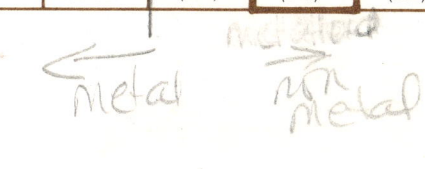

				3	4	5	6	7	8
								1 H 1.0079	2 He 4.00260
				5 **B** 10.81	6 C 12.011	7 N 14.0067	8 O 15.9994	9 F 18.998403	10 Ne 20.179
				13 **Al** 26.98154	14 **Si** 28.0855	15 P 30.97376	16 S 32.06	17 Cl 35.453	18 Ar 39.948
28 Ni 58.69	29 Cu 63.546	30 Zn 65.38	31 Ga 69.72	32 **Ge** 72.59	33 **As** 74.9216	34 Se 78.96	35 Br 79.904	36 Kr 83.80
46 Pd 106.42	47 Ag 107.868	48 Cd 112.41	49 In 114.82	50 Sn 118.69	51 **Sb** 121.75	52 **Te** 127.60	53 **I** 126.9045	54 Xe 131.29
78 Pt 195.08	79 Au 196.9665	80 Hg 200.59	81 Tl 204.383	82 Pb 207.2	83 Bi 208.9804	84 **Po** (209)	85 **At** (210)	86 Rn (222)

metal → ← non metal

62 Sm 150.36	63 Eu 151.96	64 Gd 157.25	65 Tb 158.9254	66 Dy 162.50	67 Ho 164.9304	68 Er 167.26	69 Tm 168.9342	70 Yb 173.04
94 Pu (244)	95 Am (243)	96 Cm (247)	97 Bk (247)	98 Cf (251)	99 Es (252)	100 Fm (257)	101 Md (258)	102 No (259)

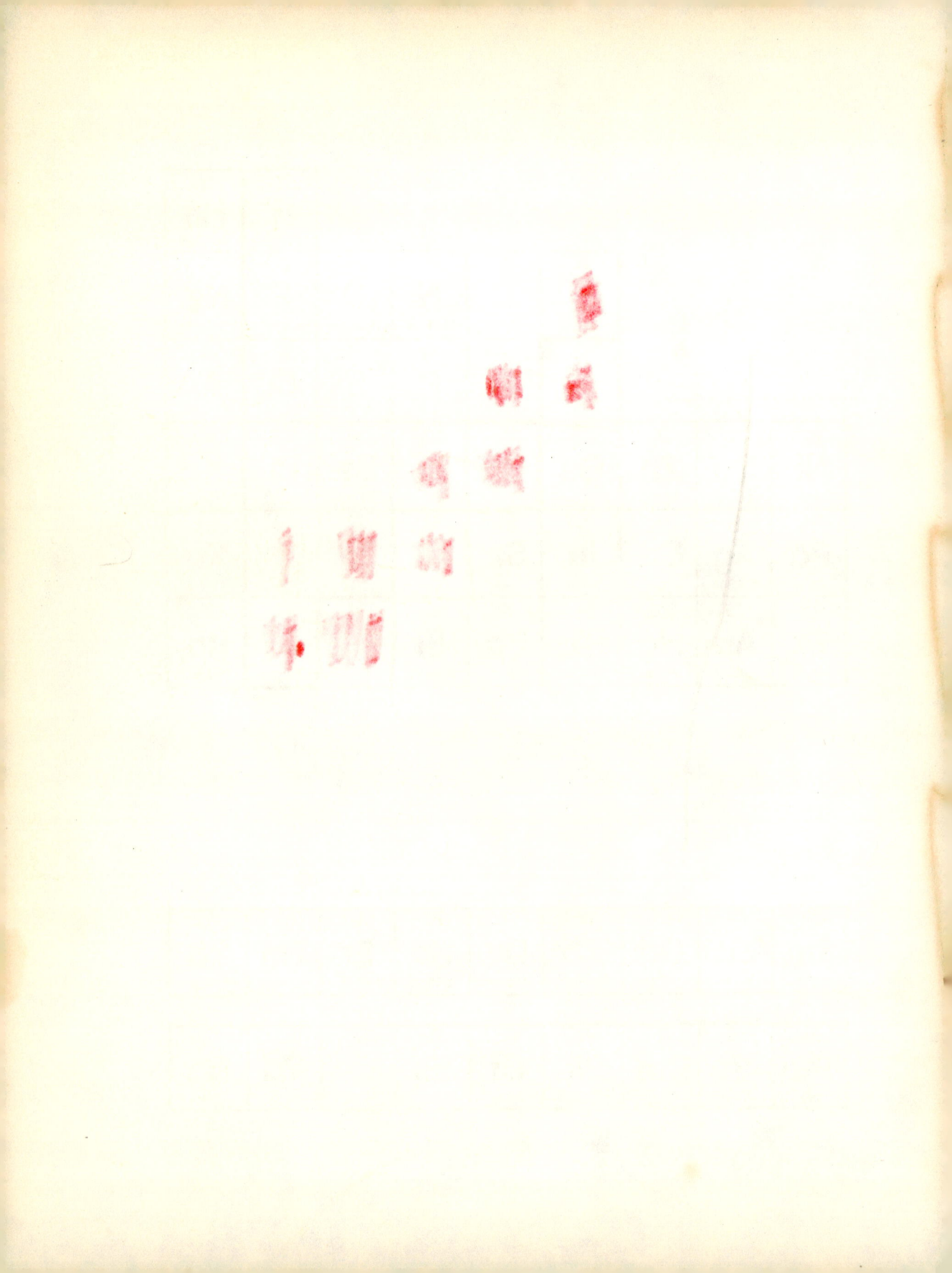

Introduction to Chemistry

William L. Masterton
Professor of Chemistry
University of Connecticut
Storrs, Connecticut

Stanley M. Cherim
Professor of Chemistry
Delaware County Community College
Media, Pennsylvania

Introduction to Chemistry

Saunders Golden Sunburst Series
Saunders College Publishing
Philadelphia New York Chicago
San Francisco Montreal Toronto
London Sydney Tokyo Mexico City
Rio de Janeiro Madrid

Address orders to:
383 Madison Avenue
New York, NY 10017

Address editorial correspondence to:
West Washington Square
Philadelphia, PA 19105

Text Typeface: 10/12 Caledonia
Compositor: Clarinda
Acquisitions Editor: John Vondeling
Project Editor: Rebecca Gruliow
Copyeditor: Cate Barnett Rzasa
Managing Editor & Art Director: Richard L. Moore
Art/Design Assistant: Virginia A. Bollard
Text Design: Caliber Design Planning, Inc.
Cover Design: William Boehm
New Text Artwork: Philadelphia Technical Art Services
Production Manager: Tim Frelick
Assistant Production Manager: Maureen Iannuzzi

Cover credit: Copyright Jay Freedman

Library of Congress Cataloging in Publication Data

Masterton, William L., 1927–
   Introduction to chemistry.

   (Sanders golden sunburst series)
   Includes index.

   1. Chemistry. I. Cherim, Stanley M. II. Title.
QD31.2.M386   1984   540   83–17244
ISBN 0-03-059676-9

INTRODUCTION TO CHEMISTRY   ISBN 0-03-059676-9

© 1984 by CBS College Publishing. All rights reserved. Printed in the United States of America.
Library of Congress catalog card number 83-17244.

7 032 9876543

CBS COLLEGE PUBLISHING
Saunders College Publishing
Holt, Rinehart and Winston
The Dryden Press

Preface

This book is written for the student taking a one-semester or one-term course in preparation for a year course in general chemistry. We assume this is the student's first exposure to chemistry. There is no attempt to be encyclopedic; this book is designed to be covered in a 12- to 14-week course.

In our writing, we have been guided by four premises.

1. *The preparatory course should emphasize the fundamentals of general chemistry.* To this end, we devote the first six chapters to such basic topics as measurements, properties of matter, formulas, stoichiometry, nomenclature, and the gas laws. With this background in concrete principles that can be tested in the laboratory, the text goes on to the more abstract ideas of atomic structure (Chapter 7) and chemical bonding (Chapter 8).

2. *Several more advanced principles and calculations can safely be postponed to the year course.* In Chapters 9–12 (Solutions, Reaction Rate and Equilibrium, Acids and Bases, Oxidation–Reduction Reactions), we have kept the discussion elementary. The concept of the equilibrium constant is introduced, but extensive algebraic manipulations are avoided. With acids and bases, the emphasis is on such fundamentals as pH, acid-base strengths, and titrations in water solution. More general acid-base models are mentioned but not stressed.

3. *The approach throughout the preparatory course should be consistent with that of the year course.* Under no circumstances should the student have to "unlearn" material from a prior course. As one might expect, we make extensive use of the conversion factor approach in solving problems. At the same time, the student is exposed to simple equations like the Ideal Gas Law. Net ionic equations are stressed throughout the chapters dealing with reactions in water solution.

4. *Students taking a first course in chemistry need all the help they can get.* A clearly written text is the most important factor here; we hope to have achieved that.

Beyond that, the following features should be helpful: Each chapter is preceded by a set of learning objectives, keyed to specific examples within the chapter. Within each chapter there are about six to ten examples worked out, that deal with basic principles. Each example is followed by an exercise that tests whether the student has grasped the principle illustrated. At the end of each chapter, there are "Questions" dealing with factual material. These are followed by "Problems," which are indexed as to type and require an understanding of principles. Finally, there are several "Multiple Choice" questions to test the student's mastery of all the material covered in the chapter.

We are indebted to a large number of chemistry teachers at all levels who have reviewed this book at various stages and made valuable suggestions. These reviewers include Stan Ashbaugh, Orange Coast College; Alice Corey, Pasedena City College; Arthur Hayes, Santa Ana College; Joseph Kanamueller, Western Michigan University; David Katz, Community College of Philadelphia; Paul Lauren, Suffolk County Community College; Steven Murov, Modesto Junior College; and Lee Pedersen, University of North Carolina, Chapel Hill.

In particular, we appreciate the candid comments of Emil Slowinski, who has a unique talent for reading a manuscript from a student's point of view. It is a pleasure to acknowledge the assistance of our friend of long standing, John Vondeling, who also happens to be chemistry editor and associate publisher at Saunders. His talented assistant, Tricia Manning, has also been very helpful, as has Don Jackson, our publisher.

William L. Masterton
Stanley M. Cherim

Contents Overview

CHAPTER 1   Measurement   1
CHAPTER 2   Matter and Energy   32
CHAPTER 3   The Building Blocks of Matter   54
CHAPTER 4   Chemical Formulas and Names   80
CHAPTER 5   Chemical Equations   106
CHAPTER 6   The Physical Behavior of Gases   134
CHAPTER 7   Electronic Structure and the Periodic Table   166
CHAPTER 8   Chemical Bonding   196
CHAPTER 9   Solutions   226
CHAPTER 10   Reaction Rate and Chemical Equilibrium   254
CHAPTER 11   Acids and Bases   286
CHAPTER 12   Oxidation–Reduction Reactions   316
CHAPTER 13   Descriptive Inorganic Chemistry: Metals and Nonmetals   346
CHAPTER 14   Introduction to Organic Chemistry   378
CHAPTER 15   Nuclear Chemistry   410
             Glossary   437
APPENDIX I   Review of Mathematics   447
APPENDIX II Answers to Problems and Multiple Choice   465
             Index   483

Contents

1 Measurement   1

     1.1 Exponential Notation   3
     1.2 Metric System   7
     1.3 Measurement of Volume and Mass   12
     1.4 Temperature   15
     1.5 Density   18
     1.6 Measurement Errors: Uncertainty, Significant Figures   21

2 Matter and Energy   32

     2.1 Types of Matter   34
     2.2 Changes in Matter   39
     2.3 Properties of Matter   41
     2.4 Energy Changes   44

3 The Building Blocks of Matter   54

     3.1 Atomic Theory   56
     3.2 Subatomic Particles   58
     3.3 Molecules and Ions   64
     3.4 Masses of Atoms   67
     3.5 The Mole Concept   71

4 Chemical Formulas and Names   80

     4.1 Types of Formulas   82
     4.2 Formulas of Ionic Compounds   84

Contents

    4.3 Names of Inorganic Compounds   87
    4.4 Formulas and Percent Composition   94

5 Chemical Equations   106

    5.1 Writing and Balancing Chemical Equations   108
    5.2 Meaning of Balanced Equations   113
    5.3 Stoichiometric Calculations   117
    5.4 Limiting Reactant and Theoretical Yield   121
    5.5 Thermochemical Equations   125

6 The Physical Behavior of Gases   134

    6.1 Kinetic Molecular Theory of Gases   137
    6.2 Volume–Pressure–Temperature Relations   139
    6.3 Molar Volumes of Gases   149
    6.4 Ideal Gas Law   152
    6.5 Dalton's Law of Partial Pressures   154
    6.6 Volumes of Gases in Reactions   157

7 Electronic Structure and the Periodic Table   166

    7.1 The Bohr Model of the Hydrogen Atom   168
    7.2 The Quantum Mechanical Atom   172
    7.3 Principal Energy Levels and Sublevels   173
    7.4 Orbitals and Electron Spin   179
    7.5 Electron Configuration and the Periodic Table   185
    7.6 Trends in the Periodic Table   189

8 Chemical Bonding   196

    8.1 Ionic Bonding   198
    8.2 Covalent Bonding   203
    8.3 Lewis Structures   206
    8.4 Molecular Geometry   214
    8.5 Polarity in Molecules   218

9 Solutions 226

- 9.1 Solutes in Water 228
- 9.2 Factors Affecting Solubility 230
- 9.3 Solubility of Ionic Compounds; Precipitation Reactions 235
- 9.4 Concentrations of Solutions 240
- 9.5 Freezing Points and Boiling Points of Solutions 246

10 Reaction Rate and Chemical Equilibrium 254

- 10.1 Rate of Reaction; Activation Energy 256
- 10.2 Effect of Changes in Conditions upon Reaction Rate 260
- 10.3 Chemical Equilibrium in a Gaseous System 264
- 10.4 Equilibrium Constants for Gaseous Systems 267
- 10.5 Equilibrium Constant and Extent of Reaction 271
- 10.6 Changes in Gaseous Equilibrium Systems 274

11. Acids and Bases 286

- 11.1 Acidic and Basic Water Solutions 289
- 11.2 Strong and Weak Acids 296
- 11.3 Strong and Weak Bases 301
- 11.4 Acid–Base Reactions 304
- 11.5 General Models of Acids and Bases 308

12. Oxidation–Reduction Reactions 316

- 12.1 Oxidation and Reduction 318
- 12.2 Balancing Redox Equations 324
- 12.3 Predicting Redox Reactions; Activity Series 328
- 12.4 Electrical Cells 332

13 Descriptive Inorganic Chemistry: Metals and Nonmetals 346

- 13.1 Alkali and Alkaline Earth Metals 348
- 13.2 Transition Metals 352

Contents

    13.3 Extraction of Metals from Their Ores: Metallurgy   357
    13.4 Nonmetallic Elements; Molecular Structure   362
    13.5 Hydrogen   367
    13.6 Oxygen   370

14 Introduction to Organic Chemistry   378

    14.1 Saturated Hydrocarbons (Alkanes)   381
    14.2 Unsaturated Hydrocarbons: Alkenes and Alkynes   390
    14.3 Aromatic Hydrocarbons: Benzene   395
    14.4 Oxygen Compounds   397

15 Nuclear Chemistry   410

    15.1 Radioactivity   413
    15.2 Rate of Radioactive Decay   420
    15.3 Nuclear Fission   425
    15.4 Nuclear Fusion   430
    15.5 Mass–Energy Relations   431

Glossary   437

Appendix 1 Review of Mathematics   447

Appendix 2 Answers to Problems and Multiple Choice   465

Index   483

Measurement

Learning Objectives

After studying this chapter, you should be able to:

1. Convert an ordinary number to an exponential number; carry out the reverse operation (Examples 1.1, 1.2).
2. Multiply or divide exponential numbers (Example 1.3).
3. State the meaning, using exponents, of the common metric prefixes.
4. Use conversion factors to convert a quantity expressed in one metric unit to a different metric unit (Examples 1.4–1.6).
5. Describe the use of instruments for measuring volume (Figures 1.3, 1.4) and mass (Figures 1.5, 1.6).
6. Given a temperature in °C, convert to °F; carry out the reverse operation (Example 1.7).
7. Given a temperature expressed in °C, convert to K; carry out the reverse operation (Example 1.8).
8. Given any two of the three quantities mass, volume, or density, calculate the other quantity (Examples 1.9, 1.10).
9. Determine the number of significant figures in a measured quantity (Example 1.11).
10. Determine the number of significant figures in a quantity obtained by multiplying or dividing measured quantities (Example 1.12).
11. Determine the uncertainty in a quantity obtained by adding or subtracting measured quantities (Example 1.13).

Science is valued for its practical advantages, it is valued because it gratifies disinterested curiosity, and it is valued because it provides the contemplative imagination with objects of great aesthetic charm.

J. W. N. SULLIVAN

Photograph by Peter L. Kresan.

CHAPTER 1

You are just beginning what is perhaps your first course in chemistry. At this point, you may well wonder what chemistry is all about. What do chemists do? Different people may answer that question in quite different ways. A young child may think of a chemist as a magician who concocts mysterious "potions." A hospital patient may idealize chemists as life-givers who furnish antibiotics, painkillers, and other "miracle drugs." In contrast, a person who lives downwind from a paint factory may think that chemists spend their time producing foul odors that pollute the air we breathe.

More generally, without making value judgments, we can say that chemists study the structure and properties of different substances. They seek answers to many basic questions. Among these are the following:

What are the "building blocks" of which substances are made? What basic particles are present in water? in table salt? in carbon monoxide?

What factors determine the properties of substances? Why is water a liquid and table salt a solid? Why is carbon monoxide a deadly poison? Why do copper and silver, but not sulfur or water, conduct electricity?

What happens when substances undergo changes? How, if at all, does the particle structure of water change when it freezes? When it boils? When it decomposes to hydrogen and oxygen?

How can one substance be converted to another? How can a drug like insulin be made from readily available starting materials? How can we ensure that carbon monoxide will *not* be formed when gasoline burns in an automobile engine?

To answer questions such as these, chemists follow an approach that blends theory with experiment. In this book, we will consider some of the basic theories and principles that form the foundation of chemistry. We will also describe a great many different experiments. These experiments involve measurements of various kinds. For that reason, it seems appropriate to begin our study of chemistry by examining the general subject of measurement.

In discussing measurements in this chapter, we will consider

—how measured quantities are expressed in metric units and how we can convert from one metric unit to another (Section 1.2).

Measurement

—how the quantities volume, mass, temperature, and density are measured (Sections 1.3–1.5).
—how the errors associated with measured quantities are expressed (Section 1.6).

As background for these topics, we begin this chapter (Section 1.1) by discussing exponential notation. This is a way of writing numbers that will be used again and again throughout the chapter and indeed throughout the entire text.

1.1 Exponential Notation

Chemists often deal with very large or very small numbers. In a drop of water, there are approximately

3,000,000,000,000,000,000,000

molecules of water. At the opposite extreme, the mass of a water molecule in grams is:

0.000 000 000 000 000 000 000 029 93 g

Numbers such as these are very awkward to work with. Neither of the above numbers could be entered directly into a calculator. To simplify operations involving large or small numbers, chemists and other scientists use what is known as **exponential** or **scientific notation**. Here a quantity is expressed as a number between one and ten (**coefficient**) times an integral power of ten (**exponential**). Examples include

1×10^4; 2.45×10^2; 3.8×10^{-3}

To understand what these quantities mean, look at Table 1.1 (p.4). This shows the values of exponentials ranging from 10^4 to 10^{-4}. We see that

$1 \times 10^4 = 1(10,000) = 10,000$

$2.45 \times 10^2 = 2.45(100) = 245$

$3.8 \times 10^{-3} = 3.8(0.001) = 0.0038$

Converting from Exponential Numbers to Ordinary Numbers

Suppose we have a quantity expressed in exponential notation,

$C \times 10^n$

where C is a number between one and ten (such as 1, 2.45, or 3.8) and n is an integer which may be positive (2, 4) or negative (−3). We wish to express

Chapter 1

TABLE 1.1 Exponentials

$$
\begin{aligned}
10^4 &= (10)(10)(10)(10) &&= 10{,}000 \\
10^3 &= (10)(10)(10) &&= 1{,}000 \\
10^2 &= (10)(10) &&= 100 \\
10^1 &= (10) &&= 10 \\
10^0 & &&= 1 \\
10^{-1} &= (0.1) &&= 0.1 \\
10^{-2} &= (0.1)(0.1) &&= 0.01 \\
10^{-3} &= (0.1)(0.1)(0.1) &&= 0.001 \\
10^{-4} &= (0.1)(0.1)(0.1)(0.1) &&= 0.0001
\end{aligned}
$$

this quantity as an ordinary number. If you look closely at the examples just presented, you may discover a simple rule for this type of conversion. Notice that in converting 2.45×10^2 we *moved the decimal point two places to the right* (from 2.45 to 245). In converting 3.8×10^{-3}, we *moved the decimal point three places to the left* (from 3.8 to 0.0038). The general rule is

To convert $C \times 10^n$ to an ordinary number, move the decimal point n places. If n is positive, move the decimal point to the right; if n is negative, move the decimal point to the left.

EXAMPLE 1.1
Convert the following to ordinary numbers
a. 2.59×10^{-4} b. 8.5×10^1 c. 2.72×10^0

Solution
a. Move the decimal point four places to the left: 0.000259.
b. Move the decimal point one place to the right: 85.
c. Since n = 0, leave the decimal point where it is: 2.72.

Exercise
Express 2×10^3 and 3×10^2 as ordinary numbers. Answer: 2000, 300.

Often you will have occasion to compare the magnitudes of two or more exponential numbers. You might, for example, want to know which is the larger

2×10^3 or 3×10^2

This can always be done by converting to ordinary numbers. In the exercise just presented, you found that $2 \times 10^3 = 2000$ and $3 \times 10^2 = 300$. Clearly, 2×10^3 is larger than 3×10^2. In another case, you might want to compare 2×10^{-3} with 3×10^{-2}. Again, converting to ordinary numbers

$2 \times 10^{-3} = 0.002$

Measurement

$3 \times 10^{-2} = 0.03$

We see that 3×10^{-2} is greater than 2×10^{-3}.
 From these examples, you may be able to deduce a general rule. **The larger exponential number is the one with the larger value of n.** Thus,

$3 > 2$, so $2 \times 10^3 > 3 \times 10^2$

$-2 > -3$, so $3 \times 10^{-2} > 2 \times 10^{-3}$

Converting Ordinary Numbers to Exponential Numbers

We have just seen how to convert from exponential to ordinary numbers. Now we consider the reverse operation, expressing an ordinary number such as

5196 or 0.0028

in exponential notation. That is, we want to express such numbers in the form

$C \times 10^n$

where C is a number between 1 and 10 and n is an integer.
 As you might suppose, the rule here is essentially the reverse of that used when we converted exponential numbers to ordinary numbers. **To find n, count the number of places the decimal point must be moved to give the coefficient C. If the decimal point must be moved to the left, n is a positive integer. If it must be moved to the right, n is a negative integer.**
 Applying this rule to the number 5196, we first note that C is 5.196. That is, to express 5196 in exponential notation, we must write it as

5.196×10^n

because 5.196 (*not* 0.5196 or 51.96) is a number between 1 and 10. To go from 5196 to 5.196, we have to move the decimal point three places to the *left*. So, n must be 3. In other words,

$5196 = 5.196 \times 10^3$

Similarly, $0.0028 = 2.8 \times 10^n$. In this case, the decimal point is moved three places to the *right* going from 0.0028 to 2.8. Hence,

$0.0028 = 2.8 \times 10^{-3}$

EXAMPLE 1.2

At the beginning of this section, it was stated that there are approximately

3,000,000,000,000,000,000,000

molecules in one drop of water. Express this number in exponential notation.

Continued

Solution

The coefficient must be 3 (a number between 1 and 10). To obtain this coefficient, the decimal point must be moved 21 places (count them!) to the left. Hence, the exponential number is

3×10^{21}

Exercise

Express the mass in grams of a water molecule, given on p. 3, in exponential notation. Answer: 2.993×10^{-23} g.

Multiplication and Division of Exponential Numbers

A major advantage of exponential notation is that it greatly simplifies the processes of multiplication and division. Here, we make use of the fact that *to multiply we add exponents*

$10^1 \times 10^2 = 10^{1+2} = 10^3$ (i.e., $10 \times 100 = 1000$)

$10^6 \times 10^{-4} = 10^{6+(-4)} = 10^2$

To divide, we subtract exponents

$10^3/10^2 = 10^{3-2} = 10^1$ (i.e., $1000/100 = 10$)

$10^6/10^{-3} = 10^{6-(-3)} = 10^9$

Using these principles, we arrive at the following rules for multiplication and division of exponential numbers.

To multiply one exponential number by another, first multiply the coefficients together in the usual way and then add exponents.

To divide one exponential number by another, divide coefficients in the usual way and subtract exponents.

EXAMPLE 1.3

Carry out the following operations
a. $(4.0 \times 10^{-3}) \times (2.0 \times 10^5)$ b. $(4.0 \times 10^{-3})/(2.0 \times 10^5)$

Solution

a. For convenience, we first separate the coefficients from the exponential terms

$(4.0 \times 2.0) \times (10^{-3} \times 10^5)$

Multiplying coefficients and adding exponents, we obtain

$8.0 \times 10^{-3+5} = 8.0 \times 10^2$

b. $\dfrac{4.0 \times 10^{-3}}{2.0 \times 10^5} = \dfrac{4.0}{2.0} \times \dfrac{10^{-3}}{10^5} = 2.0 \times 10^{-3-5} = 2.0 \times 10^{-8}$

Continued

Measurement

> **Exercise**
> Evaluate $\dfrac{(3 \times 10^4) \times (4 \times 10^{-6})}{2 \times 10^{-9}}$. Answer: 6×10^7.

Often when exponential numbers are multipled or divided, the answer is not in standard exponential notation. Consider, for example

$$(5.0 \times 10^4) \times (6.0 \times 10^3) = 30 \times 10^7$$

The product is not in standard exponential notation. The coefficient, 30, does not lie between 1 and 10. To correct this, we can rewrite the coefficient as 3.0×10^1 and then add exponents

$$30 \times 10^7 = (3.0 \times 10^1) \times 10^7 = 3.0 \times 10^8$$

In another case

$$0.50 \times 10^{-6} = (5.0 \times 10^{-1}) \times 10^{-6} = 5.0 \times 10^{-7}$$

Many arithmetical operations besides multiplication and division can be carried out with exponential numbers. These can be done most conveniently with a scientific calculator. On such a calculator, it is possible to enter numbers in exponential notation. Typically, it involves using a key labeled EXP, EE, or EEX. Check your instruction manual for the procedure to be followed with your calculator. (See also Appendix 1.)

1.2 Metric System

Chemists and other scientists always report the results of their measurements in the metric system. Here, all the units used to express a quantity are related to each other by powers of ten (Table 1.2). The metric system is in common use in virtually every country in the world except the United States. In this country, conversion from English to metric units has been slow and painful.

To illustrate the use of the prefixes listed in Table 1.2, consider the metric units of length. The basic unit here is the *meter* (abbreviation m). You

TABLE 1.2 Common Metric Prefixes

Prefix	Factor	Symbol	Prefix	Factor	Symbol
mega-	10^6	M	deci-	10^{-1}	d
kilo-	10^3	k	centi-	10^{-2}	c
			milli-	10^{-3}	m
			micro-	10^{-6}	μ
			nano-	10^{-9}	n

Figure 1.1 A long stride is a rough approximation of one meter.

can think of a meter as being roughly the length of a long stride (Figure 1.1). The large, practical unit of length, used to express highway distances, is the *kilometer* (km). One kilometer is equal to 1000 m

$1 \text{ km} = 1000 \text{ m} = 10^3 \text{m}$

Two smaller units of length are the *centimeter* (cm), which is 1/100 of a meter, and the *millimeter* (mm), which is 1/1000 of a meter.

$1 \text{ cm} = 0.01 \text{ m} = 10^{-2}\text{m}$

$1 \text{ mm} = 0.001 \text{ m} = 10^{-3}\text{m}$

A centimeter is approximately equal to the diameter of a stick of chalk or the tip of your little finger. A dime has a thickness of about 1 mm (Figure 1.2). A still smaller unit of length, used to express the dimensions of atoms and molecules, is the *nanometer* (nm)

$1 \text{ nm} = 0.000\ 000\ 001 \text{ m} = 10^{-9}\text{m}$

The same metric prefixes are used for quantities other than length. For

Figure 1.2 The width of the tip of your little finger is approximately 1 cm. A dime has a thickness of approximately 1 mm.

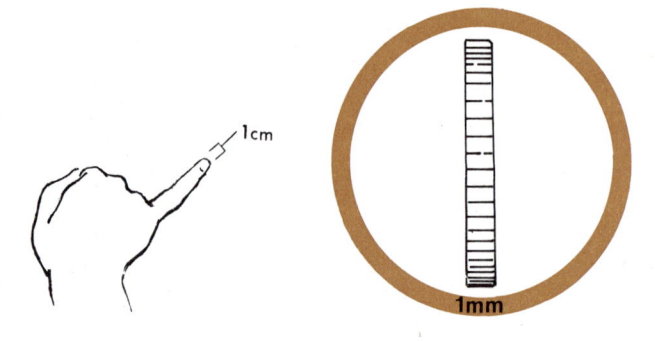

example, one *kilogram* (kg) has a mass exactly 1000 times that of one *gram* (g). A gram, in turn, has a mass of one thousand *milligrams* (mg)

1 kg = 1000 g = 10^3 g

1 g = 1000 mg = 10^3 mg

Perhaps the most common unit of volume in the metric system is the *liter* (L). A liter is one thousand *milliliters* (mL). A milliliter has a volume equal to that of a cube one centimeter on an edge (cubic centimeter or cm^3)

1 L = 1000 mL = 1000 cm^3

Conversions Between Units: Conversion Factors

Frequently you will have occasion to change the units in which a measured quantity is expressed. You might, for example, want to express a length of 1.64 m in centimeters

1.64 m = ? cm

In another case, you might want to express a length of 3.54 cm in meters

3.54 cm = ? m

To make these conversions, we start with the basic equation relating the two units, meters and centimeters

1 cm = 10^{-2} m

If we divide both sides of this equation by 10^{-2} m, we obtain a ratio that is equal to one

$$\frac{1 \text{ cm}}{10^{-2} \text{ m}} = 1$$

A similar ratio, also equal to one, is obtained when we divide both sides of the above equation by 1 cm:

$$\frac{10^{-2} \text{ m}}{1 \text{ cm}} = 1$$

The ratios 1 cm/10^{-2} m and 10^{-2} m/1 cm are referred to as **conversion factors**. We will consider many different conversion factors here and in later chapters. All of them have certain properties in common.

1. *A conversion factor is always equal to one.* Hence, a quantity can be multiplied by a conversion factor without changing the value of the quantity.
2. *A conversion factor is a ratio in which one unit appears in the numerator and another in the denominator* (centimeters and meters in the above example).

3. *Every relation between two units leads to two conversion factors.* Thus, 1 cm = 10^{-2} m leads to the factors 1 cm/10^{-2} m and 10^{-2} m/1 cm.

With this background, let's see how you would accomplish the conversions referred to earlier

1.64 m = ? cm

3.54 cm = ? m

To convert 1.64 m to centimeters, multiply by the conversion factor 1 cm/10^{-2} m

$$1.64 \text{ m} \times \frac{1 \text{ cm}}{10^{-2} \text{ m}} = 1.64 \times 10^2 \text{ cm}$$

Note that meters "cancel out," leaving centimeters in the answer. The conversion tells you that 1.64 m is equal to 1.64×10^2 cm. Multiplication by the conversion factor did not change the value of the length, but it did change the units of length.

To change 3.54 cm to meters, you need to multiply by a ratio with meters in the numerator and centimeters in the denominator. The required conversion factor is 10^{-2} m/1 cm

$$3.54 \text{ cm} \times \frac{10^{-2} \text{ m}}{1 \text{ cm}} = 3.54 \times 10^{-2} \text{ m}$$

Here, centimeters cancel, leaving the answer in meters.

These examples illustrate the general procedure used with conversion factors.

1. *A conversion factor is used to convert a given quantity, expressed in one unit, to the required quantity, expressed in a different unit.*

    given quantity × conversion factor = required quantity

2. *The conversion factor used is the one that cancels the units in the given quantity and yields the proper units in the required quantity.*

    given quantity × conversion factor = required quantity

    unit 1 × $\dfrac{\text{unit 2}}{\text{unit 1}}$ = unit 2

This procedure is further illustrated in Example 1.4.

EXAMPLE 1.4

Using the relation: 1 km = 10^3 m, convert
a. 2.50 km to meters   b. 3.68×10^4 m to kilometers

Continued

Solution

The two possible conversion factors are

$$\frac{1 \text{ km}}{10^3 \text{ m}} \quad \text{and} \quad \frac{10^3 \text{ m}}{1 \text{ km}}$$

In each case, we choose whichever conversion factor is appropriate.

a. To convert from kilometers to meters, we need a conversion factor with meters in the numerator and kilometers in the denominator. That factor is 10^3 m/1 km

$$2.50 \text{ km} \times \frac{10^3 \text{ m}}{1 \text{ km}} = 2.50 \times 10^3 \text{ m}$$

b. Here, we want to go in the opposite direction, from meters to kilometers. The required conversion factor is 1 km/10^3 m

$$3.68 \times 10^4 \text{ m} \times \frac{1 \text{ km}}{10^3 \text{ m}} = 3.68 \times 10^1 \text{ km}$$

Exercise

Express a distance of 3.94 mm in meters. Answer: 3.94×10^{-3} m.

We have illustrated the use of conversion factors with some rather simple examples. Don't be fooled into thinking that a formal procedure of this sort is unnecessary. Granted, you could probably have obtained the correct answers in Example 1.4 by "multiplying by 1000" in one case and "dividing by 1000" in the other. However, with units that are less familiar, it may be far from obvious how you should proceed. The conversion factor approach (also referred to as "dimensional analysis" or the "factor label" method) can be applied to a wide variety of problems in chemistry. It is particularly useful where more than one conversion is required (Example 1.5).

EXAMPLE 1.5

The radius of a hydrogen atom is 0.037 nm. Express the radius in centimeters.

Solution

Table 1.2 does not show a direct relation between nanometers and centimeters. However, we can relate both of these units to meters

1 nm = 10^{-9} m   1 cm = 10^{-2} m

To convert 0.037 nm to centimeters, we might follow a two-step path.
 (1) Convert nanometers to meters, using the conversion factor 10^{-9} m/1 nm.
 (2) Convert meters to centimeters, using the conversion factor 1 cm/10^{-2} m.
We set up the two conversions in a single calculation

$$0.037 \text{ nm} \times \frac{10^{-9} \text{ m}}{1 \text{ nm}} \times \frac{1 \text{ cm}}{10^{-2} \text{ m}} = 0.037 \times 10^{-7} \text{ cm} = 3.7 \times 10^{-9} \text{ cm}$$

         (1)   (2)

Continued

> **Exercise**
> Convert 1.29 cm to kilometers. Answer: 1.29×10^{-5} km.

1.3 Measurement of Volume and Mass

The two quantities that you will measure most often in the laboratory are volume and mass. In this section, we will describe the instruments you will use to make these measurements. We will also comment on the units in which the results of these measurements are most often expressed.

Volume Measurements

Of all metric units, the liter is perhaps the one most commonly used in the United States. To an increasing extent, soft drinks, liquor, and even milk are being sold in liter or 2-L containers. A liter bottle of ginger ale contains approximately 6% more than a quart bottle (more exactly, 1 L = 1.057 qt). Much of the glassware used in the general chemistry laboratory is marked in either liters or milliliters. You may work with a "1-L beaker" or a "250-mL flask." In this text, we will use the cubic centimeter more often than the milliliter. Remember that these two units have the same value: 1 cm^3 = 1 mL.

When you measure out a volume of liquid for an experiment, you will most likely use one of the pieces of glassware shown in Figure 1.3. A gradu-

Figure 1.3 A graduated cylinder is used to measure out volumes of liquids when a great degree of accuracy is not required. A measuring pipet is more accurate, a transfer pipet most accurate of all.

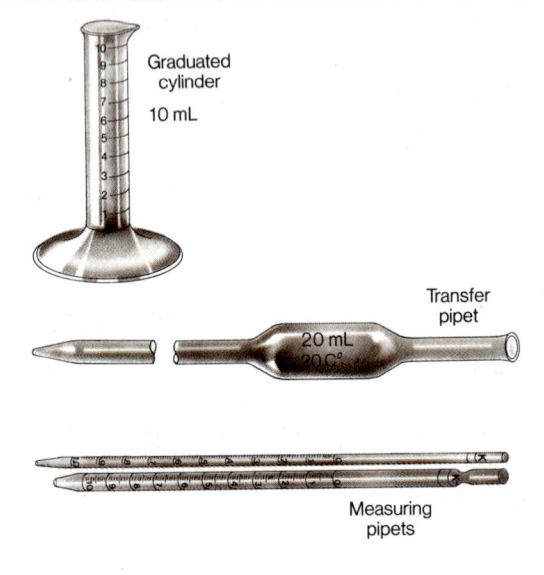

ated cylinder is used when a high degree of accuracy is not required. Using the cylinder shown, you might be able to pour out 10 cm^3 to the nearest cubic centimeter (i.e., 10 ± 1 cm^3). With a measuring pipet, you could probably do somewhat better, perhaps delivering 10.0 ± 0.1 cm^3. A transfer pipet of the type shown in Figure 1.3 is designed to deliver a specified volume of liquid, such as ten cubic centimeters, with a high degree of accuracy. Using such a pipet, you might be able to deliver 10.00 ± 0.01 cm^3.

The two pieces of volumetric glassware in Figure 1.4 are useful for special purposes. A buret measures accurately the volume of a reagent consumed in a chemical reaction. In the course of an experiment, you might deliver 16.25 cm^3 or 30.62 cm^3 or _____ of a reagent from a buret. A volumetric flask is used to make up a solution to an accurately determined volume, such as 25.00 cm^3.

Mass Measurements

The mass of a sample is a measure of the amount of matter it contains. In contrast, the weight of a sample is a measure of the gravitational force acting upon it. Mass is independent of location. You contain the same amount of matter whether you are sitting at your desk, flying in a jet plane, or orbiting the earth in a spacecraft. In contrast, your weight steadily decreases as you move out from the surface of the earth because the strength of the gravitational field decreases. Throughout the text, we will refer to mass rather than weight.

The unit of mass used most often in chemistry is the gram (g). To give you an idea of its magnitude, a dime has a mass of about 2 g. Occasionally,

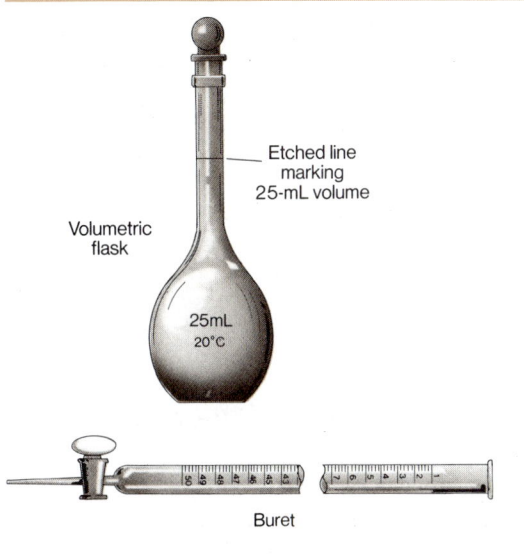

Figure 1.4 A volumetric flask is used to prepare solutions to a specified concentration (see Chapter 9). A buret is used to measure the volume of a liquid reagent required in a reaction (see Chapter 11).

we will express mass in kilograms (kg): 1 kg = 10^3 g. This book has a mass of approximately 1 kg. High-precision work often involves the use of the milligram (mg): 1 mg = 10^{-3} g. A small pinch of salt has a mass of approximately 1 mg.

Masses expressed in one metric unit are readily converted to other metric units (Example 1.6).

EXAMPLE 1.6
Assuming this book weighs 1.1 kg, express its mass in milligrams.

Solution
The required relations are

1 kg = 10^3 g   1 mg = 10^{-3} g

We first convert 1.1 kg to grams and then to milligrams. Two successive conversions are required.

$$\text{mass in milligrams} = 1.1 \text{ kg} \times \frac{10^3 \text{ g}}{1 \text{ kg}} \times \frac{1 \text{ mg}}{10^{-3} \text{ g}} = 1.1 \times 10^6 \text{ mg}$$

Exercise
What would be the mass of this book in micrograms? Answer: 1.1×10^9 μg.

In the laboratory, you will determine mass using an instrument called a balance. What you do is to balance a sample against a known mass. The process is perhaps most obvious with the double-pan balance shown at the right of Figure 1.5. Here you add metal pieces of known mass ("weights") to one pan while your sample rests on the other pan. When the scale is exactly horizontal, you can be sure that the mass of your sample is equal to that of the known masses. The same principle applies, a bit less obviously, to the other coarse balances shown in Figure 1.5.

A coarse balance is used when an accuracy of 1 g, 0.1 g, or perhaps 0.01 g is sufficient. If you are asked to weigh out a sample to the nearest milligram

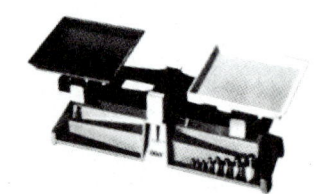

Figure 1.5 Typical coarse balances.

Figure 1.6 Analytical balances.

(± 0.001 g), you should use an *analytical balance* (Figure 1.6). A word of caution: An analytical balance is an expensive (~ $1000) and delicate instrument. Learn how to use it properly and treat it with an appropriate measure of respect.

1.4 Temperature

All of us have a feeling for relative temperatures. When we say that an object is "hot," we mean that its temperature is relatively high, most often in comparison to body temperature. A "cold" object is at a lower temperature. Following this model, we can consider temperature to be an indicator of the direction of heat flow. When you drink a cup of hot coffee, you feel warm because heat is transferred from the coffee to your body. With a bottle of cold soda, which is below body temperature, heat flows in the opposite direction. Heat *always* flows from a higher to a lower temperature.

To measure temperature, we take advantage of the fact that liquids expand when they are heated. A thermometer (Figure 1.7, p.16) contains a liquid, usually mercury, in a fine tube that widens into a bulb at the lower end. The height of the liquid depends upon the temperature around the thermometer. If the temperature increases, the liquid expands up the tube. If the temperature decreases, the liquid level falls.

To express temperature as a number, a thermometer must be marked according to some scale. In the remainder of this section, we will consider three different temperature scales (°F, °C, K) and the relations between them.

Fahrenheit and Celsius Scales

Virtually all thermometers are marked according to one or the other of two scales. The thermometer at the right of Figure 1.7 is marked in degrees Celsius (formerly called Centigrade). Here, 0°C is taken to be the temperature

Figure 1.7 Comparison between the Celsius scale *(right)* and the Fahrenheit scale *(left)*. Note that an interval of 1°C corresponds to 1.8°F.

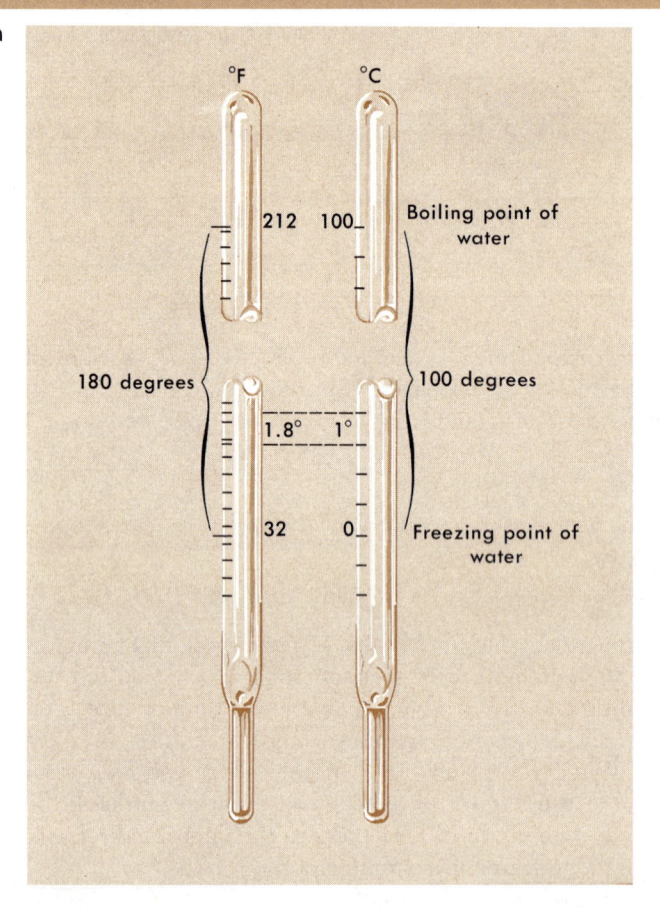

at which pure water freezes; 100°C is the temperature at which pure water boils at one atmosphere pressure. Laboratory thermometers are marked in °C. However, another temperature scale, degrees Fahrenheit, is still commonly used in the United States. The thermometer at the left of Figure 1.7 is marked in °F. On this scale, the freezing point of pure water is taken to be 32°F and the boiling point 212°F.

From this discussion and examination of Figure 1.7, it should be clear that

—the Fahrenheit degree is smaller than the Celsius degree. To be specific, 1.8°F are required to cover the same temperature interval as 1°C.
—the Fahrenheit scale "starts out" 32° ahead of the Celsius scale, since 32°F = 0°C.

Putting these two relations together, we obtain a general equation relating the two scales

$$°F = 1.8(°C) + 32° \quad (1.1)$$

Equation 1.1 is readily used to calculate a temperature in °F, given the value in °C. In that case, all that is required is a simple substitution (Example 1.7a). We can also use Equation 1.1 to go in the opposite direction, from °F to °C. There, a bit of algebra is involved (Example 1.7b).

EXAMPLE 1.7
a. The freezing point of seawater is approximately −2.0°C. Express this in °F.
b. Normal body temperature is 98.6°F. Express this in °C.

Solution
a. Substituting directly into Equation 1.1

°F = 1.8(−2.0°) + 32° = −3.6° + 32.0° = 28.4°

b. Substituting for °F, we have

98.6° = 1.8(°C) + 32°

To solve for °C, we first subtract 32° from each side of the equation

66.6° = 1.8(°C)

Then we divide each side by 1.8

°C = 66.6°/1.8 = 37.0°

Exercise
Express 30°C in °F; 0°F in °C Answers: 86°F, −18°C.

The Kelvin (Absolute) Temperature Scale

A third temperature scale, the Kelvin scale, is very useful in chemistry and the other sciences. Here the zero point, 0 K, is taken to be the lowest temperature that is possible to reach. In contrast to °F or °C, all temperatures on the Kelvin scale are positive numbers. Comparing the Kelvin and Celsius scales (Figure 1.8, p.18), we see that

—the Kelvin degree (called simply the kelvin) is the same size as the Celsius degree. Note that 100 degrees separate the freezing and boiling points of water on both scales.
—the Kelvin scale "starts out" 273° ahead of the Celsius scale, because 0 K = −273°C (more exactly, −273.15°C).

Figure 1.8
Comparison between the Kelvin and Celsius scales. On the Kelvin scale, 0 represents the lowest possible temperature (−273.15°C).

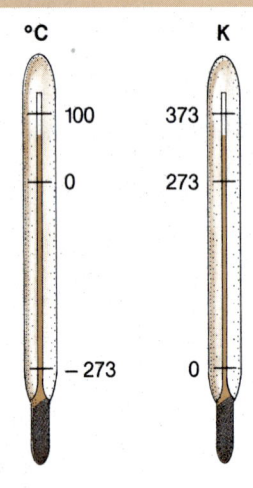

Putting these two relations together, we obtain the equation

$$K = °C + 273 \tag{1.2}$$

EXAMPLE 1.8
Express
a. 27°C in K   b. 90 K in °C

Solution
We proceed in much the same way as in Example 1.7.
a. K = 27 + 273 = 300
b. 90 = °C + 273

°C = 90 − 273 = −183

Exercise
Express 50 K in °C; 50°C in K. Answer: −223°C, 323K.

1.5 Density

In Section 1.3, we discussed mass and volume measurements. Both the mass and volume of a substance depend upon amount. The larger the sample, the larger its mass and volume. In contrast, the ratio of mass to volume, called density, is independent of amount. Samples of different sizes have the same density (Figure 1.9).

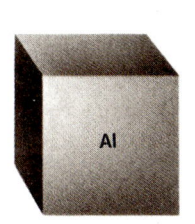

Measurement

Mass	2.70 g	5.40 g	8.10 g
Volume	1.00 cm³	2.00 cm³	3.00 cm³
Density	2.70 g/cm³	2.70 g/cm³	2.70 g/cm³

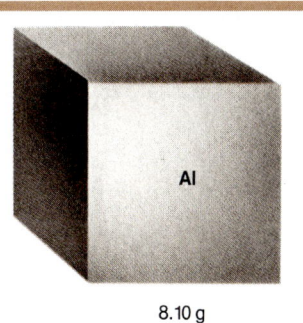

Figure 1.9 These three samples of aluminum metal have different masses and volumes. However, they all have the same density, 2.70 g/cm³. The density of aluminum is a property of that substance which can be used to identify it.

The defining equation for density is

$$\text{density} = \frac{\text{mass}}{\text{volume}} \quad \text{or} \quad d = \frac{m}{V} \tag{1.3}$$

To determine the density of a substance in the laboratory, we need to find the mass and volume of a sample. For a liquid, the measurements are quite simple. All we need do is to run out a specified volume of the liquid, using one of the volume-measuring devices described in Section 1.3. Then we weigh the liquid and use Equation 1.3 to calculate the density (Example 1.9).

EXAMPLE 1.9

To determine the density of a liquid, a student runs out a sample from a buret into a previously weighed flask. The data are as follows

mass empty flask = 16.0 g   initial buret reading = 0.2 cm³
mass flask + liquid = 45.0 g   final buret reading = 25.2 cm³

Calculate the density of the liquid.

Solution

The mass of the liquid is

45.0 g − 16.0 g = 29.0 g

The volume is found by subtracting the initial from the final buret reading

25.2 cm³ − 0.2 cm³ = 25.0 cm³

Substituting into Equation 1.3

$$d = \frac{m}{V} = \frac{29.0 \text{ g}}{25.0 \text{ cm}^3} = 1.16 \text{ g/cm}^3$$

Continued

Chapter 1

> **Exercise**
> Using a pipet, a student measures out 20.0 cm³ of a liquid. She finds that it weighs 25.2 g. What is the density of the liquid? Answer: 1.26 g/cm³.

The density of a solid is somewhat more difficult to determine than that of a liquid. The principal problem is to find the volume accurately. One approach that is commonly used is illustrated in Figure 1.10. Here, we measure the increase in volume that occurs when a solid is added to a liquid in which it is insoluble. That increase in volume is equal to the volume of the solid.

We have seen how Equation 1.3 is used to calculate the density of a sample, given its mass and volume. In some problems, you may be required to calculate the mass or volume of a sample, knowing its density. Here, two approaches are possible:

1. Solve Equation 1.3 for the required quantity (mass or volume) and substitute numbers.

$$\text{mass} = \text{density} \times \text{volume}; \quad m = d \times V \tag{1.4}$$

$$\text{volume} = \frac{\text{mass}}{\text{density}}; \quad V = \frac{m}{d}$$

2. Use density as a conversion factor to relate volume to mass.

The use of these two approaches is illustrated in Example 1.10.

EXAMPLE 1.10

a. A sample of aluminum (d = 2.70 g/cm³) has a volume of 12.6 cm³. What is its mass in grams?
b. Another sample of aluminum has a mass of 12.0 g. What is its volume?

Solution

a. Method (1): $m = d \times V = 2.70 \frac{g}{cm^3} \times 12.6 \text{ cm}^3 = 34.0 \text{ g}$

   Method (2): Since the density is 2.70 g/cm³, 2.70 g = 1 cm³

   $\text{mass} = 12.6 \text{ cm}^3 \times \frac{2.70 \text{ g}}{1 \text{ cm}^3} = 34.0 \text{ g}$

b. Method (1): $V = \frac{m}{d} = \frac{12.0 \text{ g}}{2.70 \text{ g/cm}^3} = 4.44 \text{ g/cm}^3$

   Method (2): $\text{volume} = 12.0 \text{ g} \times \frac{1 \text{ cm}^3}{2.70 \text{ g}} = 4.44 \text{ g/cm}^3$

Exercise
What would be the mass of a sample of aluminum with a volume of 1.00 cm³? The volume of an aluminum sample weighing 1.00 g? Answer: 2.70 g, 0.370 cm³

The *specific gravity* of a substance is the ratio of its density to that of a reference substance, ordinarily pure water. Since the density of water is about 1.00 g/cm³, the specific gravity of a substance is numerically equal to its density in grams per cubic centimeter. For aluminum, with a density of 2.70 g/cm³

$$\text{specific gravity} = \frac{2.70 \text{ g/cm}^3}{1.00 \text{ g/cm}^3} = 2.70$$

1.6 Measurement Errors: Uncertainty, Significant Figures

In every measurement there is a likely error or **uncertainty.** Suppose, for example, you want to find the mass of a piece of metal. Using a coarse balance that weighs to the nearest tenth of a gram, you might obtain a mass of 26.5 g:

mass = 26.5 ± 0.1 g

Here the uncertainty is 0.1 g. So far as you know, the metal could weigh as little as 26.4 g or as much as 26.6 g.

The uncertainty in a measurement can be reduced by using a more sensitive instrument. With an analytical balance (recall Figure 1.6), you could determine the mass of the metal to the nearest milligram. You might now find that:

mass = 26.517 ± 0.001 g

The uncertainty has now been reduced to 0.001 g. You know that the metal weighs between 26.516 and 26.518 g, presumably closer to 26.517 g than to either of these values.

When you report the result of a measurement, you should indicate the uncertainty. That way, other people can decide how much confidence to place in your work. You could indicate uncertainty with the "plus or minus" notation used above

measurement: 26.5 ± 0.1 g   26.517 ± 0.001 g
uncertainty: 0.1 g   0.001 g

More commonly, the measurement is simply quoted by itself with the understanding that *there is an uncertainty of one in the last digit*. If you report that a sample weighs 26.5 g, it is assumed that you weighed it to the nearest tenth of a gram. If you quote a mass of 26.517 g, an uncertainty of 1 mg is implied.

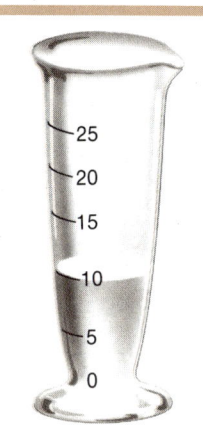

measurement: 26.5 g   26.517 g
uncertainty implied: 0.1 g   0.001 g

Figure 1.10
Measuring the density of a solid. A solid sample weighing 12.5 g was added to 10.0 cm³ of water in a cylinder. The water level went up to 15.0 cm³. This indicates that the volume of the solid is 5.0 cm³. The density of the solid, mass/volume, equals 12.5 g/5.0 cm³ = 2.5 g/cm³.

Often, the uncertainty is described by stating the number of **significant figures**. A "significant figure" is exactly what it sounds like: an experimentally meaningful digit. In 26.5 g, all three digits are significant; they have experimental meaning. There are three significant figures in 26.5 g. Similarly, there are five significant figures in 26.517 g.

measurement:	26.5 g	26.517 g
number of significant figures:	3	5

Counting Significant Figures

In many cases, as with the masses 26.5 g and 26.517 g, the number of significant figures is easily determined. *All non-zero digits in a measurement are significant.* Where zeros appear at the beginning or end of a number, there may be some question as to whether or not they are significant. In this situation, common sense is the best guide. Suppose you are told that a sample weighs

26.510 g

Is the zero at the end of this number significant? A moment's thought should convince you that it is. It was put there to indicate that the mass was determined to the nearest 0.001 g. There are five significant figures in 26.510 g, just as there are five significant figures in 26.517 g.

In another, quite different case, suppose you weigh a tiny pin on an analytical balance and find that its mass is 18 mg. Expressing this in grams, you might report that the pin weighs

0.018 g

Here, the zeros are not significant. They are written simply to fix the position of the decimal point. There are two significant figures in 0.018 g, just as there are two significant figures in 18 mg.

EXAMPLE 1.11
Give the number of significant figures in each of the following
a. 6.391 cm   b. 712 cm^3 c. 22.00 g   d. 0.002 g

Solution
In (a), (b), and (c), all the digits are meaningful. There are four significant figures in 6.391 cm, three significant figures in 712 cm^3, and four significant figures in 22.00 g. In (d), the zeros serve only to fix the position of the decimal point; there is only one significant figure.

Continued

> **Exercise**
> Which of the following has the largest number of significant figures: 0.012 g, 0.100 g, or 0.008 g? Answer: 0.100 g.

There is one surefire way to determine the number of significant figures in a measured quantity. *Express it in exponential notation, i.e., $C \times 10^n$. The number of significant figures in the quantity is that in the coefficient C.* Applying this rule to the masses in the exercise you just worked

0.012 g = 1.2×10^{-2} g   (two significant figures)

0.100 g = 1.00×10^{-1} g   (three significant figures)

0.008 g = 8×10^{-3} g   (one significant figure)

Sometimes you will encounter quantities where the number of significant figures cannot be specified exactly. Suppose you are told that a sample weighs

500 g

Without further information it is impossible to tell whether the uncertainty is 1 g, 10 g, or 100 g. If the mass is known to the nearest gram, both zeros are meaningful and there are three significant figures. If the uncertainty is 10 g, the last zero in 500 g is not meaningful; there are only two significant figures. If the mass is 500 ± 100 g, neither of the zeros is meaningful and there is only one significant figure. This ambiguous situation could have been resolved by expressing the mass in exponential notation

   5.00×10^2 g—three significant figures

or 5.0×10^2 g—two significant figures

or 5×10^2 g—one significant figure

Significant Figures in Multiplication and Division

Individual measurements are seldom end results in themselves. Instead, they are combined with other measurements, often by multiplication or division. We might find the area of a piece of tin foil by multiplying its length by its width

area = length × width

In another case, we calculate density by dividing mass by volume

$$\text{density} = \frac{\text{mass}}{\text{volume}}$$

TABLE 1.3 Significant Figures in Calculated Quantities

Sample	Mass	Volume	Density	m	V	d
1	2.70 g	1.00 cm^3	2.70 g/cm^3	3	3	3
2	2.702 g	1.001 cm^3	2.699 g/cm^3	4	4	4
3	2.7021 g	1.0 cm^3	2.7 g/cm^3	5	2	2
4	2.7 g	1.001 cm^3	2.7 g/cm^3	2	4	2

(Number of Significant Figures: m, V, d)

To express the uncertainty in a product or quotient of measured quantities, we follow a simple rule. **In multiplication or division, keep only as many significant figures in your answer as there are in the measured quantity having the fewest significant figures.** Table 1.3 applies this rule to the calculation of density from mass and volume.

Usually, when you carry out multiplication or division on your calculator, not all of the digits that appear in the display are significant. In Sample 2, Table 1.3

d = 2.702 g/1.001 cm^3

If you carry out this division on a typical calculator, the result is:

d = 2.6993007 ___ g/cm^3

All of the digits beyond 2.699 are meaningless. You should keep only four digits in the density, since the mass and volume are each known to four significant figures.

EXAMPLE 1.12
Calculate the volume of a rectangular solid that has a length of 4.16 cm, a width of 2.2 cm, and a height of 2.00 cm.

Solution

Volume = length × width × height
= 4.16 cm × 2.2 cm × 2.00 cm = 18 cm^3

You should retain only two significant figures in the volume since that is the number in the width, the dimension with the greatest uncertainty. On a calculator, you would obtain a volume of 18.304 cm^3; the last three digits are meaningless and should be discarded.

Exercise
Calculate the volume of a solid having a mass of 12.016 g and a density of 2.61 g/cm^3. Answer: 4.60 cm^3.

As we have seen, in calculations involving measurements we often drop one or more digits to obtain the correct number of significant figures. The rules followed in "rounding off" are

1. If the first digit dropped is less than five, leave the preceding digit unchanged (i.e., 3.124 → 3.12).
2. If the first digit dropped is greater than five, increase the preceding digit by one (i.e., 3.127 → 3.13).
3. If the first digit dropped is a five, round off to make the preceding digit an even number (i.e., 3.125 → 3.12; 3.135 → 3.14). The effect of this rule is that the last digit is "rounded up" half the time and "rounded down" half the time.

Uncertainties in Addition and Subtraction

Often you will have occasion to add or subtract measured quantities. You may recall that in Example 1.9 both the mass and volume were obtained by difference. That is, they were found by subtracting an initial value from a final value. To estimate the uncertainty in a sum or difference of measured quantities, we follow a common-sense rule. **In addition or subtraction, the uncertainty in your answer is that in the measured quantity with the greatest uncertainty.**

In effect, we applied this rule to find the mass and volume of the liquid referred to in Example 1.9

mass flask + liquid = 45.0 g   uncertainty = 0.1 g
mass empty flask   = 16.0 g   uncertainty = 0.1 g
mass liquid   = 29.0 g   uncertainty = 0.1 g

When both measured quantities have the same uncertainty, as in this example, the result is rather obvious and the rule is hardly necessary. Consider, however, the slightly more complex situation in Example 1.13.

EXAMPLE 1.13

What is the total mass of coffee made by mixing the following ingredients: 212 g water, 2.1 g cream, 1.88 g instant coffee?

Solution

The masses, with their uncertainties, are

212   ± 1 g
22.1   ± 0.1 g
1.88 ± 0.01 g

Continued

> On a calculator, you would obtain a total mass of 235.98 g. However, since the mass of the water is known only to the nearest gram, the total mass has an uncertainty of ± 1 g. It should be reported as 236 g
>
> 212 g + 22.1 g + 1.88 g = 236 g
>
> **Exercise**
> A salt shaker contains 214 g of salt. How much salt remains if you pour out 0.2 g onto a hamburger? Answer: 214 g.

Exact Numbers

Throughout this section, we have dealt with measured quantities, all of which are uncertain to a lesser or greater degree. However, certain numbers used in calculations are exact. In the relation

1 kg = 1000 g

the "1000" is exact. There are exactly 1000 g in 1 kg, no more, no less. Again, in the relation

°F = 1.8(°C) + 32°

the 1.8 and 32 are exact. We might say that numbers such as there have zero uncertainty or contain an infinite number of significant figures.

When an exact number is involved in a calculation, it has no effect on the uncertainty or number of significant figures. For example, if you were asked to convert 20.5°C to °F

°F = 1.8(20.5) + 32° = 68.9°

You retain three significant figures in the answer, because there are three significant figures in the measured quantity 20.5°C.

Key Words

Celsius scale
centi-
conversion factor
cubic centimeter
deci-
density
exponential notation
Fahrenheit scale
gram
Kelvin scale
kilo-
liter
mass
mega-
meter
micro-
milli-
nano-
significant figure
specific gravity

Questions

1.1 State the rule for converting an exponential number, $C \times 10^n$, to an ordinary number.
1.2 State the rule for converting an ordinary number to an exponential number.
1.3 State the rules for multiplication and division of exponential numbers.
1.4 Explain the meaning of each of the following metric prefixes
   a. kilo-   b. centi-   c. milli-
   d. nano-
1.5 What object that you can see from where you are sitting has a length of approximately 1 cm? 1 m?
1.6 Explain what is meant by a conversion factor. Describe how conversion factors are used.
1.7 When you put a spoon in a cup of hot coffee, heat flows from the _____ to the _____. If you put a spoon in a cup of iced coffee, heat flows from the _____ to the _____.
1.8 What is the freezing point of water in °C? °F? What is the boiling point of water at one atmosphere pressure in °C? °F?
1.9 Describe a method commonly used to determine the density of solids with irregular dimensions.
1.10 Explain what is meant by specific gravity.
1.11 What is meant by a "significant figure"?
1.12 State the rule used to determine the number of significant figures in multiplication or division.
1.13 State the rule used to determine the uncertainty in addition or subtraction.

Problems

Exponential Notation

1.14 Express the following as ordinary numbers
   a. 2.81×10^2 b. 4.16×10^{-4}
   c. 3.8×10^7 d. 5.71×10^{-1}
1.15 Convert the following to ordinary numbers
   a. 3.18×10^3 b. 2.9×10^{-5}
   c. 6.12×10^0 d. 0.18×10^{-2}
1.16 Express the following numbers in exponential notation
   a. 0.000 011 2   b. 3152
   c. 0.001 902   d. 286.5
1.17 Convert the following to standard exponential notation
   a. 67131   b. 0.088
   c. 31.5×10^{-3} d. 0.49×10^2
1.18 Carry out the following operations, expressing your answer in standard exponential notation
   a. $(3.1 \times 10^5) \times (3.0 \times 10^{-2})$
   b. $(6.4 \times 10^{14})/(1.6 \times 10^{-6})$
   c. $(4.2 \times 10^6) \times (3.9 \times 10^5)$
   d. $(5.0 \times 10^3)/(8.6 \times 10^{-2})$
1.19 Express the following in standard exponential notation
   a. $(1.6 \times 10^3) \times (4.0 \times 10^{-8})$
   b. $(4.0 \times 10^{-8})/(1.6 \times 10^3)$
   c. $(3.9 \times 10^{-3}) \times (6.1 \times 10^7)$
   d. $(3.9 \times 10^{-3})/(6.1 \times 10^7)$

Metric Conversions

1.20 Carry out the following length conversions
   a. 15.3 cm = ? m
   b. 0.024 km = ? m
   c. 153 cm = ? nm
1.21 Convert
   a. 8.1×10^{-4} m to nanometers
   b. 2.3×10^7 nm to meters
   c. 1.5×10^{-6} m to centimeters
   d. 3.5×10^{-3} cm to meters

1.22 Convert each of the following to centimeters
 a. 1.52 km   b. 1.52 m   c. 1.52 mm
 d. 1.52 nm

1.23 Carry out the following volume conversions
 a. 27.8 mL = ? L
 b. 0.082 L = ? mL
 c. 6.19 cm^3 = ? L

1.24 Convert each of the following to liters
 a. 352 mL   b. 26.1 cm^3
 c. 2.19 × 10^4 cm^3

1.25 Carry out the following mass conversions
 a. 0.5 kg = ? g
 b. 514 mg = ? g
 c. 4.5 mg = ? kg

1.26 Convert each of the following to grams
 a. 17.0 mg   b. 1.4 × 10^{-3} kg
 c. 62 μg

Temperature Scales

1.27 Carry out the following temperature conversions
 a. 5°F = ? °C   b. −70°F = ? °C
 c. 170°C = ? °F   d. −12°C = ? K

1.28 Convert each of the following temperatures to degrees Celsius
 a. 246 K   b. −18°F   c. 99°F
 d. 99 K

1.29 A student prepares a mixture of dry ice and acetone. A Celsius thermometer immersed in this mixture reads −76.4°C. What is the temperature in °F? K?

1.30 Benzene boils at 353 K. Express this temperature in °C; °F.

1.31 Absolute zero on the Celsius scale is −273°C. Express this in °F.

1.32 Mercury freezes at −40°F. Express this in °C; K.

Density

1.33 A student finds that a sample of benzene weighing 10.0 g has a volume of 11.4 cm^3. What is the density of benzene?

1.34 An empty graduated cylinder weighs 25.2 g. After 20.0 cm^3 of liquid is added, the total mass is 41.1 g. What is the density of the liquid?

1.35 When 4.58 g of a certain solid is added to water in a graduated cylinder, the level rises from 12.0 cm^3 to 14.0 cm^3. What is the density of the solid?

1.36 The density of ethyl alcohol is 0.790 g/cm^3. Calculate
 a. the mass of 16.0 cm^3 of ethyl alcohol
 b. the volume of 16.0 g of ethyl alcohol

1.37 The density of marble is 2.70 g/cm^3. How many grams of marble chips must be added to 32.1 cm^3 of water to increase the volume to 39.6 cm^3?

1.38 Mercury has a density of 13.6 g/cm^3. A thermometer is filled with 1.00 g of mercury. What volume does the mercury occupy?

1.39 A flask is filled when 19.6 g of water is added to it.
 a. Taking the density of water to be 1.00 g/cm^3, calculate the volume of the flask
 b. It takes 22.7 g of a certain liquid to fill the flask. What is the density of the liquid?

1.40 A block of metal weighs 85.2 g. It is 13.6 cm long, 6.12 cm wide, and 5.8 mm thick. Calculate
 a. the volume of the metal
 b. the density of the metal

Significant Figures

1.41 State the number of significant figures in each of the following measured quantities
 a. 1.6 × 10^3 g   b. 2.38 × 10^{-4} cm^3
 c. 3.90 × 10^3 L

1.42 State the number of significant figures in
 a. 0.82 kg   b. 0.011 g   c. 1.010 cm^3
 d. 29.20 L

1.43 Carry out the following operations, expressing the answers to the correct number of significant figures
 a. 6.10 cm^3/2.8 cm^2 b. 2.19 cm × 0.38 cm
 c. 92.963 g/31.5 cm^3

1.44 Find the masses in grams of samples of alu-

minum (d = 2.70 g/cm^3) having the following volumes
a. 1.5 cm^3 b. 1.50 cm^3 c. 1.500 cm^3

1.45 Find the volumes in cubic centimeters of samples of aluminum (d = 2.70 g/cm^3) with the following masses
a. 1.5 g   b. 1.50 g   c. 1.500 g

1.46 What is the total mass of a solution containing 59 g of water, 6.2 g of sugar, and 0.14 g of salt?

1.47 A bottle contains 239 g of silver nitrate. Four students withdraw samples weighing 62 g, 12.83 g, 0.36 g, and 11.9 g in that order. What mass of silver nitrate remains in the flask?

Multiple Choice

1.48 When 12.4 g of nickel is added to 12.0 cm^3 of water in a graduated cylinder, the volume increases to 13.3 cm^3. The density of nickel, in g/cm^3, is
a. 0.93   b. 1.0   c. 1.1   d. 9.5   e. 11

1.49 It is found that when 25.00 cm^3 of a certain liquid is added to a flask weighing 59.940 g, the mass increases to 89.942 g. The density of the liquid, in g/cm^3, is
a. 1.2   b. 1.20   c. 1.200   d. 1.2001   e. 1.20008

1.50 A gold atom has a radius of 0.144 nm. The radius expressed in centimeters is
a. 1.44×10^{-1} b. 1.44×10^{-7} c. 1.44×10^{-8} d. 1.44×10^{-9} e. 1.44×10^{-10}

1.51 Which of the following numbers is the smallest?
a. 1.6×10^{-2} b. 1.6×10^{3} c. 1.6×10^{-6} d. 1.6   e. 1.6×10^{15}

1.52 How many of the following numbers are equal to 1.6×10^{-4}? 16×10^{-3}, 16×10^{-4}, 16×10^{-5}, 0.00016, 0.000016
a. 1   b. 2   c. 3   d. 4   e. 5

1.53 A sample of water is heated from 20.0 to 50.0°C. The *increase* in temperature in degrees Fahrenheit is
a. 16.7   b. 22.0   c. 30.0   d. 54.0   e. 86.0

1.54 A flask has a volume of 26.1 cm^3. It contains 13.5 g of aluminum (d = 2.70 g/cm^3). What volume of a liquid must be added to the aluminum to fill the flask?
a. 5.0 cm^3 b. 12.6 cm^3 c. 21.1 cm^3 d. 26.0 cm^3 e. 26.1 cm^3

1.55 A student finds that an empty beaker weighs 12.024 g. He adds a solid and finds the total mass to be 12.108 g. The mass of the solid is known to how many significant figures?
a. 1   b. 2   c. 3   d. 4   e. 5

1.56 How many of the following will freeze when cooled to 30°F? ice (fp = 0°C), benzene (fp = 5°C), ethyl alcohol (fp = −112°C)
a. 0   b. 1   c. 2   d. 3

1.57 What is the total volume in cubic centimeters of 12.0 g of benzene (d = 0.880 g/cm^3) and 12.0 g of ethyl alcohol (d = 0.790 g/cm^3)?
a. $12.0(0.880) + 12.0(0.790)$
b. $\dfrac{12.0}{0.880} + \dfrac{12.0}{0.790}$
c. $12.0(0.880) + \dfrac{12.0}{0.790}$
d. $\dfrac{12.0 + 12.0}{0.780 + 0.890}$

1.58 Which of the following represents the largest mass?
a. 1512 g   b. 1.52 kg   c. 1.51×10^4 mg   d. 1.52×10^{-2} kg

1.59 To how many significant figures can you express the volume of a rectangular block of

metal that has the dimensions 2.14 cm × 4.6 cm × 3.181 mm?
   a. 2   b. 3   c. 4   d. 5   e. 9
1.60 The densities of iron, cobalt, nickel, copper, and zinc are 7.87, 0.892×10^1, 9.91×10^0, 0.0894×10^2, and 71.3×10^{-1} g/cm^3 in that order. You are given 10.0-g samples of each metal. Which one has the largest volume?
   a. iron   b. cobalt   c. nickel
   d. copper   e. zinc

Matter and Energy

Learning Objectives
After studying this chapter, you should be able to:

1. Distinguish between pure substances and mixtures, between coarse mixtures and solutions, and between elements and compounds (Example 2.2).
2. Explain what is meant by a period and a group in the Periodic Table.
3. Locate metals, nonmetals, and metalloids in the Periodic Table.
4. State and explain the significance of the Law of Constant Composition (Example 2.1), the Law of Conservation of Mass (Example 2.3), and the Law of Conservation of Energy.
5. Distinguish between a physical change and a chemical change and between a physical property and a chemical property. Give examples of each.
6. Explain what is meant by melting point and boiling point; show how these properties can be used to identify a substance (Example 2.4).
7. Convert energies expressed in calories to joules or vice versa (Example 2.5).
8. Distinguish between an endothermic and an exothermic reaction.
9. Relate heat of fusion to ΔH_{fus} per gram and mass of solid; do the same for heat of vaporization, ΔH_{vap} per gram, and mass of liquid (Example 2.6).

My business is to teach my aspirations to conform themselves to fact, not to try to make facts harmonise with my aspirations. Sit down before fact as a little child, be prepared to give up every preconceived notion, follow humbly wherever nature leads, or you will learn nothing.

THOMAS HUXLEY

Photograph by Richard L. Moore.

CHAPTER 2

Chemistry may be defined as the study of the properties and structure of matter and the changes that matter undergoes. Such a definition raises a number of questions. What is matter? How do we distinguish between different types of matter? What is meant by change? What are the results of change?

In this chapter, we will look for answers to these and other questions. We will first examine the different types of matter (Section 2.1). Then we will consider the changes that matter undergoes (Section 2.2) and the properties that distinguish one type of matter from another (Section 2.3). Finally, we will discuss the energy effects (Section 2.4) that accompany changes in matter.

2.1 Types of Matter

Matter is anything that has mass and occupies space. In other words, any object with a measurable density contains matter. Examples include water, air, minerals, plants, and animals. In contrast, we would not ordinarily consider heat, sunlight, or lightning to be matter. These are forms of energy.

For our purposes in chemistry, it is convenient to classify matter as shown in Table 2.1. We start by distinguishing between two broad types of matter, pure substances and mixtures. Each of these is further divided into two classes. A pure substance may be either an **element** or a **compound**. A mixture may be a solution or a coarse mixture. In the remainder of this section, we will look more closely at the characteristics of elements, compounds, and mixtures.

Elements

An element is a pure substance that cannot be broken down by chemical means into two or more pure substances. Hydrogen is an element. No matter what we do to a sample of hydrogen gas, we cannot decompose it into other substances. There are, at last count, 108 known elements. Of these, 91 occur

TABLE 2.1 Classification of Matter

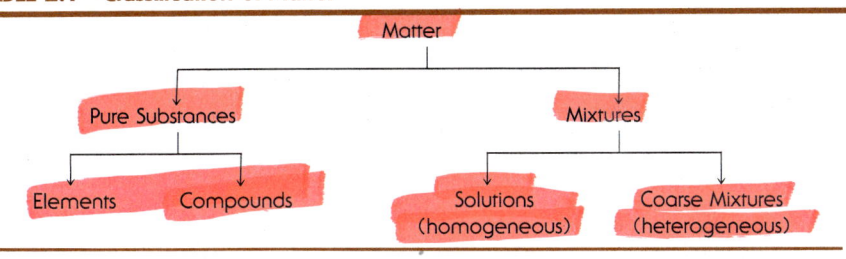

```
                        Matter
                   /              \
          Pure Substances        Mixtures
           /        \            /        \
      Elements   Compounds   Solutions    Coarse Mixtures
                             (homogeneous) (heterogeneous)
```

in nature. A few of these are familiar to everyone. The charcoal used for outdoor barbecues is nearly pure carbon. Electrical wiring is most often made from the metallic element copper. Another element, mercury, is the liquid commonly used in thermometers.

Table 2.2 lists the mass percents of the 20 most abundant elements in nature. Note that two elements, oxygen and silicon, account for approximately 75% of the total mass; ten elements account for over 99%. Perhaps the most interesting feature of the table is the number of familiar elements that do *not* appear in it. Among these are mercury, gold, and silver, which have been known since ancient times. Together, these elements make up less than 0.00001% of the earth's crust. However, they occur in concentrated ore deposits from which they are readily extracted. Mercury is obtained from cinnabar, a sulfide ore, by heating in air. Gold, which occurs as an element, can be separated from rocky impurities by mechanical means or by dissolving it in mercury.

Each element is assigned a symbol that distinguishes it from all other elements. The symbol consists of one or two letters. The first letter is capitalized; the second letter, if there is one, is not capitalized. The symbol is ordinarily derived from the name of the element. Thus carbon has the symbol C, aluminum the symbol Al. The symbols of certain elements come from Latin or Greek (Table 2.3, p.36).

TABLE 2.2 Abundance of Elements (Earth's Crust, Oceans, and Atmosphere)

Element	%		Element	%	
Oxygen	49.5%		Chlorine	0.19%	
Silicon	25.7		Phosphorus	0.12	
Aluminum	7.5		Manganese	0.09	
Iron	4.7		Carbon	0.08	
Calcium	3.4		Sulfur	0.06	
Sodium	2.6	99.2%	Barium	0.04	0.7%
Potassium	2.4		Chromium	0.033	
Magnesium	1.9		Nitrogen	0.030	
Hydrogen	0.87		Fluorine	0.027	
Titanium	0.58		Zirconium	0.023	
			All others	<0.1%	

TABLE 2.3 Symbols of Several Elements and Their Derivations

Element	Symbol	Derivation
Bromine	Br	Greek *bromos*—stench
Cobalt	Co	German *Kobald*—goblin (small amounts of cobalt in iron ore make the extraction of iron metal difficult)
Copper	Cu	Latin *cuprum*—island of Cyprus, site of ancient copper mines
Gold	Au	Latin *aurum*—gold (shining dawn)
Iron	Fe	Latin *ferrum*—iron
Lead	Pb	Latin *plumbum*—lead
Mercury	Hg	Latin *hydrargyrum*—mercury
Phosphorus	P	Greek *phosphoros*—light-bearing (white phosphorus glows in moist air because of light evolved in its oxidation)
Silver	Ag	Latin *argentum*—silver
Tin	Sn	Latin *stannum*—tin

You may be familiar with the **Periodic Table,** a device used to organize the properties of elements. A copy of the Periodic Table appears on the inside front cover of this text. We will refer to it frequently in later chapters. At the moment, we need only be concerned with three general features of the Periodic Table.

1. The horizontal rows are referred to as **periods.** The first period consists of the two elements hydrogen (H) and helium (He). The second period starts with lithium (Li) and ends with neon (Ne), and so on.
2. The vertical columns are called **groups.** The groups at the far left and right of the Table are numbered. Groups 1 and 2 are at the left, Groups 3 to 8 at the right. Elements in these groups are often referred to as **main-group elements.** As we will see later, elements within a given main group, such as Group 1, have many similar chemical properties. The elements in the center of Periods 4 through 6 are called *transition elements*. There are ten transition elements in each period (Sc through Zn in the fourth period). These groups are *not* numbered in the Periodic Table we use.
3. The sawtooth, diagonal line shown in color in the Periodic Table separates metals from nonmetals. Elements below and to the left of the line (e.g., Mg, Al, Sn) are metals. All metals possess certain properties in common (Table 2.4). In particular, they are good conductors of heat and electricity. Elements above and to the right of the diagonal line (e.g., C, O, Br) are nonmetals. Several of the elements that fall along the line, including boron (B), silicon (Si), germanium (Ge), arsenic (As), and antimony (Sb), have properties intermediate between those of metals and nonmetals. They are, for example, semiconductors. These elements are often referred to as *metalloids*.

TABLE 2.4 Properties of Metallic and Nonmetallic Elements

Metals	Nonmetals
Good to excellent conductors of heat and electricity	Poor conductors or nonconductors (graphite is an exception)
Ductile, malleable	Brittle in the solid state
Lustrous, silvery surface	Solid nonmetals generally have a dull appearance (iodine is an exception)
Solids (except mercury)	May be gases, liquids, or solids

Compounds

In contrast to an element, a compound *can* be broken down into two or more pure substances. Looking at it another way, a compound is a pure substance containing two or more elements. The most common compound is water, which consists of the elements hydrogen and oxygen chemically combined with each other. Passage of an electric current through water decomposes it to the elements (Figure 2.1).

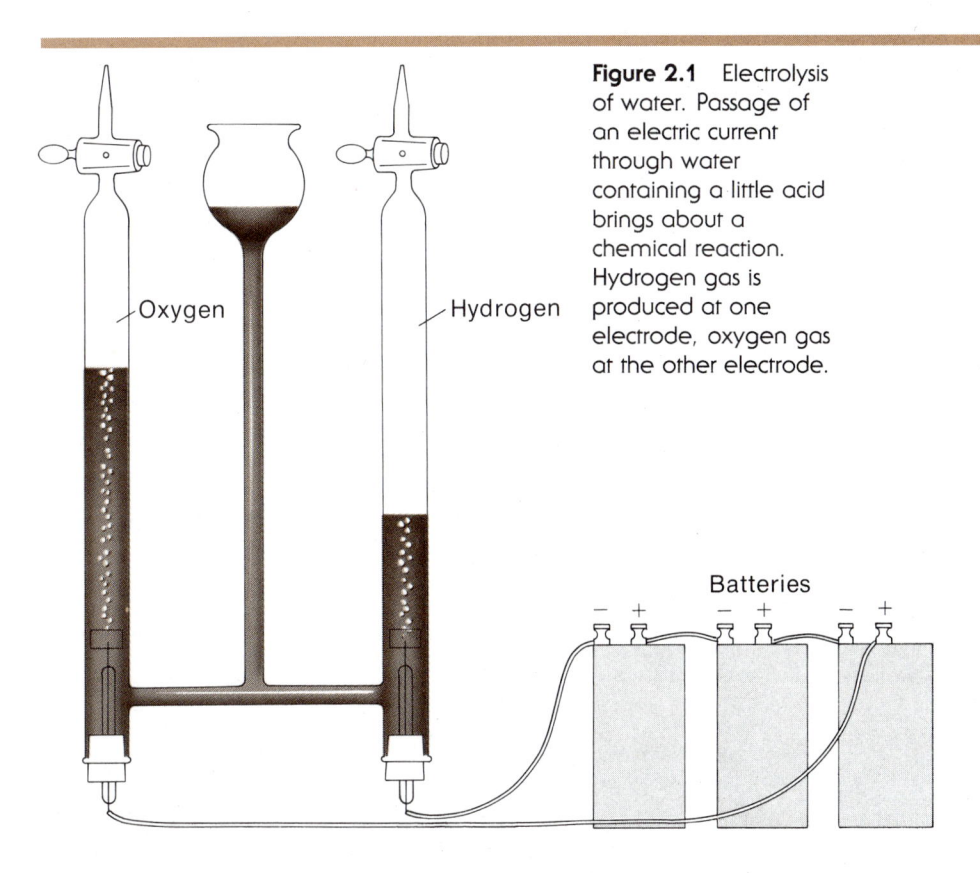

Figure 2.1 Electrolysis of water. Passage of an electric current through water containing a little acid brings about a chemical reaction. Hydrogen gas is produced at one electrode, oxygen gas at the other electrode.

Some compounds can be decomposed to the elements by heating. Among these is the red oxide of mercury, mercury(II) oxide. Heating to about 200°C breaks this compound down to mercury and oxygen. This process was first carried out by Joseph Priestley in 1774 to prepare oxygen gas. The source of heat was sunlight, focused by a magnifying glass.

Compounds have properties very different from those of the elements they contain. Water, at room temperature and atmospheric pressure, is a liquid. The two elements from which it is formed, hydrogen and oxygen, are both gases. A glowing splint inserted into a sample of oxygen gas bursts into flame. Immersed in water, it sputters and goes out. Clearly, when a compound such as water is formed from the elements, a profound change takes place.

All compounds follow the **Law of Constant Composition.** That is, a given compound always contains the same elements in the same percentages by mass. Water, regardless of where it comes from, contains 11.19% hydrogen and 88.81% oxygen. Electrolysis of a one hundred gram sample of water produces 11.19 g of hydrogen, no more and no less. The mass percents of a few other common compounds are given in Table 2.5.

EXAMPLE 2.1

Using the data in Table 2.5, calculate the mass of carbon in a sample of carbon dioxide weighing 3.618 g.

Solution

$$\text{mass carbon} = \text{mass carbon dioxide} \times \frac{\text{mass \% carbon}}{100}$$

$$= 3.618 \text{ g} \times \frac{27.29}{100} = 0.9874 \text{ g}$$

(See Appendix 1 for a further discussion of the percent concept.)

Exercise

What is the mass of oxygen in this sample of carbon dioxide? Answer: 2.631 g.

TABLE 2.5 Mass Percents of Elements in Compounds

Water	11.19% H, 88.81% O
Carbon dioxide	27.29% C, 72.71% O
Table salt (sodium chloride)	39.34% Na, 60.66% Cl
Limestone (calcium carbonate)	40.04% Ca, 12.00% C, 47.96% O
Sugar (sucrose)	42.10% C, 6.48% H, 51.42% O

Mixtures

You will recall from Table 2.1 that there are two general types of mixtures. A coarse mixture is **heterogeneous** (nonuniform). An egg is a heterogeneous mixture; we can distinguish between the shell, the white portion, and the yolk. Most of the rocks and minerals that we find in nature are coarse mixtures. In a piece of granite, you can see shiny flecks of mica and tiny bits of quartz.

A **homogeneous** mixture is referred to as a **solution.** It appears to be uniform to the eye, or indeed under an ordinary miscroscope. Air is a solution of several gases. Its composition can vary slightly depending upon location and weather conditions. The percentage of water vapor in air can vary from virtually zero over the Sahara Desert to more than 4% in Washington, D.C., on a humid summer day. The fact that air does not have a constant composition shows it to be a mixture rather than a pure compound. Other familiar solutions include seawater (water + dissolved salts) and brass (Cu + Zn).

EXAMPLE 2.2
Classify the following as elements, compounds, coarse mixtures, or solutions.
a. nickel   b. carbon monoxide   c. rainwater   d. sodium chloride
e. gasoline

Solution
a. an element (symbol Ni)
b. compound of carbon and oxygen
c. solution containing water and small amounts of impurities including dissolved gases
d. compound of sodium and chlorine
e. solution of many different organic compounds

Exercise
Garlic salt contains a small amount of garlic added to solid sodium chloride. In which of the above categories would you put garlic salt? Answer: Coarse mixture.

2.2 Changes in Matter

The changes that matter undergoes can be divided into physical and chemical changes.

Physical Change

A physical change takes place when a substance changes its form or appearance but retains its chemical identity. Mixing finely divided iron with sulfur produces a physical change. The black and yellow elements blend together to form a gray mixture. Both the iron and the sulfur retain their chemical identities. The mixture can be separated by using a magnet or by shaking with an organic solvent that will dissolve the sulfur but not the iron.

Among the most important physical changes are those in which a substance changes phase. These include

melting (solid converted to liquid)
freezing (liquid converted to solid)
sublimation (solid converted to vapor, i.e., gas)
vaporization (liquid converted to vapor)
condensation (vapor converted to liquid or solid)

Chemical Change

In a chemical change, substances change their chemical identities. If a mixture of iron and sulfur is heated, a chemical change occurs. A new substance, a compound of iron and sulfur, is formed. It does not respond to a magnet and is completely insoluble in organic solvents.

A chemical change is more frequently referred to as a *chemical reaction*, or simply a reaction. The starting materials are called *reactants*; the substances formed are called *products*. In the electrolysis of water, shown in Figure 2.1, the reaction is:

water (liquid) → hydrogen (gas) + oxygen (gas)

Here we refer to water as the reactant; hydrogen and oxygen are products. We will have a great deal more to say about chemical reactions later. In Chapter 5, we will see how they are described by chemical equations.

Conservation of Mass

There is an important law that applies to matter undergoing any kind of change. This is the **Law of Conservation of Mass**, which states that *in any ordinary physical or chemical change, the total amount of mass remains constant*. The law tells us, for example, that

—if 10.000 g of ice melts, 10.000 g of liquid water must be formed.
—if electrolysis of 10.000 g of water yields 1.119 g of hydrogen gas, then

10.000 g − 1.119 g = 8.881 g

of oxygen must be formed at the same time. In general, if we know the masses of all but one of the substances involved in a physical or chemical change, the conservation law can be used to calculate the mass of that substance.

EXAMPLE 2.3

When 1.261 g of magnesium burns in air, it forms 2.091 g of magnesium oxide. What is the mass of oxygen in the sample of magnesium oxide?

Solution

Applying the Law of Conservation of Mass

mass oxygen = mass magnesium oxide − mass magnesium

$$= 2.091 \text{ g} - 1.261 \text{ g} = 0.830 \text{ g}$$

Exercise

What is the mass percent of magnesium in magnesium oxide? Answer: 60.31%.

2.3 Properties of Matter

To identify a substance, we determine its properties and compare them with the properties of known substances. Sometimes this is done using **chemical properties**, which are properties observed when a substance undergoes a chemical change. To check whether a gas is hydrogen, you might attempt to react it with oxygen. If you find that liquid water is formed as the sole product, you can be confident that the gas is indeed hydrogen. The reaction involved is the reverse of that brought about by electrolysis of water. Here, the two elements combine with one another

hydrogen(gas) + oxygen(gas) → water(liquid)

More commonly, a substance is identified by its **physical properties.** These are properties that can be observed without changing the chemical identity of the substance. Table 2.6 lists some important physical properties of several elements and compounds. Of the properties listed, density was

TABLE 2.6 Physical Properties of Substances

Substance	Density (g/cm^3)	Melting Point (°C)	Boiling Point (°C)
Benzene	0.879	5	80
Bromine	3.12	−7	59
Ethyl alcohol	0.789	−112	78
Mercury	13.6	−39	357
Sodium chloride	2.16	800	1413
Water	1.00	0	100

Melting Point

The melting point of a pure substance is identical to its freezing point. A sample of solid benzene, warmed to 5°C, melts to liquid. Liquid benzene, upon cooling, freezes at this same temperature, 5°C. Moreover, the temperature remains constant at 5°C throughout the melting or freezing process (Figure 2.2A). Only when all the solid has melted or all the liquid has frozen, does the temperature change.

The melting point behavior of an impure solid is shown at the right in Figure 2.2. The presence of an impurity lowers the melting point. Moreover, the temperature rises steadily during the melting process. Any evidence of deviation from the horizontal in a temperature–time plot during melting suggests the presence of impurities.

Boiling Point

When a pure liquid is heated in an open container, nothing spectacular happens at first. The liquid slowly evaporates, forming vapor that escapes from the surface. Suddenly, at a particular temperature, vapor bubbles start to form in the liquid. These bubbles rise to the surface and break. When this happens, we say that the liquid is boiling; the temperature at this point is referred to as the boiling point.

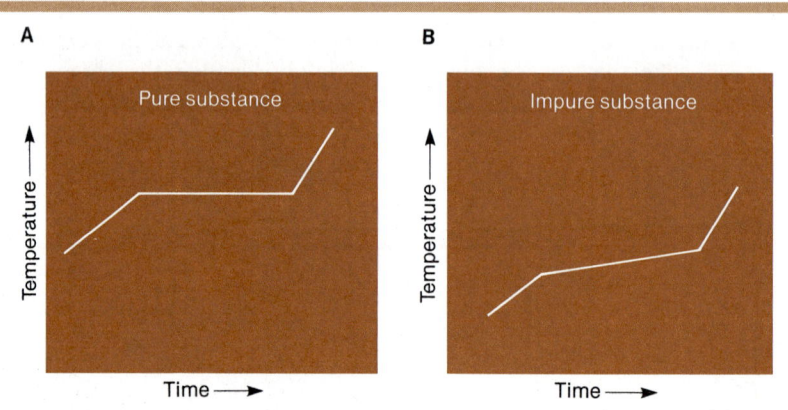

Figure 2.2 Time–temperature curves for the melting of a pure substance (A) and a mixture (B). A pure substance will melt at a constant temperature. If the substance contains an impurity, its melting point will increase as melting proceeds and will, at all stages, be lower than when pure.

A temperature–time plot for the boiling of a pure liquid looks very much like Figure 2.2A. The temperature stays constant during boiling until all the liquid has been converted to vapor. If the liquid is impure, the temperature rises during boiling.

The boiling point of a liquid depends strongly on the pressure above it. Water, heated in an open container where the pressure of the air above it is one atmosphere, boils at 100°C. Consider, however, what happens in a pressure cooker, a device used to cook foods more rapidly. Here the pressure above the water is allowed to build up. Under these conditions, the water may not boil at temperatures as high as 110°C. Conversely, at the top of a mountain where the air pressure is quite low, water may boil well below 100°C. At the top of Mount Washington in New Hampshire (elevation = 1917 m), water heated in a saucepan boils at approximately 92°C.

To understand the effect of pressure on a boiling point, consider Figure 2.3. As pointed out above, a liquid boils when vapor bubbles form below the surface. The bubbles that you see when water boils contain water vapor (steam). The pressure within such a bubble is referred to as the vapor pressure of the liquid. For a vapor bubble to form and expand, this pressure must be equal to the pressure above the liquid. (Otherwise the bubble would collapse or would not form in the first place.) This means that for boiling to occur

vapor pressure = pressure above liquid

In the case of water, we can distinguish three possibilities:

1. *The pressure above the liquid is one atmosphere.* In this case, water boils at 100°C because its vapor pressure is one atmosphere at that temperature.
2. *The pressure above the liquid is greater than one atmosphere.* Under these

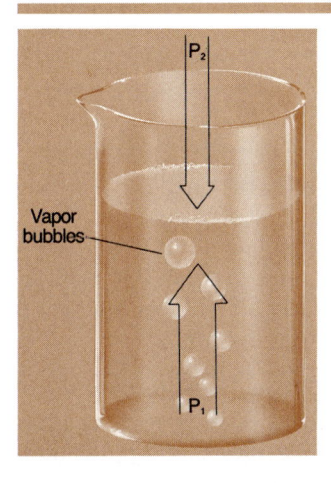

Figure 2.3 A liquid will boil when its vapor pressure becomes slightly greater than the pressure applied to it. At that point bubbles can form and remain stable until they reach the surface of the liquid.

Chapter 2

conditions water cannot boil at 100°C. The water must be heated to a higher temperature where its vapor pressure is greater than one atmosphere.

3. *The pressure above the liquid is less than one atmosphere.* Under these conditions it is possible for water to boil at temperatures below 100°C where its vapor pressure is less than one atmosphere.

The behavior we have described for water is typical of all liquids. The temperature at which the liquid boils when the temperature is one atmosphere is referred to as the *normal boiling point*. When we speak of the "boiling point" without qualifying it, we mean the normal boiling point. This is the case, for example, with the boiling points listed in Table 2.6.

EXAMPLE 2.4
A student finds that a certain substance boils when heated in an open container to about 60°C. It does not freeze when placed in an ice–water mixture at 0°C. Which one of the substances in Table 2.6 is it most likely to be?

Solution
Bromine (boiling point = 59°C, melting point = −7°C)

Exercise
How would you check further to identify this substance as bromine? Answer: Measure the density, which should be somewhat greater than 3 g/cm^3. It would also help to observe the color and odor; bromine is a dark red liquid with an acrid odor.

2.4 Energy Changes

Physical and chemical changes are accompanied by a flow of energy. Sometimes energy is absorbed from the surroundings. This happens when we supply heat from a Bunsen burner to boil water. In other changes, energy is evolved. The chemical change that occurs when methane gas burns in air is an example of such a process. This reaction, taking place within a Bunsen burner, gives off energy in the form of heat.

The energy absorbed or evolved in physical or chemical changes may appear in any of several forms. The type of energy most common in chemistry is *heat*. Most of the discussion in the rest of this section will deal with heat flow. However, there are many other forms of energy, including

—*mechanical energy*—the heat produced when gasoline burns in an automobile engine is converted to mechanical energy. This powers the car, enabling it to climb hills, accelerate when a traffic light turns yellow, and so on.

Matter and Energy 45

- *electrical energy*—through the absorption of electrical energy, water can be broken down into hydrogen and oxygen (recall Figure 2.1). The electrical energy is most often supplied by a different chemical reaction taking place within a storage battery.
- *light*—when a fuel burns, most of the energy is given off as heat. However, at least a small fraction of the energy is evolved as light. The absorption of light from the sun can bring about chemical reactions, the most important of which is photosynthesis, a process by which plants convert carbon dioxide and water into carbohydrates.

Units of Energy

A variety of different units can be used to express the amount of energy involved in a physical or chemical change. You may be most familiar with the *calorie* (cal). A calorie is the amount of energy in the form of heat required to raise the temperature of one gram of water one degree Celsius. Frequently, we use a larger unit, the *kilocalorie* (kcal)

$1 \text{ kcal} = 10^3 \text{ cal}$

The "calorie" referred to in nutrition is really a kilocalorie. When you are on a diet of "1000 calories" per day, you are supposed to eat enough food to produce 1000 kcal or 10^6 ordinary calories of energy.

Nowadays scientists are shifting to the joule (J) and kilojoules (kJ) as the base units of energy. A joule is smaller than a calorie; one calorie is a bit more than four joules

$1 \text{ cal} = 4.184 \text{ J}; \quad 1 \text{ kcal} = 4.184 \text{ kJ}$ \hfill (2.1)

Throughout the text we will ordinarily express energy changes in joules or kilojoules. Conversions from calories or kilocalories are readily made using Equation 2.1.

EXAMPLE 2.5
It takes 3.15 kcal of energy to decompose one gram of water into hydrogen and oxygen. Express this in kilojoules.

Solution
Here, as usual, we follow a conversion factor approach. We need a conversion factor to go from kilocalories to kilojoules. It should have "kJ" in the numerator and "kcal" in the denominator. The conversion factor is 4.184 kJ/1 kcal

$3.15 \text{ kcal} \times \dfrac{4.184 \text{ kJ}}{1 \text{ kcal}} = 13.2 \text{ kJ}$

Exercise
Suppose you were on a diet calling for 6.00×10^3 kJ. Express this in nutritional calories. Answer: 1.43×10^3 kcal (i.e., 1430 "calories").

Chapter 2

Conservation of Energy

To a certain extent, different forms of energy can be converted from one to another. The heat given off when coal burns can be used to vaporize water to steam. The energy in the steam can turn a turbine, producing mechanical energy. The turbine runs a generator which yields electrical energy. Electrical energy can be converted back to heat in an electric stove or to mechanical energy in a washing machine.

In energy conversions such as these, there is an exact balance between input and output; energy is neither gained nor lost. According to the **Law of Conservation of Energy,** *in any ordinary physical or chemical change, the total amount of energy remains constant.* If you burn enough coal to produce 1000 J of heat, none of that heat is "lost." Exactly 1000 J of energy appear in the surroundings in one form or another.

We have now considered two conservation laws dealing with mass and energy. There is one restriction on these individual laws. According to Einstein's theory of relativity, mass and energy are related by the equation

$$E = mc^2 \tag{2.2}$$

Albert Einstein

Here, E represents energy, m stands for mass, and c is the velocity of light (3×10^8 m/s). This equation implies that it should be possible to convert mass into energy or energy into mass. Taking this into account, it might be

Matter and Energy

best to modify the conservation laws as follows: *The total amount of energy + mass remains constant in any physical or chemical change.* In practice, mass–energy conversions are not detectable in ordinary chemical and physical changes. Only where nuclear processes are involved (see Chapter 15) does Equation 2.2 become of practical importance in chemistry.

Exothermic and Endothermic Reactions

In chemical reactions, energy effects are most often observed as heat. When a reaction is carried out in an open container at constant pressure, the heat flow is given a special name and symbol. It is referred to as the **enthalpy change, ΔH.** We can distinguish between two types of reactions:

1. **Exothermic** reactions, in which heat is evolved to the surroundings and **ΔH** is a **negative** quantity. Such reactions are very common in chemistry. The combustion of fuels always gives off heat. An example is the burning of methane (natural gas)

methane(gas) + oxygen(gas) → carbon dioxide(gas) + water(liquid); ΔH < 0

Ordinarily heat is given off when a compound is formed from the elements

hydrogen(gas) + oxygen(gas) → water(liquid); ΔH < 0

sodium(solid) + chlorine(gas) → sodium chloride(solid); ΔH < 0

2. **Endothermic** reactions, in which heat is absorbed from the surroundings and **ΔH** is a **positive** quantity. Usually, heat must be supplied to decompose a compound to elements. Such is the case with mercury(II) oxide mentioned earlier

mercury(II) oxide(solid) → mercury(liquid) + oxygen(gas); ΔH > 0

Another endothermic reaction is the decomposition of limestone (calcium carbonate) at 800°C

calcium carbonate(solid) → calcium oxide(solid) + carbon dioxide(gas); ΔH > 0

We will return to the topic of ΔH in chemical reactions in Chapter 5, after chemical equations have been introduced.

Heats of Fusion and Vaporization

In order to melt a solid, heat must be absorbed. When an ice cube melts in a glass of water, it absorbs heat from the water. This explains why the temperature of the water drops, approaching 0°C.

The heat required to melt a solid is referred to as the heat of fusion and given the symbol ΔH_{fus}. The fact that the **heat of fusion** is a positive quantity

TABLE 2.7 Heats of Fusion and Vaporization

Substance	ΔH_{fus}	ΔH_{vap}*
Benzene	126 J/g	395 J/g
Bromine	67.8 J/g	187 J/g
Ethyl alcohol	104 J/g	854 J/g
Mercury	11.6 J/g	296 J/g
Sodium chloride	517 J/g	3100 J/g
Water	333 J/g	2257 J/g

*ΔH_{vap} varies somewhat with temperature. Values given here are at the normal boiling point.

explains, at least in part, why the temperature remains constant while a pure solid is melting. The heat supplied by a Bunsen burner or other source is absorbed in melting the solid. As a result, it does not bring about an increase in temperature.

The heat of fusion of a substance is directly proportional to mass. It takes 333 J of heat to melt 1 g of ice, 666 J to melt 2 g, 3330 J to melt 10 g and so on. For this reason, heats of fusion must be quoted for a specific amount of solid, most often one gram. We would say that the heat of fusion of ice is 333 J/g. That quantity, 333 J/g, is a physical property of ice, similar in nature to density or melting point. Every pure solid has its own characteristic heat of fusion (Table 2.7).

Vaporization, like fusion, is endothermic. You have to supply heat, perhaps from a Bunsen burner or electric range, to boil water. When water vaporizes at room temperature, heat is also absorbed. This explains why evaporation is a cooling process. Perspiration evaporating on a hot summer day absorbs heat from your skin, lowering its surface temperature.

The heat required to vaporize a liquid is referred to as the **heat of vaporization, ΔH_{vap}**. Heat of vaporization, like heat of fusion, is directly proportional to mass. It is ordinarily quoted in joules per gram. For water, the heat of vaporization at 100°C is 2257 J/g. As you can see from Table 2.7, heat of vaporization per gram varies considerably from one substance to another. Like heat of fusion per gram, it is a characteristic physical property that can be used to identify a substance.

The fact that ΔH_{fus} and ΔH_{vap} are directly proportional to mass is useful in calculations. It allows us to relate the amount of substance melted or vaporized to the amount of heat absorbed in these processes (Example 2.6).

EXAMPLE 2.6
Using Table 2.7, calculate
a. the amount of heat that must be absorbed to melt 2.50 g of benzene
b. the mass of benzene that can be vaporized by 2.50 kJ of heat

Continued

Solution

a. Since the heat of fusion is 126 J/g, we can write the relation

$$126 \text{ J} = 1 \text{ g}$$

Following the conversion factor approach to go from grams to joules

$$\Delta H_{fus} = 2.50 \text{ g} \times \frac{126 \text{ J}}{1 \text{ g}} = 315 \text{ J}$$

b. According to Table 2.7, 395 J must be absorbed per gram of benzene vaporized. Hence

$$\text{mass benzene} = 2500 \text{ J} \times \frac{1 \text{ g}}{395 \text{ J}} = 6.33 \text{ g}$$

Exercise
How much heat is absorbed in vaporizing 0.200 g of Hg? Answer: 59.2 J.

Just as melting and vaporization are endothermic, the reverse processes, freezing and condensaton, are exothermic (Figure 2.4). When one gram of water freezes, 333 J of heat are given off, i.e., $\Delta H = -333$ J. When one gram of steam condenses, 2257 J of heat are evolved, i.e., $\Delta H = -2257$ J. A steam-heating system takes advantage of this effect. Fuel is burned to furnish the heat required to vaporize water in a steam boiler. The steam passes into radiators, where it condenses, giving off heat to warm a home, a classroom, or a factory.

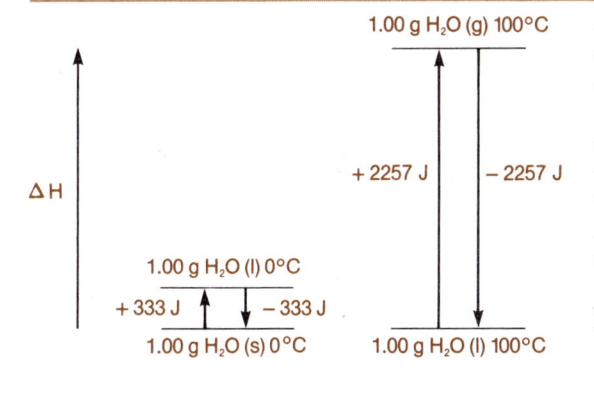

Figure 2.4 The amount of heat absorbed when one gram of ice melts ($\Delta H = +333$ J) is exactly equal to the amount of heat evolved when one gram of liquid water freezes ($\Delta H = -333$ J). Similarly, when one gram of liquid water vaporizes, $\Delta H = +2257$ J; for the reverse process, $\Delta H = -2257$ J.

Key Words

boiling point
calorie
chemical change

chemical property
compound
element

Continued

Chapter 2

endothermic
enthalpy change (ΔH)
exothermic
freezing point
group
heat of fusion
heat of vaporization
heterogeneous
homogeneous
joule
Law of Conservation of Energy
Law of Conservation of Mass
Law of Constant Composition
main-group element
matter

melting point
metal
metalloid
nonmetal
period
Periodic Table
physical change
physical property
product
reactant
solution
symbol
transition element
vapor pressure

Questions

2.1 Distinguish between an element and a compound; a pure substance and a mixture.

2.2 What is the most abundant element in the earth's crust? the most common compound?

2.3 Give the symbols of the following elements
 a. carbon b. lead c. chlorine
 d. potassium e. iron

2.4 Name the elements with the following symbols
 a. He b. Na c. Cu d. P
 e. Ag

2.5 In the Periodic Table, the horizontal rows are called _____; the vertical columns are referred to as _____.

In answering Questions 2.6–2.10, you may refer to a Periodic Table.

2.6 Classify each of the following as a main-group or transition element
 a. Ca b. Cr c. Al d. Fe e. W
 f. I

2.7 Classify each of the following as a metal, metalloid, or nonmetal
 a. Sr b. Br c. Ge d. V e. Ar

2.8 Give the names of the Group 4 elements that are
 a. metals b. metalloids c. nonmetals

2.9 Classify each of the following as a metal, metalloid, or nonmetal
 a. antimony b. mercury
 c. phosphorus d. neon e. sodium

2.10 Which of the following groups contain elements classified as metalloids?
 a. Group 1 b. Group 2 c. Group 3
 d. Group 4 e. Group 5

2.11 List several properties of metals that distinguish them from nonmetals.

2.12 List several physical properties that can be used to identify a substance.

2.13 The melting points of a pure substance is identical to its _____ _____.

2.14 Explain briefly why the boiling point of a liquid depends upon the pressure above it.

2.15 A certain substance is a solid at room temperature. Assuming it is one of those listed in Table 2.6, identify the substance.

2.16 A certain substance is a liquid in the temperature range 25–200°C. Which one of the

Matter and Energy

substances in Table 2.6 is it most likely to be? How would you prove that it is that substance?
2.17 List four different types of energy.
2.18 State the Law of Conservation of Mass; the Law of Conservation of Energy.
2.19 What is meant by an exothermic process? an endothermic process?
2.20 What is the sign of ΔH for an exothermic process? an endothermic process?

Problems

Types of Substances
2.21 Identify each of the following as an element, compound, solution, or heterogeneous mixture
  a. water   b. squash   c. blood
  d. hydrogen   e. copper
2.22 Identify each of the following as an element, compound, or mixture
  a. lead   b. air   c. cranberry sauce
  d. sodium chloride   e. sulfur

Percent Composition of Compounds
2.23 Using Table 2.5, calculate the masses of Na and Cl in a sample of sodium chloride weighing 2.601 g.
2.24 Using Table 2.5, determine the mass of carbon in one kilogram of carbon dioxide.
2.25 Using Table 2.5, calculate the mass of sucrose that contains 1.00 g of carbon.
2.26 It is found that a sample of potassium chloride weighing 2.124 g contains 1.010 g of chlorine.
  a. What is the percent of chlorine in potassium chloride?
  b. What is the percent of potassium in potassium chloride?

Types of Changes
2.27 Label the following changes as physical or chemical
  a. freezing milk
  b. formation of rust on an iron shovel
  c. formation of vinegar from wine
  d. distillation of water
  e. dissolving alcohol in water
2.28 Two elements, copper and sulfur, are heated together. Describe one or more experiments that would help you decide whether a physical or a chemical change has taken place.
2.29 Which of the following involve a physical change? a chemical change?
  a. burning charcoal
  b. dissolving instant coffee in water
  c. sublimation of dry ice
  d. electrolysis of water

Conservation of Mass
2.30 When 1.000 g of carbon burns in air, 3.664 g of carbon dioxide is formed. What is the mass of oxygen in this sample of carbon dioxide?
2.31 When 1.0000 g of water is electrolyzed, 0.8881 g of oxygen is formed. What mass of hydrogen is formed at the same time?

Energy Units
2.32 Carry out the following energy conversions (1 cal = 4.184 J)
  a. 619 cal = ? J
  b. 8.7×10^6 J = ? cal
  c. 42.3 kcal = ? kJ
  d. 5.7×10^{-2} kcal = ? J
2.33 Convert each of the following to kilojoules
  a. 34.6 cal   b. 2.467×10^{12} J
  c. 1.098 kcal

Chapter 2

2.34 The heat of fusion of ice is 333 J/g. Express this in calories per gram.

2.35 When one gram of carbon burns in air, 7.83 kcal of heat is evolved. Express this in kilojoules.

Heats of Fusion and Vaporization

2.36 Using Table 2.7, calculate the amount of heat that must be absorbed to
  a. melt 6.02 g of ice
  b. boil 1.87 g of water

2.37 Using Table 2.7, calculate the amount of heat evolved when 0.1245 g of steam condenses.

2.38 Using Table 2.7, calculate the mass of mercury that can be
  a. melted by 1.00 kJ of heat
  b. boiled by 1.00 kJ of heat

2.39 When 128 g of naphthalene melts, 19.3 kJ of heat is absorbed. Calculate the heat of fusion of naphthalene in joules per gram.

2.40 Using Table 2.7, calculate the number of calories of heat that must be absorbed to melt one kilogram of sodium chloride.

Multiple Choice

2.41 When a certain liquid is heated, it begins to boil at 62°C. The temperature rises steadily during the boiling process, eventually reaching 68°C. The liquid is most likely
  a. an element   b. a compound
  c. a solution   d. a metal

2.42 The normal boiling point of ethyl alcohol is 78°C. At a pressure below one atmosphere, the boiling point of ethyl alcohol is most likely
  a. 75°C   b. 78°C   c. 80°C
  d. 100°C

2.43 Water is most readily decomposed to elements in the laboratory by supplying
  a. heat   b. electrical energy
  c. mechanical energy   d. light

2.44 Which of the following energy units is the largest?
  a. calorie   b. joule   c. kilocalorie
  d. kilojoule

2.45 In order to heat 2.00 g of water from 15.0°C to 20.0°C, 10.0 cal of heat must be absorbed. This is equivalent to how many kilojoules?
  a. 41.8   b. 0.0418   c. 0.418
  d. 2.39   e. 0.00239

2.46 The heat of fusion of benzene is 126 J/g. The amount of heat required to melt 3.00 g of benzene is, in kilojoules
  a. 0.0420   b. 0.378   c. 42.0   d. 378

2.47 The heat of vaporization of bromine is 187 J/g. The mass, in kilograms, of bromine that can be vaporized by one kilojoule of heat is
  a. 5.35×10^{-6} b. 5.35×10^{-3}
  c. 5.35   d. 5.35×10^3 e. 5.35×10^6

2.48 How many of the following are pure substances?

   seawater, granite, bromine, carbon dioxide, petroleum

  a. 1   b. 2   c. 3   d. 4   e. 5

2.49 Which one of the following would boil at a constant temperature?
  a. gasoline   b. coffee   c. seawater
  d. benzene

To answer Questions 2.50–2.53, you may refer to a Periodic Table.

2.50 How many metals are there in Group 8?
  a. 0   b. 1   c. 2   d. 3   e. 6

2.51 How many of the following are transition metals?

   Ba, Os, Sn, Pb, Hg

  a. 1   b. 2   c. 3   d. 4   e. 5

2.52 How many elements are in the third period of the Periodic Table?
a. 2   b. 6   c. 8   d. 18   e. 32

2.53 What is the symbol of the element manganese?
a. M   b. Ma   c. Mn   d. Mg

2.54 A solid that is a good conductor of electricity is most likely a
a. metal   b. nonmetal   c. metalloid   d. compound

2.55 Potassium chlorate contains the three elements potassium, chlorine, and oxygen. The mass percents of K and Cl are 31.9 and 28.9, respectively. The mass of oxygen in 2.00 g of potassium chlorate is
a. 0.578 g   b. 0.638 g   c. 0.784 g   d. 2.00 g

2.56 The mass percent of carbon in carbon dioxide is 27.3. The mass of carbon dioxide that contains 1.00 g of carbon is
a. 0.273 g   b. 0.727 g   c. 1.00 g   d. 1.38 g   e. 3.66 g

2.57 The process by which a solid is converted directly to a vapor is called
a. condensation   b. freezing
c. melting   d. sublimation
e. vaporization

2.58 Which one of the following processes is most likely to result in the formation of a new compound?
a. condensation   b. freezing
c. chemical change   d. stirring
e. heating

2.59 Which one of the following is *not* a physical property of hydrogen?
a. colorless   b. odorless
c. combustible   d. density = 0.0823 g/L at 25°C, 1 atmosphere

2.60 The normal boiling point of a liquid is the temperature at which
a. boiling occurs when heated normally
b. bubbles form in the liquid
c. the liquid boils when the pressure is reduced to zero
d. the liquid boils when the pressure is one atmosphere

The Building Blocks of Matter

Learning Objectives
After studying this chapter, you should be able to:

1. State the major postulates of the atomic theory of matter.
2. List three subatomic particles and indicate their charges, relative masses, and locations within the atom.
3. Define atomic number, mass number, and isotope (Example 3.1).
4. Given the number of protons and neutrons in an atom, write its nuclear symbol. Carry out the reverse operation (Example 3.2).
5. Define and give examples of: molecule, cation, anion.
6. Given the symbol of an ion, state the number of protons and electrons present (Example 3.3).
7. Given their atomic masses, calculate the ratio of the masses of two different atoms (Example 3.4).
8. Given the atomic mass of an element, calculate the mass in grams of a specified number of atoms, or the number of atoms in a sample of known mass (Example 3.5).
9. Given the formula of a substance, calculate its molar mass (Examples 3.6, 3.7).
10. Knowing or having calculated the molar mass of a substance, convert moles to grams or vice versa (Examples 3.8, 3.9).

It has been said that science has no ethical basis, that is no more than a cold, impersonal way of arriving at the objective truth about natural phenomena. This view I wish to challenge, since it is my belief that by examining critically the nature, origins, and methods of science we may logically arrive at a conclusion that science is ineluctably involved in questions of values, is inescapably committed to standards of right and wrong, and unavoidably moves in the large toward social aims.

Bentley Glass

Photograph by Peter L. Kresan.

CHAPTER 3

In Chapters 1 and 2, we took what might be called a macroscopic view of matter. We dealt with the properties of bulk samples of matter, samples that can be seen and weighed. In this chapter, we will look at matter from a microscopic point of view. Our emphasis will be on the particle structure of matter, starting with that tiny, invisible, unweighable particle, the atom.

In studying the atom, we will be looking for answers to such questions as: What is meant by an atom? How do atoms differ from one another? Of what are atoms composed? How do atoms combine with one another to form elements and compounds? How much do atoms weigh?

To answer these and other questions about atoms, we start with a basic survey of atomic theory (Section 3.1). The nature of subatomic particles (electrons, protons, and neutrons) is investigated in Section 3.2. This is followed (Section 3.3) by a brief introduction to two important types of particles derived from atoms: molecules and ions. The relative and actual masses of different atoms are considered in Section 3.4. The chapter concludes (Section 3.5) with a discussion of the most important counting unit in chemistry, the mole. We will find the mole concept useful in dealing with all kinds of substances, whether they are made up of atoms, molecules, or ions.

3.1 Atomic Theory

The concept of the atom as the ultimate particle of matter goes back to the Greeks. It appears to have originated with Leucippus, a Greek philosopher, in about 450 B.C. His disciple Democritus expanded on the rather vague ideas of Leucippus. He suggested that all matter was made up of small particles that could not be broken down further. Democritus named this particle "atomos," meaning indivisible.

Nothing came of the ideas of Leucippus and Democritus, in part because they were rejected by Plato and Aristotle, who were much more influential. Aristotle believed that matter could be subdivided indefinitely and that there could be no such thing as an ultimate particle. His ideas prevailed for about 2000 years.

Modern atomic theory is based on the work of John Dalton, an English schoolteacher and chemist. He revived the idea of the atom in 1808 by publishing a treatise entitled a *New System of Chemical Philosophy*. Most of Dalton's ideas have stood the test of time; some have been modified over the years. The major features of present-day atomic theory are summarized in the following statements.

John Dalton

1. **An element is composed of tiny particles called atoms.** To get an idea of just how tiny an atom is, consider the hydrogen atom. We know that a hydrogen atom, on the average, weighs 1.67×10^{-24} g. The smallest sample of hydrogen that can be weighed on an analytical balance contains a billion billion atoms (10^{18} atoms).
2. **All the atoms of a given element have the same chemical properties; they differ chemically from atoms of all other elements.** Hydrogen atoms have a unique set of chemical properties that distinguish them from all other atoms
3. **Atoms retain their identity in an ordinary chemical or physical change.** Atoms are not created, destroyed, or converted into other kinds of atoms. In a reaction involving hydrogen, there are the same number of hydrogen atoms in the products as in the reactants.

4. **Compounds contain atoms of two or more elements in a fixed ratio.** This ratio can be expressed by small whole numbers. The compound water contains hydrogen and oxygen atoms. In every sample of water, there are two hydrogen atoms for every oxygen atom.
5. **When elements form more than one compound, the atom ratios differ.** Carbon and oxygen form two different compounds. In one, called carbon monoxide, there is one oxygen atom for every carbon atom. In the other, called carbon dioxide, there are two oxygen atoms for every carbon atom.

The atomic theory offers a simple explanation for two of the basic laws of chemistry discussed in Chapter 2. The third statement leads directly to the *Law of Conservation of Mass.* If atoms are "conserved" in a reaction, there can be no change in mass. The fourth statement leads to the *Law of Constant Composition.* If the atom ratio in a compound is fixed, the mass ratio or mass percents must also be constant.

3.2 Subatomic Particles

Dalton believed that atoms were the ultimate particles of matter. In his words, "Thou canst not split an atom." This idea prevailed through much of the 19th century. Then, as experimental techniques improved, several subatomic particles were discovered. In this section we will consider the properties of three such particles and the way in which they are combined in atoms.

Electrons

The English physicist J. J. Thomson discovered the first subatomic particle in 1897. Using the apparatus shown in Figure 3.1, he found that, at very high voltages and low gas pressures, rays of light became visible. These are called cathode rays. Thomson studied the behavior of cathode rays in magnetic and electrical fields. He concluded that they consist of a stream of tiny, negatively charged particles. Thomson named this new particle the **electron.** It has a unit negative charge (-1). This same particle is produced regardless of the identity of the gas in the tube. This suggests that the electron is a fundamental particle, present in the atoms of all elements.

Thomson was able to determine the charge-to-mass ratio of the electron. A few years later, in 1909, Robert Millikan at the University of Chicago, determined the charge on the electron. Putting these two measurements together, it was possible to calculate the mass of the electron. This turns out to be 9.11×10^{-28} g, a very small quantity indeed. The electron weighs only a little more than 1/2000 as much as a hydrogen atom.

The Building Blocks of Matter

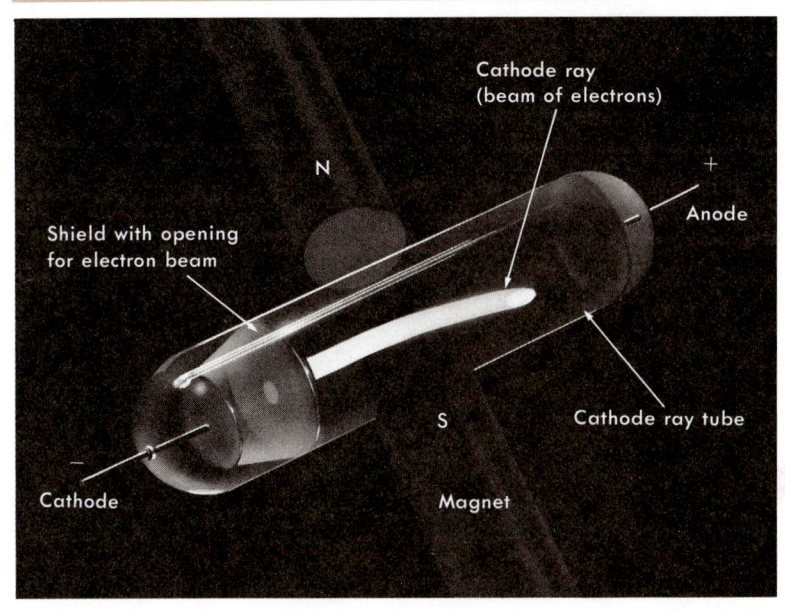

Sir Joseph John Thomson

Figure 3.1 J. J. Thomson (1856–1940) found that, at high voltages and low gas pressures, rays of light became visible in a tube such as this. These cathode rays were bent by electrical or magnetic fields. By studying the deflection of the rays, Thomson showed that they consist of a stream of tiny, negatively charged particles (electrons).

Chapter 3

The Nucleus: Protons and Neutrons

In 1911, Ernest Rutherford in England carried out an experiment that was crucial to the understanding of the structure of the atom. He bombarded a thin gold foil (Figure 3.2) with alpha particles (helium atoms stripped of their electrons). Most of the alpha particles went through the foil undeflected. However, about one of every 8000 bounced back at an angle of more than 90°. This was entirely unexpected. In Rutherford's words, "It was about as credible as if you had fired a 15-inch shell at a piece of tissue paper and it came back and hit you."

Ernest Rutherford

On the basis of his experiments, Rutherford proposed a model of the atom that we use today. Atoms have a very small center called the **nucleus.** The nucleus is positively charged and contains nearly all of the mass of the atom. Electons are located in the region outside the nucleus. The negative charge of the electrons just balances the positive charge of the nucleus, making atoms neutral.

Rutherford's nucleus was later shown to contain two different kinds of particles. One of these, the **proton,** has a positive charge (+1), equal in magnitude but opposite in sign to that of the electron. The other nuclear particle, the **neutron,** has no charge. The proton and neutron have about the same mass, nearly equal to that of an ordinary hydrogen atom. this means that protons and neutrons are about 2000 times as heavy as electrons. The properties of these subatomic particles are summarized in Table 3.1.

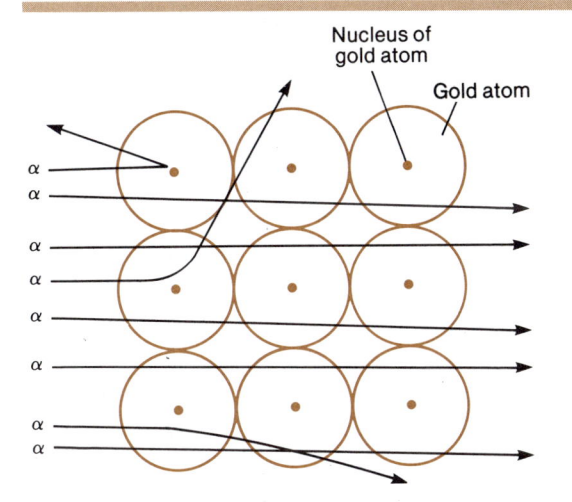

Figure 3.2 Rutherford directed a beam of alpha particles at a gold foil. Most of the foil undeflected. A few, such as the one at the upper left of the figure, were reflected back toward the source. This showed that there must be a tiny, positively charged nucleus in the gold foil.

TABLE 3.1 Components of the Atom

Particle	Charge	Mass*	Location
Electron	−1	1/2000	outside nucleus
Proton	+1	1	in the nucleus
Neutron	0	1	in the nucleus

*Relative to a hydrogen atom (approximate)

Atomic Number

From a chemical standpoint, the most important property of an atom is the number of protons in the nucleus. This quantity is referred to as the **atomic number** (at. no.)

atomic number = number of protons in the nucleus (3.1)

Each element has a characteristic number of protons in the nucleus and hence has its own unique atomic number. All hydrogen atoms have one proton in the nucleus; the atomic number of hydrogen is one. All helium atoms have two protons in the nucleus; the atomic number of helium is two.

In the Periodic Table, the atomic number of an element is shown in color above its symbol. Looking at the Periodic Table inside the front cover of this text you should find that:

—potassium (K) has an atomic number of 19; there must then be 19 protons in the nucleus of every potassium atom.
—uranium (U) has an atomic number of 92; the nucleus of every uranium atom contains 92 protons.

Potassium is the only atom whose nucleus contains 19 protons. Any atom with 92 protons in its nucleus must be a uranium atom.

Isolated atoms are electrically neutral. This means that the number of electrons (−1 charge) outside the nucleus must be exactly equal to the number of protons (+1 charge) in the nucleus. Putting it another way, the number of electrons in a neutral atom is equal to the atomic number of the element. There is one electron in a hydrogen atom (at. no. H = 1), two electrons in a helium atom (at. no. He = 2), 19 electrons in a potassium atom (at. no. K = 19), and so on.

Mass Number and Isotopes

Since an electron is so much lighter than a proton or neutron, the mass of an atom is determined mainly by the number of protons and neutrons in its nucleus. This quantity is known as the **mass number**

$$\text{mass number} = \text{number of protons} + \text{number of neutrons} \tag{3.2}$$

An atom with one proton and zero neutrons in the nucleus has a mass number of 1 + 0 = 1. An atom with one proton and one neutron has a mass number of 1 + 1 = 2.

EXAMPLE 3.1
Give the number of protons, electrons, and neutrons in an atom of atomic number 17 and mass number 35.

Solution
no. protons = no. electrons = atomic no. = 17
no. neutrons = mass no. − no. protons = 35 − 17 = 18

Exercise
Give the symbol of this atom, using the Periodic Table. Answer: Cl.

All atoms of a given element must have the same number of protons in the nucleus. However, atoms of an element may, and often do, differ in the number of neutrons in the nucleus. This is the case with hydrogen. Most hydrogen atoms found in nature (99.985%) have no neutrons in the nucleus. A few (0.015%) have one neutron. These two different kinds of atoms are sometimes referred to as "light" hydrogen and "heavy" hydrogen (or deuterium).

light hydrogen: 1 proton, 0 neutrons; at. no. = 1, mass no. = 1

heavy hydrogen: 1 proton, 1 neutron; at. no. = 1, mass no. = 2

The Building Blocks of Matter

This situation is typical of most elements. They exist in nature as a mixture of **isotopes.** *Isotopes are atoms that have the same number of protons but a different number of neutrons.* There are two isotopes of hydrogen and two isotopes of boron

"B-10" : 5 protons, 5 neutrons; at. no. = 5, mass no. = 10

"B-11" : 5 protons, 6 neutrons; at. no. = 5, mass no. = 11

The uranium found in nature is made up of three isotopes

"U-234" : 92 protons, 142 neutrons; at. no. = 92, mass no. = 234

"U-235" : 92 protons, 143 neutrons; at. no. = 92, mass no. = 235

"U-238" : 92 protons, 146 neutrons; at. no. = 92, mass no. = 238

Isotopes of an element have two important characteristics.

1. *They have identical or nearly identical chemical properties.* The three isotopes of uranium are chemically identical. The same is true of the two isotopes of boron; they behave the same way in reactions. The two isotopes of hydrogen are very similar chemically, even though they differ in mass by a factor of two. In a few instances, it is possible to detect small differences in the chemical behavior of "light" and "heavy" hydrogen.
2. *They nearly always occur in nature in a fixed ratio.* Hydrogen contains 99.985% of H-1 and 0.015% of H-2. In samples of the element boron, or boron compounds, the percents of the two elements are fixed. About 19% (more exactly, 18.83%) of the boron atoms are of the B-10 variety while the remainder are B-11. Lead is one of the few elements where the percents of the different isotopes can vary slightly depending upon the source.

Nuclear Symbols

To identify a nucleus, we specify the atomic number, the mass number, and the symbol of the element. In a nuclear symbol

—the atomic number appears as a subscript at the lower left of the symbol.
—the mass number appears as a superscript at the upper left of the symbol.

The isotopes of hydrogen, boron, and uranium that we have discussed have the following nuclear symbols

$^{1}_{1}H$, $^{2}_{1}H$; $^{10}_{5}B$, $^{11}_{5}B$; $^{234}_{92}U$, $^{235}_{92}U$, $^{238}_{92}U$

EXAMPLE 3.2
a. Give the nuclear symbols of the isotopes of carbon (at. no. = 6) which have 6, 7, and 8 neutrons in the nucleus.
b. Give the number of protons and neutrons in an atom of $^{23}_{11}Na$; $^{122}_{50}Sn$

Continued

Solution
a. The atomic number in each case is 6. The mass numbers are

$6 + 6 = 12; 6 + 7 = 13; 6 + 8 = 14$

The nuclear symbols are

$^{12}_{6}C$, $^{13}_{6}C$, $^{14}_{6}C$

(These isotopes are often referred to as C-12, C-13, and C-14.)

b. The number of protons is the atomic number. The number of neutrons is found by subtracting the atomic number from the mass number.

$^{23}_{11}Na$: 11 protons, 12 neutrons

$^{122}_{50}Sn$: 50 protons, 72 neutrons

Exercise
Give the number of protons, neutrons, and electrons in a potassium atom of mass number 39. Answer: 19, 20, 19.

3.3 Molecules and Ions

Isolated atoms rarely occur as such in nature. Only among the *noble gases* in Group 8 of the Periodic Table do we find individual atoms with very weak forces between them. In gaseous helium, the basic "building block" is the He atom; in argon, it is the Ar atom, and so on.

In most elements and all compounds, atoms interact with each other to form more complex structural units. One such unit, the **molecule,** is found in several nonmetallic elements and in many compounds formed between nonmetals. In contrast, charged **ions** are common in compounds formed by metals with nonmetals.

Molecules

A molecule is a group of two or more atoms held to one another by strong forces called chemical bonds. One of the simplest molecules is that of the element hydrogen. It consists of two hydrogen atoms joined by a chemical bond. The structure of the molecule can be indicated as:

H—H

with the understanding that the straight line between the two atoms represents a chemical bond. More commonly, we write the **molecular formula,** H_2. The subscript "2" in the formula indicates that there are two hydrogen atoms in the molecule. The H_2 molecule is the basic structural unit in gaseous hydrogen. There are no isolated hydrogen atoms as such in this substance; instead they are always found in pairs. Many nonmetallic elements besides hydrogen form diatomic molecules. This is the case with all of the Group 7 elements *(halogens)*: F_2, Cl_2, Br_2, I_2. The two elements that make up most of the atmosphere, nitrogen and oxygen, exist as molecules of N_2 and O_2.

Compounds formed by two different nonmetals are most often molecular. An example in the gaseous compound hydrogen chloride. Here, the basic structural unit is a molecule in which a hydrogen atom is bonded to a chlorine atom. We can represent the hydrogen chloride molecule as:

H—Cl

More commonly, we use the molecular formula, HCl. (Note that no subscript is used in the molecular formula when only one atom of a given type is present in the molecule.) Probably the best-known molecule to most people is that of the compound water. In the H_2O molecule, two hydrogen atoms are bonded to a central oxygen atom

H—O—H

The geometries of several simple molecules are shown in Figure 3.3 (the small spheres represent hydrogen atoms). Many more complex molecules are known, particularly among organic (carbon) compounds. For example, aspirin is a molecular substance with the formula $C_9H_8O_4$. The structural unit in ordinary sugar (sucrose) is a molecule with the formula $C_{12}H_{22}O_{11}$.

Ions

Compounds formed by metals with nonmetals seldom contain molecules. Instead, they are built up of charged particles called ions. An example of an ionic compound is sodium chloride, NaCl, formed when sodium metal reacts with the nonmetal chlorine. In sodium chloride, there are two kinds of ions

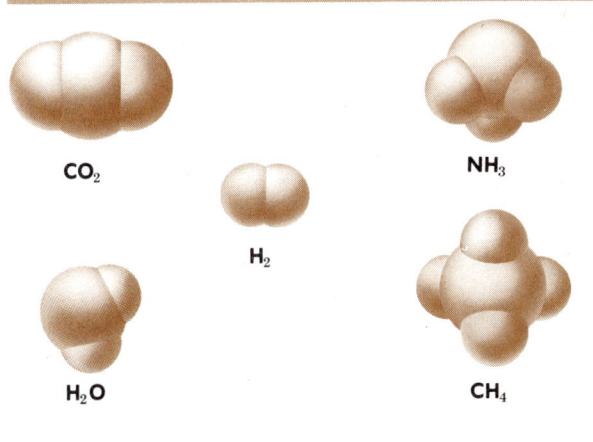

Figure 3.3 Examples of molecules.

Chapter 3

1. An Na$^+$ ion, formed when an Na atom loses an electron

   Na(11 p$^+$, 11 e$^-$) → Na$^+$(11 p$^+$, 10 e$^-$) + e$^-$

2. A Cl$^-$ ion, formed when a Cl atom gains an electron

   Cl(17 p$^+$, 17 e$^-$) + e$^-$ → Cl$^-$(17 p$^+$, 18 e$^-$)

Notice that in forming an ion from an atom, there is a loss or gain of *electrons;* the number of protons, and hence the identify of the element, stays the same. When an atom loses electrons, it forms a positive ion (**cation**). A cation such as Na$^+$ contains more protons than electrons. When an atom gains electrons, it forms a negative ion (**anion**). An anion such as Cl$^-$ contains more electrons than protons.

A small portion of a sodium chloride crystal is shown in Figure 3.4. The crystal consists of alternating Na$^+$ and Cl$^-$ ions held together by strong electrical attractive forces. There are no small discrete molecules in sodium chloride. We write its formula as NaCl to indicate that there are an equal number of Na$^+$ and Cl$^-$ ions in the crystal. This must be the case since the charges are equal in magnitude (+1 and −1) and the crystal, like all bulk samples of matter, must be electrically neutral.

Figure 3.4 The crystal structure of sodium chloride, NaCl. Notice that each sodium ion, Na$^+$, is surrounded by chloride ions, Cl$^-$, and vice versa.

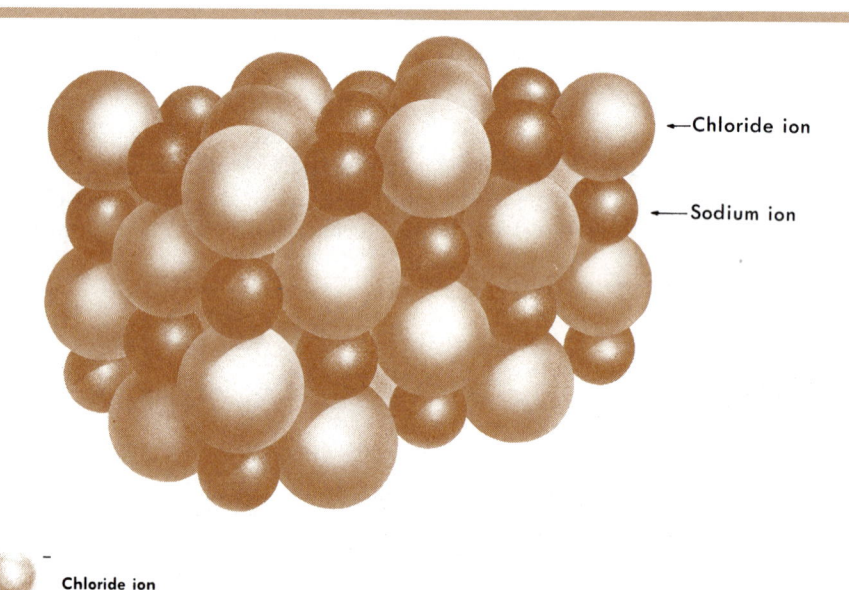

← Chloride ion

← Sodium ion

− Chloride ion

+ Sodium ion

Another ionic compound, calcium oxide (CaO), contains the Ca^{2+} and O^{2-} ions

$Ca\ (20\ p^+, 20\ e^-) \rightarrow Ca^{2+}\ (20\ p^+, 18\ e^-) + 2\ e^-$

$O\ (8\ p^+, 8\ e^-) + 2\ e^- \rightarrow O^{2-}\ (8\ p^+, 10\ e^-)$

Calcium oxide has a structure similar to that of sodium chloride. Its formula is written CaO, since equal numbers of Ca^{2+} and O^{2-} ions are present. One $+2$ ion (Ca^{2+}) is required to balance one -2 ion (O^{2-}).

EXAMPLE 3.3
How many protons and electrons are there in an
a. Al^{3+} ion?   b. N^{3-} ion?

Solution
a. The atomic number of aluminum is 13. The Al atom contains 13 p^+ and 13 e^-. To form a $+3$ ion, three electrons must be lost. The Al^{3+} ion contains 13 p^+ and 10 e^-.
b. The N atom (at. no. = 7) contains 7 p^+ and 7 e^-. The N^{3-} ion contains three extra electrons: 7 p^+, 10 e^-.

Exercise
Give the symbol and charge of the ion that has 21 p^+ and 18 e^-; 16 p^+ and 18 e^-.
Answer: Sc^{3+}, S^{2-}.

3.4 Masses of Atoms

As we pointed out earlier, individual atoms are too small to be seen, let alone weighed. However, it is possible to determine the relative masses of different atoms. That is, we can find out how heavy an atom is compared with an atom of a different element. We will now consider how this is done. Later in this section, we will show how it is possible to calculate the masses of individual atoms, tiny as they are.

Relative Atomic Masses

Referring back to the Periodic Table on the inside front cover, you will note a number in black below the symbol of each element. These numbers (1.0079 for H, 15.9994 for O, 32.06 for S, etc.) are referred to as **atomic masses**. The atomic mass of an element tells us how heavy, on the average, an atom of that element is relative to a $^{12}_{6}C$ (i.e., C-12) atom, which is assigned an

atomic mass of exactly 12. Since the atomic masses of hydrogen, oxygen, and sulfur are about 1, 16, and 32, respectively, we can say that

An H atom is about $\frac{1}{12}$ as heavy as C-12 atom.

An O atom is about $\frac{16}{12}$ as heavy as C-12 atom.

An S atom is about $\frac{32}{12}$ as heavy as a C-12 atom.

The atomic mass scale is set up using C-12 as a reference. However, we need not restrict mass comparisons to that atom. Atomic masses can be used to find the relative masses of any and all kinds of atoms. For example, taking the atomic masses of H, O, and S to be 1, 16, and 32, we can say that

An H atom is $\frac{1}{16}$ as heavy as an O atom.

An H atom is $\frac{1}{32}$ as heavy as an S atom.

An O atom is 16 times as heavy as an H atom.

An O atom is $\frac{16}{32}$ or $\frac{1}{2}$ as heavy as an S atom.

An S atom is 32 times as heavy as an H atom.

An S atom is $\frac{32}{16}$ or 2 times as heavy as an O atom.

EXAMPLE 3.4
Using the Periodic Table as a source of atomic masses, find the ratio of the mass of an iron atom to that of a
a. chlorine atom   b. mercury atom

Solution
The atomic masses of Fe, Cl, and Hg are 55.847, 35.453, and 200.59, respectively.
a. $\frac{\text{mass Fe atom}}{\text{mass Cl atom}} = \frac{55.847}{35.453} = 1.5752$

An iron atom is a bit more than 3/2 as heavy as a chlorine atom.
b. $\frac{\text{mass Fe atom}}{\text{mass Hg atom}} = \frac{55.847}{200.59} = 0.27841$

An iron atom weighs somewhat more than 1/4 as much as a mercury atom.

Continued

> **Exercise**
> How many nitrogen atoms would be required to have the same mass as 1.00 iron atom? Answer: 3.99 atoms.

The proton and neutron, both of which have nearly the same mass as a hydrogen atom, have atomic masses very close to one. This means that individual isotopes such as $^{1}_{1}H$ and $^{16}_{8}O$ have atomic masses very nearly equal to their mass numbers (1, 16). As a result, elements that consist entirely or almost entirely of a single isotope have atomic masses that are very close to whole numbers. Hydrogen, which consists almost entirely of $^{1}_{1}H$ (99.985%), has an atomic mass only slightly greater than one (i.e., 1.0079). Oxygen, which is mostly $^{16}_{8}O$ (99.759%), has an atomic mass of almost exactly 16 (i.e., 15.9994). The atomic mass of the element carbon is slightly greater than 12 (i.e., 12.011). This reflects the fact that the principal isotope, $^{12}_{6}C$, is mixed with a small amount (about 1%) of the heavier isotope, $^{13}_{6}C$.

Elements whose atomic masses are not close to whole numbers usually contain appreciable amounts of more than one isotope. Consider, for example, the element boron, which has an atomic mass of 10.81. As you might guess, naturally occurring boron is a mixture of two isotopes, $^{10}_{5}B$ and $^{11}_{5}B$. Most of the boron is in the form of the heavier isotope (81%), so the atomic mass is closer to 11 than 10. As another example, neon (atomic mass = 20.179) consists mostly of $^{20}_{10}Ne$. Small amounts of two heavier isotopes (0.3% $^{21}_{10}Ne$ and 8.8% $^{22}_{10}Ne$) raise the atomic mass somewhat above 20.

Absolute Masses of Atoms: Avogadro's Number

For our purposes, it is usually sufficient to know the relative masses of different atoms. These, as we have seen, can be calculated from a table of atomic masses. We would like, though, to go one step further and obtain the actual masses of atoms in grams. We expect these masses to be extremely small; the question is, just how small are they?

To answer this question, we start with an important observation. *In X grams of an element, where X is its atomic mass, there is a certain definite number of atoms that we will call N. This number, N, is the same for all elements.* For example, the number of hydrogen atoms in 1.0079 g of H is the same as the number of oxygen atoms in 15.9994 g of O or the number of sulfur atoms in 32.06 g of sulfur.

The italicized statement just made is by no means obvious. However, we can demonstrate that it is correct by a simple exercise in logic. Suppose we take N to be the number of atoms in twelve grams of C-12. Since a hydrogen atom weighs 1.0079/12 as much as C-12 atom

N hydrogen atoms weigh $\dfrac{1.0079}{12} \times 12 \text{ g} = 1.0079 \text{ g}$

By the same reasoning

$$\text{N oxygen atoms weigh } \frac{15.9994}{12} \times 12 \text{ g} = 15.9994 \text{ g}$$

$$\text{N sulfur atoms weigh } \frac{32.06}{12} \times 12 \text{ g} = 32.06 \text{ g}$$

In other words, 1.0079 g of hydrogen, 15.994 g of oxygen, and 32.06 g of sulfur all contain the same number of atoms, N. This is the relationship we set out to prove.

At this point, we could calculate the masses of individual atoms if we knew one important quantity: the value of N. As you can imagine, chemists and other scientists over the years have devoted a great deal of attention to the evaluation of this important number. It is beyond the scope of this text to describe the experiments that lead to accurate values of N. Suffice it to say that N, often called **Avogadro's number,** is, to four significant figures, 6.022×10^{23}

$$N = \text{Avogadro's number} = 6.022 \times 10^{23} \tag{3.3}$$

In other words, **there are 6.022×10^{23} atoms in X grams of any element, where X is the atomic mass.** This means that 1.0079 g of H contains 6.022×10^{23} hydrogen atoms, 15.9994 g of O contains 6.022×10^{23} oxygen atoms, and 32.06 g of S contains 6.022×10^{23} sulfur atoms.

We think you'll agree that 6.022×10^{23} is a very large number. To get some idea of just how big it is, suppose you had 6.022×10^{23} marbles. If these marbles were piled up on the surface of the earth, they would cover it to a depth of about 1 kilometer. Of course, you will never see Avogadro's number of marbles, potato chips, or any other visible objects. However, when you look at 32.06 g of sulfur or 63.55 g of copper, you are in effect observing 6.022×10^{23} atoms.

Knowing the value of Avogadro's number, we can now calculate

—the masses of individual atoms or specified numbers of atoms (Example 3.5a)
—the number of atoms in a sample of known mass (Example 3.5b)

EXAMPLE 3.5
Taking Avogadro's number to be 6.022×10^{23} and the atomic mass of potassium to be 39.098, calculate
a. the mass of a K atom.
b. the number of K atoms in a sample of potassium weighing 1.00 g.

Continued

Solution
For potassium: 6.022×10^{23} atoms $= 39.098$ g
This relation gives us the conversion factors we need to solve the problem.
a. Here we want to "convert" a potassium atom to grams. We start with one atom and multiply by the conversion factor 39.098 g/6.022×10^{23} atoms

$$\text{mass of 1 atom} = 1 \text{ atom} \times \frac{39.098 \text{ g}}{6.022 \times 10^{23} \text{ atoms}} = 6.493 \times 10^{-23} \text{ g}$$

This is a very small mass, as we would expect. In general, atoms have masses ranging from a little more than 10^{-24} g (mass H atom $= 1.674 \times 10^{-24}$ g) to somewhat greater than 10^{-22} g (mass U atom $= 3.953 \times 10^{-22}$ g).

b. This time we want to go in the opposite direction, from grams to atoms. We use the conversion factor 6.022×10^{23} atoms/39.098 g. The answer, as we would expect, is huge; it takes a lot of atoms to weigh a gram.

$$\text{number of atoms} = 1.00 \text{ g} \times \frac{6.022 \times 10^{23} \text{ atoms}}{39.098 \text{ g}} = 1.54 \times 10^{22} \text{ atoms}$$

Exercise
What is the mass in grams of 213 Na atoms? Answer: 8.13×10^{-21} g.

3.5 The Mole Concept

Eggs are commonly sold by the dozen. Paper clips can be purchased by the gross. The dozen (12 units) is a convenient counting unit for poultry farmers. The gross (144 units) is a more useful counting unit for small items like paper clips. Neither of these is particularly useful to a chemist counting atoms or molecules. Here, a much larger unit is required since the items to be counted are so small. Chemists use Avogadro's number, 6.022×10^{23}. This counting unit is so important in chemistry that it is given a special name, the **mole** (abbreviated as **mol**). **A mole refers to Avogadro's number of items,** whether they be atoms, molecules, or anything else.

One mole of H atoms $= 6.022 \times 10^{23}$ H atoms
One mole of H_2O molecules $= 6.022 \times 10^{23}$ H_2O molecules
One mole of neutrons $= 6.022 \times 10^{23}$ neutrons
One mole of marbles $= 6.022 \times 10^{23}$ marbles

You will recall that Avogadro's number of atoms of an element has a mass in grams that is numerically equal to the atomic mass of the element. Since one mole represents Avogadro's number of atoms, it follows that the

molar mass of an element (grams per mole) is numerically equal to its atomic mass (Figure 3.5).

1 mol C = 12.01 g C; molar mass C = 12.01 g/mol

1 mol S = 32.06 g S; molar mass S = 32.06 g/mol

1 mol Cu = 63.55 g Cu; molar mass Cu = 63.55 g/mol

It is possible to extend this reasoning to obtain the molar mass of any substance, element or compound, molecular or ionic. The rule is a simple one. **The molar mass of a substance (grams per mole) is numerically equal to its formula mass. The formula mass is the sum of the atomic masses of the atoms in the formula.** To illustrate how this rule is applied, consider water, H_2O. Rounding off the atomic masses of H and O to four significant figures

formula mass H_2O = 2(1.008) + 16.00 = 18.02

one mole of H_2O weighs 18.02 g

molar mass H_2O = 18.02 g/mol

EXAMPLE 3.6
Calculate the molar mass of
a. monatomic oxygen, O   b. O_2 c. O_3 (ozone)

Solution
The atomic mass of oxygen is, to four significant figures, 16.00.
a. formula mass O = 16.00; molar mass O = 16.00 g/mol
b. formula mass O_2 = 32.00; molar mass O_2 = 32.00 g/mol
c. formula mass O_3 = 48.00; molar mass O_3 = 48.00 g/mol

Exercise
What is the molar mass of P_4? Answer: 123.88 g/mol.

Figure 3.5 One mole of carbon (12.01 g), sulfur (32.06 g), and copper (63.55 g). Each sample contains the same number of atoms, 6.022×10^{23}.

EXAMPLE 3.7

Calculate the molar mass (g/mol) of
a. magnesium nitride, Mg_3N_2 b. sugar, $C_{12}H_{22}O_{11}$
(Round off atomic masses to four significant figures.)

Solution
a. formula mass = 3(atomic mass Mg) + 2(atomic mass N)
   = 3(24.30) + 2(14.01) = 100.92
   molar mass = 100.92 g/mol
b. formula mass = 12(atomic mass C) + 22(atomic mass H) + 11(atomic mass O)
   = 12(12.01) + 22(1.008) + 11(16.00) = 342.30
   molar mass = 342.30 g/mol

Exercise
What is the molar mass of aspirin, $C_9H_8O_4$? Answer: 180.15 g/mol.

Mole–Gram Conversions

Very often in chemistry, we want to convert from moles to grams or vice versa. The molar mass, in effect, gives us the conversion factors we need to make this type of conversion. The procedure followed is illustrated in Examples 3.8 and 3.9.

EXAMPLE 3.8

Calculate the number of grams in 1.20 mol of
a. Mg_3N_2 b. $C_{12}H_{22}O_{11}$

Solution
The molar masses of Mg_3N_2 and $C_{12}H_{22}O_{11}$ were found in Example 3.7 to be 100.92 and 342.30 g/mol, respectively. Since we want to go from moles to grams, these are just the conversion factors we need

a. mass of Mg_3N_2 = 1.20 mol × $\dfrac{100.92 \text{ g}}{1 \text{ mol}}$ = 121 g

b. mass of $C_{12}H_{22}O_{11}$ = 1.20 mol × $\dfrac{342.30 \text{ g}}{1 \text{ mol}}$ = 411 g

Exercise
What is the mass in grams of 1.20 mol of H_2O? Answer: 21.6 g.

EXAMPLE 3.9

Calculate the number of moles in 1.60 g of $CaCl_2$.

Continued

Solution

Let us first find the molar mass of $CaCl_2$.

formula mass = 40.08 + 2(35.45) = 110.98

molar mass = 110.98 g/mol

Since we want to go from grams to moles, the conversion factor we need is 1 mol/110.98 g

$$\text{no. moles} = 1.60 \text{ g} \times \frac{1 \text{ mol}}{110.98 \text{ g}} = 0.0144 \text{ mol}$$

Exercise

How many moles are there in 20.0 g of H_2O? Answer: 1.11 mol.

So far our discussion of the mole concept has dealt entirely with pure substances. However, many of the experiments that you are carrying out in the laboratory involve solutions. Most of the solutions you work with contain two components. One of these, the *solvent*, does the dissolving; in the general chemistry laboratory, this is usually water. The other component, which is dissolved by the solvent, is called the *solute*. Frequently, the solute is a water-soluble solid such as sodium chloride, NaCl, or sugar, $C_{12}H_{22}O_{11}$.

To indicate how much solute there is in a given amount of solution, we most often specify its *molarity*. Molarity (M) is defined by the equation

$$\text{molarity (M)} = \frac{\text{no. moles solute}}{\text{no. liters solution}} \tag{3.4}$$

Thus a solution labeled "1.00 M NaCl" would contain 1.00 mol (58.4 g) of NaCl per liter. Another solution labeled "0.100 M NaCl" would contain 0.100 mol (5.84 g) of NaCl per liter (see also Figure 3.6).

We will have more to say about molarity when we discuss general aspects of solutions in Chapter 9. In the meantime, you should know what molarity means in case you come across it in the laboratory.

Figure 3.6 A bottle of 6 M HCl (dilute hydrochloric acid) contains 6 mol of HCl per liter, or 3 mol HCl per 500 cm³, or 0.6 mol HCl per 100 cm³, and so on.

The Building Blocks of Matter　　75

Key Words

anion
atom
atomic mass
atomic number
Avogadro's number
cation
electron
formula mass
ion
isotope

mass number
molar mass
molarity
mole
molecular formula
molecule
neutron
nuclear symbol
nucleus
proton

$^{40}_{20}Ca$

Questions

3.1 State, in your own words, the postulates of the atomic theory.

3.2 How does the atomic theory explain the Law of Conservation of Mass? the Law of Constant Composition?

3.3 Give the charges and relative masses of the electron, proton, and neutron.

3.4 Define atomic number, mass number, and isotope.

3.5 Explain what is meant by a molecule and give the formulas of at least five different molecules containing hydrogen.

3.6 Explain what is meant by an ion, a cation, and an anion.

3.7 When an ion is formed from an atom, the number of _____ changes. In contrast, the number of _____ in the _____ stays the same.

3.8 "The atomic mass of magnesium is 24.305." Explain in your own words what this statement means.

3.9 Using the Periodic Table, complete the following table.

Element	Atomic Number	Atomic Mass
Potassium	19	39
Iron	26	56
As	33	74.92
I	53	127

3.10 The atomic mass of chlorine is 35.453. The element consists of two isotopes of mass numbers 35 and 37. Which isotope is the more abundant? Explain your reasoning.

3.11 A mole represents how many items?

3.12 A laboratory reagent bottle is labeled "1.0 M KCl." Explain in your own words what the label means.

3.13 What is the difference between a mole and a molecule?

3.14 How is molar mass related to formula mass?

Problems

Nuclear Symbols

(You may use the Periodic Table to work out these problems.)

3.15 Write nuclear symbols for atoms in which
a. the atomic number is 26 and the mass number is 55
b. the atomic number is 27 and the mass number is 60
c. the atomic number is 15 and the mass number is 30
d. the atomic number is 94 and the mass number is 242

3.16 Write nuclear symbols for
a. carbon-14   b. fluorine-19
c. strontium-90   d. barium-137

3.17 Write nuclear symbols for atoms in which there are
a. 7 protons and 8 neutrons
b. 19 protons and 21 neutrons
c. 47 protons and 61 neutrons
d. 82 protons and 124 neutrons

3.18 Give the number of protons and neutrons in each of the atoms in Problem 3.15.

3.19 Complete the following table

Element	Atomic Number	Mass Number	Number of Neutrons
$^{14}_{7}N$	7	14	7
	82	207	
$^{27}_{13}Al$			
		45	24

3.20 Complete the following table

Nuclear Symbol	Atomic Number	Mass Number	Number of Neutrons
	38	88	
$^{39}_{19}K$			
		75	42
$^{16}_{8}O$			

Ions
(You may use the Periodic Table to work out these problems.)

3.21 Give the number of protons and electrons in
a. Mg^{2+} b. Br^- c. Sc^{3+}
d. Cu^+

3.22 Give the symbol and charge of an ion having 10 electrons and
a. 8 protons   b. 9 protons
c. 11 protons   d. 12 protons

3.23 Complete the following table

Symbol	No. of Protons	No. of Electrons	No. of Neutrons	Net Charge
	7	10	7	
	11	10	12	
$^{80}_{35}Br^-$				
$^{56}_{26}Fe^{3+}$				

3.24 Complete the following table

Symbol	No. of Protons	No. of Electrons	No. of Neutrons	Net Charge
$^{64}_{29}Cu^+$				
	36		48	0
	15	18	16	

Relative Atomic Masses
(Use the Periodic Table to find atomic masses.)

3.25 Find the ratio of the mass of an iron atom to an atom of
a. Mn   b. Co   c. Ni

3.26 Find the ratio of the mass of a Li atom to
a. a hydrogen atom   b. a helium atom
c. a beryllium atom

3.27 How many Ca atoms are required to equal the mass of a Pt atom?

3.28 If the atomic mass of carbon-12 had been taken to be 6.00 . . . instead of 12.00 . . ., what would have been the atomic mass of magnesium?

Masses of Atoms
(Take Avogadro's number to be 6.022×10^{23} and use the Periodic Table for atomic masses.)

3.29 Calculate the mass, in grams, of an atom of
a. H   b. He   c. Li

3.30 Calculate the mass in grams of
a. 126 oxygen atoms
b. 2.549×10^{18} sulfur atoms
c. 6.022×10^{22} Ar atoms

3.31 How many atoms are there in 1.00 g of
   a. Na   b. Fe   c. Pt
3.32 How many atoms are there in
   a. 1.00 kg of nitrogen
   b. 1.20×10^{-6} g of neon
   c. 2.496 g of Ar
3.33 Complete the following table

Element	Number of Atoms	Mass in Grams
Al	12	___
Si	___	2.02 g
P	1.6×10^{19}	___

3.34 Complete the following table.

Element	Number of Atoms	Mass in Grams
Gold	2.92×10^4	___
Mercury	1.00×10^{18}	___
Lead	___	1.00 g

Molar Masses
(Use the Periodic Table for atomic masses.)
3.35 What is the molar mass of
   a. Ca   b. CaO   c. $CaCl_2$
   d. $CaCO_3$
3.36 Calculate the molar mass of
   a. Na_2CrO_4 b. $C_6H_{12}O_6$ c. Bi_2O_3
3.37 What is the ratio of the molar mass of Na_3PO_4 to that of
   a. H_3PO_4 b. NaH_2PO_4 c. Na_2HPO_4
3.38 What is the molar mass of a compound if 0.0500 mol weighs 3.69 g?

Mole–Gram Conversions
(Use the Periodic Table for atomic masses.)
3.39 What is the mass in grams of 0.250 mol of
   a. H_2 b. O_2 c. H_2O d. H_2O_2
3.40 How many moles are there in 1.00 g of
   a. H_2 b. O_2 c. H_2O d. H_2O_2
3.41 Determine the number of moles in
   a. 6.02 g of Mg   b. 400.0 g of Cu
   c. 1.62 kg of $CuSO_4$
3.42 Calculate the mass in grams of
   a. 0.22 mol Fe   b. 1.86 mol O_2
   c. 0.054 mol NO_2
3.43 A bottle contains 25.0 cm^3 of mercury (d = 13.6 g/cm^3). What is the
   a. mass of Hg in grams?
   b. number of moles of Hg?
   c. number of atoms of Hg?
3.44 How many moles of oxygen atoms are there in 50.0 g of $CaCO_3$?
3.45 Arrange the following in order of increasing mass in grams
   a. 2.00 mol H_2O b. 0.900 mol CO_2
   c. 0.750 mol C_2H_6O

Multiple Choice

3.46 In a chemical reaction, which one of the following is most likely to change?
   a. number of atoms
   b. number of grams
   c. number of molecules
   d. number of protons
   e. number of neutrons

3.47 Which one of the following particles is the heaviest?
   a. proton   b. neutron   c. ^2_1H nucleus
   d. ^2_1H atom
3.48 What is the total number of protons and electrons in a K^+ ion (at. no. = 19)?
   a. 18   b. 19   c. 37   d. 38
   e. some other number

3.49 The atomic masses of cerium and lead are 140.1 and 207.2, respectively. The number of atoms of lead required to balance two atoms of cerium is
a. 0.3381 b. 0.6762 c. 1.352
d. 1.479 e. 2.958

3.50 The mass of an atom of vanadium (at. mass = 50.94) is
a. 1.182×10^{-24} g b. 8.459×10^{-23} g
c. 50.94 g d. 3.068×10^{25} g

3.51 How many atoms are there in 8.00 g of oxygen (at. mass O = 16.00)?
a. 0.500 b. 2.00 c. 3.01×10^{23}
d. 6.02×10^{23} e. 1.20×10^{24}

3.52 In an atom of $^{235}_{92}U$, there are
a. 92 protons, 92 electrons, 235 neutrons
b. 92 protons, 143 electrons, 235 neutrons
c. 92 protons, 143 electrons, 92 neutrons
d. 143 protons, 143 electrons, 92 neutrons
e. none of the above

3.53 Which one of the following atoms contains the largest number of neutrons?
a. $^{35}_{17}Cl$ b. $^{40}_{18}Ar$ c. $^{39}_{19}K$ d. $^{40}_{20}Ca$

3.54 The formula of aspirin is $C_9H_8O_4$. In 104 aspirin molecules, the number of oxygen atoms is
a. 26 b. 104 c. 416 d. 6.02×10^{23}
e. 6.26×10^{25}

3.55 How many of the following ions contain 18 electrons?
$_{16}S^{2-}$, $_{17}Cl^{-}$, $_{19}K^{+}$, $_{20}Ca^{2+}$, $_{21}Sc^{3+}$
a. 1 b. 2 c. 3 d. 4 e. 5

3.56 Which one of the following has the largest molar mass?
a. $NaClO_2$ b. $NaClO$ c. $NaCl$
d. $NaClO_4$ e. $NaClO_3$

3.57 Which one of the following contains the smallest number of molecules (at. mass H = 1.0, O = 16.0)?
a. 1.00 g of H_2 b. 1.00 g of O_2
c. 1.00 g of H_2O d. 1.00 g of H_2O_2

3.58 What is the mass in grams of 1.20 mol of $C_6H_{12}O_6$ (molar mass = 1.80×10^2 g/mol)?
a. 1.50×10^2 b. 2.16×10^2 c. 1.99×10^{-24} d. 7.22×10^{23}

3.59 How many moles of atoms are there in 24.5 g of H_2SO_4 (molar mass = 98.0 g/mol)?
a. 0.250 b. 0.500 c. 1.75 d. 4.00
e. 98.0

3.60 Which one of the following has the largest mass (at. mass Na = 23.0, Fe = 55.8)?
a. Na^+ ion b. Na atom c. Fe atom
d. Fe^{2+} ion e. Fe^{3+} ion

Chemical Formulas and Names

Learning Objectives

After studying this chapter, you should be able to:

1. Distinguish between simplest and molecular formulas; between molecular, structural, and condensed structural formulas.
2. Predict the formulas of ionic compounds, using Table 4.1 and Figure 4.1 (Examples 4.1–4.3).
3. Given the corresponding formula, name an ionic compound (Examples 4.4, 4.5) or a binary molecular compound (Example 4.6).
4. Give the names and formulas of the acids listed in Tables 4.4 and 4.5; identify and name the corresponding oxyanions and their compounds (Example 4.7).
5. Given the formula of a compound, determine the mass percents of the elements present (Example 4.8); carry out the reverse calculation (Examples 4.11, 4.12).
6. Knowing or having calculated the mass percent of an element in a compound, calculate the mass of that element in a given mass of the compound (Example 4.10).
7. Given the formula of a hydrate, determine the mass percent of water present (Example 4.9); carry out the reverse calculation (Example 4.13).
8. Given the simplest formula of a compound and its molar mass, determine its molecular formula (Example 4.14).

> *Science is built of facts the way a house is built of bricks; but an accumulation of facts is no more science than a pile of bricks is a house.*
>
> **HENRI POINCARÉ**

Photograph by Peter L. Kresan.

CHAPTER 4

In Chapter 3, we mentioned chemical formulas briefly in connection with ionic and molecular substances. In this chapter, we will discuss this topic in greater depth. We will look for answers to such questions as: How are formulas such as N_2O_4, $Ca(OH)_2$, and $CuSO_4 \cdot 5H_2O$ interpreted? How can we predict the formulas of simple compounds? How are formulas of compounds determined in the laboratory? How do we associate names with chemical formulas?

To answer these and other questions, we start this chapter with an overview of the different types of formulas (Section 4.1). In Section 4.2, we will see how it is possible to predict the formulas of ionic compounds. The rules for naming ionic compounds, binary molecular compounds, and acids are considered in Section 4.3. Finally, in Section 4.4, we will examine the relationship between the formula of a compound and the mass percents of the elements present.

4.1 Types of Formulas

Many different types of formulas are used to represent compounds. The **simplest (empirical) formula** gives the simplest ratio of atoms of the elements present. An example is the simplest formula of water, H_2O. This formula tells us that there are two hydrogens atoms for every oxygen atom in water. Ionic compounds, such as calcium chloride, $CaCl_2$, or potassium chlorate, $KClO_3$, are almost always represented by their simplest formulas. The formula, $CaCl_2$, tells us that the atom ratio of calcium to chlorine in calcium chloride is 1:2. From the formula $KClO_3$, we can conclude that in this compound the atom ratio is 1 K : 1 Cl : 3 O.

Molecular substances are most often represented by their **molecular formula,** which shows the number of atoms of each type in the molecule. Sometimes the molecular formula is the same as the simplest formula. This is the case with water, H_2O; there are two hydrogen atoms and one oxygen atom in a water molecule. With other compounds, the molecular formula is a whole-number multiple of the simplest formula. This is the case with dinitrogen tetroxide, N_2O_4. The simplest formula of this compound is NO_2. The molecular formula, N_2O_4, indicates that the molecule contains two nitrogen and four oxygen atoms.

Sometimes we go one step beyond the molecular formula and show the *structural formula* of a molecular substance. A structural formula indicates how the atoms are bonded in the molecule. For water, we would write

H—O—H

The straight lines represent chemical bonds joining the central oxygen to the two hydrogen atoms. As another example, consider the compound methyl alcohol, molecular formula CH_4O. The structural formula is

$$\begin{array}{c} H \\ | \\ H-C-O-H \\ | \\ H \end{array}$$

Although structural formulas give more information than molecular formulas, they have the disadvantage of consuming much more space. For that reason, we will seldom use structural formulas, except for organic compounds (Chapter 14). However, from time to time we will use what might be called *condensed structural formulas* such as

HOH (water

CH_3OH (methyl alcohol)

Such formulas imply the way in which atoms are linked in a molecule without occupying much more space than an ordinary molecular formula.

Still another kind of formula is used to represent compounds called *hydrates*. These are solid ionic compounds that contain water molecules within the crystal lattice. An example is hydrated calcium chloride, which contains six moles of water, H_2O, for every mole of calcium chloride, $CaCl_2$. The formula of the hydrate is

$CaCl_2 \cdot 6H_2O$

The dot serves to separate the two formulas. The 6 indicates that there are six H_2O's for every $CaCl_2$. In another case, we interpret the formula

$CuSO_4 \cdot 5H_2O$

to mean that in this hydrate there are five moles of water for every mole of copper(II) sulfate, $CuSO_4$.

4.2 Formulas of Ionic Compounds

As pointed out in Chapter 3, compounds formed by metals with nonmetals are ordinarily ionic. The formula of an ionic compound can be predicted if you know the charges of the two ions involved. To illustrate, consider sodium chloride. The ions present are Na^+ and Cl^-. Since sodium chloride, like all pure substances, must be electrically neutral, it follows that there must be one Na^+ ion for every Cl^- ion. Hence the formula is simply NaCl. In another case, calcium chloride is composed of Ca^{2+} and Cl^- ions. Two Cl^- ions are required to electrically balance a Ca^{2+} ion, so the formula of calcium chloride is $CaCl_2$. Notice that in NaCl and $CaCl_2$, and indeed *in all ionic compounds, the symbol of the metal is written first.*

We will consider many applications of the principle of electroneutrality in this section. To apply it, you must know the charges of the ions involved. For that reason, we will organize this section around ionic charges. First, we will consider how to obtain the charges of *monatomic ions*, which are derived from single atoms by the gain or loss of electrons. Then we will look at the names, formulas, and charges of *polyatomic ions*, charged particles containing two or more atoms.

Monatomic Ions

The charges of many common ions can be predicted by applying a simple principle. **Elements that are close to a noble gas (Group 8) in the Periodic Table tend to form ions that have the same number of electrons as the noble gas atom.** To illustrate this rule consider the noble gas, neon (Ne). Since it has an atomic number of 10, a neon atom has 10 electrons. The three nonmetals that precede neon in the Periodic Table (N, O, F) each gain electrons to form negative ions with 10 electrons. Thus we have

nitride ion (7 p^+, 10 e^-); charge = -3; N^{3-} ion

oxide ion (8 p^+, 10 e^-); charge = -2; O^{2-} ion

fluoride ion (9 p^+, 10 e^-); charge = -1; F^- ion

The three metals that follow neon in the Periodic Table (Na, Mg, Al) each lose electrons to form positive ions with 10 electrons.

sodium ion (11 p^+, 10 e^-); charge = $+1$; Na^+ ion

magnesium ion (12 p^+, 10 e^-); charge = $+2$; Mg^{2+} ion

aluminum ion (13 $p+$, 10 e^-); charge = $+3$; Al^{3+} ion

Figure 4.1 lists ions, superimposed on the Periodic Table, that have the same number of electrons as the neighboring noble gas atom. Note that

Chemical Formulas and Names

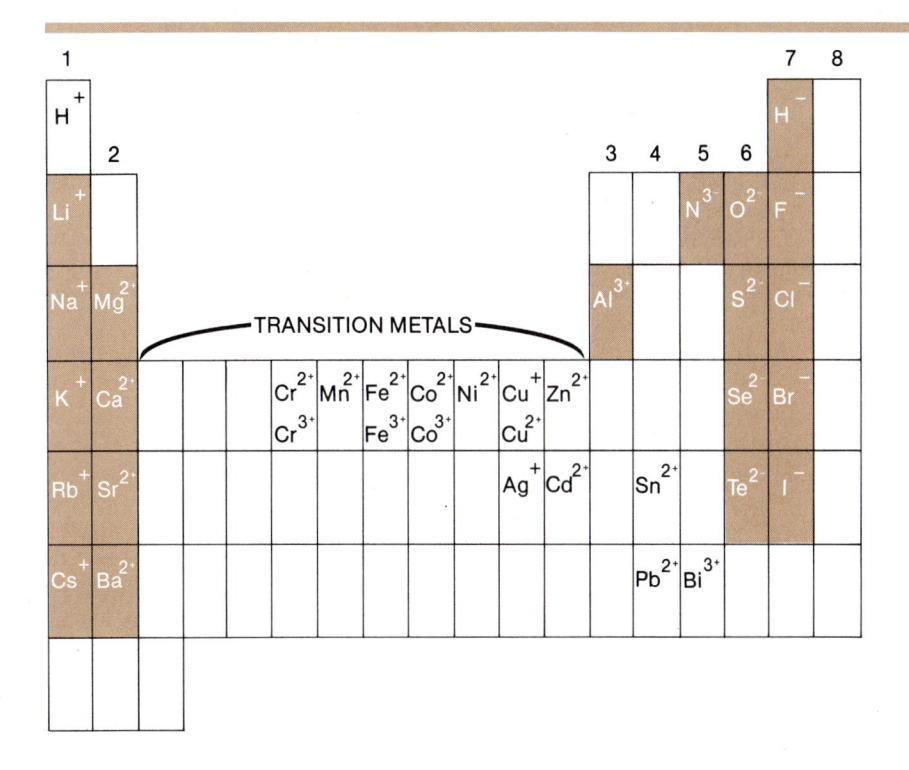

Figure 4.1 Charges of monatomic ions superimposed on the Periodic Table. Ions shown in color have noble gas structures.

1	2							3	4	5	6	7	8
H^+													H^-
Li^+										N^{3-}	O^{2-}	F^-	
Na^+	Mg^{2+}		←TRANSITION METALS→					Al^{3+}			S^{2-}	Cl^-	
K^+	Ca^{2+}		Cr^{2+}, Cr^{3+}	Mn^{2+}	Fe^{2+}, Fe^{3+}	Co^{2+}, Co^{3+}	Ni^{2+}	Cu^+, Cu^{2+}	Zn^{2+}		Se^{2-}	Br^-	
Rb^+	Sr^{2+}					Ag^+	Cd^{2+}		Sn^{2+}		Te^{2-}	I^-	
Cs^+	Ba^{2+}							Pb^{2+}	Bi^{3+}				

The ions of Group 1 metals have +1 charges.
The ions of Group 2 metals have +2 charges.
The ions of Group 6 nonmetals have −2 charges.
The ions of Group 7 nonmetals have −1 charges.

EXAMPLE 4.1

Using only the Periodic Table (*not* Figure 4.1) and assuming that all ions involved have the same number of electrons as the nearest noble gas atom (Group 8), predict the formulas of the ionic compounds formed by
a. lithium and sulfur   b. magnesium and iodine   c. scandium and oxygen

Solution

In each case, we first use the Periodic Table to deduce the charges of the ions. Then we use the principle of electroneutrality to predict the formula of the compound.

a. Li is in Group 1, so loses one electron to form a +1 ion, Li^+. S is in Group 6, so forms a −2 ion, S^{2-}, by gaining two electrons. Two Li^+ ions are required to balance one S^{2-} ion. Formula: Li_2S.

Continued

b. Mg, in Group 2, forms the Mg^{2+} ion. Iodine, in Group 7, forms the I^- ion. Formula: MgI_2.
c. Sc, by losing three electrons, acquires the same number of electrons as Ar, 18. It forms the Sc^{3+} ion. Oxygen, in Group 6, forms the O^{2-} ion. For electroneutrality, two Sc^{3+} ions are required to balance three O^{2-} ions. Formula: Sc_2O_3.

Exercise
What are the formulas of aluminum fluoride, aluminum oxide, and aluminum nitride, assuming they are all ionic? Answer: AlF_3, Al_2O_3, AlN.

Several metals that are far removed from the noble gases in the Periodic Table form positive ions. These include the *transition metals* in the center of the Table and the heavier metals in Groups 4 and 5, sometimes referred to as *posttransition metals*. The charges of some of the more common ions of this type are listed in Figure 4.1 (in black). Note that

1. There is no simple way to predict the charges of these ions. The most common charge is +2 (e.g., Mn^{2+}, Ni^{2+}, Zn^{2+}, Cd^{2+}, Sn^{2+}, Pb^{2+}, --), but some of these metals form +3 (e.g., Cr^{3+}, Fe^{3+}) or +1 (Ag^+) ions.
2. In several cases, a metal forms more than one positive ion. For example, iron forms both Fe^{2+} and Fe^{3+}, copper both Cu^+ and Cu^{2+}, and so on.

EXAMPLE 4.2
Using Figure 4.1, predict the formulas of
a. the ionic compound formed by nickel with iodine
b. the ionic compound formed by silver with oxygen
c. two ionic compounds formed by iron with chlorine

Solution
a. Ni^{2+} and I^- ions; NiI_2 b. Ag^+ and O^{2-} ions; Ag_2O
c. Iron can form either Fe^{2+} or Fe^{3+}. Chlorine (Group 7) forms Cl^-. $FeCl_2$, $FeCl_3$

Exercise
Copper forms two different ionic oxides. What are their formulas? Answer: Cu_2O, CuO.

Polyatomic Ions

Table 4.1 lists some of the more common polyatomic ions, along with their names and charges. (Note that all but one of the ions listed is an anion containing oxygen atoms, an *oxyanion*.) The formulas of compounds containing these ions are predicted in the usual way, using the principle of electroneutrality. A small complication arises when there are two or more polyatomic

TABLE 4.1   Polyatomic Ions

+1	−1	−2	−3
NH_4^+ (ammonium)	OH^- (hydroxide)	CO_3^{2-} (carbonate)	PO_4^{3-} (phosphate)
	NO_3^- (nitrate)	SO_4^{2-} (sulfate)	
	NO_2^- (nitrite)	SO_3^{2-} (sulfite)	
	ClO_4^- (perchlorate)	CrO_4^{2-} (chromate)	
	ClO_3^- (chlorate)	$Cr_2O_7^{2-}$ (dichromate)	
	ClO_2^- (chlorite)	O_2^{2-} (peroxide)	
	ClO^- (hypochlorite)		
	MnO_4^- (permanganate)		

ions of a given type in the formula. In that case, the polyatomic ion is enclosed in parentheses, followed by the proper subscript. Thus,

$Ca(OH)_2$ (1 Ca^{2+} ion, 2 OH^- ions) $Fe(NO_3)_3$ (1 Fe^{3+} ion, 3 NO_3^- ions)

EXAMPLE 4.3
Using Figure 4.1 and Table 4.1, predict the formulas of
a. barium nitrate   b. ammonium sulfate   c. magnesium phosphate

Solution
a. $Ba(NO_3)_2$: 1 Ba^{2+} ion required to balance 2 NO_3^- ions
b. $(NH_4)_2SO_4$: 2 NH_4^+ ions required to balance 1 SO_4^{2-} ion
c. $Mg_3(PO_4)_2$: 3 Mg^{2+} ions required to balance 2 PO_4^{3-} ions

Exercise
Give the formulas of ammonium nitrate, ammonium phosphate, and ammonium bromide. Answer: NH_4NO_3, $(NH_4)_3PO_4$, NH_4Br.

4.3   Names of Inorganic Compounds

A compound can be designated by formula or by name. You should be able to translate back and forth between the two designations. Given the formula of a substance, you should be able to name it. Conversely, if you come across the name of a compound, you should be able to associate a formula with it. Today, the relation between the formula of an inorganic compound and its name is a systematic one. This was not always true (Table 4.2). Students a few generations ago had to learn a great many common names.

In this section, we will consider the relation between names and formulas for the three types of compounds most common in general chemistry.

Chapter 4

TABLE 4.2 Common (Trivial) Names of Various Chemicals

Formula	Name	Derivation of Name
$AgNO_3$	Lunar caustic	Alchemists associated silver with the moon (Latin *luna*)
CaO	Quicklime (lime)	Made by heating limestone, $CaCO_3$
$Ca(OH)_2$	Slaked lime	Made by mixing (slaking) CaO with water
$CaSO_4 \cdot \frac{1}{2}H_2O$	Plaster of Paris	First used near Paris to make plaster, cement
$CaSO_4 \cdot 2H_2O$	Gypsum (alabaster)	Derived from Greek word for plaster
$CuSO_4 \cdot 5H_2O$	Blue vitriol	Blue, glassy crystals (Latin *vitreus*)
FeS_2	Fool's gold (pyrites)	Resembles gold in appearance
$FeSO_4 \cdot 7H_2O$	Green vitriol	Green, glassy crystals
HCl	Muriatic acid	Chlorides—muriates (Latin *muria* = brine)
$HCl + HNO_3$	Aqua regia	Dissolves Au, Pt ("royal water")
HNO_3	Aqua fortis	"Strong water"
H_2SO_4	Oil of vitriol	Concentrated H_2SO_4 (oily) used to make metal sulfates (vitriols)
$HgCl_2$	Corrosive sublimate	Poisonous, readily sublimes
K_2MnO_4	Chameleon mineral	Green in water, turns red in acid
KNO_3	Saltpeter	Latin: salt obtained from rock
$Mg(OH)_2$	Milk of magnesia	Forms cloudy suspension with water
$MgSO_4 \cdot 7H_2O$	Epsom salts	First isolated from mineral springs in Epsom, England
NH_4Cl	Sal ammoniac	Salt made by reacting ammonia with HCl
$Na_2CO_3 \cdot 10H_2O$	Washing soda	Used to make soap
$NaHCO_3$	Baking soda	Used in baking as source of $CO_2(g)$
$NaOH$	Caustic soda (lye)	Strong base
$Na_2SO_4 \cdot 10H_2O$	Glauber's salt	First made by alchemist Johann Glauber, approximately 1650
$Na_2S_2O_3$	Hypo	Old name was sodium hyposulfite

They are

1. Ionic compounds of the type discussed in Section 4.2 (e.g., NaCl, $Ca(OH)_2$).
2. Binary molecular compounds, which contain atoms of two different nonmetals. Examples include N_2O, and P_2O_5.
3. Molecular acids, which contain a nonmetal atom, one or more hydrogen atoms, and, in most cases, oxygen atoms as well. Examples include HCl, H_2SO_4 and HNO_3. In water solution, these compounds form ions; for example, HCl forms H^+ and Cl^- ions.

Ionic Compounds

The name of an ionic compound consists of two words. The first word is the name of the positive ion *(cation)*, which appears first in the formula. The

Chemical Formulas and Names

second word is the name of the negative ion *(anion)*, which appears last in the formula.

Compound	Cation	Anion	Name of Compound
NaCl	Na^+ (sodium)	Cl^- (chloride)	Sodium chloride
Ca(OH)$_2$	Ca^{2+} (calcium)	OH^- (hydroxide)	Calcium hydroxide

Clearly, to name an ionic compound all we need do is name the individual ions in succession. There are a few simple rules for naming ions.

1. Monatomic positive ions that have noble gas structures (shown in color in Figure 4.1) take the names of the metals from which they are derived:

   Na^+ sodium Ca^{2+} calcium Al^{3+} aluminum

2. When a metal forms more than one ion, these ions must be distinguished in some way. The approved method* is to indicate the charge of the ion by a Roman numeral in parentheses after the name of the metal:

   Fe^{2+} iron(II) Fe^{3+} iron(III)

   In practice, we often follow this system with all transition and post-transition cations. Thus we may write

   Ni^{2+} nickel(II) Sn^{2+} tin(II)

3. Monatomic negative ions are named by adding the suffix *-ide* to the stem of the name of the corresponding nonmetal:

   N^{3-} nitride O^{2-} oxide H^- hydride
                      S^{2-} sulfide F^- fluoride
                      Se^{2-} selenide Cl^- chloride
                      Te^{2-} telluride Br^- bromide
                                          I^- iodide

4. Polyatomic ions are given special names (see Table 4.1 p. 87). When a nonmetal combines with oxygen to form two different polyatomic ions, the suffixes *-ate* and *-ite* are used to distinguish between them. The *ate* ion is the one containing the greater number of oxygen atoms:

   SO_4^{2-} sulfate NO_3^- nitrate
   SO_3^{2-} sulfite NO_2^- nitrite

---

*An older system used the suffixes -ous and -ic for the ions of lower charge and higher charge, respectively. These were added to the Latin stem of the name of the metal. For example, Fe^{2+} = ferrous and Fe^{3+} = ferric. The approved system is simpler, more informative, and will be used throughout this book.

With an element like chlorine which forms more than two polyatomic ions with oxygen, the prefixes *per-* and *hypo-* also are used:

ClO_4^- perchlorate

ClO_3^- chlorate

ClO_2^- chlorite

ClO^- hypochlorite

Using these four rules, you should be able to name ionic compounds (Example 4.4) or assign formulas corresponding to names (Example 4.5).

EXAMPLE 4.4
Name the following ionic compounds:
a. BaI_2 b. $Al_2(SO_4)_3$ c. $Cr(NO_3)_3$ d. $PbCl_2$

Solution
In each case, decide what ions are present and name them in order (cation first, then anion, as in the formula).
a. Ba^{2+} (barium), I^- (iodide) ions; barium iodide
b. Al^{3+} (aluminum), SO_4^{2-} (sulfate) ions; aluminum sulfate
c. Cr^{3+}, NO_3^- ions; chromium(III) nitrate. The Roman numeral III is used because Cr is a transition metal capable of forming more than one cation (+2 as well as +3).
d. Pb^{2+}, Cl^- ions; lead(II) chloride. In common usage, the Roman numeral II is usually dropped because Pb^{2+} is the only common cation derived from lead.

Exercise
Name $KClO_2$ and $Cu_3(PO_4)_2$. Answer: potassium chlorite, copper(II) phosphate.

EXAMPLE 4.5
Give the formulas of the following ionic compounds:
a. calcium nitrite   b. iron(III) sulfate   c. copper(II) perchlorate

Solution
In each case, deduce from the name what ions are present. Then write the formula, using the principle of electroneutrality.
a. Ca^{2+}, NO_2^- ions; $Ca(NO_2)_2$
b. Fe^{3+}, SO_4^{2-} ions; $Fe_2(SO_4)_3$
c. Cu^{2+}, ClO_4^- ions; $Cu(ClO_4)_2$

Continued

> **Exercise**
> Write formulas for potassium sulfide, potassium sulfite, and potassium sulfate. Answer: K$_2$S, K$_2$SO$_3$, K$_2$SO$_4$.

Binary Molecular Compounds

When a compound contains two different nonmetallic elements, it is ordinarily molecular in nature. The names of such compounds consist of two words.

1. The first word is the name of the nonmetal that appears first in the formula.* A Greek prefix (Table 4.3) is used to indicate the number of atoms of the nonmetal in the molecule.
2. The second word is derived by adding the suffix *-ide* to the stem of the name of the second nonmetal. The appropriate prefix is used to indicate the number of atoms of the second nonmetal in the molecule.

To illustrate how this system works, consider the names of the several oxides of nitrogen:

N$_2$O$_5$ dinitrogen pentoxide
N$_2$O$_4$ dinitrogen tetroxide
NO$_2$ nitrogen dioxide
N$_2$O$_3$ dinitrogen trioxide
N$_2$O$_2$ dinitrogen dioxide
NO nitrogen oxide
N$_2$O dinitrogen oxide

Notice that

—the final "a" in the prefix is dropped when it would precede a vowel. This makes the word easier to say. Thus we write "pentoxide" instead of "pentaoxide."

TABLE 4.3 Greek Prefixes Used in Names

Atoms per Molecule	Prefix	Atoms per Molecule	Prefix
2	di-	7	hepta-
3	tri-	8	octa-
4	tetra-	9	nona-
5	penta-	10	deca-
6	hexa-		

*Ordinarily, in a binary molecular compound, the symbol of the more metallic element is written first. The more nonmetallic element (the one closer to the upper right corner of the Periodic Table) is written last.

—when there is only one atom of a particular element in a molecule, we do not ordinarily use a prefix. (A common exception is CO, carbon monoxide.)

EXAMPLE 4.6
Name the following molecular compounds:
a. HF   b. NI$_3$ c. P$_4$O$_6$

Solution
a. No prefixes are needed, since there is only one atom of each type. The correct name is hydrogen fluoride.
b. Nitrogen triiodide
c. Tetraphosphorus hexoxide

Exercise
Write the formulas of phosphorus pentachloride and sulfur hexafluoride. Answer: PCl$_5$, SF$_6$.

Many of the best known binary molecular compounds have common names that are widely used. Indeed, the common name is often the only one used. You will never find water referred to as "dihydrogen oxide." Other common names include

H$_2$O$_2$ hydrogen peroxide
N$_2$H$_4$ hydrazine
NH$_3$ ammonia
PH$_3$ phosphine
NO nitric oxide
N$_2$O nitrous oxide

Acids

One of the most important acids in the general chemistry laboratory is hydrochloric acid, a water solution of hydrogen chloride, HCl. Although HCl is molecular in the gas state, it ionizes in aqueous solution to form H$^+$ and Cl$^-$ ions. The hydrogen compounds of the other Group 7 elements (HF, HBr, HI) behave similarly. We will have more to say about the acidic properties of these species in Chapter 11. At the moment, we are only interested in the names assigned to their water solutions (Table 4.4).

Most acids, with the exception of those listed in Table 4.4, contain oxygen in addition to hydrogen and other nonmetal atoms. The names and formulas of the more common inorganic acids of this type, often referred to as

Chemical Formulas and Names

TABLE 4.4 Acids Formed by HF, HCl, HBr, and HI

Compound	Name	Water Solution	Name
HF(g)	Hydrogen fluoride	HF(aq)	Hydrofluoric acid
HCl(g)	Hydrogen chloride	HCl(aq)	Hydrochloric acid
HBr(g)	Hydrogen bromide	HBr(aq)	Hydrobromic acid
HI(g)	Hydrogen iodide	HI(aq)	Hydriodic acid

oxyacids, are listed in Table 4.5. We also include the names and formulas of the oxyanions formed, along with H^+, when the acids ionize in water. For example, nitric acid, HNO_3, ionizes to form H^+ ions and nitrate ions, NO_3^-. Sulfuric acid, H_2SO_4, ionizes to form H^+ ions and hydrogen sulfate ions, HSO_4^-. Further ionization of HSO_4^- gives another H^+ ion and the sulfate ion, SO_4^{2-}.

As you can perhaps see from Table 4.5, the names of oxyanions are closely related to those of the acids from which they are derived. In particular,

1. Oxyanions whose names end in *-ate* are derived from acids whose names end in *-ic*. Thus,
   HNO_3 nitric acid $\qquad$ NO_3^- nitrate ion
   $HClO_4$ perchloric acid $\qquad$ ClO_4^- perchlorate ion
2. Oxyanions whose names end in *-ite* are derived from acids whose names end in *-ous*. Examples include
   HNO_2 nitrous acid $\qquad$ NO_2^- nitrite ion
   $HClO$ hypochlorous acid $\qquad$ ClO^- hypochlorite ion

Notice also the system used to name oxyanions that contain a hydrogen atom. To distinguish HCO_3^- from CO_3^{2-}, the former is called the *hydrogen carbonate* ion. (Commonly, the HCO_3^- ion is often referred to by an older name, bicarbonate). A similar system is used to distinguish the two anions

TABLE 4.5 Oxyacids and Oxyanions

Acid Formula	Name	Anion Formula	Name	Anion Formula	Name
HNO_3	Nitric acid	NO_3^-	Nitrate		
HNO_2	Nitrous acid	NO_2^-	Nitrite		
$HClO_4$	Perchloric acid	ClO_4^-	Perchlorate		
$HClO_3$	Chloric acid	ClO_3^-	Chlorate		
$HClO_2$	Chlorous acid	ClO_2^-	Chlorite		
$HClO$	Hypochlorous acid	ClO^-	Hypochlorite		
H_2CO_3	Carbonic acid	HCO_3^-	Hydrogen carbonate	CO_3^{2-}	Carbonate
H_2SO_4	Sulfuric acid	HSO_4^-	Hydrogen sulfate	SO_4^{2-}	Sulfate
H_2SO_3	Sulfurous acid	HSO_3^-	Hydrogen sulfite	SO_3^{2-}	Sulfite

formed by H_2SO_4 and by H_2SO_3. With H_3PO_4, three different anions are possible. The names assigned here are

$H_2PO_4^-$ dihydrogen phosphate ion
HPO_4^{2-} hydrogen phosphate ion (or monohydrogen phosphate ion)
PO_4^{3-} phosphate ion

Compounds containing these anions are named in the usual way (Example 4.7).

EXAMPLE 4.7
Name the following ionic compounds
a. $NaHCO_3$ b. $Ca(HSO_4)_2$ c. NaH_2PO_4

Solution
As usual, the first step is to decide what ions are present. They are then named in order, starting with the cation.
a. Na^+, HCO_3^- ions; sodium hydrogen carbonate (the common name is sodium bicarbonate)
b. Ca^{2+}, HSO_4^- ions; calcium hydrogen sulfate
c. Na^+, $H_2PO_4^-$ ions; sodium dihydrogen phosphate

Exercise
What are the formulas of potassium phosphate, potassium hydrogen phosphate, and potassium dihydrogen phosphate? Answer: K_3PO_4, K_2HPO_4, KH_2PO_4.

4.4 Formulas and Percent Composition

Chemical formulas are useful for many purposes. As we will see in Chapter 5, they are a prerequisite for balancing chemical equations. A formula can also be used to calculate the *percent composition* of a compound, i.e., the mass percents of the elements present. Conversely, if the percent composition is known, the *simplest formula* of the compound can be determined. With one more piece of information, the molar mass, the *molecular formula* can be found.

Percent Composition from Formula

Knowing the formula of a compound, it is possible to calculate the mass percents of the elements it contains. To do this, we make use of the fact that
   The subscripts in a formula represent, not only the atom ratio, but also the mole ratio in which the elements are combined.

Chemical Formulas and Names

Thus, given the formulas $CaCl_2$ and $KClO_3$, we can say that

—in $CaCl_2$, 1 atom of Ca is combined with 2 atoms of Cl, or 1 mol of Ca (40.08 g) is combined with 2 mol of Cl (70.91 g).

—in $KClO_3$, 1 atom of K is combined with 1 atom of Cl and three atoms of O or 1 mol of K (39.10 g) is combined with 1 mol of Cl (35.45 g) and 3 mol of O (48.00 g). The use of this principle in determining percent composition is shown in Example 4.8.

EXAMPLE 4.8

Determine the mass percents of the elements in
a. iron(II) sulfate, $FeSO_4$ b. iron(III) sulfate, $Fe_2(SO_4)_3$

Solution

a. In one mole of $FeSO_4$, there is 1 mol of Fe, 1 mol S, and 4 mol O.

   1 mol Fe = 55.85 g
   1 mol S   = 32.06 g
   4 mol O   = 64.00 g
                ───────
                151.91 g

We see that 1 mol of $FeSO_4$ weighs 151.91 g and contains 55.85 g Fe, 32.06 g S, and 64.00 g of O. Therefore,

$$\% \text{ Fe} = \frac{55.85 \text{ g}}{151.91 \text{ g}} \times 100 = 36.77; \quad \% \text{ S} = \frac{32.06 \text{ g}}{151.91 \text{ g}} \times 100 = 21.10$$

$$\% \text{ O} = \frac{64.00 \text{ g}}{151.91 \text{ g}} \times 100 = 42.13$$

b. In one mole of $Fe_2(SO_4)_3$, there are 2 mol of Fe, 3 mol of S, and $3 \times 4 = 12$ mol of O.

   2 mol Fe = 2(55.85 g) = 111.70 g
   3 mol S = 3(32.06 g) = 96.18 g
   12 mol O = 12(16.00 g) = 192.00 g
                            ────────
                            399.88 g

Proceeding as in (a)

$$\% \text{ Fe} = \frac{111.70 \text{ g}}{399.88 \text{ g}} \times 100 = 27.93; \quad \% \text{ S} = \frac{96.18 \text{ g}}{399.88 \text{ g}} \times 100 = 24.05$$

$$\% \text{ O} = \frac{192.00 \text{ g}}{399.88 \text{ g}} \times 100 = 48.01$$

Exercise

Which would contain the greater mass of iron, 100 g of $FeSO_4$ or 100 g of $Fe_2(SO_4)_3$? Answer: 100 g of $FeSO_4$.

The procedure we have used to determine the mass percents of the elements in a compound consists essentially of three steps.

1. *Find the mass in grams of each element in one mole of the compound* (e.g., 55.85 g of Fe, 32.06 g of S, 64.00 g of O in $FeSO_4$).
2. *Find the mass of one mole of the compound by adding the masses of the elements obtained in Step 1* (e.g., mass of one mole $FeSO_4$ = 55.85 g + 32.06 g + 64.00 g = 151.91 g).
3. *Use the relation*

$$\text{mass \% X} = \frac{\text{mass of X}}{\text{total mass compound}} \times 100 \quad (4.1)$$

to find the various mass percents.

This same approach can be used to find the mass percent of water in a hydrate (Example 4.9).

EXAMPLE 4.9
Consider the hydrate of magnesium sulfate called Epsom salts, $MgSO_4 \cdot 7H_2O$. What is the percent of water in this hydrate?

Solution
One mole of hydrate contains 1 mol of Mg, 1 mol of S, 4 mol of O, and 7 mol of H_2O (molar mass = 18.02 g/mol).

1 mol Mg	=	24.30 g
1 mol S	=	32.06 g
4 mol O	= 4(16.00 g) =	64.00 g
7 mol H_2O	= 7(18.02 g) =	126.14 g
		246.50 g

$$\% \ H_2O = \frac{126.14 \text{ g}}{246.50 \text{ g}} \times 100 = 51.17$$

Exercise
What is the percent of $MgSO_4$ in $MgSO_4 \cdot 7H_2O$? Answer: 48.83%

Once we know the mass percent of an element in a compound, we can readily find the mass of that element in a given mass of the compound. To do this, it is convenient to rearrange Equation 4.1 to obtain

$$\text{mass of X} = \text{total mass compound} \times \frac{\text{mass \% X}}{100} \quad (4.2)$$

Chemical Formulas and Names

EXAMPLE 4.10
Knowing from Example 4.8 that the mass percent of iron in $FeSO_4$ is 36.77, calculate the mass of iron in a sample of $FeSO_4$ weighing 6.782 g.

Solution
Applying Equation 4.2

$$\text{mass Fe} = \text{total mass FeSO}_4 \times \frac{\text{mass \% Fe}}{100}$$

$$= 6.782 \text{ g} \times \frac{36.77}{100} = 2.494 \text{ g}$$

Exercise
What is the mass of iron in a sample of $Fe_2(SO_4)_3$ weighing 6.782 g? Answer: 1.894 g.

Simplest Formula from Percent Composition

We have just seen that it is possible to determine the percent composition of a compound if its formula is known. As you might guess, this calculation can be reversed; the formula can be deduced if the percent composition is known. Indeed, this is the way formulas of "unknown" compounds are ordinarily determined in the laboratory. The pure compound is first analyzed to find the percents by mass of the different elements. From these data, the *simplest formula* of the compound is obtained. The logic involved is illustrated in Example 4.11.

EXAMPLE 4.11
Determine the simplest formulas of compounds containing
a. 58.92% Na, 41.08% S
b. 32.38% Na, 22.57% S, 45.05% O

Solution
a. Let's start with a fixed mass of the compound, 100 g for convenience. In 100 g of this compound there are

58.92 g Na, 41.08 g S.

We now calculate the number of moles of each element in the 100-g sample. Since the atomic masses of Na and S are 22.99 and 32.06,

1 mol Na = 22.99 g Na; 1 mol S = 32.06 g S

Continued

Using the conversion factor approach,

$$\text{moles Na} = 58.92 \text{ g Na} \times \frac{1 \text{ mol Na}}{22.99 \text{ g Na}} = 2.563 \text{ mol Na}$$

$$\text{moles S} = 41.08 \text{ g S} \times \frac{1 \text{ mol S}}{32.06 \text{ g S}} = 1.281 \text{ mol S}$$

Looking at the two numbers just calculated, we see that the mole ratio of Na to S is

$$\frac{2.563 \text{ mol Na}}{1.281 \text{ mol S}} = 2.00 \text{ mol Na/mol S}$$

But, as pointed out earlier, the mole ratio and the atom ratio are identical. In other words, the atom ratio of Na to S must be 2:1. The simplest formula is Na_2S.

b. We start, as in (a), with a 100-g sample, which must contain

32.38 g Na, 22.57 g S, 45.05 g O.

Again, we calculate the number of moles of each element in the 100-g sample

$$\text{moles Na} = 32.38 \text{ g Na} \times \frac{1 \text{ mol Na}}{22.99 \text{ g Na}} = 1.408 \text{ mol Na}$$

$$\text{moles S} = 22.57 \text{ g S} \times \frac{1 \text{ mol S}}{32.06 \text{ g S}} = 0.7040 \text{ mol S}$$

$$\text{moles O} = 45.05 \text{ g O} \times \frac{1 \text{ mol O}}{16.00 \text{ g O}} = 2.816 \text{ mol O}$$

To find the simplest mole ratio, we compare Na and O to S

$$\frac{1.408 \text{ mol Na}}{0.704 \text{ mol S}} = 2.00 \text{ mol Na/mol S}$$

$$\frac{2.816 \text{ mol O}}{0.704 \text{ mol S}} = 4.00 \text{ mol O/mol S}$$

The simplest formula is Na_2SO_4 (the compound is sodium sulfate).

Exercise
What is the simplest formula of a compound that contains 50 mass percent S and 50 mass percent O? Answer: SO_2.

Summarizing the three-step approach used to obtain the simplest formula of a compound from percent composition:

1. *Using the percent composition, write down the masses of each element in a 100-g sample.*
2. *Convert these masses in grams to moles.*
3. *Determine the simplest mole ratio between the elements, which is also the*

atom ratio in the formula. To do this, divide by the smallest of the numbers obtained in (2).

Sometimes, the "simplest mole ratio" may not be obvious, even after you have done the division just called for. Example 4.12 illustrates such a case.

EXAMPLE 4.12

A certain compound contains 26.58% K, 35.35% Cr, and 38.07% O. What is its simplest formula?

Solution

Following the three-step procedure
(1) In a 100-g sample, there are

26.58 g K, 35.35 g Cr, 38.07 g O

(2) The numbers of moles of each element are

$$\text{moles K} = 26.58 \text{ g K} \times \frac{1 \text{ mol K}}{39.10 \text{ g K}} = 0.6798 \text{ mol K}$$

$$\text{moles Cr} = 35.35 \text{ g Cr} \times \frac{1 \text{ mol Cr}}{52.00 \text{ g Cr}} = 0.6798 \text{ mol Cr}$$

$$\text{moles O} = 38.07 \text{ g O} \times \frac{1 \text{ mol O}}{16.00 \text{ g O}} = 2.379 \text{ mol O}$$

(3) $\frac{0.6798 \text{ mol Cr}}{0.6798 \text{ mol K}} = 1.00 \text{ mol Cr/mol K}$

$\frac{2.379 \text{ mol O}}{0.6798 \text{ mol K}} = 3.50 \text{ mol O/mol K}$

Here, the oxygen to potassium mole ratio is not a whole number. To obtain a whole number ratio, we multiply numerator and denominator by 2 to obtain

7 mol O/2 mol K

The simplest formula is $K_2Cr_2O_7$ (potassium dichromate).

Exercise

For a certain oxide of iron, the mole ratio is calculated to be 1.33 mol O/mol Fe. What is the simplest formula of the compound? Answer: Fe_3O_4.

The three-step procedure used to obtain simplest formulas is readily adapted to find the formula of a hydrate. Starting with the percent of water in the hydrate, we can find the number of moles of water per mole of anhydrous compound (Example 4.13).

> **EXAMPLE 4.13**
> A certain hydrate of copper(II) sulfate, $CuSO_4 \cdot xH_2O$, is found to contain 36.08% by mass of water. What is the formula of the hydrate?
>
> **Solution**
> In 100 g of the hydrate, there are
>
> 36.08 g H_2O and $(100.00 - 36.08)$g $= 63.92$ g $CuSO_4$
>
> We next find the numbers of moles of H_2O and $CuSO_4$ in the 100-g sample. To do this, note that
>
> 1 mol $H_2O = 2(1.01 \text{ g}) + 16.00 \text{ g} = 18.02$ g H_2O
>
> 1 mol $CuSO_4 = 63.55 \text{ g} + 32.06 \text{ g} + 64.00 \text{ g} = 159.61$ g $CuSO_4$
>
> $$\text{moles } H_2O = 36.08 \text{ g } H_2O \times \frac{1 \text{ mol } H_2O}{18.02 \text{ g } H_2O} = 2.002 \text{ mol } H_2O$$
>
> $$\text{moles } CuSO_4 = 63.92 \text{ g } CuSO_4 \times \frac{1 \text{ mol } CuSO_4}{159.61 \text{ g } CuSO_4} = 0.4005 \text{ mol } CuSO_4$$
>
> The mole ratio is: $\dfrac{2.002 \text{ mol } H_2O}{0.4005 \text{ mol } CuSO_4} = 5.00$ mol H_2O/mol $CuSO_4$
>
> The formula must be $CuSO_4 \cdot 5H_2O$.
>
> **Exercise**
> In another hydrate of $CuSO_4$, the percent of water is 10.14. What is the formula of the hydrate? Answer: $CuSO_4 \cdot H_2O$.

Molecular Formula from Simplest Formula

From the percent composition of a compound, we determine only the simplest formula. For a molecular substance, the molecular formula may or may not be the same as the simplest formula. With water, H_2O, the two formulas are the same. For hydrogen peroxide, H_2O_2, the molecular formula is twice the simplest formula, HO. In general,

molecular formula = n(simplest formula)   (4.3)

where n = 1 or 2 or 3 or some larger whole number

To find the molecular formula from the simplest formula, we need to obtain the value of n in Equation 4.3. This is readily done if the molar mass is known (Example 4.14).

EXAMPLE 4.14

Two common organic compounds have the simplest formula CH. One of these is a gas, acetylene, with a molar mass of 26 g/mol. The other is a liquid, benzene, with a molar mass of 78 g/mol. What are the molecular formulas of acetylene and benzene?

Solution

The molar mass of CH is

$$(12 + 1) \text{ g/mol} = 13 \text{ g/mol}$$

Acetylene has a molar mass just twice that of CH (26/13 = 2). Hence, for acetylene, n in Equation 4.3 must be 2 and the molecular formula must be C_2H_2. We proceed similarly with benzene

$$n = \frac{78}{13} = 6$$

The molecular formula is C_6H_6.

Exercise

Two different gases both have the simplest formula NO_2. One has a molar mass of 46 g/mol, the other a molar mass of 92 g/mol. What are their molecular formulas?
Answer: NO_2, N_2O_4.

Key Words

acid
hydrate
molar mass
mole
molecular formula
noble gas
oxyacid

oxyanion
percent composition
polyatomic ion
posttransition metal
simplest formula
structural formula
transition metal

Questions

4.1 Explain the difference between simplest formula and molecular formula; between molecular formula and structural formula.

4.2 Explain in your own words what the formula $CuSO_4 \cdot 5H_2O$ means.

4.3 Give the names of the following ions
    a. NH_4^+ b. OH^- c. CO_3^{2-}
    d. PO_4^{3-}

4.4 State the names of the following polyatomic ions
    a. NO_3^- b. NO_2^-
    c. ClO_4^- d. ClO_3^-
    e. ClO^-

4.5 Give the names of the following ions
    a. MnO_4^- b. CrO_4^{2-} c. CO_3^{2-}
    d. SO_4^{2-} e. SO_3^{2-}

Chapter 4

4.6 Name the following cations
   a. Na^+ b. Ca^{2+} c. K^+ d. Ba^{2+}

4.7 Name the following cations, indicating the charge by a Roman numeral
   a. Fe^{2+} b. Fe^{3+} c. Cu^+ d. Cu^{2+}

4.8 Name the following anions
   a. N^{3-} b. O^{2-} c. H^- d. F^-

4.9 Name the following anions
   a. S^{2-} b. Se^{2-} c. Cl^- d. Br^-

4.10 State the Greek prefixes used to indicate the following numbers of atoms per molecule
   a. 2   b. 3   c. 4   d. 5   e. 6

4.11 Give the common names of the following compounds
   a. H_2O b. H_2O_2 c. NH_3
   d. N_2H_4

4.12 Name the following acids
   a. HCl b. $HClO$ c. $HClO_4$
   d. HNO_3 e. H_2SO_4

4.13 Name the following acids
   a. HF b. HNO_2 c. $HClO_3$
   d. $HClO_2$ e. H_2CO_3

4.14 Name the following anions
   a. HCO_3^- b. HSO_4^- c. HSO_3^-

4.15 Name the following anions
   a. $H_2PO_4^-$ b. HPO_4^{2-} c. PO_4^{3-}

4.16 Describe the three-step approach used to determine the mass percents of the elements of a compound of known formula.

4.17 Describe the three-step approach used to obtain the simplest formula of a compound from percent composition.

4.18 In order to determine the molecular formula of a compound, you need one piece of information in addition to the simplest formula. What is that piece of information?

Problems

Formulas of Ionic Compounds
(You may use the Periodic Table in these problems.)

4.19 Predict the formulas of the ionic compounds formed by
   a. lithium and iodine
   b. magnesium and sulfur
   c. aluminum and oxygen

4.20 Predict the formulas of
   a. calcium bromide
   b. sodium oxide
   c. barium sulfide

4.21 Predict the formulas of the ionic compounds formed by
   a. zinc and iodine
   b. manganese and sulfur
   c. silver and oxygen

4.22 Predict the formulas of two different ionic compounds formed by chlorine with
   a. iron   b. copper   c. cobalt

4.23 Predict the formulas of
   a. ammonium chloride
   b. potassium carbonate
   c. sodium phosphate

4.24 Predict the formulas of
   a. ammonium sulfate
   b. magnesium chlorate
   c. lithium sulfate

4.25 Predict the formulas of the ionic compounds formed by iodine with the following elements
   a. sodium   b. strontium   c. zinc
   d. silver   e. cadmium

4.26 Predict the formulas of the nitrates and sulfates of the following metals
   a. calcium   b. aluminum   c. nickel
   d. silver

Names of Ionic Compounds

4.27 Name the following ionic compounds
   a. $NaBr$ b. K_2S c. $CaCl_2$

Chemical Formulas and Names

4.28 Name the following ionic compounds
   a. CuSO₄   b. Cu₂SO₄   c. FeI₂
   d. FeBr₃

4.29 Name the following ionic compounds
   a. Al₂(SO₄)₃   b. Ca(NO₃)₂   c. K₂CO₃
   d. MgCrO₄

4.30 Give the formulas of the following ionic compounds
   a. chromium(II) chloride
   b. chromium(III) iodide
   c. cobalt(III) nitrate
   d. cobalt(II) nitrite

4.31 Name the following ionic compounds
   a. NaHCO₃   b. Mg(HCO₃)₂
   c. KHSO₃   d. Ca(HSO₄)₂

4.32 Give the formulas of the following ionic compounds
   a. sodium dihydrogen phosphate
   b. copper(II) phosphate
   c. calcium hydrogen carbonate
   d. cobalt(III) sulfate

Names of Molecular Compounds

4.33 Name the following molecular compounds
   a. HCl   b. NCl₃   c. P₄O₁₀

4.34 Name the following molecular compounds
   a. N₂O   b. N₂O₃   c. N₂O₄
   d. N₂O₅

4.35 Write the molecular formulas of
   a. xenon tetrafluoride
   b. uranium hexafluoride
   c. nitrogen oxide
   d. tellurium dioxide

4.36 Write the molecular formula of
   a. selenium trioxide
   b. diiodine pentoxide
   c. boron trifluoride
   d. sulfur tetrafluoride

Percent Composition from Formula
(Use the Periodic Table for atomic masses.)

4.37 Determine the mass percents of the elements in
   a. KOH   b. (NH₄)₂SO₄   c. C₂H₅OH

4.38 What is the mass percent of water in CuSO₄ · 5H₂O?

4.39 Determine the percent compositions of the following compounds and the mass of copper in a 5.00-g sample of each compound.
   a. CuCO₃   b. Cu(NO₃)₂
   c. CuSO₄·3H₂O

4.40 What is the mass percent of water in Na₂CO₃ · 10H₂O? the mass percent of Na₂CO₃? the mass of water formed when two grams of the hydrate is heated until all the water is driven off?

4.41 Determine the mass of metal in a 1.264-g sample of
   a. Fe(NO₃)₃   b. FeSO₄
   c. CoCl₂ · 6H₂O

Simplest Formula from Percent Composition
(Use the Periodic Table for atomic masses.)

4.42 Find the simplest formulas of compounds with the following compositions
   a. 75.0% C, 25.0% H
   b. 21.8% Mg, 27.9% P, 50.3% O

4.43 Determine the simplest formulas of compounds with the following compositions
   a. 45.9% K, 16.5% N, 37.6% O
   b. 74.0% C, 8.7% H, 17.3% N

4.44 A 20.0-g sample of a compound is found to contain 8.42 g of Na, 3.78 g of P, and 7.80 g of oxygen. Determine
   a. the mass percents of Na, P, and O in the compound
   b. the simplest formula of the compound

4.45 A sample of an oxide of tin weighing 3.014 g is heated in hydrogen to form 2.374 g of Sn.
   a. What are the mass percents of tin and oxygen in the compound?
   b. What is the simplest formula of the compound?

4.46 A certain hydrate of nickel chloride, NiCl₂ · xH₂O, contains 35.8% of H₂O and 64.2% of NiCl₂. What is the value of x in the formula of the hydrate?

4.47 In analyzing a different hydrate of nickel

chloride, a student finds that 1.390 g of the hydrate gives, after heating, 0.758 g of NiCl$_2$. How many moles of H$_2$O are there per mole of NiCl$_2$ in the hydrate?

Molecular Formula from Simplest Formula
(Use the Periodic Table for atomic masses.)

4.48 In a certain hydrocarbon, there are equal numbers of C and H atoms. The molar mass is approximately 65 g/mol. What is the molecular formula?

4.49 A certain hydrocarbon contains 92.3% carbon and 7.70% hydrogen. What is its simplest formula? If the molar mass is 78.0 g/mol, what is the molecular formula?

4.50 Vitamin C contains 40.92% C, 4.58% H, and 54.51% O. The molar mass of Vitamin C is 176.1 g/mol. What is the molecular formula of Vitamin C?

Multiple Choice

4.51 The atomic masses of Cr, Cl, and Br are 52.00, 35.45, and 79.90, respectively. Of the following, which compound contains the largest mass percent of chromium?
  a. chromium(II) chloride
  b. chromium(III) chloride
  c. chromium(II) bromide
  d. chromium(III) bromide

4.52 The mass percents of iron and sulfur in FeSO$_4$ are 36.77 and 21.10, respectively. The mass of oxygen in 2.625 g of FeSO$_4$ is
  a. 0.5539 g   b. 0.9652 g   c. 1.106 g
  d. 2.625 g

4.53 The mass percent of chromium (at. mass = 52.00) in a certain compound is 24.8. The number of moles of chromium in 10.0 g of the compound is
  a. 0.0477   b. 0.192   c. 0.403
  d. 2.48

4.54 The number of moles of water in 1.000 g of MgSO$_4$·7H$_2$O (molar mass = 246.5 g/mol) is
  a. 0.004057   b. 0.02840   c. 0.07310
  d. 0.5117   e. 7

4.55 The simplest formula of a certain hydrocarbon is CH$_2$. Which one of the following (g/mol) could *not* represent the molar mass of the compound?
  a. 14   b. 28   c. 42   d. 60   e. 84
  (at. mass C = 12.0, H = 1.0)

4.56 The formula of chromium(II) chlorite is
  a. CrCl$_2$ b. Cr(ClO)$_2$ c. Cr(ClO$_2$)$_2$
  d. Cr(ClO$_3$)$_2$ e. Cr(ClO$_4$)$_2$

4.57 Using the Periodic Table, predict the formula of the ionic compound formed by magnesium with nitrogen
  a. MgN   b. MgN$_2$ c. MgN$_3$
  d. Mg$_2$N   e. none of these

4.58 How many moles of potassium are there in 3.00 mol of potassium sulfate?
  a. 1.50   b. 3.00   c. 4.50   d. 6.00
  e. 9.00

4.59 The systematic name of N$_2$O$_2$ is
  a. nitrogen oxide
  b. nitric oxide
  c. nitrogen dioxide
  d. dinitrogen oxide
  e. dinitrogen dioxide

4.60 The name of the H$_2$PO$_4^-$ ion is
  a. phosphate
  b. phosphoric acid
  c. hydrogen phosphate
  d. dihydrogen phosphate

Chemical Equations

Learning Objectives
After studying this chapter, you should be able to:

1. Write a balanced chemical equation to represent a reaction, knowing the formulas and physical states of reactants and products (Example 5.1).
2. Write a balanced equation for the reaction of a metal with a nonmetal to form an ionic compound (Example 5.2).
3. Interpret a chemical equation in terms of moles or grams of substances taking part in the reaction (Example 5.3).
4. Relate, using a balanced equation:
   —moles of one substance to moles of another substance taking part in the reaction (Example 5.4).
   —moles of one substance to grams of another substance (Examples 5.5, 5.6).
   —grams of one substance to grams of another substance (Example 5.7).
5. Determine, knowing the amounts in moles or grams of two reactants, the limiting reactant and the theoretical yield of product (Examples 5.8, 5.9).
6. Calculate one of the three quantities: theoretical yield, actual yield, percent yield, given the other two quantities (Example 5.10).
7. Write a balanced thermochemical equation for the combustion of a substance or the formation of a compound from the elements (Example 5.11).

*Thoughts, like fleas,
jump from man to man.
But they don't bite everybody.*

STANISLAW LEC

Photograph by Peter L. Kresan.

CHAPTER 5

In Chapter 2, we referred to a few chemical reactions. Among these was the electrolysis of water to form hydrogen and oxygen. This reaction could be described in words as

water (liquid) → hydrogen (gas) + oxygen (gas)

A more convenient and useful way to describe the reaction is by means of a **chemical equation**

$$2\ H_2O(l) \rightarrow 2\ H_2(g) + O_2(g)$$

Chemical equations, such as the one just written, are the subject of this chapter. We will be interested in the meaning of equations. For example, what do the coefficients (2 for H_2O, 2 for H_2) mean in the above equation? We will also look at the procedure used to translate word descriptions of reactions into *balanced* equations. How do we go about writing an equation and how do we balance it? Most important, we will study the uses of chemical equations. How does a balanced equation help us to relate amounts of reactants and products? How can an equation be used to show the amount of heat absorbed or evolved in a reaction?

To answer questions such as these, we start with a discussion of how chemical equations are written and balanced (Section 5.1). Then we consider what a balanced equation tells us about the relative numbers of moles or grams of reactants or products (Sections 5.2–5.4). Finally, in Section 5.5, we consider thermochemical equations, which specify the heat flow, ΔH, in a reaction.

5.1 Writing and Balancing Chemical Equations

A chemical equation summarizes what happens in a reaction. To translate a reaction carried out in the laboratory into an equation on paper, we follow certain rules. To illustrate these rules, we will apply them to a familiar reaction, that which occurs when natural gas burns in air.

Chemical Equations

The principal component of natural gas is methane, CH_4. It combines with oxygen, O_2, in the air. A study of the reaction shows that there are two products. One of these is a gas, carbon dioxide, CO_2. The other is liquid water, H_2O. To write a balanced equation for this reaction, we follow a three-step path.

1. **Write an unbalanced equation.** To do this, we write the formulas of reactants (starting materials) on the left of the equation. These are separated by an arrow from the formulas of the products of the reaction, which appear on the right of the equation. In this case, we write

   $CH_4 + O_2 \rightarrow CO_2 + H_2O$

   When we are dealing with molecular substances, as we are here, we use molecular formulas. We represent oxygen as O_2 rather than O because we know that the element consists of diatomic molecules.

2. **Balance the equation.** To balance an equation, we make it conform to the Law of Conservation of Mass. This requires that there be the same number of atoms of each element on both sides of the equation. Equations are balanced by writing numbers called *coefficients* in front of the appropriate formulas.

   Examination of the equation we have written for the reaction of CH_4 with O_2 shows that we have

   1 C atom on the left, 1 C atom on the right

   4 H atoms on the left, 2 H atoms on the right

   2 O atoms on the left, 3 O atoms on the right

   Clearly, hydrogen and oxygen atoms are not balanced. To balance hydrogen, we write a coefficient of 2 in front of H_2O

   $CH_4 + O_2 \rightarrow CO_2 + 2\ H_2O$

   The understanding is that the coefficient, 2 in this case, multiplies the entire formula. Hence we now have

   1 C atom on the left, 1 C atom on the right

   4 H atoms on the left, 4 H atoms on the right

   2 O atoms on the left, 4 O atoms on the right

   Oxygen is still unbalanced. This is easily remedied by writing a coefficient of 2 in front of O_2

   $CH_4 + 2\ O_2 \rightarrow CO_2 + 2\ H_2O$

   The equation is now balanced. We have 1 carbon atom, 4 hydrogen atoms, and 4 oxygen atoms on both sides.

   Notice that equations are balanced by adjusting coefficients, *never* by changing subscripts in formulas. For example, we balanced oxygen by writing 2 O_2, not O_4. Oxygen exists as diatomic molecules, O_2; there is no such thing as an O_4 molecule. Remember, chemical equations describe chemical reactions; they must tell us what really happens.

Another point concerning equation balancing is worth commenting about. You will recall that, in the equation we wrote originally for the combustion of methane, both hydrogen and oxygen were out of balance. We chose to balance hydrogen first for a very good reason. It appears in only one substance on both sides of the equation (CH_4 on the left, H_2O on the right). Oxygen, on the other hand, appears in two different substances on the same side of the equation: CO_2 and H_2O on the right. It would have been difficult, to say the least, to start by balancing oxygen. In general, we always prefer to start the balancing process with an element that appears in only one place on each side of the equation.

3. **Indicate the physical states of reactants and products.** We will do this by using

(g) for substances in the gas state

(s) for solids

(l) for pure liquids

(aq) for species in water, i.e., aqueous solution

For this reaction, we have

$$CH_4(g) + 2\ O_2(g) \rightarrow CO_2(g) + 2\ H_2O(l) \tag{5.1}$$

This is the final balanced equation for the combustion of methane.

EXAMPLE 5.1

When gaseous ethane, molecular formula C_2H_6, burns in air, a reaction similar to the one in Equation 5.1 occurs. The products are carbon dioxide gas and liquid water. Write a balanced equation for the reaction.

Solution

We follow the three-step path described in connection with the reaction shown in Equation 5.1.

(1) The unbalanced equation is

$$C_2H_6 + O_2 \rightarrow CO_2 + H_2O$$

(2) To balance the equation, we might start with carbon. At the moment, we have 2 C atoms on the left and 1 on the right. We can bring carbon into balance by writing a coefficient of 2 in front of CO_2

$$C_2H_6 + O_2 \rightarrow 2\ CO_2 + H_2O$$

Continued

Now we balance hydrogen, leaving oxygen until last. We write a coefficient of 3 for H_2O, so as to get six hydrogen atoms on both sides

$$C_2H_6 + O_2 \rightarrow 2\ CO_2 + 3\ H_2O$$

At this stage, we have

$2 \times 2 + 3 \times 1 = 7$ oxygen atoms on the right

To get an equal number of oxygen atoms on the left, we could write a coefficient of 7/2 in front of O_2

$$C_2H_6 + \frac{7}{2} O_2 \rightarrow 2\ CO_2 + 3\ H_2O$$

To get whole-number coefficients, we multiply both sides by 2

$$2\ C_2H_6 + 7\ O_2 \rightarrow 4\ CO_2 + 6\ H_2O$$

(3) The final balanced equation, including physical states, is

$$2\ C_2H_6(g) + 7\ O_2(g) \rightarrow 4\ CO_2(g) + 6\ H_2O(l)$$

Exercise
Which of the following equations are balanced?

$$3\ C_2H_6(g) + \frac{21}{2} O_2(g) \rightarrow 6\ CO_2(g) + 9\ H_2O(l)$$

$$4\ C_2H_6(g) + 14\ O_2(g) \rightarrow 8\ CO_2(g) + 12\ H_2O(l)$$

$$6\ C_2H_6(g) + 21\ O_2(g) \rightarrow 12\ CO_2(g) + 18\ H_2O(l)$$

Answer: All of them have the same number of atoms of each type on both sides.

Example 5.1 and the exercise that follows illustrate an important point. *A chemical equation remains valid (i.e., balanced) if each coefficient is multiplied by the same number.* That number might be 2, ³⁄₂, 3, etc. This means that there are an infinite number of balanced equations for a given reaction. In practice, we ordinarily choose to work with the equation that has the *simplest whole-number coefficients*. Thus for the combustion of ethane we would write

$$2\ C_2H_6(g) + 7\ O_2(g) \rightarrow 4\ CO_2(g) + 6\ H_2O(l) \quad (5.2)$$

rather than any of the other equations shown in Example 5.1.

Another point is sometimes overlooked in connection with chemical equations. *You cannot write an equation to describe a reaction unless you know what happens in the reaction.* In particular, you must know the formulas and physical states of all reactants and products. For the combustion of methane (Equation 5.1) and ethane (Equation 5.2), you were in essence given this information. Sometimes, though, you will be expected to know the formulas and physical states of reactants and products. For

Chapter 5

TABLE 5.1 States and Formulas of the Elements (Chemical Equations)

All elements may be shown as monatomic solids, e.g., C(s), Mg(s), *except:*
Hg(l)
$N_2(g)$, $O_2(g)$, $P_4(s)$
$H_2(g)$, $F_2(g)$, $Cl_2(g)$, $Br_2(l)$, $I_2(s)$
He(g), Ne(g), Ar(g), Kr(g), Xe(g)

example, at this stage you should be able to write a balanced equation for the reaction of aluminum with oxygen without any further information. This requires that you know that

—aluminum is a solid element, symbol Al.
—oxygen is a gaseous element built up of diatomic molecules, O_2.
—the two elements react to form an ionic solid containing Al^{3+} and O^{2-} ions, which has the formula Al_2O_3 (recall the discussion in Chapter 4).

With this information, you can write the unbalanced equation

$$Al + O_2 \rightarrow Al_2O_3$$

This equation is readily balanced, giving

$$4\ Al + 3\ O_2 \rightarrow 2\ Al_2O_3$$

Finally we indicate the physical states, arriving at

$$4\ Al(s) + 3\ O_2(g) \rightarrow 2\ Al_2O_3(s) \tag{5.3}$$

At this stage, you should be able to write balanced equations for the formation of simple ionic compounds from the elements. To do this, you need to know

1. The symbols or formulas and physical states of the elements involved. Here, the information in Table 5.1 will be helpful.
2. The charges of the ions formed. These, you will recall, were listed in Figure 4.1, p. 85. Knowing the charges of both cation and anion, you can use the principle of electroneutrality to predict the formula of the ionic solid formed.
3. How to balance a simple chemical equation, as described in this section.

Example 5.2 illustrates how equations of this sort are written and balanced.

EXAMPLE 5.2
Write a balanced equation for the reaction of
a. aluminum with fluorine   b. magnesium with nitrogen

Continued

Solution

In each case we (1) write an unbalanced equation. To do that, we deduce the formula of the product, using Figure 4.1, and the formulas of the reactants, using Table 5.1. Then (2) we balance the equation in the usual way. Finally (3) we indicate the physical states of reactants and products, again making use of Table 5.1 (note that *all ionic compounds are solids*).

a. (1) The product, aluminum fluoride, contains Al^{3+} and F^- ions. It must then have the formula AlF_3. Taking the formula of the element fluorine from Table 5.1, the unbalanced equation is

$$Al + F_2 \rightarrow AlF_3$$

(2) To balance fluorine, we use a coefficient of 3 for F_2 and a coefficient of 2 for AlF_3

$$Al + 3\,F_2 \rightarrow 2\,AlF_3$$

To balance aluminum, we insert a coefficient of 2 for Al

$$2\,Al + 3\,F_2 \rightarrow 2\,AlF_3$$

(3) Aluminum is a solid, fluorine a gas (Table 5.1), and aluminum fluoride a solid. The final equation is

$$2\,Al(s) + 3\,F_2(g) \rightarrow 2\,AlF_3(s)$$

b. (1) The ions formed are Mg^{2+} and N^{3-}; magnesium nitride must have the simplest formula Mg_3N_2. The unbalanced equation is

$$Mg + N_2 \rightarrow Mg_3N_2$$

(2) To balance magnesium, we write a coefficient of 3 for Mg

$$3\,Mg + N_2 \rightarrow Mg_3N_2$$

(3) Magnesium is a solid, nitrogen a gas, and magnesium nitride a solid. The final balanced equation is

$$3\,Mg(s) + N_2(g) \rightarrow Mg_3N_2(s)$$

Exercise

Write a balanced equation for the reaction of silver with chlorine. Answer: $2\,Ag(s) + Cl_2(g) \rightarrow 2\,AgCl(s)$.

5.2 Meaning of Balanced Equations

The simplest way to interpret a balanced equation is in terms of the relative numbers of particles of reactants and products. Consider for example

$$CH_4(g) + 2\,O_2(g) \rightarrow CO_2(g) + 2\,H_2O(l) \tag{5.1}$$

The coefficients in this equation tell us the relative numbers of molecules reacting and produced. Specifically,

1 molecule CH_4 + 2 molecules $O_2 \rightarrow$ 1 molecule CO_2 + 2 molecules H_2O

In other words, one molecule of CH_4 (methane) reacts with two molecules of O_2 (molecular oxygen) to form one molecule of CO_2 (carbon dioxide) and two molecules of H_2O (water). We sometimes express these relationships in a slightly different way, writing:

1 molecule $CH_4 \simeq$ 2 molecules $O_2 \simeq$ 1 molecule $CO_2 \simeq$ 2 molecules H_2O

The symbol $\simeq$ means "is chemically equivalent to." In this reaction

1 molecule of CH_4 is chemically equivalent to (reacts with) 2 molecules of O_2

1 molecule of CH_4 is chemically equivalent to (reacts to form) 1 molecule of CO_2 and so on.

Mole Relations

You will recall (p. 111) that a chemical equation remains valid if we multiply each coefficient by the same number. This number could be 2, 3/2, – – – or, in particular, Avogadro's number, 6×10^{23}. Consider what happens if we multiply each coefficient in Equation 5.1 by Avogadro's number

6×10^{23} $CH_4(g)$ + 12×10^{23} $O_2(g) \rightarrow 6 \times 10^{23}$ $CO_2(g)$ + 12×10^{23} $H_2O(l)$

This tells us that

6×10^{23} molecules CH_4 + 12×10^{23} molecules $O_2 \rightarrow 6 \times 10^{23}$ molecules CO_2 + 12×10^{23} molecules H_2O

But, 6×10^{23} molecules represents one mole of any molecular substance. Therefore

1 mol CH_4 + 2 mol $O_2 \rightarrow$ 1 mol CO_2 + 2 mol H_2O

Comparing the equation just written with Equation 5.1, we see that the coefficients (1 for CH_4, 2 for O_2, 1 for CO_2, 2 for H_2O) give us not only the relative numbers of molecules but also *the relative numbers of moles of reactants and product*. We conclude that 1 mol of CH_4 reacts with 2 mol of O_2 to form 1 mol of CO_2 and 2 mol of H_2O. In other words, in Equation 5.1,

1 mol $CH_4 \simeq$ 2 mol $O_2 \simeq$ 1 mol $CO_2 \simeq$ 2 mol H_2O

1 mol of CH_4 is chemically equivalent to (reacts with) 2 mol of O_2

1 mol of CH_4 is chemically equivalent to (reacts to form) 1 mol of CO_2

and so on.

Chemical Equations

This observation concerning the equation for Reaction 5.1 is valid for any reaction. **The coefficients of a balanced equation represent the relative numbers of moles of reactants and products.** In the reaction

$$4\ Al(s) + 3\ O_2(g) \rightarrow 2\ Al_2O_3(s)$$

4 mol Al + 3 mol O_2 → 2 mol Al_2O_3

or

4 mol Al ≎ 3 mol O_2 ≎ 2 mol Al_2O_3

which means that

4 mol Al reacts with 3 mol O_2

4 mol Al reacts to form 2 mol Al_2O_3

3 mol O_2 reacts to form 2 mol Al_2O_3

Mass Relations

We have just seen that the coefficients of a balanced equation allow us to relate numbers of moles of reactants and products. We can carry this development one step further to relate masses in grams of reactants and products. To see how this is done, consider again the combustion of methane

$$CH_4(g) + 2\ O_2(g) \rightarrow CO_2(g) + 2\ H_2O(l) \tag{5.1}$$

We have just seen that

1 mol CH_4 + 2 mol O_2 → 1 mol CO_2 + 2 mol H_2O

We can readily convert moles to grams for each species. Taking the atomic masses of carbon, hydrogen, and oxygen to be 12.01, 1.008, and 16.00, we find that

1 mol CH_4 weighs 12.01 g + 4(1.008 g) = 16.04 g

1 mol O_2 weighs 2(16.00 g) = 32.00 g

1 mol CO_2 weighs 12.01 g + 2(16.00 g) = 44.01 g

1 mol H_2O weighs 2(1.008 g) + 16.00 g = 18.02 g

Referring back to Equation 5.1, we can say that

16.04 g CH_4 + 2(32.00 g) O_2 → 44.01 g CO_2 + 2(18.02 g) H_2O

or

16.04 g CH_4 + 64.00 g O_2 → 44.01 g CO_2 + 36.04 g H_2O

We see that 16.04 g of CH_4 reacts with 64.00 g of O_2 to form 44.01 g of CO_2 and 36.04 g of H_2O. In terms of equivalences

16.04 g CH_4 ≃ 64.00 g O_2 ≃ 44.01 g CO_2 ≃ 36.04 g H_2O

This development can be carried out with any balanced equation. Example 5.3 applies it in a slightly more complex reaction.

EXAMPLE 5.3

In Example 5.1, we found the balanced equation for the combustion of ethane to be

$$2\ C_2H_6(g) + 7\ O_2(g) \rightarrow 4\ CO_2(g) + 6\ H_2O(l)$$

For this reaction, obtain a relation between
a. moles of C_2H_6 and moles of O_2
b. grams of C_2H_6 and grams of O_2
c. grams of C_2H_6 and grams of CO_2

Solution
a. The coefficients of the balanced equation give us directly the mole relation

2 mol C_2H_6 ≃ 7 mol O_2

b. To relate masses in grams of C_2H_6 and O_2, we start with the relation in (a) and convert moles to grams, taking the molar masses of C_2H_6 and O_2 to be 30.07 and 32.00 g/mol, respectively

2(30.07 g) C_2H_6 ≃ 7(32.00 g) O_2

60.14 g C_2H_6 ≃ 224.00 g O_2

We interpret this relation to mean that 60.14 g of ethane will react with 224.00 g of oxygen

c. We first write down the mole relation, using the coefficients of the balanced equation

2 mol C_2H_6 ≃ 4 mol CO_2

Now we translate into grams (molar mass C_2H_6 = 30.07 g/mol, CO_2 = 44.01 g/mol)

2(30.07 g) C_2H_6 ≃ 4(44.01 g) CO_2

60.14 g C_2H_6 ≃ 176.04 g CO_2

In other words, 60.14 g of C_2H_6 burns to form 176.04 g of CO_2.

Exercise
For the reaction of aluminum with oxygen (Equation 5.3), obtain a relation between grams of aluminum and grams of oxygen. Answer: 107.92 g Al ≃ 96.00 g O_2.

Chemical Equations

5.3 Stoichiometric Calculations

The study of the relationships between amounts of substances produced and consumed in reactions is referred to as *stoichiometry*. In this section, we will be interested in four types of stoichiometric calculations.

1. Given the amount in moles of one substance, X, determine the equivalent number of moles of another substance, Y, involved in the reaction. Calculations of this type are often referred to as *mole to mole conversions*.
2. Given the amount in moles of X, determine the equivalent number of grams of Y. We call this a *mole to mass conversion*.
3. Given the amount in grams of X, determine the equivalent number of moles of Y *(mass to mole conversion)*.
4. Given the amount of grams in X, determine the equivalent number of grams of Y *(mass to mass conversion)*.

Although we treat these as four separate calculations, they have a great deal in common. In particular, they all require that you

—work with the balanced equation for the reaction
—recognize that the coefficients of that equation represent moles of each species

Mole to Mole Conversions

Frequently we want to know the number of moles of a substance that reacts with or is formed from a given number of moles of a different substance. The conversion factor required for this type of calculation follows directly from the coefficients of the balanced equation. The conversion

moles X → moles Y

can be carried out in a single step (Example 5.4).

EXAMPLE 5.4

In the combustion of ethane

$2 \, C_2H_6(g) + 7 \, O_2(g) \rightarrow 4 \, CO_2(g) + 6 \, H_2O(l)$

calculate
a. the moles of O_2 required to react with 1.42 mol C_2H_6
b. the moles of CO_2 formed from 2.85 mol C_2H_6

Continued

Solution

a. As we pointed out earlier

$$2 \text{ mol } C_2H_6 \simeq 7 \text{ mol } O_2$$

For calculation purposes, the equivalence sign can be treated as an "equals" sign. This relation gives us the conversion factor we need to go from moles of C_2H_6 (given) to moles of O_2 (required). That conversion factor is 7 mol O_2/2 mol C_2H_6

$$\text{moles } O_2 = 1.42 \text{ mol } C_2H_6 \times \frac{7 \text{ mol } O_2}{2 \text{ mol } C_2H_6} = 4.97 \text{ mol } O_2$$

b. Again, the relation we need is given directly by the coefficients of the balanced equation

$$2 \text{ mol } C_2H_6 \simeq 4 \text{ mol } CO_2$$

Setting up the conversion from moles of C_2H_6 to moles of CO_2

$$2.85 \text{ mol } C_2H_6 \times \frac{4 \text{ mol } CO_2}{2 \text{ mol } C_2H_6} = 5.70 \text{ mol } CO_2$$

Exercise

In the reaction between aluminum and oxygen (Equation 5.3), how many moles of O_2 are required to react with 5.29 mol Al? Answer: 3.97 mol O_2.

Mole to Mass Conversions

Occasionally, you may be asked to calculate the mass in grams of a substance Y that reacts with or is produced from a given number of moles of another substance X. This involves a two-step conversion

moles X → moles Y → grams Y

The first step is carried out using the coefficients of the balanced equation, as in Example 5.4. Having found the number of moles of Y, you then convert to grams, using the molar mass. In practice, the two steps are carried out in one continuous operation (Example 5.5).

EXAMPLE 5.5

For the combustion of methane

$$CH_4(g) + 2 O_2(g) \rightarrow CO_2(g) + 2 H_2O(l)$$

calculate the mass in grams of water formed from 1.82 mol of CH_4.

Continued

Solution
The indicated path is

1.82 mol CH$_4$ → ? moles H$_2$O → ? grams H$_2$O
                  (1)   (2)

The relations and conversion factors required are

relation	conversion factor
(1) 1 mol CH$_4$ ≏ 2 mol H$_2$O	2 mol H$_2$O/1 mol CH$_4$
(2) 1 mol H$_2$O = 18.02 g H$_2$O	18.02 g H$_2$O/1 mol H$_2$O

The set-up is

$$\text{mass H}_2\text{O} = 1.82 \text{ mol CH}_4 \times \frac{2 \text{ mol H}_2\text{O}}{1 \text{ mol CH}_4} \times \frac{18.02 \text{ g H}_2\text{O}}{1 \text{ mol H}_2\text{O}} = 65.6 \text{ g H}_2\text{O}$$

Exercise
For the reaction between aluminum and fluorine (Example 5.2), calculate the mass in grams of fluorine required to react with 2.00 mol Al. Answer: 114 g F$_2$.

Mass to Mole Conversions

This type of calculation, like that in Example 5.5, involves a two-step conversion

mass X → moles X → moles Y

In the first step, grams of X are converted to moles. Then, using the coefficients of the balanced equation, we convert moles of X to moles of Y.

EXAMPLE 5.6
For the reaction

4 Al(s) + 3 O$_2$(g) → 2 Al$_2$O$_3$(s)

how many moles of Al$_2$O$_3$ are formed from 5.18 g of Al?

Solution
The indicated path is

5.18 g Al → ? mol Al → ? mol Al$_2$O$_3$
              (1)   (2)

To find the conversion factors required, we note that
(1) 1 mol Al = 26.98 g Al
(2) 4 mol Al ≏ 2 mol Al$_2$O$_3$
The conversion is

Continued

Chapter 5

$$\text{moles Al}_2\text{O}_3 = 5.18 \text{ g Al} \times \frac{1 \text{ mol Al}}{26.98 \text{ g Al}} \times \frac{2 \text{ mol Al}_2\text{O}_3}{4 \text{ mol Al}} = 0.0960 \text{ mol Al}_2\text{O}_3$$

Exercise
For the reaction of magnesium with nitrogen (Example 5.2), calculate the number of moles of N_2 required to react with 12.15 g Mg. Answer: 0.1667 mol N_2.

Mass to Mass Conversions

Perhaps the most common type of stoichiometric calculation requires that you calculate the mass in grams of one species involved in a reaction, knowing the mass of another species

mass X → mass Y

Here, a three-step conversion is required.

1. Convert grams of X to moles of X, using the molar mass of X.
2. Convert moles of X to moles of Y, using the coefficients of the balanced equation.
3. Convert moles of Y to grams of Y, using the molar mass of Y.

In other words, we follow a three-step path

mass X → moles X → moles Y → mass Y
        (1)   (2)   (3)

As usual, the arithmetic is carried out in a single operation (Example 5.7).

EXAMPLE 5.7
For the combustion of ethane

$2 \text{ C}_2\text{H}_6(g) + 7 \text{ O}_2(g) \rightarrow 4 \text{ CO}_2(g) + 6 \text{ H}_2\text{O}(l)$

calculate
a. the mass in grams of O_2 required to react with 12.0 g of C_2H_6
b. the mass in grams of CO_2 formed from 12.0 g of C_2H_6

Solution
a. The indicated path is

12.0 g C_2H_6 → ? moles C_2H_6 → ? mol O_2 → ? g O_2
                (1)   (2)   (3)

The relations required are

(1) 1 mol C_2H_6 = 30.07 g C_2H_6
(2) 2 mol C_2H_6 ≃ 7 mol O_2
(3) 1 mol O_2 = 32.00 g O_2

Continued

The conversion is

$$\text{mass } O_2 = 12.0 \text{ g } C_2H_6 \times \frac{1 \text{ mol } C_2H_6}{30.07 \text{ g } C_2H_6} \times \frac{7 \text{ mol } O_2}{2 \text{ mol } C_2H_6} \times \frac{32.00 \text{ g } O_2}{1 \text{ mol } O_2}$$

$$= 44.7 \text{ g } O_2$$

b. This is similar to (a), except that we want the mass of CO_2 instead of O_2. The path to be followed is

12.0 g C_2H_6 → ? mol C_2H_6 → ? mol CO_2 → ? g CO_2

Note from the balanced equation that 2 mol C_2H_6 forms 4 mol CO_2. The molar masses of C_2H_6 and CO_2 are 30.07 and 44.01 g/mol, respectively. The conversion is

$$\text{mass } CO_2 = 12.0 \text{ g } C_2H_6 \times \frac{1 \text{ mol } C_2H_6}{30.07 \text{ g } C_2H_6} \times \frac{4 \text{ mol } CO_2}{2 \text{ mol } C_2H_6} \times \frac{44.01 \text{ g } CO_2}{1 \text{ mol } CO_2}$$

$$= 35.1 \text{ g } CO_2$$

Exercise

For the combustion of methane (Equation 5.1), how many grams of CO_2 are formed from 32.0 g of CH_4? Answer: 88.0 g.

These examples should convince you of the importance of

—*balanced equations*. All stoichiometric calculations require that you start with a balanced equation.
—*the mole concept*. The balanced equation allows you to relate moles of one substance to moles of another. You learned in Chapter 3 how to convert moles of a given substance to grams or vice versa; that skill is required here as well.
—*the conversion factor approach*. If you're not using conversion factors, chances are you're lost at this point. If so, go back and work through the various examples until you can set them up properly.

5.4 Limiting Reactant and Theoretical Yield

Let's consider once again the reaction of aluminum with oxygen

4 Al(s) + 3 O_2(g) → 2 Al_2O_3(s)

As we have seen, the coefficients in this equation represent relative numbers of moles. They tell us that 4 mol of Al will react with 3 mol of O_2 to form 2 mol of Al_2O_3. If we were to mix 4.00 mol Al (108 g) with 3.00 mol O_2 (96 g), we should produce 2.00 mol (204 g) of Al_2O_3. Both reactants would be completely consumed. The Al_2O_3 produced should be pure, uncontaminated by either of the starting materials, aluminum metal or oxygen gas.

Usually when we carry out a reaction in the laboratory, the reactants are not in exactly the mole ratio required for reaction. We might, for example, burn 4.00 mol of Al (108 g) in a room containing 20.00 mol of O_2 (640 g). In that case, there would be an excess of oxygen, since only 3.00 mol of O_2 is required to react with 4.00 mol of Al. After the reaction is over, we should have

$(20.00 - 3.00)$ mol $O_2 = 17.00$ mol O_2 left over

All the aluminum should be used up, forming 2.00 mol of Al_2O_3

$$4.00 \text{ mol Al} \times \frac{2 \text{ mol Al}_2O_3}{4 \text{ mol Al}} = 2.00 \text{ mol Al}_2O_3$$

In other words, the reaction would form 2.00 mol (204 g) of Al_2O_3 mixed with 17.00 mol (544 g) of unreacted O_2.

In this situation, we distinguish between the reactant in excess (O_2) and the other reactant (Al), which is called the **limiting reactant.** The amount of product formed is determined (limited) by the amount of limiting reactant. With 4.00 mol of Al, we cannot get more than 2.00 mol of Al_2O_3, no matter how large an excess of O_2 we use. The amount of product that would be formed if all the limiting reactant were consumed is called the **theoretical yield** of product. If we mix 4.00 mol of Al with excess O_2, the theoretical yield of Al_2O_3 is 2.00 mol.

Often you will need to determine the limiting reactant in a reaction and then find the theoretical yield of product. To do this, it helps to follow a systematic procedure. The one that we will use involves three steps.

1. Calculate the amount of product that would be formed if the first reactant were completely consumed.
2. Repeat this calculation for the second reactant. That is, calculate how much product would be formed if all of that reactant were consumed.
3. Choose the **smaller** of the two amounts calculated in Steps 1 and 2. This is the theoretical yield of product. The reactant that produces the smaller amount is the limiting reactant.

This procedure is illustrated in Example 5.8.

EXAMPLE 5.8

Consider the reaction

$3 \text{ Mg}(s) + N_2(g) \rightarrow Mg_3N_2(s)$

Suppose 2.00 mol of Mg is mixed with 2.00 mol of N_2. What is the theoretical yield of Mg_3N_2 in moles? Which is the limiting reactant, magnesium or nitrogen?

Continued

Solution

Following the three steps described above

(1) Assuming all the magnesium is used up

$$\text{moles } Mg_3N_2 = 2.00 \text{ mol Mg} \times \frac{1 \text{ mol } Mg_3N_2}{3 \text{ mol Mg}} = 0.667 \text{ mol } Mg_3N_2$$

(2) Assuming all the nitrogen is used up

$$\text{moles } Mg_3N_2 = 2.00 \text{ mol } N_2 \times \frac{1 \text{ mol } Mg_3N_2}{1 \text{ mol } N_2} = 2.00 \text{ mol } Mg_3N_2$$

(3) Choosing the *smaller* amount of Mg_3N_2, we decide that

—the theoretical yield of Mg_3N_2 is 0.667 mol
—the limiting reactant is Mg

There is an excess of N_2 in this reaction mixture, far more than is required to react with all the magnesium.

Exercise

How many moles of N_2 are required to react with the 2.00 mol of Mg? Of the 2.00 mol of N_2 used, how much is left unreacted? Answer: 0.667 mol; 1.33 mol.

If the amounts of reactants are expressed in grams, the calculation of theoretical yield is a bit more tedious. However, the procedure followed is exactly the same as before (Example 5.9).

EXAMPLE 5.9

Consider the reaction

$$2 \text{ Al}(s) + 3 \text{ F}_2(g) \rightarrow 2 \text{ AlF}_3(s)$$

Suppose we start with a mixture of 10.0 g of Al and 10.0 g of F_2. What is the theoretical yield, in grams, of AlF_3? Which reactant is limiting?

Solution

We proceed as in Example 5.8, except that gram–gram conversions are required in Steps 1 and 2. Recall (Example 5.7) that these are three-step conversions. The molar masses of Al, F_2, and AlF_3 are

Al: 27.0 g/mol F_2: 38.0 g/mol AlF_3: 84.0 g/mol

(1) If all the aluminum is consumed

$$\text{mass } AlF_3 = 10.0 \text{ g Al} \times \frac{1 \text{ mol Al}}{27.0 \text{ g Al}} \times \frac{2 \text{ mol } AlF_3}{2 \text{ mol Al}} \times \frac{84.0 \text{ g } AlF_3}{1 \text{ mol } AlF_3}$$

$$= 31.1 \text{ g } AlF_3$$

Continued

(2) If all the fluorine is consumed

$$\text{mass AlF}_3 = 10.0 \text{ g F}_2 \times \frac{1 \text{ mol F}_2}{38.0 \text{ g F}_2} \times \frac{2 \text{ mol AlF}_3}{3 \text{ mol F}_2} \times \frac{84.0 \text{ g AlF}_3}{1 \text{ mol AlF}_3}$$
$$= 14.7 \text{ g AlF}_3$$

(3) The theoretical yield is 14.7 g of AlF_3; the limiting reactant is F_2. Some of the aluminum is left over.

Exercise
How many grams of Al are required to react with 10.0 g of F_2? How many grams of Al remain unreacted? Answer: 4.74 g; 5.3 g.

Remember that, in deciding upon the theoretical yield of product, you choose the *smaller* of the two amounts calculated in Steps 1 and 2. To see why this must be the case, it may be helpful to refer back to Example 5.9. There, we started with 10.0 g of Al and 10.0 g of F_2. We decided that the theoretical yield of AlF_3 was 14.7 g and that 5.3 g of Al was left over. Thus we have

10.0 g Al + 10.0 g $F_2 \rightarrow$ 14.7 g AlF_3 + 5.3 g Al

This makes sense; 20.0 g of reactants yields a total of 20.0 g of "products," including 5.3 g of unreacted aluminum. Suppose though that we had chosen 31.1 g of AlF_3 as the theoretical yield. We would then have had the nonsensical situation

10.0 g Al + 10.0 g $F_2 \rightarrow$ 31.7 g AlF_3

This violates the Law of Conservation of Mass. There is no way we can get 31.7 g of product from 20.0 g of reactants.

In calculating the theoretical yield, we assume that the limiting reactant is completely (i.e., 100%) converted to product. In practice, this almost never happens. Some of the limiting reactant may be consumed in other reactions. Some product may be lost in separating it from the reaction mixture. In general, the **actual yield** in a reaction is expected to be less than the theoretical yield. Putting it another way, the **percent yield** is ordinarily less than 100.

$$\% \text{ yield} = \frac{\text{actual yield}}{\text{theoretical yield}} \times 100 \tag{5.4}$$

EXAMPLE 5.10
If, with the mixture referred to in Example 5.9, the actual yield of AlF_3 is 12.6 g, what is the percent yield?

Continued

Solution
Recall that the theoretical yield was found to be 14.7 g. Hence,

% yield $= \dfrac{12.6 \text{ g}}{14.7 \text{ g}} \times 100 = 85.7$

Exercise
If the percent yield were 90.0, what would be the actual yield, taking the theoretical yield to be 14.7 g? Answer: 13.2 g.

5.5 Thermochemical Equations

In Chapter 2, we referred briefly to the flow of heat in chemical reactions. We pointed out that, from the standpoint of heat flow, there are two different kinds of reactions.

1. Exothermic reactions, in which heat is given off by the reaction mixture to the surroundings. For such reactions, the heat flow, ΔH, is a negative quantity.

   Exothermic: Heat evolved, $\Delta H < 0$ \hfill (5.5)

2. Endothermic reactions, in which heat is absorbed by the reaction mixture from the surroundings. For such reactions, the heat flow, ΔH, is a positive quantity.

   Endothermic: Heat absorbed, $\Delta H > 0$ \hfill (5.6)

We can interpret the heat flow, ΔH, as a difference in enthalpy or "heat content" between products and reactants

$$\Delta H = H_{products} - H_{reactants} \quad (5.7)$$

From this point of view, an endothermic reaction is one in which the heat content of the products is greater than that of the reactants (Figure 5.1, *left*). In an exothermic reaction, the products have a lower heat content than the reactants (Figure 5.1, *right*, p. 126).

In a thermochemical equation, we cite the value of ΔH at the right of the chemical equation. The thermochemical equation for the combustion of methane is

$$CH_4(g) + 2\ O_2(g) \rightarrow CO_2(g) + 2\ H_2O(l);\ \Delta H = -890.3 \text{ kJ} \quad (5.8)$$

This equation tells us that 890.3 kJ of heat is given off when 1 mol of CH_4 reacts with 2 mol of O_2 to form 1 mol of CO_2 and 2 mol of H_2O. Similarly, we interpret the thermochemical equation

$$Sn(s) + O_2(g) \rightarrow SnO_2(s);\ \Delta H = -580.7 \text{ kJ} \quad (5.9)$$

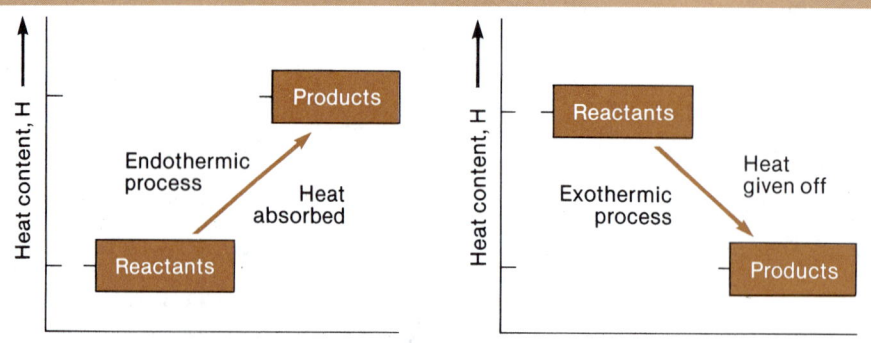

Figure 5.1 In an endothermic process, the heat content of the products is higher than that of the reactants. The increase in heat content comes from the heat absorbed from the surroundings. In an exothermic process, the heat content of the products is less than that of the reactants. The loss in heat content occurs as heat is given off to the surroundings.

to mean that 580.7 kJ of heat is given off when one mole of SnO_2 is formed from the elements.

Equations 5.8 and 5.9 represent two different types of reactions that are quite common in chemistry. The ΔH's associated with these reactions are given special names.

1. The **heat of combustion** is ΔH when one mole of a substance burns with excess oxygen. The heat of combustion of $CH_4(g)$ is -890.3 kJ/mol. Other heats of combusion are listed in Table 5.2. Note that heats of combustion are always negative; heat is given off when a substance burns.
2. The **heat of formation** is ΔH when one mole of a substance is formed from the elements. The heat of formation of $SnO_2(s)$ is -580.7 kJ/mol. Note from Table 5.2 that heats of formation are usually, although not always, negative. Heat is ordinarily given off when a compound is formed from the elements.

EXAMPLE 5.11
Using Table 5.2, write balanced thermochemical equations for the
a. formation of $MgO(s)$ b. combustion of $C_2H_2(g)$

Solution
a. The value given for the heat of formation in Table 5.2 is for one mole of MgO. The balanced equation is

$$Mg(s) + \frac{1}{2} O_2(g) \rightarrow MgO(s); \Delta H = -601.8 \text{ kJ}$$

Continued

Chemical Equations

TABLE 5.2 Heats of Formation and Combustion (kJ/mol)

Heat of Formation		Heat of Combustion to $CO_2(g)$ and $H_2O(l)$	
$Al_2O_3(s)$	−1669.8	$H_2(g)$	−285.8
$CaCl_2(s)$	−795.0	$C(s)$	−393.5
$Ca(OH)_2(s)$	−986.6	$CO(g)$	−283.0
$CO(g)$	−110.5	$CH_4(g)$	−890.3
$CO_2(g)$	−393.5	$CH_3OH(l)$	−726.8
$CuO(s)$	−155.2	$C_2H_2(g)$	−1299.6
$Fe_2O_3(s)$	−822.2	$C_2H_4(g)$	−1410.8
$HCl(g)$	−92.3	$C_2H_6(g)$	−1559.8
$HNO_3(l)$	−173.2	$C_2H_5OH(l)$	−1366.9
$H_2O(l)$	−285.8	$C_3H_8(g)$	−2220.0
$H_2SO_4(l)$	−811.3	$C_4H_{10}(g)$	−2878.6
$HgO(s)$	−90.7	$C_5H_{12}(l)$	−3510.0
$MgO(s)$	−601.8	$C_6H_6(l)$	−3268.1
$NaCl(s)$	−411.0	$C_6H_{14}(l)$	−4163.1
$NO(g)$	+90.4	$C_7H_{16}(l)$	−4817.0
$NO_2(g)$	+33.9	$C_8H_{18}(l)$	−5470.6
$PbO(s)$	−217.9	$C_{10}H_8(s)$	−5156.8
$SiO_2(s)$	−859.4		
$SnO_2(s)$	−580.7		
$SO_3(g)$	−395.2		

b. The reactants are $C_2H_2(g)$ and $O_2(g)$; the products are $CO_2(g)$ and $H_2O(l)$. The unbalanced equation is

$$C_2H_2(g) + O_2(g) \rightarrow CO_2(g) + H_2O(l)$$

The balanced equation for the combustion of one mole of acetylene, C_2H_2, is

$$C_2H_2(g) + \frac{5}{2}O_2(g) \rightarrow 2\,CO_2(g) + H_2O(l); \Delta H = -1299.6 \text{ kJ}$$

Exercise
Write the balanced thermochemical equation for the formation of NO(g). Answer:

$$\frac{1}{2}N_2(g) + \frac{1}{2}O_2(g) \rightarrow NO(g); \Delta H = +90.4 \text{ kJ}$$

Notice that, in writing thermochemical equations, it is permissible and often most convenient to use fractional coefficients.

Key Words

actual yield
balanced equation
coefficient

conversion factor
enthalpy change, ΔH
heat of combustion

Continued

heat of formation
limiting reactant
mole
percent yield

stoichiometry
theoretical yield
thermochemical equation

Questions

5.1 Explain what is meant by a "balanced" equation.

5.2 What does each of the following symbols mean in a balanced equation

(s), (l), (g), (aq)

5.3 Why, in balancing an equation, do we adjust coefficients, not subscripts?

5.4 The balanced equation

$C_3H_8(g) + 5\ O_2(g) \rightarrow 3\ CO_2(g) + 4\ H_2O(l)$

tells us that

1 mol $C_3H_8 \simeq$? mol $O_2 \simeq$? mol $CO_2 \simeq$? mol H_2O

5.5 The balanced equation

$C(s) + O_2(g) \rightarrow CO_2(g)$

tells us that

12.0 g C $\simeq$? g $O_2 \simeq$? g CO_2

5.6 What is meant by the term "limiting reactant"; "theoretical yield"?

5.7 Explain in your own words how you decide which reactant is limiting.

5.8 In an exothermic reaction, ΔH has a _____ sign; in an endothermic reaction, ΔH has a _____ sign.

5.9 State in words what the following thermochemical equation means

$CH_4(g) + 2\ O_2(g) \rightarrow CO_2(g) + 2\ H_2O(l)$; $\Delta H = -890.3$ kJ

5.10 What is meant by "heat of formation"; "heat of combustion"?

Problems

Balancing Equations

5.11 Balance the following equations
  a. $Na(s) + O_2(g) \rightarrow Na_2O_2(s)$
  b. $Na(s) + H_2O(l) \rightarrow NaOH(s) + H_2(g)$
  c. $SnO_2(s) + H_2(g) \rightarrow Sn(s) + H_2O(l)$

5.12 Balance the following equations
  a. $MnO_2(s) \rightarrow Mn_3O_4(s) + O_2(g)$
  b. $SiO_2(s) + HF(g) \rightarrow SiF_4(g) + H_2O(l)$
  c. $NaClO_3(s) \rightarrow NaCl(s) + O_2(g)$

5.13 Write balanced equations for the following reactions
  a. iron powder reacts with sulfur to form solid iron(II) sulfide
  b. two gases, xenon and fluorine, react to form crystals of xenon tetrafluoride
  c. The two elements hydrogen and nitrogen react to form $NH_3(g)$
  d. When copper(I) oxide is heated, it decomposes to the elements

5.14 Write a balanced equation for the formation of an ionic compound when
  a. aluminum reacts with oxygen
  b. calcium reacts with fluorine
  c. barium reacts with oxygen
  d. sodium reacts with sulfur

5.15 Write a balanced equation for the formation of an ionic compound by the reaction of
  a. potassium with nitrogen
  b. zinc with bromine
  c. nickel with oxygen
  d. silver with chlorine

5.16 When a hydrocarbon burns in air, the products are $CO_2(g)$ and $H_2O(l)$. Write balanced equations for the burning of the following hydrocarbons
  a. $C_2H_4(g)$ b. $C_2H_2(g)$ c. $C_6H_6(l)$

5.17 When iron reacts with oxygen, the product is iron(III) oxide. What is wrong with each of the following equations written to represent that reaction?
  a. $2\ Fe(s) + O_3(g) \rightarrow Fe_2O_3(s)$
  b. $8\ Fe(s) + 6\ O_2(g) \rightarrow 4\ Fe_2O_3(s)$
  c. $Fe(s) + O_2(g) \rightarrow Fe_2O_3(s)$
  d. $Fe(s) + O_2(g) \rightarrow FeO_2(s)$

5.18 Balance the following equations
  a. $Sr(s) + H_2O(l) \rightarrow Sr(OH)_2(s) + H_2(g)$
  b. $(NH_4)_2CO_3(s) \rightarrow NH_3(g) + CO_2(g) + H_2O(l)$
  c. $Al(s) + H_2C_2O_4(aq) \rightarrow Al_2(C_2O_4)_3(s) + H_2(g)$
  d. $Co_3O_4(s) + CO(g) \rightarrow Co(s) + CO_2(g)$

Mole to Mole Conversions

5.19 The balanced equation for the combustion of propane is

$$C_3H_8(g) + 5\ O_2(g) \rightarrow 3\ CO_2(g) + 4\ H_2O(l)$$

Calculate
  a. the number of moles of CO_2 formed from 1.26 mol of C_3H_8
  b. the number of moles of C_3H_8 required to react with 6.18 mol of O_2
  c. the number of moles of H_2O formed from 1.62×10^{-3} mol of O_2

5.20 The balanced equation for the reaction of iron with chlorine is

$$2\ Fe(s) + 3\ Cl_2(g) \rightarrow 2\ FeCl_3(s)$$

Calculate
  a. the number of moles of Fe required to form 0.585 mol of $FeCl_3$
  b. the number of moles of Cl_2 required to form 0.585 mol of $FeCl_3$
  c. the number of moles of Fe required to react with 1.18 mol of Cl_2

5.21 The balanced equation for the reaction of nitric oxide with chlorine is

$$2\ NO(g) + Cl_2(g) \rightarrow 2\ NOCl(g)$$

Complete the following table

_____ mol NO + 1.64 mol $Cl_2 \rightarrow$ _____ mol NOCl

_____ mol NO + _____ mol $Cl_2 \rightarrow$ 3.18 mol NOCl

5.22 The balanced equation for the reaction of phosphorus with oxygen is

$$P_4(s) + 5\ O_2(g) \rightarrow P_4O_{10}(s)$$

Complete the following table

_____ mol P_4 + 2.19 mol $O_2 \rightarrow$ _____ mol P_4O_{10}

7.76 mol P_4 + _____ mol $O_2 \rightarrow$ _____ mol P_4O_{10}

Mole to Mass Conversions

(Use the Periodic Table for atomic masses.)

5.23 For the reaction in Problem 5.19, calculate
  a. the mass in grams of O_2 required to react with 1.00 mol of C_3H_8
  b. the mass in grams of CO_2 formed from 1.68 mol of C_3H_8
  c. the mass in grams of H_2O formed from 1.68 mol of C_3H_8

5.24 For the reaction in Problem 5.20, calculate
  a. the mass in grams of $FeCl_3$ formed from 6.02 mol of Fe
  b. the mass in grams of Cl_2 required to react with 6.02 mol of Fe
  c. the mass in kilograms of Fe required to react with 1.00 mol of Cl_2

5.25 For the reaction in Problem 5.21, complete the following table

_____ g NO + _____ g Cl₂ →
1.64 mol NOCl

1.64 mol NO + _____ g Cl₂ →
_____ g NOCl

5.26 For the reaction in Problem 5.22, complete the following table

_____ g P₄ + _____ g O₂ →
3.77 mol P₄O₁₀

2.69 mol P₄ + _____ g O₂ →
_____ g P₄O₁₀

Mass to Mole Conversions
(Use the Periodic Table for atomic masses.)

5.27 For the reaction in Problem 5.19, calculate
   a. the number of moles of O_2 required to react with 4.19 g of C_3H_8
   b. the number of moles of C_3H_8 required to react with 3.42×10^2 g of O_2
   c. the number of moles of H_2O formed from 4.91×10^1 g of C_3H_8

5.28 For the reaction in Problem 5.20, calculate
   a. the number of moles of $FeCl_3$ formed from 1.00×10^2 g of Fe
   b. the number of moles of Cl_2 required to react with 1.00×10^2 g of Fe
   c. the number of moles of Fe required to react with 212 g of chlorine

5.29 For the reaction in Problem 5.21, complete the following table

_____ mol NO + 3.74 g Cl₂ →
_____ mol NOCl

51.9 g NO + _____ mol Cl₂ →
_____ mol NOCl

5.30 For the reaction in Problem 5.22, complete the following table

_____ mol P₄ + _____ mol O₂ →
519 g P₄O₁₀

_____ mol P₄ + 16.4 g O₂ →
_____ mol P₄O₁₀

Mass to Mass Conversions
(Use the Periodic Table for atomic masses.)

5.31 Oxygen gas is sometimes prepared in the laboratory by heating potassium chlorate

$$2 \text{ KClO}_3(s) \rightarrow 2 \text{ KCl}(s) + 3 \text{ O}_2(g)$$

What mass of potassium chlorate is required to produce 0.0448 g of oxygen?

5.32 When magnesium is added to a water solution of hydrochloric acid, the following reaction occurs

$$\text{Mg}(s) + 2 \text{ H}^+(aq) \rightarrow \text{Mg}^{2+}(aq) + \text{H}_2(g)$$

What mass of magnesium is required to produce 2.16 g of H_2?

5.33 When sodium hydrogen carbonate is heated it decomposes

$$2 \text{ NaHCO}_3(s) \rightarrow \text{Na}_2\text{CO}_3(s) + \text{H}_2\text{O}(g) + \text{CO}_2(g)$$

What mass of solid is formed when 1.000 g of $NaHCO_3$ decomposes?

5.34 Ammonia in the presence of a platinum catalyst burns to form NO

$$4 \text{ NH}_3(g) + 5 \text{ O}_2(g) \rightarrow 4 \text{ NO}(g) + 6 \text{ H}_2\text{O}(l)$$

Complete the following table

_____ g NH₃ + 162 g O₂ →
_____ g NO + _____ g H₂O

18.0 g NH₃ + _____ g O₂ →
_____ g NO + _____ g H₂O

5.35 The combustion of the rocket fuel hydrazine occurs via the reaction

$$\text{N}_2\text{H}_4(l) + \text{O}_2(g) \rightarrow \text{N}_2(g) + 2 \text{ H}_2\text{O}(l)$$

Complete the following table

_____ g N₂H₄ + _____ g O₂ →
3.19 g N₂ + _____ g H₂O

86.2 g N$_2$H$_4$ + _____ g O$_2$ →
_____ g N$_2$ + _____ g H$_2$O

5.36 Aluminum reacts with HCl(g) to produce solid aluminum chloride, AlCl$_3$, and hydrogen gas
   a. Write a balanced equation for the reaction
   b. Calculate the mass of AlCl$_3$ formed from 1.00 g of Al

5.37 Aluminum carbide, Al$_4$C$_3$, is an abrasive made by reacting aluminum with carbon
   a. Write a balanced equation for the reaction
   b. What mass of aluminum is required to make 45.5 kg of Al$_4$C$_3$?

5.38 Write a balanced equation for the reaction of potassium with iodine and calculate the mass of product formed from 1.86 g of K

5.39 Write a balanced equation for the reaction of zinc with oxygen and calculate the mass of product formed from 1.00 kg of Zn

5.40 In the production of beer by fermentation, glucose, C$_6$H$_{12}$O$_6$, is converted to ethyl alcohol, C$_2$H$_5$OH, and CO$_2$

C$_6$H$_{12}$O$_6$(aq) → 2 C$_2$H$_5$OH(aq) + 2 CO$_2$(g)

What mass of glucose is required to make 2.00 × 10^3 kg of beer which contains 4.20% by mass of ethyl alcohol?

5.41 Photosynthesis can be described by the equation

6 CO$_2$(g) + 6 H$_2$O(l) → C$_6$H$_{12}$O$_6$(s) + 6 O$_2$(g)

It has been estimated that land plants use 25 billion metric tons of CO$_2$ in this reaction each year (1 metric ton = 1000 kg). How many grams of C$_6$H$_{12}$O$_6$ and O$_2$ are produced each year?

Limiting Reactant, Theoretical Yield
(Use the Periodic Table for atomic masses.)

5.42 Consider the reaction in Problem 5.19. Suppose we start with 1.00 mol of C$_3$H$_8$ and 2.00 mol of O$_2$. Which reactant is limiting? What is the theoretical yield of CO$_2$ in moles?

5.43 Suppose that, in the reaction referred to in Problem 5.19, we start with 1.00 g of C$_3$H$_8$ and 2.00 g of O$_2$. Which reactant is limiting? What is the theoretical yield of CO$_2$ in grams?

5.44 The reaction of potassium with bromine is

2 K(s) + Br$_2$(l) → 2 KBr(s)

If 4.0 g of K is mixed with 10.0 g of Br$_2$, what is the theoretical yield of KBr?

5.45 To produce iron(II) sulfide, 142 g of sulfur is heated with 182 g of iron. Which reactant is limiting? How many grams of the other reactant are left when the reaction is over?

5.46 Suppose that, in the reaction referred to in Problem 5.21, we start with 6.00 g of NO and 7.00 g of Cl$_2$. What is the theoretical yield of NOCl? If the actual yield is 11.2 g, what is the percent yield?

5.47 For the reaction referred to in Problem 5.20, suppose you react 1.00 mol of Fe with 40.0 g of Cl$_2$. What is the theoretical yield in grams of FeCl$_3$? If the actual yield is 50.0 g, what is the percent yield?

5.48 What mass of 80.0% pure gold(III) oxide is required to yield 1.00 g of pure gold upon decomposition to the elements?

Multiple Choice

5.49 For which one of the following equations would the value of ΔH be equal to the heat of formation of magnesium oxide, in kilojoules per mole?
a. $2\ Mg(s) + O_2(g) \rightarrow 2\ MgO(s)$
b. $Mg(s) + \frac{1}{2} O_2(g) \rightarrow MgO(s)$
c. $MgO(s) \rightarrow Mg(s) + \frac{1}{2} O_2(g)$
d. $2\ MgO(s) \rightarrow 2\ Mg(s) + O_2(g)$

5.50 For which one of the following equations would the value of ΔH represent the heat of combustion of C_2H_2 in kilojoules per mole?
a. $2\ C(s) + H_2(g) \rightarrow C_2H_2(g)$
b. $C_2H_2(g) \rightarrow 2\ C(s) + H_2(g)$
c. $C_2H_2(g) + \frac{5}{2} O_2(g) \rightarrow 2\ CO_2(g) + H_2O(l)$
d. $2\ C_2H_2(g) + 5\ O_2(g) \rightarrow 4\ CO_2(g) + 2\ H_2O(l)$

5.51 In the balanced equation (with the smallest whole-number coefficients) for the formation of Cr_2O_3 from the elements, the coefficient of oxygen is
a. 1   b. 2   c. 3   d. 4   e. 5

5.52 In the balanced equation (with the smallest whole-number coefficients) for the combustion of C_4H_{10} to form CO_2 and H_2O, the coefficient of oxygen is
a. 2   b. 5   c. 8   d. 13   e. 26

5.53 In the reaction of aluminum with fluorine, the number of moles of fluorine required to react with 1.41 mol of Al is
a. 0.470   b. 0.940   c. 1.41   d. 2.11   e. 4.23

Questions 5.54–5.60 refer to the reaction of nitrogen with hydrogen

$$N_2(g) + 3\ H_2(g) \rightarrow 2\ NH_3(g)$$

(atomic mass N = 14.0, H = 1.0)

5.54 The mass in grams of N_2 required to react with 2.00 mol of H_2 is
a. 0.667   b. 4.67   c. 9.33   d. 14.0   e. 18.7

5.55 The number of moles of H_2 required to form 1.00 g of NH_3 is
a. 0.0882   b. 0.176   c. 0.333   d. 1.00   e. 1.50

5.56 The mass in grams of NH_3 formed from 1.00 g of N_2 is
a. 0.824   b. 1.00   c. 1.21   d. 2.00   e. 5.67

5.57 If one starts with 2.00 mol of N_2 and 2.00 mol of H_2, the theoretical yield of NH_3 in moles is
a. 0.667   b. 1.00   c. 1.33   d. 2.00   e. 4.00

5.58 A student calculates that the theoretical yield of NH_3 in his reaction is 12.00 g of NH_3. He actually obtains 6.92 g of NH_3. The percent yield is
a. 42.3   b. 57.7   c. 90.6   d. 173

5.59 Another student carrying out this reaction uses a mixture of H_2 and helium containing 60.0 mass percent H_2. How many grams of N_2 are required to react with 1.00 g of this mixture (the helium does not react)?
a. 0.200   b. 0.600   c. 2.80   d. 4.67   e. 7.78

5.60 When this reaction takes place, the number of moles
a. decreases   b. increases   c. remains the same

The Physical Behavior of Gases

Learning Objectives
After studying this chapter you should be able to:

1. State and explain the meaning of each postulate of the kinetic molecular theory of gases.
2. Convert between different units of volume, temperature, and pressure (Example 6.1).
3. Use the relation $P_1V_1 = P_2V_2$ to calculate one of the four quantities (P_1, V_1, P_2, V_2), knowing the values of the other three (Example 6.2).
4. Use the relation $V_1/T_1 = V_2/T_2$ to calculate one of the four quantities (V_1, T_1, V_2, T_2), knowing the values of the other three (Example 6.3).
5. Use the relation $P_1V_1/T_1 = P_2V_2/T_2$ to calculate one of the six quantities (V_1, P_1, T_1, V_2, P_2, T_2), knowing the values of the other five (Example 6.4).
6. Using the molar volume of a gas (22.4 L at standard temperature and pressure), relate density to molar mass (Examples 6.5, 6.6).
7. Given or having calculated the values of three of the four variables P, V, n, T, use the Ideal Gas Law to calculate the other variable (Examples 6.7, 6.8).
8. Apply Dalton's Law to calculate the partial pressure or number of moles (given V and T) of a gas collected over water (Example 6.9).
9. Relate the mass in grams or amount in moles of one substance to the volume of another gaseous substance involved in a reaction (Examples 6.10, 6.11).
10. Relate the volumes of two different gases (same T and P) involved in a reaction (Example 6.12).

The meaning of things lies not in the things themselves but in our attitude toward them.

SAINT-EXUPÉRY

Photograph by Peter L. Kresan.

CHAPTER 6

All substances in the gas state resemble one another in their physical behavior. Among the properties that distinguish gases from liquids or solids are the following.

1. *Gases are readily compressible.* In operating a tire pump, you compress air to perhaps ⅓ of its original volume. A scuba diver uses compressed air that has a volume less than 10% of that of the same air at atmospheric pressure. Liquids and solids are almost incompressible. You become painfully aware of this when you do a "belly flop" off a diving board into the water of a swimming pool.
2. *Gases expand readily when heated.* The volume of a balloon filled with air at atmospheric pressure increases by nearly 40% when the temperature rises from 0 to 100°C (Figure 6.1). Over the same temperature range, liquid water increases in volume by only about 4%.
3. *Gases have low densities.* A flask filled with air weighs much less than the same flask filled with a liquid or solid. Gas densities, at room temperature and atmospheric pressure, range from about 0.00008 g/cm^3 (H_2) to about 0.014 g/cm^3 (UF_6). In contrast, liquids and solids have much higher densities (d_{water} = 1.00 g/cm^3, d_{ice} = 0.91 g/cm^3).
4. *Gases diffuse readily into each other.* If you generate chlorine gas at your lab bench, your neighbor soon becomes aware of its odor. The chlorine molecules quickly mix with those of the air. Liquids mix more slowly; under ordinary conditions, solids do not diffuse at all.

Figure 6.1 The effect of temperature on gas volume. When the temperature rises, the volume of air inside the balloon increases.

cool balloon

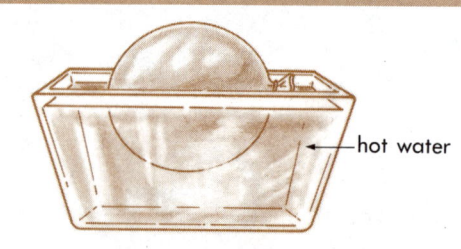

expanded balloon

hot water

In this chapter, we will search for explanations of these and other properties of gases. We will seek to find out why gases are so compressible, why they expand so much on heating, and why they have such low densities. Questions such as these can be answered qualitatively in terms of a simple model of the gas state known as the kinetic molecular theory (Section 6.1). Later in this chapter, we will become more quantitative and seek to answer the question "how much?" To what extent does the volume of a gas change when the pressure or temperature changes by a given amount? What is the molar volume of a gas at a given temperature and pressure? Calculations required to answer questions such as these use the so-called "gas laws," discussed in Sections 6.2–6.5. Finally, in Section 6.6, we will see how the gas laws allow us to make stoichiometric calculations involving the volumes of gases taking part in reactions.

6.1 Kinetic Molecular Theory of Gases

One of the greatest accomplishments of 19th century science was the development of a successful model of the gas state. This model was derived by James Maxwell in England and Ludwig Boltzmann in Austria, among others. It explains the properties of gases in terms of the motion of their molecules and hence is called the kinetic molecular theory. The main features of this theory are listed below.

1. **At ordinary temperatures and pressures, gas molecules are far removed from each other.** At 25°C and 1 atm, gas molecules are, on the average, about ten molecular diameters apart. This means that a gas is mostly empty space. Only about 0.1% of the total volume is occupied by the molecules themselves.

    This simple feature explains many of the properties of gases. Their low densities reflect the fact that gases are mostly empty space. Compressing a gas is relatively easy because all one has to do is move the molecules closer together. When a gas is heated, the molecules move farther apart and the volume occupied by the gas increases.

    The large distance between gas molecules has another important consequence. Attractive forces between molecules are quite weak when they are far apart. At ordinary temperatures and pressures, the effect of these forces in gases is very small. Under these conditions, gas molecules act as completely independent particles, unaffected by each other's existence. This situation changes when a gas is cooled and compressed. As the molecules come close together, attractive forces become more important. Eventually, they become strong enough for the gas to condense to a liquid, where the molecules touch one another.

Chapter 6

2. **Gas molecules are moving very rapidly.** At 25°C, the average speed of an O_2 molecule in a sample of gaseous oxygen is 482 m/s. In more familiar terms, this amounts to

$$482 \text{ m} \times \frac{1 \text{ km}}{1000 \text{ m}} \times \frac{3600 \text{ s}}{1 \text{ h}} = 1740 \text{ km/h}$$

or 1080 miles per hour! The high speed of gas molecules helps to explain why diffusion occurs so rapidly.

You will note that we referred to the *average* speed of an O_2 molecule. The word "average" reflects the fact that not all the molecules in a gas sample are travelling at the same speed. At any given instant, a few molecules are moving very slowly, others are moving very rapidly, and still others have intermediate speeds. It turns out that for most molecules the speed is reasonably close to the average value. In a sample of oxygen at 25°C, nearly 90% of the O_2 molecules have speeds in the range 200–800 m/s. Fewer than 1% have speeds in excess of 1000 m/s.

3. **An increase in temperature increases both the average speed and the fraction of molecules that have a very high speed.** These effects are shown for oxygen in Figure 6.2. When the temperature increases from 25

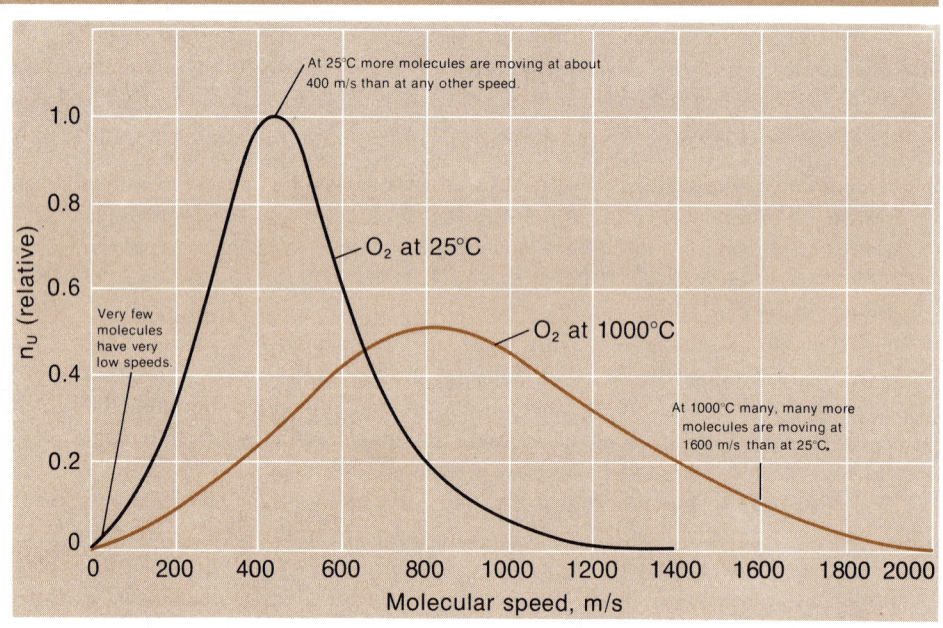

Figure 6.2 The graph shows the fraction of O_2 molecules having a given speed of 25°C and at 1000°C. At 25°C most of the molecules are moving at a speed close to 400 m/s. At 1000°C, the average speed is about twice as great. More important, the fraction of molecules moving at very high speeds is much greater at the higher temperature.

The Physical Behavior of Gases   139

Figure 6.3a When two molecules collide, the individual speeds and energies may change. However, the total kinetic energy after collision is the same as it was before collision.

to 1000°C, the average speed approximately doubles. The fraction of molecules having speeds greater than 1000 m/s increases from less than 1% to about 50%, a factor of 50.

4. **Collisions between molecules occur frequently and are perfectly elastic** (Figure 6.3). In oxygen gas at 25°C and 1 atm, an O_2 molecule collides

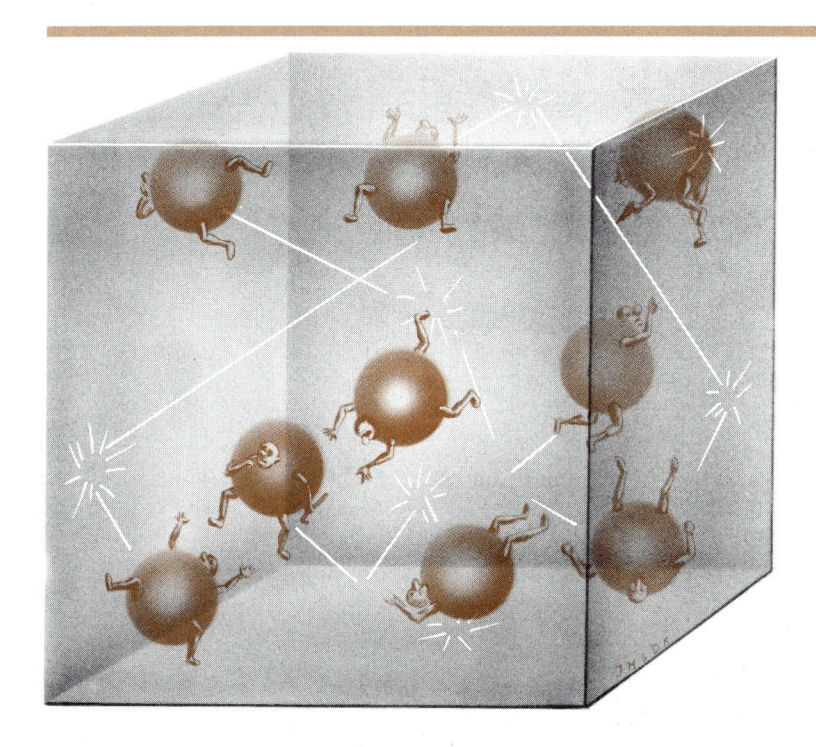

Figure 6.3b Gas molecules collide frequently with the walls of their container. These collisions are responsible for the pressure exerted by the gas.

with about three billion other molecules each second. These collisions are *elastic* in the sense that there is no overall loss or gain of kinetic energy (energy of motion). As the result of a collision, one molecule may speed up while another slows down. However, the total kinetic energy after collision is exactly the same as it was before the collision occurred (Figure 6.3a).

Gas molecules collide frequently not only with each other but also with the walls of their container (Figure 6.3b). These collisions are responsible for the pressure exerted by the gas. Every time a molecule strikes a wall and rebounds from it, it exerts a force on the wall. The sum of these forces over the area of the walls defines the gas pressure.

6.2 Volume–Pressure–Temperature Relations

The kinetic molecular theory of gases leads to some of the basic relations governing the physical behavior of gases. These relations, known as the gas laws, will occupy our attention for the rest of this chapter. In this section, we will look at two such laws. One, which relates the volume of a gas to its pressure, was first proposed by an English scientist, Robert Boyle, in 1660. The other, which describes the temperature dependence of volume, was developed by two French scientists, Jacques Charles and Joseph Gay-Lussac in 1800. Before discussing these laws, it will be helpful to consider the units in which volume, pressure, and temperature are expressed.

Units of Volume, Pressure, and Temperature

The units of volume and temperature were discussed in Chapter 1. Volumes of gases are most often expressed in *liters* (L) or *cubic centimeters* (cm^3)

$$1 \text{ L} = 1000 \text{ cm}^3 \tag{6.1}$$

In Chapter 1, we considered three different temperature scales. **In all calculations involving gases, temperature is expressed in K.** Temperatures in °C are readily converted to K by using the relation

$$K = °C + 273 \tag{6.2}$$

Pressure is defined as force per unit area. In the English system, pressure may be expressed in pounds per square inch (lb/in^2). A mass of one pound resting on an area of one square inch exerts a pressure of 1 lb/in^2. A metric pressure unit that is being used more and more nowadays is the *kilopascal* (kPa). A mass of 10 g resting on a surface 1 cm^2 in area exerts a pressure of approximately 1 kPa (Figure 6.4).

Atmospheric pressure can be measured accurately by using a device

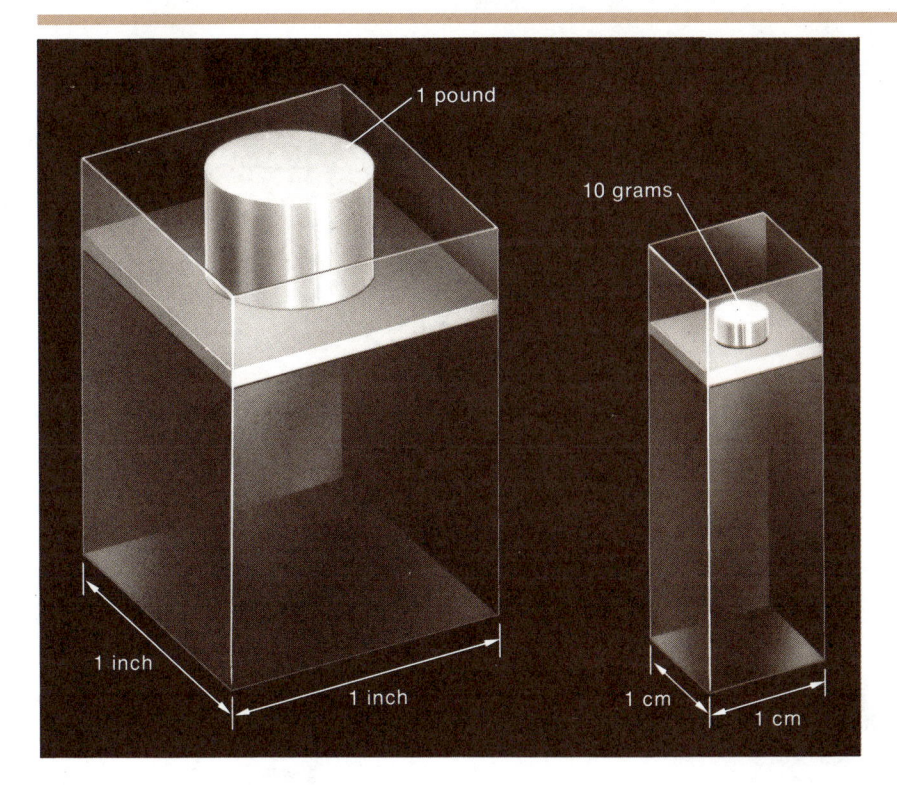

Figure 6.4 Pressure is equal to force divided by area. A mass of one pound resting on one square inch of area will exert a pressure downward due to gravity of one pound per square inch. Ten grams on a surface of 1 cm² will exert a pressure of about one kilopascal on that surface.

called a *barometer* (Figure 6.5, p. 142). Here, we measure the height of a column of mercury that exerts a pressure equal to that of the air in the room. The pressure of air, and indeed of any gas, is often expressed in *millimeters of mercury*. When we say that a gas exerts a pressure of "742 mm Hg" we mean that its pressure is the same as that exerted by a column of mercury 742 mm high.

Another unit commonly used to express pressure is the *standard atmosphere*, usually referred to simply as an "atmosphere." An atmosphere is defined as the pressure exerted by a column of mercury 760 mm high

1 atm = 760 mm Hg   (6.3)

In relation to the pressure units mentioned earlier

1 atm = 14.7 lb/in² = 101.3 kPa   (6.4)

EXAMPLE 6.1

A certain gas exerts a pressure of 729 mm Hg. Express this pressure in atmospheres.

Continued

Chapter 6

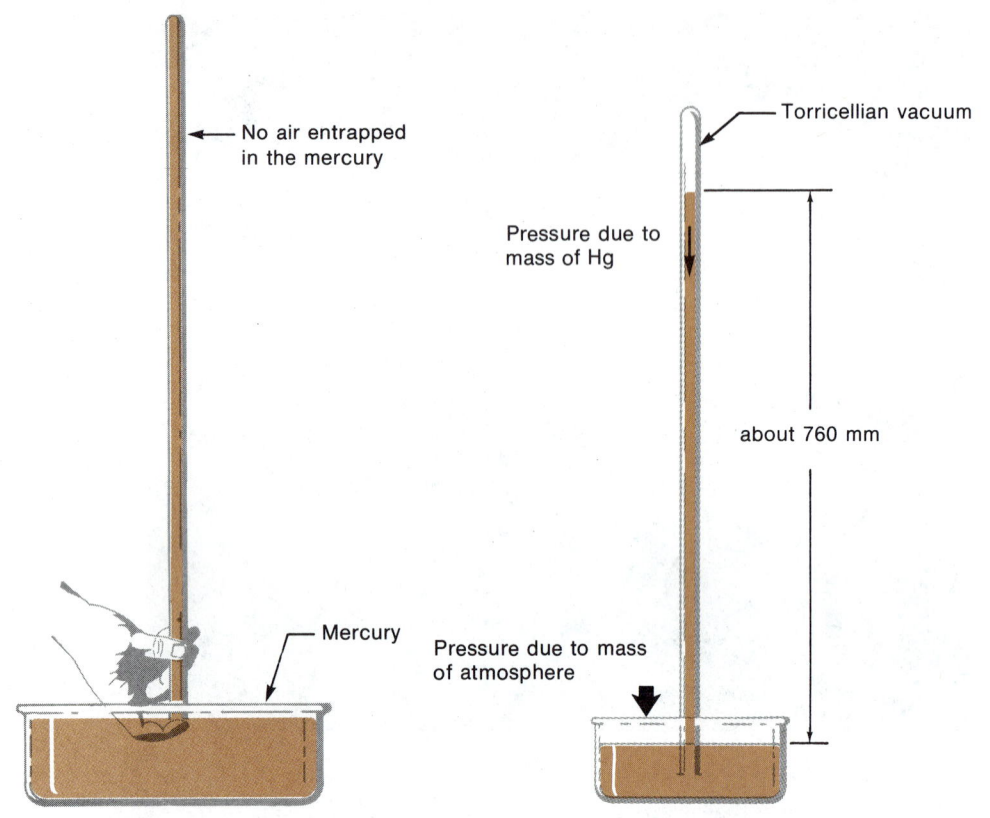

Figure 6.5 Making a barometer. The mercury is held up in the tube by the pressure of the atmosphere. The height of the column is a direct measure of the air pressure, because at the lower mercury surface, the pressure both inside and outside the tube must have the same value or else the mercury would flow. That pressure is the pressure of the air.

Solution
The conversion factor is obtained directly from Equation 6.3. Since we want to go from millimeters of mercury to atmospheres, we should have atm in the numerator and mm Hg in the denominator. The factor is 1 atm/760 mm Hg

$$\text{pressure in atm} = 729 \text{ mm Hg} \times \frac{1 \text{ atm}}{760 \text{ mm Hg}} = 0.959 \text{ atm}$$

Exercise
Express a pressure of 0.120 atm in millimeters of mercury. Answer: 91.2 mm Hg.

The Physical Behavior of Gases   143

Atmospheric pressure is ordinarily greatest at sea level. There, it averages to be approximately 760 mm Hg. At higher altitudes, where there is a smaller mass of air overhead, the pressure is lower. At Denver, Colorado (elevation = 1600 m), the normal air pressure is only about 650 mm Hg. Air pressure also varies with weather conditions. It is highest on bright sunny days and drops as a storm approaches. In a hurricane or tornado, the atmospheric pressure may be as low as 650 mm Hg at sea level.

Volume–Pressure Relation

The volume of a gas sample at constant temperature is *inversely* related to its pressure. When you increase the pressure on the air in a tire pump by pushing down on the piston, the volume of air decreases. This inverse relationship is readily explained by the kinetic molecular theory (Figure 6.6). When the gas molecules are crowded into a smaller volume, they strike the walls more frequently. Since gas pressure is caused by wall collisions, it must increase when the volume is reduced.

Table 6.1 (p. 144) lists volume data obtained by increasing the pressure on a sample of gas at constant temperature. These data are plotted in Figure 6.7. From the table or the figure, it should be clear that

The volume of a gas sample at constant temperature is inversely proportional to pressure.

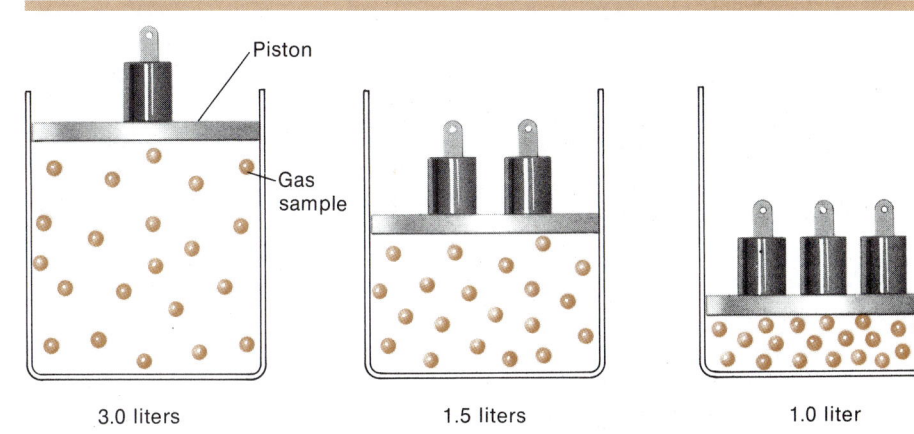

3.0 liters   1.5 liters   1.0 liter

Figure 6.6 If the volume of a gas is decreased, its pressure goes up. If the volume is cut in half, there will be twice as many molecules in a given volume as there were before. This produces twice as many collisions on any surface of the container and increases the pressure by a factor of two.

Chapter 6

TABLE 6.1 Relation Between P and V for 1.00 Mol of H$_2$ at 0°C

P (atmospheres)	V (liters)	PV (atm·L)
1.00	22.4	22.4
2.00	11.2	22.4
3.00	7.47	22.4
4.00	5.60	22.4
5.00	4.48	22.4

Mathematically,* this means that

$$V = \frac{k_1}{P} \quad \text{(T and amount of gas held constant)}$$

In this equation, V is the volume and P the pressure of the gas. The quantity k_1 is a "constant," i.e., it is a number that does not change when V and P change. If we multiply both sides of this equation by P, we obtain

$$PV = k_1 \quad \text{(T and amount of gas held constant)} \tag{6.5}$$

To see what Equation 6.5 means, consider the data in Table 6.1. The table gives the volume occupied by one mole of hydrogen gas at 0°C and a series of different pressures. Notice that, regardless of the individual values of P and V, their product, PV remains constant at 22.4 atm·L. We would say that, under these conditions, k_1 in Equation 6.5 is 22.4 atm·L.

Equation 6.5 is known as Boyle's Law. For use in calculations, it is convenient to express the law in a somewhat different form. We need a relation between "final" and "initial" conditions that does not involve k_1. To ob-

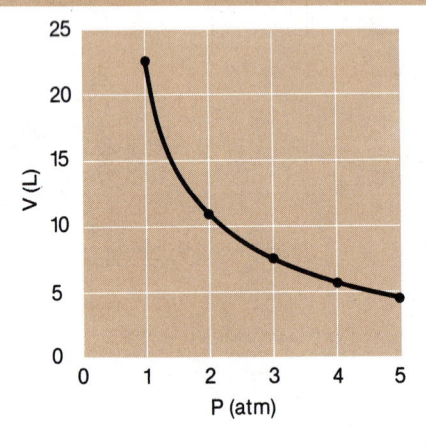

Figure 6.7 Boyle's Law. The volume of a gas sample, maintained at constant temperature, is inversely proportional to the pressure, P.

*Inverse and direct proportionalities are discussed further in Appendix 1.

tain such a relation, we write Equation 6.5 twice, once for initial conditions (subscript 1) and once for final conditions (subscript 2). Thus we have

$P_1 V_1 = k_1$ and $P_2 V_2 = k_1$

Since $P_1 V_1$ and $P_2 V_2$ are both equal to k_1, they are equal to each other. The relation we want is

$P_2 V_2 = P_1 V_1$ (T and amount of gas held constant) (6.6)

The use of Equation 6.6 is illustrated in Example 6.2

EXAMPLE 6.2
A tank of compressed air contains 20.0 L of air at a pressure of 96.5 atm. What volume would this air occupy if the pressure were reduced to 1.00 atm at constant temperature?

Solution
It helps to organize the data so that you are clear as to what is given and what is asked for. In this case

given: $P_1 = 96.5$ atm, $V_1 = 20.0$ L, $P_2 = 1.00$ atm

required: $V_2 = ?$

Clearly, we need to solve Equation 6.6 for V_2. To do that, we divide both sides by P_2:

$$V_2 = V_1 \times \frac{P_1}{P_2} = 20.0 \text{ L} \times \frac{96.5 \text{ atm}}{1.00 \text{ atm}} = 1.93 \times 10^3 \text{ L}$$

Exercise
A sample of gas has a volume of 25.0 cm^3 at 724 mm Hg. If the pressure increases, at what pressure (mm Hg) does the volume become 20.0 cm^3? Answer: 905 mm Hg.

Problems of this type can also be solved by a slightly different approach. Instead of substituting into Equation 6.6, you might reason as follows. The final volume can be obtained from the initial volume by multiplying by an appropriate factor:

$V_{final} = V_{initial} \times$ (pressure factor)

The pressure decreases from 96.5 to 1.00 atm. The effect is to *increase* the volume since V and P are inversely related. Hence the pressure factor must be a ratio greater than one, i.e., 96.5 atm/1.00 atm (rather than 1.00 atm/96.5 atm).

$$V_{final} = V_{initial} \times \frac{96.5 \text{ atm}}{1.00 \text{ atm}} = 20.0 \text{ L} \times \frac{96.5 \text{ atm}}{1.00 \text{ atm}} = 1.93 \times 10^3 \text{ L}$$

One further point concerning Equation 6.6 is worth mentioning. It doesn't matter what units are used for P and V, provided initial and final values have the same units. Thus P_2 and P_1 could both be in atmospheres or both in millimeters of mercury. We could not, however, express P_2 in atmospheres and P_1 in millimeters of mercury. If we did that, the units in the relation

$$V_2 = V_1 \times \frac{P_1}{P_2}$$

would not cancel. We would have to change the units of either P_2 or P_1, using the relation

1 atm = 760 mm Hg

Volume—Temperature Relation

You will recall (Figure 6.1) that the volume of a gas sample at constant pressure is directly related to temperature. Volume increases when temperature rises and decreases when temperature falls. This effect can be explained in terms of the kinetic molecular theory. You will recall that an increase in temperature raises the speed of gas molecules. As a result, they strike the walls more often and with greater force. If the volume were held constant, the pressure of the gas would increase. To hold the pressure constant at the higher temperature, the volume would have to increase. In this way, the number of collisions with the wall per unit time would be back to its original value.

Table 6.2 lists volume data obtained by increasing the temperature of a gas sample at constant pressure. The data are plotted in Figure 6.8. The graph is a straight line which, if extended, would pass through the origin. That is, V = 0 when T = 0. From the table or the figure, it should be clear that

The volume of a gas sample at constant pressure is directly proportional to its absolute temperature (K). (Charles' law)
Mathematically, this means that

$V = k_2 T$ (P and amount of gas held constant)

where k_2 is a constant, independent of the values of V and T. If we divide both sides of this equation by T, we obtain

$$\frac{V}{T} = k_2 \quad \text{(P and amount of gas held constant)} \tag{6.7}$$

The validity of this relation is indicated by the last column of data in Table 6.2, where the quotient V/T has the same value, 0.0821 L/K, at every point.

Equation 6.7 is commonly called Charles' Law. To obtain a "two-point"

The Physical Behavior of Gases

TABLE 6.2 Relation Between V and T for 1.00 Mol of H$_2$ at 1.00 atm

t (°C)	T (K)	V (L)	V/T (L/K)
0	273	22.4	0.0821
50	323	26.5	0.0821
100	373	30.6	0.0821
150	423	34.7	0.0821
200	473	38.8	0.0821

equation relating volume to temperature, we write Equation 6.7 twice, at initial conditions (V$_1$, T$_1$) and final conditions (V$_2$, T$_2$)

$$\frac{V_1}{T_1} = k_2 \quad \text{and} \quad \frac{V_2}{T_2} = k_2$$

or

$$\frac{V_1}{T_1} = \frac{V_2}{T_2} \quad \text{(P and amount of gas held constant)} \tag{6.8}$$

EXAMPLE 6.3

A sealed balloon filled with air has a volume of 125 cm^3 at 0°C. If the balloon is warmed to 20°C at constant pressure, what does its volume become?

Solution
Again, start by organizing the data

given: V$_1$ = 125 cm^3, T$_1$ = 0 + 273 = 273 K, T$_2$ = 20 + 273 = 293 K

required: V$_2$ = ?

Continued

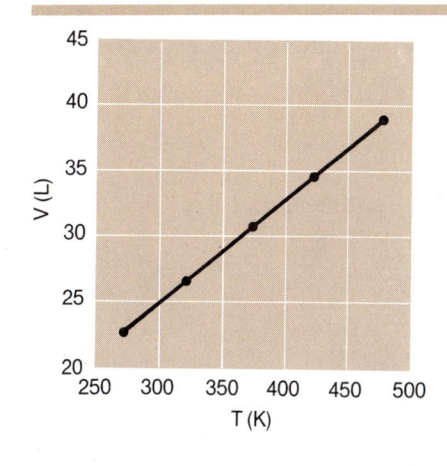

Figure 6.8 Charles' Law. The volume of a gas sample, maintained at constant pressure, is directly proportional to the absolute temperature, T.

We must solve Equation 6.8 for V_2; to do that, we multiply both sides by T_2

$$V_2 = V_1 \times \frac{T_2}{T_1} = 125 \text{ cm}^3 \times \frac{293 \text{ K}}{273 \text{ K}} = 134 \text{ cm}^3$$

We could have arrived at this same answer using the "reasoning" approach:

$V_{final} = V_{initial} \times$ (temperature factor)

Temperature increases from 273 to 293 K; volume is directly proportional to temperature. Hence volume must *increase*, so the temperature factor must be greater than 1, i.e., 293 K/273 K (rather than 273 K/293 K)

$$V_{final} = V_{initial} \times \frac{293 \text{ K}}{273 \text{ K}} = 125 \text{ cm}^3 \times \frac{293 \text{ K}}{273 \text{ K}} = 134 \text{ cm}^3$$

Exercise
A gas occupies a volume of 125 cm³ at 100°C. To what temperature must the gas be cooled at constant pressure to reduce the volume to 101 cm³? Answer: 28°C.

Notice that in using Charles' Law, **you must use absolute temperature** (Kelvin degrees). Volume is directly proportional to temperature in K but *not* in °C.

Volume–Pressure–Temperature Relations

We have seen that the volume of a sample of gas, V, is *inversely* proportional to pressure, P, and *directly* proportional to absolute temperature, T. We can express both of these relations in a single equation:

$$V = k_3 \frac{T}{P} \quad \text{(constant amount of gas)} \tag{6.9}$$

where k_3, like k_1 and k_2, is a constant in the sense that its value is independent of those of the other quantities (V, T, P) in the equation.

To obtain a "two-point" equation relating V, T, and P, we start by solving Equation 6.9 for k_3. To do this, we first multiply both sides of the equation by P and then divide by T

$$PV = k_3 T \; ; \quad \frac{PV}{T} = k_3 \tag{6.10}$$

As usual, we use the subscript 1 for initial conditions and 2 for final conditions.

$$\frac{P_1 V_1}{T_1} = k_3 \quad \text{and} \quad \frac{P_2 V_2}{T_2} = k_3$$

The final equation is

$$\frac{P_2 V_2}{T_2} = \frac{P_1 V_1}{T_1} \tag{6.11}$$

The Physical Behavior of Gases   149

Notice that Equation 6.11 reduces to Boyle's Law ($P_2V_2 = P_1V_1$) when T is constant ($T_2 = T_1$). It also reduces to Charles' Law ($V_2/T_2 = V_1/T_1$) when P is constant ($P_2 = P_1$).

EXAMPLE 6.4
A gas sample has a volume of 1.62 L at 20°C and 716 mm Hg pressure. What will its volume be at a temperature of $-10°C$ and a pressure of 760 mm Hg?

Solution

given: $V_1 = 1.62$ L, $P_1 = 716$ mm Hg, $T_1 = 20 + 273 = 293$ K
$P_2 = 760$ mm Hg, $T_2 = -10 + 273 = 263$ K

required: $V_2 = ?$

We must solve Equation 6.11 for V_2. To do that, we first multiply both sides by T_2 and divide by P_2

$$P_2V_2 = \frac{P_1V_1T_2}{T_1}$$

$$V_2 = V_1 \times \frac{T_2}{T_1} \times \frac{P_1}{P_2}$$

Substituting numbers

$$V_2 = 1.62 \text{ L} \times \frac{263 \text{ K}}{293 \text{ K}} \times \frac{716 \text{ mm Hg}}{760 \text{ mm Hg}} = 1.37 \text{ L}$$

Note that both the temperature factor (263 K/293 K) and the pressure factor (716 mm Hg/760 mm Hg) are less than one. This makes sense. The decrease in temperature (from 293 to 263 K) and the increase in pressure (from 716 to 760 mm Hg) both have the effect of *decreasing* the volume.

Exercise

Solve Equation 6.11 for T_2. Answer: $T_2 = T_1 \times \frac{P_2}{P_1} \times \frac{V_2}{V_1}$.

6.3 Molar Volumes of Gases

In 1813, Amedeo Avogadro, an Italian scientist, suggested a basic principle governing the behavior of gases. Avogadro's Law states that *equal volumes of all gases, at the same temperature and pressure, contain the same number of molecules.* This means, for example, that at 0°C and 1 atm, the number of O_2 molecules in one liter of oxygen is the same as the number of H_2 molecules in one liter of hydrogen or the number of N_2 molecules in one liter of nitrogen, etc.

Amedeo Avogadro

From Avogadro's Law, it follows that one mole (6.022×10^{23} molecules) of any gas must occupy a certain definite volume at a given temperature and pressure. For example, the volume occupied by 1 mol of O_2 at 0°C and 1 atm is the same as that occupied by 1 mol of H_2 at 0°C and 1 atm or 1 mol of N_2 at 0°C and 1 atm, etc. That is

$$V_m \, O_2 = V_m \, H_2 = V_m \, N_2 = \text{----} \quad \text{(T and P held constant)} \tag{6.12}$$

where V_m represents the molar volume. We can say that **all gases, at the same temperature and pressure, have the same molar volume.**

The molar volumes of gases are most often quoted at 0°C and 1 atm. These are often referred to as "standard conditions" and represented as STP (standard temperature and pressure). At 0°C and 1 atm, the molar volumes of all gases are very close to 22.4 L/mol.

$$V_m = 22.4 \text{ L/mol at STP} \tag{6.13}$$

The fact that a mole of any gas occupies 22.4 L at 0°C and 1 atm allows us to calculate gas densities at STP. To do this, we need only know the molar mass of the gas, which follows directly from its formula (Example 6.5).

EXAMPLE 6.5
What is the density of O_2 at STP?

Solution
The molar volume of O_2, like that of all gases, is 22.4 L/mol at STP. The molar mass is 32.00 g/mol. Density is the ratio of mass of volume

$$d = \frac{\text{molar mass}}{\text{molar volume}} = \frac{32.00 \text{ g/mol}}{22.4 \text{ L/mol}} = 1.43 \text{ g/L}$$

Exercise
What is the density of N_2 at 0°C and 1 atm? Answer: 1.25 g/L.

As we have just seen, the density of a gas at STP can be calculated if its molar mass is known. We can turn this calculation around. From the measured density of a gas, we can find its molar mass (Example 6.6). This is an important experimental approach to the determination of molar masses of volatile substances.

EXAMPLE 6.6
A student finds that a sample of natural gas has a density of 0.737 g/L at STP. What is the molar mass of the sample?

Solution
We start with the relation

$$d = \frac{\text{molar mass}}{\text{molar volume}}$$

Since we want molar mass, we solve for that quantity. Multiplying both sides of the equation by molar volume, we have

molar mass = d × molar volume

We are given the density (0.737 g/L); the molar volume is 22.4 L/mol. Hence;

$$\text{molar mass} = 0.737 \frac{\text{g}}{\text{L}} \times 22.4 \frac{\text{L}}{\text{mol}} = 16.5 \text{ g/mol}$$

This answer is reasonable. Natural gas is mostly methane, CH_4, which has a molar mass of 16.0 g/mol. Small amounts of heavier gases, mostly C_2H_6, raise the molar mass slightly.

Exercise
What is the molar mass of a gas that has a density twice that of O_2 at the same temperature and pressure? Answer: 64.0 g/mol.

We have seen that Equation 6.13 is a very useful one. Remember, though, that the concept of molar volume is not limited to STP. We can readily calculate molar volumes of gases under other conditions. For example, at 25°C and 1 atm, we have

$$V_m \text{ at } 25°C, 1 \text{ atm} = (V_m \text{ at } 0°C, 1 \text{ atm}) \times \frac{298 \text{ K}}{273 \text{ K}}$$

$$= 22.4 \frac{L}{mol} \times \frac{298 \text{ K}}{273 \text{ K}} = 24.5 \text{ L/mol}$$

This molar volume can be used in much the same way that we used 22.4 L/mol in Examples 6.5 and 6.6. Thus, to find the density of $O_2(g)$ at 25°C and 1 atm, we would write

$$d = \frac{32.00 \text{ g/mol}}{24.5 \text{ L/mol}} = 1.31 \text{ g/L} \quad (\text{at } 25°C, 1 \text{ atm})$$

To find the molar mass of a gas, given a density of 0.737 g/L at 25°C, 1 atm

$$\text{molar mass} = 0.737 \frac{g}{L} \times 24.5 \frac{L}{mol} = 18.1 \text{ g/L}$$

6.4 Ideal Gas Law

The equations that we have written in this chapter for the volume of a gas apply strictly to an *ideal gas*. An ideal gas is one for which the volume, V, is

—directly proportional to the number of moles, n
—directly proportional to the absolute temperature, T
—inversely proportional to the pressure, P.

Combining these relations, we see that, for an ideal gas

$$V = \frac{\text{constant} \times n \times T}{P}$$

This equation is ordinarily written in a different form. Both sides of the equation are multiplied by P to clear fractions. The constant in the equation, which is the same for all gases, is given the symbol R. With these changes, we obtain the **Ideal Gas Law**

$$PV = nRT \tag{6.14}$$

To evaluate R, we use the fact that 1 mol of any gas (n = 1) occupies a volume of 22.4 L (V = 22.4 L) at STP (T = 273 K, P = 1 atm). Solving Equation 6.14 for R and substituting for P, V, n, and T

$$R = \frac{PV}{nT} = \frac{(1 \text{ atm})(22.4 \text{ L})}{(1 \text{ mol})(273 \text{ K})} = 0.0821 \frac{\text{L} \cdot \text{atm}}{\text{mol} \cdot \text{K}} \qquad (6.15)$$

With this value of R, it is possible to calculate any one of the variables in Equation 6.14 (P, V, n, T), knowing the values of the other three variables. Examples 6.7 and 6.8 show how this is done.

EXAMPLE 6.7
What volume is occupied by 0.100 mol of a gas at 25°C and 720 mm Hg?

Solution
Solving the Ideal Gas Law for V

$$V = \frac{nRT}{P}$$

Here

n = 0.100 mol

R = 0.0821 L · atm/(mol · K)

T = 25 + 273 = 298 K

With R in L · atm/(mol · K), *pressure must be expressed in atmospheres*

$$P = 720 \text{ mm Hg} \times \frac{1 \text{ atm}}{760 \text{ mm Hg}} = 0.947 \text{ atm}$$

Substituting numbers

$$V = \frac{(0.100 \text{ mol})\left(0.0821 \frac{\text{L} \cdot \text{atm}}{\text{mol} \cdot \text{K}}\right)(298 \text{ K})}{0.947 \text{ atm}} = 2.58 \text{ L}$$

Exercise
What volume is occupied by 6.40 g of O_2 at 25°C and 720 mm Hg? Answer: 5.17 L.

EXAMPLE 6.8
Calculate the pressure exerted by 1.56 g of Cl_2 in a container with a volume of 1.00×10^3 cm³ at 20°C.

Solution
Solving the Ideal Gas Law for P

$$P = \frac{nRT}{V}$$

Continued

> To find the number of moles, n, we divide the mass in grams by the molar mass of Cl_2, 70.91 g/mol
>
> $$n = \frac{1.56 \text{ g}}{70.91 \text{ g/mol}} = 0.0220 \text{ mol}$$
>
> R = 0.0821 L · atm/(mol · K)
>
> V = 1.00 × 10³ cm³ = 1.00 L; with this value of R, *volume must be in liters*
>
> T = 273 + 20 = 293 K
>
> Substituting numbers
>
> $$P = \frac{(0.0220 \text{ mol})\left(0.0821 \frac{\text{L} \cdot \text{atm}}{\text{mol} \cdot \text{K}}\right)(293 \text{ K})}{1.00 \text{ L}} = 0.529 \text{ atm}$$
>
> **Exercise**
> How many moles of O_2 are in a 2.00 L cylinder if the oxygen exerts a pressure of 125 atm at 27°C? Answer: 10.2 mol.

In all calculations in this text, we will assume that gases behave ideally. At room temperature and atmospheric pressure, gases follow Equation 6.14 quite closely. At most, deviations amount to a few percent. Deviations from the Ideal Gas Law become more significant at **high pressures** and **low temperatures**. Under these conditions, the molecules are relatively close to one another and two factors that are negligible under ordinary conditions become significant. One of these is the volume occupied by the molecules themselves. The other is the strength of attractive forces between molecules. These two effects cause gases to behave nonideally.

6.5 Dalton's Law of Partial Pressures

So far in this chapter, our discussion has focused on the behavior of pure gases. Frequently, though, we deal with mixtures containing more than one gas. Here a relation discovered by John Dalton in 1807 is useful. Dalton's Law states that **The total pressure of a gas mixture is the sum of the partial pressures of the individual gases in the mixture.** For a mixture of two gases, A and B, Dalton's Law has the form

$$P_{tot} = P_A + P_B \tag{6.16}$$

In this equation, P_{tot} is the total (measured) pressure of the mixture. The quantities P_A and P_B are partial pressures. *The partial pressure of a gas is the pressure it would exert if it occupied the entire volume by itself.*

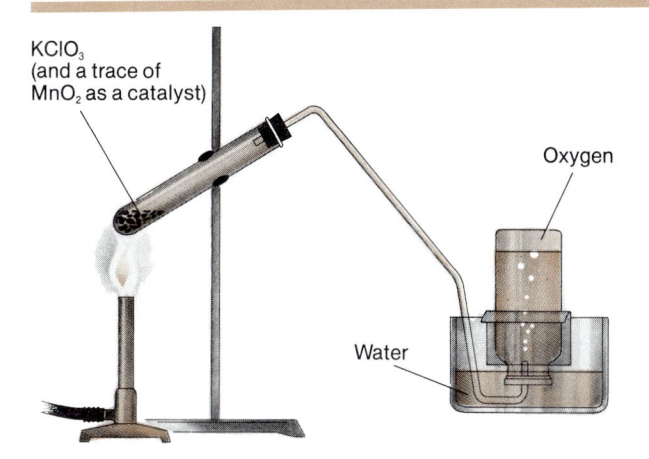

Figure 6.9 Laboratory preparation of oxygen. The oxygen gas is collected over water and is saturated with water vapor.

One of the most important applications of Dalton's Law involves the collection of gases over water (Figure 6.9). Here, the gas being collected is mixed with water vapor. If the collected gas is oxygen, Dalton's Law takes the form

$$P_{tot} = P_{O_2} + P_{H_2O} \tag{6.17}$$

The total pressure in Equation 6.17 is the measured pressure, as read on a barometer. The quantity P_{H_2O} is the vapor pressure of water, referred to in Chapter 2. It has a fixed value at a given temperature, independent of the volume of the container. The value of P_{H_2O} can be obtained from Table 6.3. Knowing P_{tot} and P_{H_2O}, we can readily calculate P_{O_2} (Example 6.9).

TABLE 6.3 Vapor Pressure of Water

T (°C)	VP (mm Hg)	T (°C)	VP (mm Hg)
0	4.6	28	28.3
5	6.5	29	30.0
10	9.2	30	31.8
15	12.8	31	33.7
16	13.6	32	35.7
17	14.5	33	37.7
18	15.5	34	39.9
19	16.5	35	42.2
20	17.5	40	55.3
21	18.6	50	92.5
22	19.8	60	149.4
23	21.1	70	233.7
24	22.4	80	355.1
25	23.8	90	525.8
26	25.2	100	760.0
27	26.7		

EXAMPLE 6.9

A sample of oxygen is collected over water as indicated in Figure 6.9. The total (barometric) pressure is 748 mm Hg. The volume of gas collected is 316 cm^3; the temperature is 23°C. Calculate

a. the partial pressure of O_2, using Dalton's Law
b. the number of moles of O_2, using the Ideal Gas Law.

Solution

a. Solving Equation 6.17 for P_{O_2}, we have

$$P_{O_2} = P_{tot} - P_{H_2O}$$

P_{tot} is given as 748 mm Hg. From Table 6.3, we see that the vapor pressure of water at 23°C is 21.1 mm Hg. Hence;

$$P_{O_2} = 748 \text{ mm Hg} - 21.1 \text{ mm Hg} = 727 \text{ mm Hg}$$

b. We first solve the Ideal Gas Law for n

$$n = \frac{PV}{RT}$$

To find the number of moles of oxygen, we use the partial pressure of oxygen

$$P_{O_2} = 727 \text{ mm Hg} \times \frac{1 \text{ atm}}{760 \text{ mm Hg}} = 0.957 \text{ atm}$$

$$V = 316 \text{ cm}^3 \times \frac{1 \text{ L}}{1000 \text{ cm}^3} = 0.316 \text{ L}$$

$$R = 0.0821 \text{ L} \cdot \text{atm/(mol} \cdot \text{K)}$$

$$T = 273 + 23 = 296 \text{ K}$$

Hence:

$$n = \frac{(0.957 \text{ atm})(0.316 \text{ L})}{\left(0.0821 \frac{\text{L} \cdot \text{atm}}{\text{mol} \cdot \text{K}}\right)(296 \text{ K})} = 0.0124 \text{ mol}$$

Note that in using the Ideal Gas Law with $R = 0.0821$ L · atm/(mol · K), *volume must be expressed in liters, pressure in atmospheres, and temperature in Kelvin degrees.*

Exercise

Suppose the temperature were 15°C instead of 23°C. If P_{tot} and V remained the same, what would be the partial pressure and number of moles of O_2? Answer: 735 mm Hg; 0.0129 mol.

6.6 Volumes of Gases in Reactions

In Chapter 5, we dealt with a variety of stoichiometric calculations. Starting with a balanced equation, we carried out mole–to–mole, mole–to–mass, mass–to–mole, and mass–to–mass conversions. Using the Ideal Gas Law, we can extend stoichiometric calculations to include the volumes of gases involved in reactions. We will consider three calculations of this type.

1. Given the amount in grams (or moles) of one substance, X, determine the volume of another, gaseous substance, Y, involved in a reaction at a known temperature and pressure. This kind of calculation is referred to as a *mass to volume* conversion.
2. Given the volume of a gas, X, at a known temperature and pressure, calculate the amount in grams (or moles) of another substance Y involved in a reaction. We call this a *volume to mass* conversion.
3. Given the volume of one gas, X, calculate the volume of another gas, Y, involved in a reaction, taking the temperature and pressure to be the same for the two gases. This is a *volume to volume* conversion.

Mass to Volume Conversions

Some reactions in the general chemistry laboratory form gaseous products. Here you will often need to know the volume of a gas, Y, produced from a known mass of reactant, X. The calculation follows a three-step path

mass X $\xrightarrow{(1)}$ moles X $\xrightarrow{(2)}$ moles Y $\xrightarrow{(3)}$ volume Y

The first step is readily carried out, knowing the molar mass of X. The second step uses the coefficients of the balanced equation, as discussed in Chapter 5. In the third step, we use the Ideal Gas Law in the form

$$V = \frac{nRT}{P}$$

to convert moles to volume. The calculations are illustrated in Example 6.10.

EXAMPLE 6.10

Oxygen is produced by heating potassium chlorate

$2\ KClO_3(s) \rightarrow 2\ KCl(s) + 3\ O_2(g)$

What volume of O_2 at 25°C and 1.00 atm is produced by the decomposition of 1.00 g of $KClO_3$?

Continued

Solution
The indicated path is

1.00 g KClO$_3$ → ? moles KClO$_3$ → ? moles O$_2$ → ? volume O$_2$
              (1)   (2)   (3)

The first two conversions are readily carried out by methods described in Chapter 5.
(1) 1 mole KClO$_3$ = 122.55 g KClO$_3$
(2) 2 mol KClO$_3$ ≃ 3 mol O$_2$

$$\text{moles O}_2 = 1.00 \text{ g KClO}_3 \times \frac{1 \text{ mol KClO}_3}{122.55 \text{ g KClO}_3} \times \frac{3 \text{ mol O}_2}{2 \text{ mol KClO}_3}$$

$$= 0.0122 \text{ mol O}_2$$

To find the volume of oxygen, we use the Ideal Gas Law

$$V = \frac{nRT}{P} = \frac{(0.0122 \text{ mol})\left(0.0821 \frac{\text{L} \cdot \text{atm}}{\text{mol} \cdot \text{K}}\right)(298 \text{ K})}{1.00 \text{ atm}} = 0.298 \text{ L}$$

Exercise
What volume of O$_2$ at 20°C and 750 mm Hg would be produced from 1.00 g of KClO$_3$? Answer: 0.297 L.

Volume to Mass Conversions

This type of calculation is, in essence, the reverse of that just gone through. It involves three steps

volume X → moles X → moles Y → mass Y
          (1)   (2)   (3)

The first step uses the Ideal Gal Law in the form

$$n = \frac{PV}{RT}$$

to convert volume of gas to moles. Then we use the coefficients of the balanced equation to make the mole to mole conversion. Finally, knowing the molar mass of Y, we can calculate the mass in grams of Y.

EXAMPLE 6.11
Hydrogen gas can be produced by the reaction of zinc with acid (H$^+$ ions)

Zn(s) + 2 H$^+$(aq) → Zn^{2+}(aq) + H$_2$(g)

What mass of Zn is required to produce 1.00 L of H$_2$ at 20°C and 750 mm Hg?

Continued

Solution
The path is

1.00 L H_2 → ? moles H_2 → ? moles Zn → ? grams Zn
  (1)   (2)   (3)

The first conversion is made using the Ideal Gas Law. Note that P = 750/760 atm, V = 1.00 L, T = 293 K. Hence;

$$n = \frac{PV}{RT} = \frac{(750/760 \text{ atm})(1.00 \text{ L})}{\left(0.0821 \frac{\text{L} \cdot \text{atm}}{\text{mol} \cdot \text{K}}\right)(293 \text{ K})} = 0.0410 \text{ mol}$$

The second and third conversions are of the type discussed in Chapter 5. Noting that 1 mol of H_2 is formed from 1 mol of Zn and that the molar mass of Zn is 65.38 g/mol, we have

$$\text{mass Zn} = 0.0410 \text{ mol } H_2 \times \frac{1 \text{ mol Zn}}{1 \text{ mol } H_2} \times \frac{65.38 \text{ g Zn}}{1 \text{ mol Zn}} = 2.68 \text{ g Zn}$$

Exercise
How many moles of Zn are required to form 2.00 L of H_2 at STP? Answer: 0.0893 mol.

Volume to Volume Conversions

When two different gases are involved in a reaction, there is a simple relationship between their volumes, measured at the same temperature and pressure. To find this relationship, consider two gases, 1 and 2, both at T and P. Applying the Ideal Gas Law to both gases, we have

$$V_2 = \frac{n_2 RT}{P}$$

$$V_1 = \frac{n_1 RT}{P}$$

Dividing the first equation by the second, the factor RT/P cancels and we obtain

$$V_2/V_1 = n_2/n_1 \tag{6.18}$$

This tells us that *the volume ratio in which two gases react at T and P is the same as the reacting mole ratio.*

To see what Equation 6.18 implies, consider the reaction

$$N_2(g) + 3 H_2(g) \rightarrow 2 NH_3(g)$$

We found in Chapter 5 that the coefficients of the balanced equation can be interpreted in terms of moles. That is, 1 mol of N_2 reacts with 3 mol of H_2 to form 2 mol of NH_3.

1 mol $N_2 \simeq$ 3 mol $H_2 \simeq$ 2 mol NH_3

But since the reacting volume ratio, at a given temperature and pressure, is the same as the mole ratio, we conclude that 1 L of N_2 reacts with 3 L of H_2 to form 2 L of NH_3 (Figure 6.10).

1 L $N_2 \simeq$ 3 L $H_2 \simeq$ 2 L NH_3 (same T, P)

We see that the reacting volume ratio is given by the coefficients of the balanced equation. More generally, we can say that

The volumes of different gases involved in a reaction, if measured at the same temperature and pressure, are in the same ratio as the coefficients in the balanced equation.

This relation, known as the Law of Combining Volumes, was first proposed in a somewhat different form by Gay-Lussac in 1808. Its use is illustrated in Example 6.12.

EXAMPLE 6.12

Consider the combustion of ethane gas

2 $C_2H_6(g)$ + 7 $O_2(g) \rightarrow$ 4 $CO_2(g)$ + 6 $H_2O(l)$

What volume of O_2, at 25°C and 720 mm Hg, is required to react with 6.18 L of C_2H_6, measured at the same temperature and pressure?

Solution

According to the Law of Combining Volumes

2 L $C_2H_6 \simeq$ 7 L O_2

This gives us the conversion factor we need to go from volume of C_2H_6 to volume of O_2.

Continued

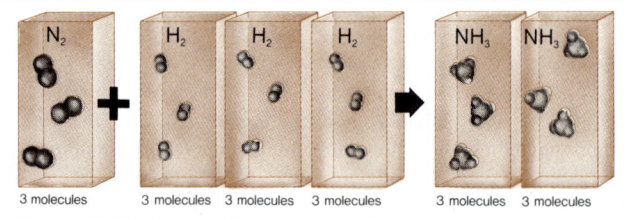

Figure 6.10 Law of Combining Volumes applied to the reaction $N_2(g)$ + 3 $H_2(g) \rightarrow$ 2 $NH_3(g)$. The reacting volumes are in the same ratio as the coefficients in the balanced equation.

$$\text{Volume } O_2 = 6.18 \text{ L } C_2H_6 \times \frac{7 \text{ L } O_2}{2 \text{ L } C_2H_6} = 21.6 \text{ L } O_2$$

Exercise
What volume of CO_2, also at 25°C and 720 mm Hg, is produced in this reaction?
Answer: 12.4 L.

You should realize that the Law of Combining Volumes applies only to *gases* at the same temperature and pressure. In Example 6.12, you could not have used the law directly to obtain

—the volume of O_2 required at conditions other than 25°C and 720 mm Hg, the temperature and pressure at which the volume of C_2H_6 is measured.
—the volume of liquid water produced in the reaction.

Key Words

atmosphere
Avogadro's Law
barometer
Boyle's Law
Charles' Law
Dalton's Law
Ideal Gas Law

kilopascal
kinetic molecular theory
Law of Combining Volumes
millimeter of mercury
molar volume
partial pressure
STP

Questions

6.1 List at least four ways in which the properties of gases differ from those of liquids and solids.

6.2 State four postulates of the kinetic theory of gases.

6.3 Which of the following pressure units is the largest? the smallest?

millimeter of Hg; atmosphere; kilopascal

6.4 State in your own words, Boyle's Law; Charles' Law; Avogadro's Law; Dalton's Law.

6.5 Explain what, if anything, is wrong with each of the following equations.
a. $P_1V_2 = P_2V_1$ b. $P_1V_1/T_2 = P_2V_2/T_1$
c. $P = nRT/V$

6.6 Solve each of the following equations for T_2.
a. $V_1/T_1 = V_2/T_2$ b. $P_1V_1/T_1 = P_2V_2/T_2$
c. $P_2V_2 = nRT_2$

6.7 The molar volume of a gas at 0°C and 1 atm (STP) is _____ L; at 25°C and 1 atm the molar volume is _____ L.

6.8 State the value of the gas constant R with volume in liters, pressure in atmospheres, and temperature in kelvins.

6.9 Explain in words how you would find the partial pressure of a gas collected over water at 25°C and a known total pressure.

6.10 State the Law of Combining Volumes for gases involved in reactions.

6.11 Explain, in terms of the properties of gases, why
   a. a light bulb explodes when it is incinerated
   b. a balloon filled with hot air rises
   c. water vapor at 100°C and 1 atm has a molar volume more than 1000 times greater than that of liquid water.

Problems

Pressure Units

6.12 Using the relations: 1 atm = 760 mm Hg = 101.3 kPa, make the following conversions
   a. 1.02 atm = ? mm Hg
   b. 712 mm Hg = ? atm
   c. 96.2 kPa = ? atm

6.13 The barometric pressure on a certain day is reported as 30.02 inches of mercury. Convert this to
   a. millimeters of Hg (1 in = 25.40 mm)
   b. atmospheres

6.14 On a Canadian radio station, the announcer gives the barometric pressure as 100.6 kPa. Convert this to
   a. millimeters of mercury
   b. atmospheres

Boyle's Law
(Assume constant temperature and amount of gas.)

6.15 A gas occupies 2.30 L at 740 mm Hg. What volume will it have at a pressure of 760 mm Hg?

6.16 A gas with a volume of 1.00 L exerts a pressure of 2.10 atm. What pressure will it exert when the volume is decreased to 1.00 cm^3?

6.17 What volume of argon must be withdrawn from a cylinder at a pressure of 96.5 atm to fill a light bulb with a volume of 226 cm^3 to a pressure of 610 mm Hg?

6.18 A gas occupies a volume of 6.18 L at a pressure of 812 mm Hg. At what pressure, in atmospheres, will it have a volume of 5.82 L?

Charles' Law
(Assume constant pressure and amount of gas.)

6.19 A gas occupies a volume of 502 cm^3 at 20°C. What will be its volume at 106°C?

6.20 A gas is at a temperature of 27°C. To what temperature in °C must it be cooled to reduce its volume to one half the original value?

6.21 The volume of a gas is 206 cm^3 at -20°C. To what temperature must it be heated for its volume to become 1.00 L?

6.22 If a gas is heated from 0 to 25°C, by what percent does its volume increase?

Volume–Pressure–Temperature Relations
(Assume constant amount of gas.)

6.23 An aerosol can at 20°C contains a gas at 2.16 atm. What pressure is exerted by the gas at 100°C, assuming the volume remains constant?

6.24 A flask full of air at 25°C and 752 mm Hg is stoppered and put in a refrigerator at 5°C. What pressure is exerted by the cold gas (constant volume)?

6.25 A gas occupies a volume of 1.41 L at 25°C and 718 mm Hg. What is its volume at STP?

6.26 A gas has a volume of 1.00 cm^3 at STP. What will be its volume at 350°C and 0.592 atm?

6.27 The air pressure inside an automobile tire is 3.16 atm when the temperature is 20°C. After the car has been driven for some time, the temperature of the air rises to 40°C. The volume of the tire at the higher temperature is 1.01 times its original value. What is the pressure inside the tire at 40°C?

6.28 A gas sample has a volume of 1.69 L at 25°C and 754 mm Hg. At what temperature will it occupy a volume of 1.25 L at 724 mm Hg?

Molar Volumes and Densities

(Use the Periodic Table for atomic masses where necessary.)

6.29 What are the densities of the following gases at STP?
   a. N_2 b. He   c. SF_6

6.30 What is the molar mass of a gas with a density of 2.69 g/L at STP?

6.31 Calculate the mass of 2.69 L of O_2 at STP.

6.32 What volume is occupied by 1.00 g of the following gases at STP?
   a. N_2 b. He   c. SF_6

Ideal Gas Law

(Use the Periodic Table for atomic masses where necessary.)

6.33 What volume is occupied by 1.38 mol of a gas at 25°C and 652 mm Hg?

6.34 How many moles of gas are there in a 0.750-L container at 19°C and 0.420 atm?

6.35 What pressure is exerted by 0.100 mol of a gas confined in a 2.00 L flask at a temperature of 19°C?

6.36 At what temperature is the molar volume of a gas 25.0 L at a pressure of 2.00 atm?

6.37 How many grams of ammonia, NH_3, are there in a 125-cm³ flask at a pressure of 1406 mm Hg and a temperature of −13°C?

6.38 What volume is occupied by 1.00 g of O_2 at 29°C and 712 mm Hg?

6.39 A student finds that 1.23 g of He occupies a volume of 6.75 L at 21°C and 1.10 atm. Use these data to calculate the gas constant R in L · atm/mol · K.

6.40 How many kilograms of N_2 are in a 50.0 L cylinder at a pressure of 50.0 atm and a temperature of 50.0°C?

Dalton's Law

(Use Table 6.3 for water vapor pressures.)

6.41 A sample of a gas is collected over water at 20°C. The total pressure is measured to be 759 mm Hg. What is the partial pressure of the dry gas?

6.42 A sample of O_2 collected over water at 21°C has a volume of 219 cm³. The total (barometric) pressure is 739 mm Hg. Determine
   a. the partial pressure of O_2
   b. the number of moles of O_2

6.43 A sample of 416 cm³ of wet H_2 collected over water at 27°C exerts a pressure of 752 mm Hg.
   a. What is the partial pressure of the H_2?
   b. What is the mass of H_2 in the wet gas?

6.44 A gas is collected over water at 18°C. The measured volume is 129 cm³ at a total pressure of 746 mm Hg.
   a. What is the partial pressure of the gas?
   b. What is the volume of the dry gas at STP?

Volumes of Gases in Reactions

(Use the Periodic Table for atomic masses.)

6.45 In the reaction: $2\ KClO_3(s) \rightarrow 2\ KCl(s) + 3\ O_2(g)$ determine the volume of oxygen produced at 20°C and 750 mm Hg from
   a. 0.100 mol $KClO_3$ b. 2.00 g $KClO_3$

6.46 The combustion of acetylene can be represented by the equation

$$2\ C_2H_2(g) + 5\ O_2(g) \rightarrow 4\ CO_2(g) + 2\ H_2O(g)$$

If 1.00 g of acetylene is burned to form products at 1.00 atm and 1150°C, what volume of CO_2 is produced?

6.47 For the reaction in Problem 6.46, what volume of O_2 at 25°C and 1.00 atm is required to react with 1.00 g of C_2H_2?

6.48 How many grams of aluminum are required to produce 1.00 L of H_2 at 20°C and 712 mm Hg by the reaction

$$2\ Al(s) + 6\ H^+(aq) \rightarrow 2\ Al^{3+}(aq) + 3\ H_2(g)$$

6.49 Numbers and letters can be etched on glassware by exposing it to hydrogen fluoride. The reaction involved is

$$SiO_2(s) + 4 HF(g) \rightarrow SiF_4(g) + 2 H_2O(l)$$

What mass of SiO_2 reacts with 1.00 L of HF at 1.00 atm and 70°C?

6.50 Consider the reaction $N_2(g) + 3 H_2(g) \rightarrow 2 NH_3(g)$. What volume of N_2 is required to react with 1.60 L of H_2 at the same temperature and pressure?

Multiple Choice

6.51 In the reaction; $4 NH_3(g) + 5 O_2(g) \rightarrow 4 NO(g) + 6 H_2O(l)$, what is the total volume of reactants required to form 1.20 L of NO at the same temperature and pressure?
  a. 1.08 L b. 1.20 L c. 1.50 L
  d. 2.70 L e. 10.8 L

6.52 Which one of the following gases has the highest density at STP?
  a. H_2 b. CH_4 c. C_2H_6 d. C_2H_4
  e. C_2H_2

6.53 An increase in temperature increases all but one of the following quantities. Choose that quantity.
  a. average molecular speed
  b. fraction of very slow molecules
  c. gas pressure (constant volume)
  d. gas volume (constant pressure)

6.54 A sample of air, originally at a pressure of 754 mm Hg, is saturated with ether, a volatile liquid. The temperature, volume, and amount of air remain constant. The final pressure of the air will be
  a. 754 mm Hg
  b. less than 754 mm Hg
  c. greater than 754 mm Hg

6.55 Which one of the following pairs are inversely proportional?
  a. V vs T (constant P, n)
  b. V vs n (constant P, T)
  c. P vs T (constant n, V)
  d. n vs T (constant P, V)

6.56 The molar volume of a gas at 100°C and 1.00 atm is 30.6 L. The density of O_2 at 100°C and 1.00 atm (at. mass O = 16.0) is, in grams per liter
  a. 0.523 b. 0.956 c. 1.05 d. 1.43

6.57 In reaction; $2 H_2(g) + O_2(g) \rightarrow 2 H_2O(l)$, the volume of H_2 at 25°C and 2.00 atm required to react with 1.60 L of O_2 at 25°C and 4.00 atm is
  a. 0.80 L b. 1.60 L c. 3.20 L
  d. 6.40 L

6.58 The number of moles of a gas in 22.4 L at 27°C and 1.00 atm is
  a. 1.00 b. less than 1.00 c. more than 1.00

6.59 At room temperature, the average velocity of an O_2 molecule, in miles per hour, is approximately
  a. 10^{-3} b. 10^{-1} c. 10^1 d. 10^3
  e. 10^5

6.60 Gases deviate most widely from the Ideal Gas Law at
  a. high pressures and high temperatures
  b. high pressures and low temperatures
  c. low pressures and low temperatures
  d. low pressures and high temperatures

Electronic Structure and the Periodic Table

Learning Objectives
After studying this chapter, you should be able to:

1. Describe the main features of the Bohr model of the hydrogen atom and explain how it can be checked experimentally.
2. Explain how the quantum mechanical model differs from the Bohr model.
3. State the capacity for electrons of each principal energy level.
4. State the capacity for electrons of each type of sublevel (s, p, d, f); state what sublevels are found in a given principal level (Example 7.1).
5. Write the electron configuration of any atom from H (at. no. = 1) to Kr (at. no. = 36) (Example 7.2).
6. State the number of orbitals in each sublevel (Example 7.3).
7. Write an orbital diagram for any atom from H (at. no. = 1) to Kr (at. no. = 36) (Example 7.4).
8. Using the Periodic Table, give the outer electron configuration of any main-group element (Example 7.5).
9. State how metallic character, ionization energy, and electronegativity vary as one moves down or across in the Periodic Table (Examples 7.6, 7.7).

> Two sorts of truth: trivialities, where opposites are obviously absurd, and profound truths, recognized by the fact that the opposite is also a profound truth.
>
> **NIELS BOHR**

An early version of Mendeleev's Periodic Table. From Brescia, F., et al.: *Chemistry: A Modern Introduction*. Philadelphia, W. B. Saunders Co., 1978.

	Ti = 50	Zr = 90	? = 180
	V = 51	Nb = 94	Ta = 182
	Cr = 52	Mo = 96	W = 186
	Mn = 55	Rh = 104.4	Pt = 197.4
	Fe = 56	Ru = 104.4	Ir = 198
	Ni = Co = 59	Pl = 106.6	Os = 199
	Cu = 63.4	Ag = 108	Hg = 200
Mg = 24	Zn = 65.2	Cd = 112	
Al = 27.4	? = 68	Ur = 116	Au = 197?
Si = 28	? = 70	Sn = 118	
P = 31	As = 75	Sb = 122	Bi = 210
S = 32	Se = 79.4	Te = 128?	
Cl = 35.5	Br = 80	I = 127	
K = 39	Rb = 85.4	Cs = 133	Tl = 204
Ca = 40	Sr = 87.6	Ba = 137	Pb = 207
? = 45	Ce = 92		
?Er = 56	La = 94		
?Yt = 60	Di = 95		
?In = 75.6	Th = 118?		

CHAPTER 7

In Chapter 3, we pointed out that the number of electrons in a neutral atom is equal to the atomic number of the element. These electrons, we noted, are located outside the nucleus. In this chapter, we will look more closely at the electronic structure of atoms. We will be interested in finding answers to two questions. How are the electrons arranged around the nucleus; what are their positions relative to one another? Perhaps more important, from the standpoint of experiment, what are the relative energies of these electrons? To answer these questions, we start by examining a model of the simplest atom, that of hydrogen, in which there is only one electron (Section 7.1). Then we will develop a somewhat more complex model that can be applied to multi-electron atoms (Section 7.2). Using this model, we will discuss in some detail the electronic structures of atoms up to atomic number 36 (Sections 7.3, 7.4).

The Periodic Table was introduced in Chapter 2 and has been referred to occasionally since. In the last sections of this chapter, we will look more closely at this arrangement of elements, seeking answers to two questions. What is the theoretical basis of the Periodic Table, i.e., why do elements in the same vertical group have similar chemical properties (Section 7.5)? How do atomic properties vary as we move horizontally or vertically in the Periodic Table (Section 7.6)?

7.1 The Bohr Model of the Hydrogen Atom

In 1913, the Danish physicist Niels Bohr proposed a simple model to describe the behavior of the single electron in the hydrogen atom. He assumed that normally the electron moves about the nucleus in a circular orbit of fixed radius. In that sense, the electron would resemble the earth revolving about the sun. According to Bohr, as long as the electron remains in this orbit its energy is constant. We would say that under these conditions, the electron is in its *ground state* and has its lowest energy.

It is possible for the hydrogen electron to absorb energy from an outside

Niels Bohr

source. Bohr believed that when this happens the electron jumps to a different orbit of higher energy, farther from the nucleus. In this way the electron reaches what we call an *excited state*. There are several excited states of successively higher energies. Ordinarily, an electron does not stay in any one of these states very long. Instead, it gives off energy and drops back to a lower energy state, eventually to the ground state.

Bohr's greatest contribution was to apply the quantum theory, first proposed by Max Planck in a quite different context, to the hydrogen atom. Bohr stated, and we continue to believe today that **The electron in the hydrogen atom can only exist in certain states (energy levels) associated with definite amounts of energy. When an electron changes its state, it must absorb or emit the exact amount of energy required to bring it from the initial to the final state.**

We commonly describe this situation by saying that the energy of the electron is *quantized*. It can change only by certain definite amounts, "quantum jumps." There is an analogy to a person climbing a ladder (Figure 7.1). He is restricted to certain distances from the ground, defined by the rungs of the ladder. When he moves to a higher rung, his gravitational energy in-

Figure 7.1 A man on a ladder changes his energy by fixed amounts as he moves from one rung to another. An atom has energy levels similar to those fixed by the ladder rungs. In an atom the energy levels (rungs) are not equally spaced.

creases by a certain amount. Likewise, his energy changes by finite amounts when he descends from one rung to another, finally reaching the ground.

Bohr was able to calculate the energies that are available to a hydrogen electron. He derived the following equation

$$E = \frac{-1312}{n^2} \text{ kJ/mol} \tag{7.1}$$

Here, E is the energy of one mole of electrons in kilojoules and n, called a *quantum number*, can be any whole number

n = 1, 2, 3, 4, --

When n = 1, Equation 7.1 gives the molar energy of the hydrogen electron in the ground state

E (n = 1) = −1312 kJ/mol

Taking n = 2, we obtain the molar energy in the first excited state

$$E (n = 2) = \frac{-1312}{4} \text{ kJ/mol} = -328 \text{ kJ/mol}$$

Several energy levels calculated from Equation 7.1 are shown in Figure 7.2. Notice that as n increases, E becomes a smaller and smaller negative number. In other words, as n becomes very large, approaching infinity, E approaches zero. Physically, this corresponds to the situation where the electron is completely removed from the atom. Under these conditions, there is no longer any attractive force between the electron and the proton and E = 0.

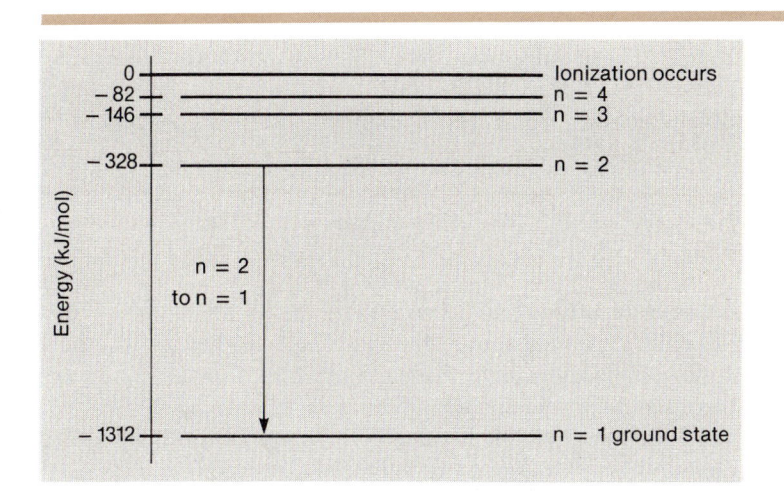

Figure 7.2 Energy levels in the hydrogen atom, calculated from Equation 7.1. In the normal hydrogen atom, the electron is in the ground state (n = 1). By absorbing energy, the electron can move to an excited state (n = 2, 3, – –). As n increases, E approaches zero; when E = 0, ionization has occurred.

There are several ways to check the Bohr model against experiment. One is to measure the *ionization energy* of the hydrogen atom.

$$H(g) \rightarrow H^+(g) + e^-; \Delta E = \text{ionization energy}$$

This is the energy that must be absorbed to remove an electron, originally in the ground state, from the hydrogen atom. By experiment, the ionization energy of the hydrogen atom is found to be 1317 kJ/mol. The value calculated from the Bohr model is 1312 kJ/mol. The agreement here is quite good; the two values are within less than 0.5% of each other.

Another way to test the Bohr model makes use of the *atomic spectrum* of hydrogen (Figure 7.3). When hydrogen atoms are "excited" by exposure to a spark or electrical discharge, they give off light at a series of discrete wave-

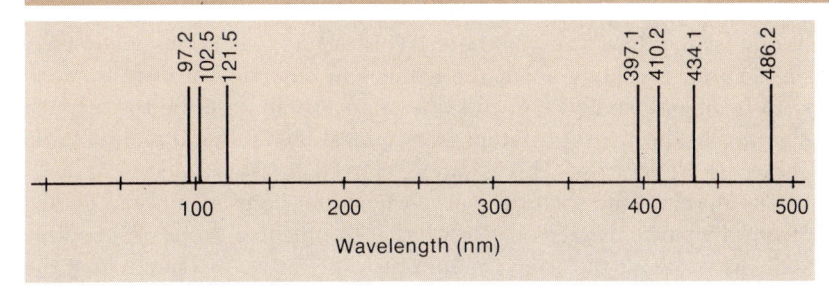

Figure 7.3 Some lines in the atomic spectrum of hydrogen. The lines appear in groups. The group at the left (Lyman series) is in the ultraviolet. The group at the right (Balmer series) is in the visible region. The lines are formed when the electron drops from a higher to a lower energy level. For example, the line at 121.5 nm is produced when the electron moves from n = 2 to n = 1 (Figure 7.2).

lengths. The energies of the spectral lines shown in Figure 7.3 can be measured very accurately. These values can then be compared with those calculated from Equation 7.1. When this is done, the agreement is excellent, within 0.1% or less.

7.2 The Quantum Mechanical Atom

The Bohr model was brilliantly successful in explaining the atomic properties of hydrogen. However, it proved impossible to extend the model to other, more complex atoms, containing more than one electron. There, the agreement between theory and experiment was poor, to say the least.

Scientists in the 1920's gradually became convinced that the Bohr model had to be abandoned. They developed instead a more complex model based on a whole new branch of the physical sciences called *quantum mechanics*. The quantum mechanical model of the atom is a highly mathematical one. It attempts to describe the energy of electrons in terms of rather complex mathematical equations. We will not attempt to state these equations, let alone solve them. Instead, we will concentrate upon certain qualitative aspects of the quantum mechanical atom.

Quantum mechanics tells us that small particles confined to very small regions of space behave quite differently from large, visible objects. The motion of electrons in atoms differs in a very basic way from that of a baseball in flight or an automobile moving down the highway. Perhaps surprisingly, *we cannot specify precisely where an electron is, relative to the nucleus, at a given instant. The best we can do is to quote the odds or the "probability" that the electron will be found in a particular region.* For example, we might say that the chances are five in ten that the electron will be within 0.05 nm of the nucleus, or nine in ten that it will be within 0.14 nm of the nucleus.

Since we can't specify exactly the position of an electron at any given time, we can't say what path it follows (If I don't know where you are, I certainly don't know how you got there.) This means that Bohr's idea of the hydrogen electron revolving about the nucleus in a circular orbit of fixed radius had to be abandoned. We need some other way to describe the location of the electron in the hydrogen atom. A graphical model that attempts to do this is shown in Figure 7.4. This is often referred to as an "electron cloud" diagram. The depth of the shading is proportional to the probability of finding the electron in a particular region. Notice that the shading drops off smoothly and steadily as we move out from the nucleus. The farther we move from the nucleus in any given direction, the less likely we are to find the electron.

Quantum mechanics, unlike the Bohr model, can be applied to multielectron atoms. The main concern here is with the relative energies of the different electrons in these atoms. The quantization of electron energies,

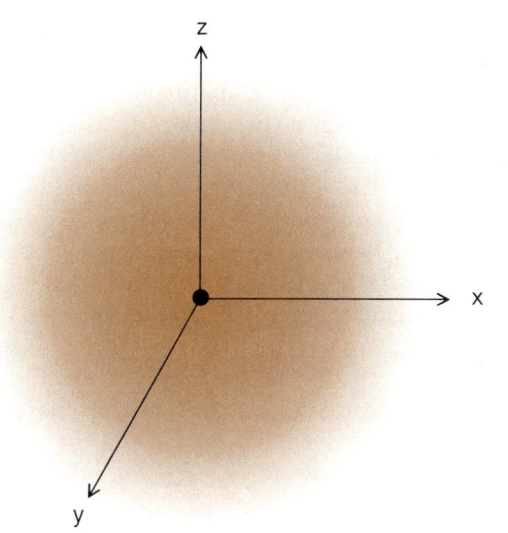

Figure 7.4 Electron cloud surrounding an atomic nucleus. The depth of color is proportional to the probability of finding an electron in a particular region. Notice that the probability decreases rapidly and smoothly as one moves out from the nucleus.

which was a central concept of the Bohr model, is retained. However, there is a complication. To completely specify the energy of an electron in an atom, we must cite four quantum numbers rather than one, as in Equation 7.1. We will not discuss these quantum numbers as such. Instead, we will concentrate upon the quantities they describe. These are

—the *principal energy level* in which the electron is located (Section 7.3).
—the *sublevel* in which the electron is located (Section 7.3).
—the *orbital* occupied by the electron (Section 7.4).
—the *spin* of the electron (Section 7.4).

7.3 Principal Energy Levels and Sublevels

The quantum mechanical model of the multi-electron atom (two or more electrons) considers that these electrons are distributed among different principal energy levels. These levels are quite similar to those proposed by Bohr for the hydrogen atom. However, each principal level (except the first) is divided into sublevels, which differ at least slightly from one another in energy.

In discussing electron energy levels, we start by considering what principal levels are available and how they are designated. Then we look at the sublevel structure of the various principal levels. With this background, we will describe how electrons in atoms of atomic number 1–36 are distributed among principal levels and sublevels. Here and throughout the rest of this chapter, we will be talking about *gaseous atoms in their ground states*.

Principal Energy Levels

Recall that the Bohr model of the hydrogen atom established a series of energy levels designated by the quantum number n (Equation 7.1). Under normal conditions, the hydrogen electron is in the ground state (n = 1). Higher levels (n = 2, 3, - -) are also available. The quantum mechanical model retains these levels, now described as principal energy levels. As with the Bohr model, energy increases with increasing value of n. An electron in the level n = 2 has a higher energy than an electron in the first principal level (n = 1). An electron in the level n = 3 is higher in energy than an electron in the level n = 2, and so on.

Each principal level has a limited capacity for electrons. In general, as n increases, the number of electrons that can fit into a given principal level is $2n^2$, where n is the quantum number of the level. This means that the capacity for electrons of various principal energy levels is

quantum number of level (n)	1	2	3	4
capacity of level ($2n^2$)	$2e^-$	$8e^-$	$18e^-$	$32e^-$

Notice that the first principal level can hold only two electrons. Atoms with more than two electrons, starting with lithium (at. no. = 3) must put some of these electrons into higher levels.

In a general way, the distance of electrons from the nucleus can be related to the quantum number n of the principal level. As n increases, an electron is more likely to be found farther out from the nucleus. On the average, an electron in the second level is farther from the nucleus than one in the first level. An electron in the n = 3 level is still more distant from the nucleus, and so on. Qualitatively, this explains why the energy of the electron increases as n increases. Energy has to be absorbed for an electron (negative charge) to move farther out from the nucleus (positive charge).

Sublevels

All the electrons in a given principal energy level need not have the same energy. Within a principal level, there are sublevels which differ in energy. We might think of these as analogous to different floors in an apartment building. Energy must be absorbed to move an electron from a lower to a higher sublevel, just as it takes energy to climb stairs leading from one apartment to another on a higher floor.

For historical reasons, sublevels are designated by the letters s, p, d, and f. Two important principles apply to sublevels.

1. The energies of sublevels increase in the order
   s < p < d < f

   Within a given principal level, the s sublevel is always lowest in energy. The p sublevel is somewhat higher in energy, the d sublevel still higher, and the f sublevel highest of all. Following our apartment-house analogy, the s sublevel might correspond to the basement and the f sublevel to the top floor.

2. Each sublevel always has a fixed capacity for electrons. The maximum number of electrons that can fit into each sublevel is

sublevel	s	p	d	f
capacity for electrons	2	6	10	14

With these ideas in mind, we arrive at the sublevel structure for the principal levels n = 1 through n = 4 shown in Table 7.1. Notice that **the total number of different sublevels in a given principal level is equal to the quantum number, n, of that level.** Thus,

—for the first principal level (n = 1), there is only one sublevel, designated as 1s. That sublevel can hold two electrons, which is the total capacity of the first principal level.

—for the second principal level (n = 2), there are two sublevels, designated as 2s and 2p. Of the two sublevels, the 2s is lower in energy and can hold two electrons. The higher sublevel, 2p, can hold six electrons. The total capacity of the second principal level is 2 + 6 = 8.

—for the third principal level (n = 3), there are three sublevels. These increase in energy in the order 3s < 3p < 3d. Of the 18 electrons that can be accomodated in the third level, two can fit into the 3s level, six into the 3p, and ten into the 3d.

—for the fourth principal level (n = 4), there is a complete set of four sublevels, 4s, 4p, 4d, 4f. These increase in energy in the order 4s < 4p < 4d < 4f. In an atom where the fourth level is completely filled with 32 electrons, 2 will be in the 4s sublevel, 6 in the 4p, 10 in the 4d, and 14 in the 4f.

Higher principal levels beyond those listed in Table 7.1 also have s, p, d, and f sublevels. Thus we can have a 5s, 5p, 5d, or 5f sublevel. The general principles listed above apply to these sublevels as well (Example 7.1).

TABLE 7.1 Sublevel Structure of Principal Levels n = 1 through n = 4

Principal Level	Sublevels (Electron Capacities)	Total Capacity
1	1s (2e$^-$)	2
2	2s (2e$^-$) 2p (6e$^-$)	8
3	3s (2e$^-$) 3p (6e$^-$) 3d (10e$^-$)	18
4	4s (2e$^-$) 4p (6e$^-$) 4d (10e$^-$) 4f (14e$^-$)	32

> **EXAMPLE 7.1**
> a. How many electrons can go into a 5d sublevel? a 5f?
> b. Which sublevel is higher in energy, 5d or 5f?
>
> **Solution**
> a. 10e⁻ in 5d, 14e⁻ in 5f.
> b. 5f; an f sublevel is always higher in energy than a d sublevel in the same principal level.
>
> **Exercise**
> What is the total capacity for electrons of the 6s, 6p, and 6d sublevels? Which of these sublevels has the lowest energy? Answer: 18; 6s.

Electron Configurations

The electronic structure of an atom is commonly described by giving the number of electrons in each energy sublevel. To do this, we write the **electron configuration** of the atom. Here the number of electrons in each sublevel is indicated by a superscript to the right of the sublevel symbol. For example, an atom that has two electrons in the 1s sublevel would have the electron configuration

$1s^2$

In another case, we would interpret the electron configuration

$1s^2 2s^2 2p^1$

to mean that the atom involved has two 1s electrons, two 2s electrons, and one 2p electron.

We can obtain the electron configurations of atoms by a quite simple process. **As atomic number increases, electrons are added to sublevels of increasingly higher energy, filling each sublevel to capacity before moving on to the next one.** We will now follow this process to obtain the electron configurations of the elements in the first four periods of the Periodic Table.

First Period ($_1$H and $_2$He) As you might expect, the sublevel of lowest energy is the 1s. This sublevel is half filled with hydrogen (1e⁻) and completely filled with helium (2e⁻). The electron configurations of these atoms are

$_1$H $1s^1$ $_2$He $1s^2$

Second Period ($_3$Li → $_{10}$Ne) Since the 1s sublevel has a capacity of only two electrons, the third electron in lithium must go into the next highest sublevel. There are no more sublevels for n = 1, so we move to n = 2. Here, there are two sublevels, the 2s and the 2p. Of these, the 2s is lower in energy; the third electron in lithium enters the 2s sublevel

$_3$Li $1s^2 2s^1$

The 2s sublevel fills with beryllium (at. no. = 4). The next element, boron, puts an electron into the 2p sublevel

$_5$B $1s^2 2s^2 2p^1$

Electrons continue to enter the 2p sublevel until it is filled with neon (at. no. = 10).

$_{10}$Ne $1s^2 2s^2 2p^6$

Third Period ($_{11}$Na → $_{18}$Ar) In this period, we repeat the process just described for the second period. The only difference is that the 3s and 3p sublevels are filling instead of the 2s and 2p. Thus we have the following electron configurations

$_{11}$Na $1s^2 2s^2 2p^6 3s^1$

$_{12}$Mg $1s^2 2s^2 2p^6 3s^2$

$_{13}$Al $1s^2 2s^2 2p^6 3s^2 3p^1$

$_{18}$Ar $1s^2 2s^2 2p^6 3s^2 3p^6$

Fourth Period ($_{19}$K → $_{36}$Kr) Potassium (at. no. = 19) has to put its outermost electron into the next higher sublevel beyond the 3p. You will recall from Table 7.1 that there is one sublevel left in the third principal level, 3d. However, it turns out that, so far as the atoms in the fourth period are concerned, the 4s sublevel is lower in energy than the 3d. That is, the *lowest* sublevel for n = 4, 4s, fills before the *highest* sublevel for n = 3, 3d. The 19th electron of potassium enters the 4s sublevel; that sublevel fills with calcium (at. no. = 20).

$_{19}$K $1s^2 2s^2 2p^6 3s^2 3p^6 4s^1$

$_{20}$Ca $1s^2 2s^2 2p^6 3s^2 3p^6 4s^2$

Once the 4s sublevel is filled, the 3d starts to fill, with scandium (at. no. = 21). The 3d sublevel continues to fill through zinc (at. no. = 30).

$_{21}$Sc $1s^2 2s^2 2p^6 3s^2 3p^6 4s^2 3d^1$

$_{30}$Zn $1s^2 2s^2 2p^6 3s^2 3p^6 4s^2 3d^{10}$

The last six elements in the fourth period put electrons into the next highest sublevel, 4p

$_{31}$Ga $1s^2 2s^2 2p^6 3s^2 3p^6 4s^2 3d^{10} 4p^1$

$_{36}$Kr $1s^2 2s^2 2p^6 3s^2 3p^6 4s^2 3d^{10} 4p^6$

In Figure 7.5, we indicate the relative energies of the different sublevels we have considered. Note the "overlap" between 4s and 3d. Using this figure, if necessary, and following the process just described, you should be able to deduce the electron configuration of any atom of atomic number 1–36 (Example 7.2). For elements of higher atomic number, electron configurations are most readily obtained by interpreting the Periodic Table (Section 7.5).

EXAMPLE 7.2
What is the electron configuration of nickel (at. no. = 28)?

Solution
There are 28 electrons to account for. The first 18 fill the 1s through 3p sublevels, as in argon. The next two electrons fill the 4s sublevel. That leaves eight electrons (28 − 18 − 2 = 8) to go into the 3d sublevel. The electron configuration of nickel is

$_{28}$Ni $1s^2 2s^2 2p^6 3s^2 3p^6 4s^2 3d^8$

Continued

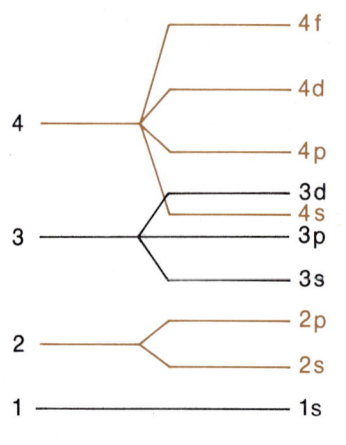

Figure 7.5 In general, energy increases with the principal quantum number, **n**. However, it is possible for the lowest sublevel of **n** = 4 (i.e., 4s) to be below the highest sublevel of **n** = 3 (i.e., 3d). This appears to be the situation in the potassium and calcium atoms, where successive electrons enter the 4s rather than the 3d sublevel.

Exercise

What is the electron configuration of As (at. no. = 33)? Answer: $1s^2 2s^2 2p^6 3s^2 3p^6 4s^2 3d^{10} 4p^3$.

In Table 7.2, we show the electron configurations of the first 36 elements in the Periodic Table. To save space, we have used the symbols [$_2$He], [$_{10}$Ne] and [$_{18}$Ar] as abbreviations

for [$_2$He] read $1s^2$

for [$_{10}$Ne] read $1s^2 2s^2 2p^6$

for [$_{18}$Ar] read $1s^2 2s^2 2p^6 3s^2 3p^6$

You will note that with two minor exceptions, Cr and Cu, the electron configurations listed in the table follow the order we have described.

7.4 Orbitals and Electron Spin

As we have seen, it is not enough to specify the principal level in which an electron in an atom is located. The energy of the electron also depends upon the sublevel it occupies. It turns out that to completely specify the energy of the electron, we must go two steps further. We must state the *orbital* occupied by the electron. All sublevels, except s sublevels, contain more than one orbital. Then we must indicate the *spin* of the electron within the orbital.

In this section we will first consider the orbital structure of different sublevels. Then we look at the possibilities so far as electron spin is con-

TABLE 7.2 Electron Configurations of the First 36 Elements

$_1$H $1s^1$	$_{13}$Al [$_{10}$Ne] $3s^2 3p^1$	$_{25}$Mn [$_{18}$Ar] $4s^2 3d^5$
$_2$He $1s^2$	$_{14}$Si [$_{10}$Ne] $3s^2 3p^2$	$_{26}$Fe [$_{18}$Ar] $4s^2 3d^6$
$_3$Li [$_2$He] $2s^1$	$_{15}$P [$_{10}$Ne] $3s^2 3p^3$	$_{27}$Co [$_{18}$Ar] $4s^2 3d^7$
$_4$Be [$_2$He] $2s^2$	$_{16}$S [$_{10}$Ne] $3s^2 3p^4$	$_{28}$Ni [$_{18}$Ar] $4s^2 3d^8$
$_5$B [$_2$He] $2s^2 2p^1$	$_{17}$Cl [$_{10}$Ne] $3s^2 3p^5$	*$_{29}$Cu [$_{18}$Ar] $4s^1 3d^{10}$
$_6$C [$_2$He] $2s^2 2p^2$	$_{18}$Ar [$_{10}$Ne] $3s^2 3p^6$	$_{30}$Zn [$_{18}$Ar] $4s^2 3d^{10}$
$_7$N [$_2$He] $2s^2 2p^3$	$_{19}$K [$_{18}$Ar] $4s^1$	$_{31}$Ga [$_{18}$Ar] $4s^2 3d^{10} 4p^1$
$_8$O [$_2$He] $2s^2 2p^4$	$_{20}$Ca [$_{18}$Ar] $4s^2$	$_{32}$Ge [$_{18}$Ar] $4s^2 3d^{10} 4p^2$
$_9$F [$_2$He] $2s^2 2p^5$	$_{21}$Sc [$_{18}$Ar] $4s^2 3d^1$	$_{33}$As [$_{18}$Ar] $4s^2 3d^{10} 4p^3$
$_{10}$Ne [$_2$He] $2s^2 2p^6$	$_{22}$Ti [$_{18}$Ar] $4s^2 3d^2$	$_{34}$Se [$_{18}$Ar] $4s^2 3d^{10} 4p^4$
$_{11}$Na [$_{10}$Ne] $3s^1$	$_{23}$V [$_{18}$Ar] $4s^2 3d^3$	$_{35}$Br [$_{18}$Ar] $4s^2 3d^{10} 4p^5$
$_{12}$Mg [$_{10}$Ne] $3s^2$	*$_{24}$Cr [$_{18}$Ar] $4s^1 3d^5$	$_{36}$Kr [$_{18}$Ar] $4s^2 3d^{10} 4p^6$

*The unusual configurations of Cr and Cu are believed to reflect the special stability of half-filled sublevels (s^1, d^5) or completely filled sublevels (d^{10}).

cerned. With that background, we will see how to use orbital diagrams as a means of describing the complete electronic structure of atoms.

Orbitals

In Figure 7.4, we showed the nature of the electron cloud associated with the single electron in the hydrogen atom. In principle, we could draw similar diagrams for electron clouds in other atoms. This is seldom done because, as the number of electrons increases, such diagrams quickly become complex and difficult to interpret. Instead, we draw simple three-dimensional figures that enclose the region in which there is a 90% chance of finding a particular electron. These figures are referred to as **orbitals.** The shape of an orbital differs depending upon the sublevel (s, p, d, or f) in which the electron is located.

Figure 7.6 shows the nature of the 1s and 2s orbitals. Note that they are *spherical*, as are all s orbitals. This means that an electron in an s orbital is as likely to be found "north" of the nucleus as it is "south" (or east or west, or – –). In other words the density of the electron cloud is symmetrical about the nucleus; there is no preferred direction in space. Notice also that the volume of the orbital increases with the principal quantum number (2s > 1s). An electron in a 2s orbital is, on the average, farther out from the nucleus than an electron in a 1s orbital.

The orbitals associated with p sublevels have a more complex shape than those of s sublevels. In Figure 7.7, we show the geometries of the three p orbitals that are available to electrons in each p sublevel. (The size of p orbitals, like that of s orbitals, increases with the principal quantum number, n). Notice that

—a p orbital has a "dumbbell" or "figure eight" shape. The dumbbell has its center at the nucleus. It consists of two identical segments called "lobes" directed at a 180° angle to one another.

—the three p orbitals are oriented at 90° angles to one another. We can con-

Figure 7.6 Shapes of 1s and 2s orbitals. Both are spherical, as are all s orbitals. The radius of the 2s orbital is larger, indicating that an electron in that orbital is more likely to be found farther out from the nucleus.

1s

2s

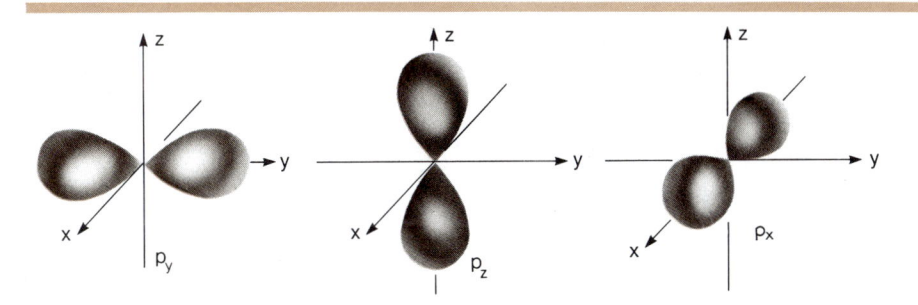

Figure 7.7 Three p orbitals available to an electron. Electrons in different p orbitals are relatively far apart. Their charge clouds are concentrated along perpendicular directions.

sider one of them (the one shown at the left of Figure 7.7) to be centered along the y axis. This is referred to as the p_y orbital. Another p orbital (p_z) is oriented along the z axis. The third orbital, p_x, is directed along the x axis.

This picture can be extended to describe the geometries of d and f orbitals. We will not attempt to do this; the three-dimensional shapes of these orbitals are rather complex. At this point, we turn our attention to the electron capacities of orbitals. Here the rule is a very simple one:

Each orbital has a capacity of two electrons.

Recall that an s sublevel has a capacity of $2e^-$, a p sublevel a capacity of $6e^-$, a d sublevel a capacity of $10e^-$, and an f sublevel a capacity of $14e^-$. This means that there must be

$\dfrac{2}{2}$ = 1 s orbital within each s sublevel (1s, 2s, – –)

$\dfrac{6}{2}$ = 3 p orbitals within each p sublevel (2p, 3p, – –)

$\dfrac{10}{2}$ = 5 d orbitals within each d sublevel (3d, 4d, – –)

$\dfrac{14}{2}$ = 7 f orbitals within each f sublevel (4f, 5f, – –)

This situation is summarized in Table 7.3.

TABLE 7.3 Orbital Structure of Sublevels

Sublevel	Number of Orbitals	Capacity of Each Orbital	Total Capacity Sublevel
s	1	$2e^-$	$2e^-$
p	3	$2e^-$	$6e^-$
d	5	$2e^-$	$10e^-$
f	7	$2e^-$	$14e^-$

Chapter 7

> **EXAMPLE 7.3**
> What is the total number of orbitals
> a. in the third principal energy level?
> b. occupied by electrons in neon, at. no. = 10?
>
> **Solution**
> a. The 3s sublevel contains one orbital, the 3p contains three orbitals, and the 3d contains five orbitals
>
> $$1 + 3 + 5 = 9$$
>
> b. The electron configuration of neon is $1s^2 2s^2 2p^6$. The orbitals filled are the 1s, the 2s, and the three 2p orbitals
>
> $$1 + 1 + 3 = 5$$
>
> **Exercise**
> How many orbitals are occupied in the argon atom (at. no. = 18)? Answer: 9.

Electron Spin

To describe an electron in an atom completely, we need to consider one more factor. This is the **spin** of the electron about its axis (Figure 7.8). Electrons in atoms can be thought of as having either of two different spins. We might describe these as "clockwise" versus "counterclockwise." More commonly, we represent the spins by arrows drawn up or down:

↑ or ↓

Figure 7.8 In some respects, an electron behaves as if it were a spherical particle spinning about its axis. There is an analogy between the alignment of electron spins and the alignment of bar magnets (*top* of figure). Within an orbital, the more stable arrangement is the one in which the two electrons have opposed spins (*lower right*).

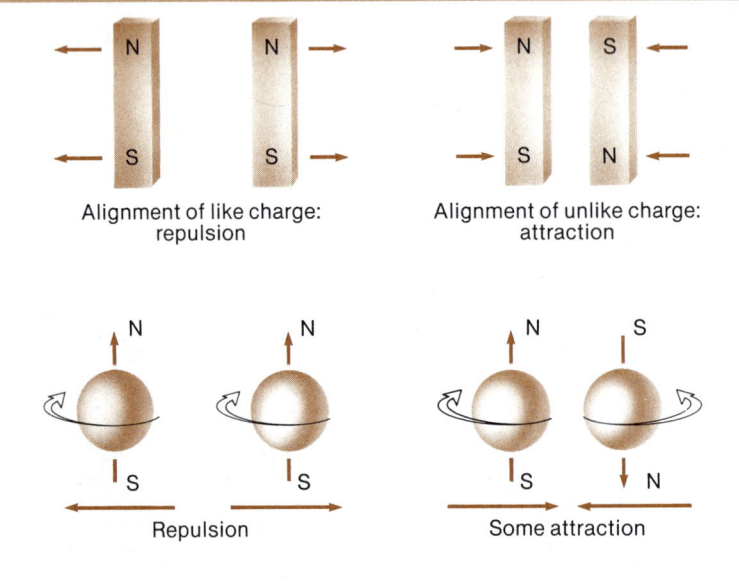

Alignment of like charge: repulsion

Alignment of unlike charge: attraction

Repulsion

Some attraction

Electronic Structure and the Periodic Table

183

Under ordinary conditions, the energy of a single isolated electron is independent of its spin. We can equally well write the arrow up or down for such an electron. However, with two electrons, the situation changes. It makes a difference whether the spins are in the same direction *(parallel)* or in opposite directions *(opposed)*. This explains why we need to specify the spins of electrons in multi-electron atoms.

Orbital Diagrams

Perhaps the most complete description of the electronic structure of an atom uses what are known as orbital diagrams. Here, we start with the electron configuration and go two steps further. We show the orbitals in which the electrons are located within sublevels. To do that, we will use parentheses ( ) to represent an orbital. Then we show the spin of each electron within an orbital, indicating it as ↑ or ↓.

A simple principle allows us to go from electron configuration to orbital diagram for the elements of atomic number 1–5: **Two electrons in the same orbital have opposed spins, ↑ and ↓.** Hence, we have

Element	Electron Configuration	Orbital Diagram		
		1s	2s	2p
$_1$H	$1s^1$	(↑)		
$_2$He	$1s^2$	(↑↓)		
$_3$Li	$1s^2 2s^1$	(↑↓)	(↑)	
$_4$Be	$1s^2 2s^2$	(↑↓)	(↑↓)	
$_5$B	$1s^2 2s^2 2p^1$	(↑↓)	(↑↓)	(↑)( )( )

Note that in the hydrogen atom, it makes no difference whether we show the electron spin as ↑ or ↓; for no reason in particular, we show it as ↑. The same situation applies to the single 2s electron in lithium and the single 2p electron in boron. Along the same line, it doesn't matter which of the 2p orbitals we choose to receive the outermost electron in boron. The p_x, p_y, and p_z orbitals are of equal energy at that point. It seems reasonable to start filling from the left, so we do that.

With the next element, carbon, a complication arises. It has two 2p electrons. This leads to three possibilities. In principle, these two electrons could be

—in the same 2p orbital, in which case they should have opposed spins, ↑ and ↓.
—in two different 2p orbitals with opposed spins, ↑ and ↓.
—in two different 2p orbitals with parallel spins, ↑ and ↑.

Experimental evidence indicates that the carbon atom prefers the last choice. More generally, we can say that

Chapter 7

When more than one orbital is available within a given sublevel (p, d, or f), electrons enter singly, giving the maximum number of half-filled orbitals. The single electrons in these orbitals have parallel spins.

Following this principle, we can now write orbital diagrams for elements of atomic number 6–10.

Element	Electron Configuration	1s	2s	2p
$_6$C	$1s^2 2s^2 2p^2$	(↑↓)	(↑↓)	(↑)(↑)( )
$_7$N	$1s^2 2s^2 2p^3$	(↑↓)	(↑↓)	(↑)(↑)(↑)
$_8$O	$1s^2 2s^2 2p^4$	(↑↓)	(↑↓)	(↑↓)(↑)(↑)
$_9$F	$1s^2 2s^2 2p^5$	(↑↓)	(↑↓)	(↑↓)(↑↓)(↑)
$_{10}$Ne	$1s^2 2s^2 2p^6$	(↑↓)	(↑↓)	(↑↓)(↑↓)(↑↓)

The same principle is readily applied to elements of higher atomic number (Example 7.4).

EXAMPLE 7.4

The electron configuration of nickel was derived in Example 7.2

$_{28}$Ni $1s^2 2s^2 2p^6 3s^2 3p^6 4s^2 3d^8$

Write the orbital diagram of nickel.

Solution

In the completed sublevels, all the orbitals are filled with electron pairs of opposed spins. The 3d sublevel is incomplete. The eight 3d electrons are distributed so as to give the maximum number of half-filled orbitals. The result is

1s	2s	2p	3s	3p	4s	3d
(↑↓)	(↑↓)	(↑↓)(↑↓)(↑↓)	(↑↓)	(↑↓)(↑↓)(↑↓)	(↑↓)	(↑↓)(↑↓)(↑↓)(↑)(↑)

Note that the electrons in the two half-filled 3d orbitals have parallel spins.

Exercise

For As (at. no. = 33), give the orbital diagram for the 15 electrons beyond Ar.
Answer:

(↑↓) (↑↓)(↑↓)(↑↓)(↑↓)(↑↓) (↑)(↑)(↑)
4s   3d   4p

7.5 Electron Configuration and the Periodic Table

The Periodic Table was first proposed by Dmitri Mendeleev, a Russian chemist, in 1869. Fifty years passed before the electron configurations of the elements were worked out. These structures provided the theoretical basis for the Periodic Table. The reason why elements in the same vertical group show similar chemical properties is that: **Elements in the same group of the Periodic Table have the same outer electron configuration.**

Dmitri Mendeleev

To understand what this statement means, consider the (abbreviated) configurations of the *alkali metals* (Group 1) and the *alkaline earth metals* (Group 2)

Group 1	Group 2
$_3$Li [$_2$He] $2s^1$	$_4$Be [$_2$He] $2s^2$
$_{11}$Na [$_{10}$Ne] $3s^1$	$_{12}$Mg [$_{10}$Ne] $3s^2$
$_{19}$K [$_{18}$Ar] $4s^1$	$_{20}$Ca [$_{18}$Ar] $4s^2$
$_{37}$Rb [$_{36}$Kr] $5s^1$	$_{38}$Sr [$_{36}$Kr] $5s^2$
$_{55}$Cs [$_{54}$Xe] $6s^1$	$_{56}$Ba [$_{54}$Xe] $6s^2$

Notice that all of the alkali metals have a single s electron in the outermost principal energy level. All of the alkaline earths have two outer s electrons. In this sense, we would say that all five alkali metals have the same outer electron configuration. The same is true of all five alkaline earth metals.

As we will see in Chapter 8, it is the outer electrons that are involved in chemical reactions. This explains why elements in the same group of the Periodic Table have similar chemical properties. With metals, the outer electrons are lost in reactions with nonmetals. On this basis, we see why all the alkali metals form +1 cations (Li^+, Na^+, K^+, Rb^+, Cs^+) and all the alkaline earth metals form +2 cations (Be^{2+}, Mg^{2+}, Ca^{2+}, Sr^{2+}, Ba^{2+}).

In Figure 7.9, we indicate the regions of the Periodic Table where various sublevels are filling.

1. The elements in Groups 1 and 2 are filling an s sublevel. Lithium puts a single electron into the 2s sublevel; Be fills that sublevel with two electrons. In the third period, Na starts to fill the 3s sublevel and Mg completes the job.
2. The elements in Groups 3 through 8 (six elements in each period beyond the first) fill p sublevels which, you will recall, have a capacity of six electrons. In the second period, the 2p sublevel starts to fill with B (at. no. = 5) and is completed by Ne (at. no. = 10). In the third period, the 3p sublevel fills starting with Al (at. no. = 13) and ending with argon (at. no. = 18).

    Throughout the remainder of this chapter and indeed in Chapter 8, our discussion will concentrate upon the *main-group elements*, those filling s and p sublevels. As we will see, the properties of these elements vary in a regular way as we move down or across in the Periodic Table.
3. The *transition metals*, which appear for the first time in the middle of the fourth period, fill d sublevels. Remember that a d sublevel has a capacity of ten electrons. Small wonder that there are ten elements in a transition series. In the fourth period, the elements Sc (at. no. = 21) through Zn (at. no. = 30) fill the 3d sublevel. In the fifth period, the 4d sublevel is filled, starting with Y (at. no. = 39) and ending with Cd (at. no. = 48). The ten transition metals in the sixth period ($_{71}$Lu → $_{80}$Hg) fill the 5d sublevel. Finally, in the seventh period there is still another series of tran-

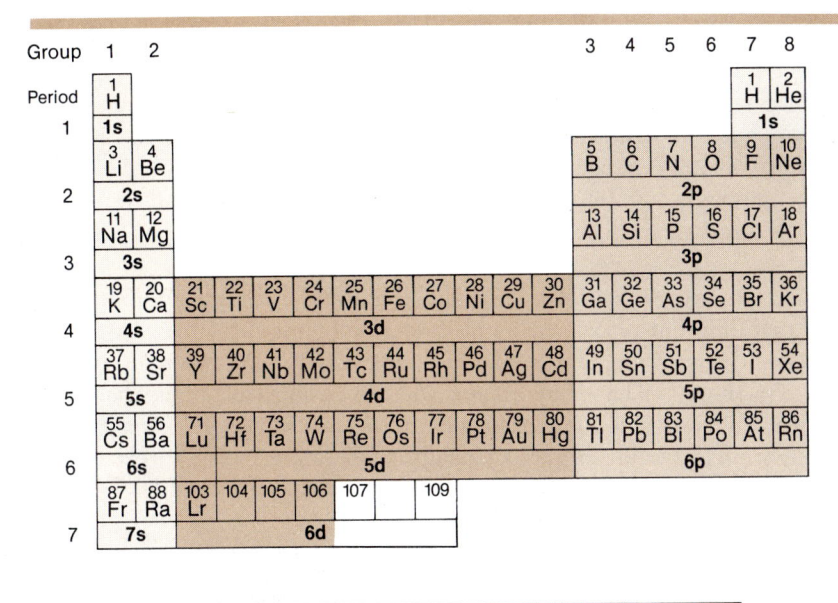

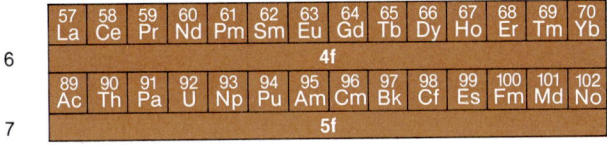

Electronic Structure and the Periodic Table

Figure 7.9 The Periodic Table can be used to deduce the electron configurations of atoms. Elements in Groups 1 and 2 are filling an ns sublevel, where n is the number of the period. Elements in Groups 3 through 8 (except for H and He) fill an np sublevel. The transition metals fill an (n − 1)d sublevel. For example, the elements Sc through Zn in the fourth period fill the 3d sublevel. The lanthanides fill the 4f sublevel, while the actinides fill the 5f sublevel.

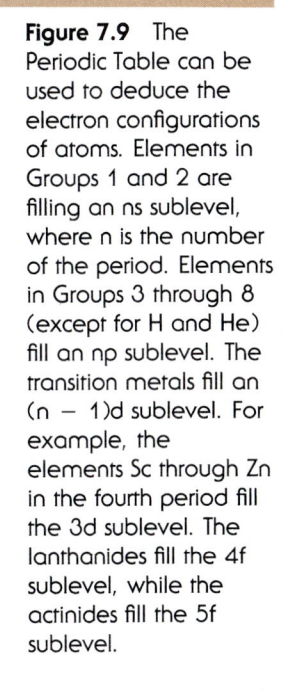

sition metals, this one incomplete. Elements 103–109, none of which occurs in nature, are believed to be putting electrons in the 6d sublevel.

Many of our most familiar elements are transition metals. Of the five metals produced in the greatest quantity in the United States, four (iron, manganese, copper, and zinc) are in the first transition series. We will have more to say about the physical and chemical properties of the transition metals in Chapter 13.

4. The f sublevels, with a capacity of 14 electrons, are filled by two series of elements listed separately below the main body of the Table. The 14 elements in the sixth period fill the 4f sublevel; they are commonly referred to as *lanthanides* or rare earths. These elements (lanthanum, atomic number 57, through ytterbium, atomic number 70) resemble each other closely in their chemical properties. They occur together in nature and for many years were very difficult to obtain in pure form. Modern separation methods have made it possible to produce many of the lanthanides and their compounds at a reasonable cost. They have a variety of uses. A brilliant red phosphor used in TV receivers contains europium oxide, Eu_2O_3. The oxides of lanthanum, praseodymium, and neodymium (La_2O_3, Pr_2O_3,

Nd$_2$O$_3$) are used in tinted sunglasses. They absorb ultraviolet radiation and reduce the intensity of sunlight.

In the seventh period, the 5f sublevel fills with the 14 elements called *actinides*. These start with atomic number 89 (actinium) and go through atomic number 102 (nobelium). All of these elements are radioactive. Only two, thorium (at. no. = 90) and uranium (at. no. = 92) are found in nature. Uranium is used in nuclear reactors to generate electrical energy (Chapter 15).

You will notice that in the Periodic Table one element, hydrogen, appears in two different places. In its electron configuration, 1s^1, hydrogen resembles the Group 1 elements *(alkali metals)* in that it has a single, outer s electron. On the other hand, hydrogen, like the Group 7 elements (halogens) is one electron short of a noble gas configuration. In its properties, hydrogen behaves more like a halogen than an alkali metal. In particular, it reacts readily with active metals to form a -1 ion, H$^-$, similar to the *halide* ions F$^-$, Cl$^-$, Br$^-$, I$^-$.

Figure 7.9 could be used to deduce the complete electron configuration of a element. More commonly, we use the Periodic Table to predict the outer electron configurations of *main-group elements* (the numbered groups). To do this, we note that the outer electron configurations of these elements are:

Group	1	2	3	4	5	6	7	8
Outer Configuration	ns^1	ns^2	ns^{2}np^1	ns^{2}np^2	ns^{2}np^3	ns^{2}np^4	ns^{2}np^5	ns^{2}np^6

Here, n is the number of the period in which the element is located. That is n = 2 for the second period, n = 3 for the 3rd period, and so on.

EXAMPLE 7.5
Using the Periodic Table, give the outer electron configuration of
a. Ga (at. no. = 31)   b. I (at. no. = 53)

Solution
a. Gallium is in the fourth period, in Group 3. Its outer electron configuration is 4s^{2}4p^1.
b. Iodine is in Period 5, Group 7. Outer electron configuration: 5s^{2}5p^5.

Exercise
What is the symbol of the element with the outer electron configuration 6s^{2}6p^3?
Answer: Bi.

7.6 Trends in the Periodic Table

Many of the properties of atoms depend in a systematic way upon their position in the Periodic Table. In this section, we will look at three such properties. One of them, *metallic character*, was touched upon briefly in Chapter 2. The other two properties, *ionization energy* and *electronegativity*, are important so far as chemical bonding is concerned (Chapter 8). Ionization energy measures the ease (or difficulty) of removing an electron from an atom. Electronegativity measures the tendency of an atom to attract electrons to itself.

Metallic Character

As we pointed out in Chapter 2, metals are located at the lower left of the Periodic Table, nonmetals at the upper right. Putting it another way, we might say that **metallic character increases as we move down in the Periodic Table and decreases as we move across from left to right.**

To illustrate the vertical trend, consider the elements in Group 4. The first element, carbon, is clearly nonmetallic. The two elements below carbon, silicon and germanium, are metalloids. Their properties are intermediate between those of metals and nonmetals; both Si and Ge are electrical semiconductors. The last two elements in Group 4, tin and lead, are metals. Both elements are good electrical conductors and have other typical metallic properties, including luster and malleability.

The horizontal trend in metallic character is shown by the elements in the third period, $_{11}$Na through $_{18}$Ar. The first three elements, sodium, magnesium, and aluminum, are clearly metals. All of them are excellent conductors. As we just pointed out, the next element, silicon, is a metalloid. Silicon is followed by four nonmetals (phosphorus, sulfur, chlorine, and argon). All four of these elements are nonconductors; none of them shows any of the properties we associate with metals.

Ionization Energy

The ionization energy of an element is the energy change, usually expressed in kilojoules per mole, when an electron is removed from a gaseous atom. Using the letter M to stand for the element

$$M(g) \rightarrow M^+(g) + e^-; \quad \Delta E = \text{ionization energy}$$

As you might expect, ionization energy is always positive (i.e., $\Delta E > 0$). There is a strong attraction between an electron (negative charge) and the nucleus (positive charge). Energy must be absorbed to separate them from each other.

Ionization energy is a measure of how readily an atom gives up electrons. Metals, which lose electrons rather easily, tend to have relatively low ionization energies. Nonmetals, on the other hand, have high ionization energies. As you might expect on this basis, **ionization energy generally decreases as we move down in the Periodic Table and increases as we move across from left to right.** Notice from Figure 7.10 that the element with the lowest ionization energy is Cs ($\Delta E = 377$ kJ/mol), located at the lower left corner of the Table. Cesium is also the most metallic of all the common elements. The element with the highest ionization energy is helium ($\Delta E = 2377$ kJ/mol), located at the upper right corner.

EXAMPLE 7.6

Using only the Periodic Table, arrange the following elements in order of decreasing ionization energy: Na, Mg, K.

Solution

These elements are arranged in the Periodic Table in the pattern

Na   Mg

K

Continued

Figure 7.10 First ionization energies of the main-group elements (in kilojoules per mole). In general, ionization energy decreases as one moves down in the Periodic Table and increases as one moves across from left to right. Cesium has the smallest ionization energy and helium the largest.

	1	2		3	4	5	6	7	8
								H	He
								1318	2377
	Li	Be		B	C	N	O	F	Ne
	527	904		808	1092	1410	1322	1686	2088
	Na	Mg		Al	Si	P	S	Cl	Ar
	502	745		586	791	1021	1004	1264	1527
	K	Ca		Ga	Ge	As	Se	Br	Kr
	427	594		586	770	954	946	1146	1356
	Rb	Sr		In	Sn	Sb	Te	I	Xe
	410	556		565	715	841	879	1017	1176
	Cs	Ba		Tl	Pb	Bi	Po	At	Rn
	377	510		594	724	711	820		1042

Since Na is to the left of Mg, it should have a lower ionization energy. Since K is below Na, it should have a still lower ionization energy. The predicted order is

Mg > Na > K

(Note that this is also the order of increasing metallic character.) From Figure 7.10 we see that the predicted order is correct: ΔE decreases from 745 kJ/mol for magnesium to 502 kJ/mol for sodium to 427 kJ/mol for potassium.

Exercise
Which of the three elements O, F, or Cl should have the largest ionization energy?
Answer: F.

Electronegativity

There are various ways to describe the tendency of an atom to gain electrons. The only one we will consider is called electronegativity. The more electronegative an atom is, the greater its attraction for electrons. Unlike ionization energy, electronegativity cannot be measured directly. The numbers quoted in Figure 7.11 are based on a relative scale first proposed by Linus Pauling. They range from 4.0 for the most electronegative element, fluorine, to 0.7 for the element having the least attraction for electrons, cesium.

H 2.1						
Li 1.0	Be 1.5	B 2.0	C 2.5	N 3.0	O 3.5	F 4.0
Na 0.9	Mg 1.2	Al 1.5	Si 1.8	P 2.1	S 2.5	Cl 3.0
K 0.8	Ca 1.0	Ga 1.6	Ge 1.8	As 2.0	Se 2.4	Br 2.8
Rb 0.8	Sr 1.0	In 1.7	Sn 1.8	Sb 1.9	Te 2.1	I 2.5
Cs 0.7	Ba 0.9	Tl 1.8	Pb 1.8	Bi 1.9	Po 2.0	At 2.2

Figure 7.11
Electronegatives of the main-group elements. As metallic character goes down, electronegativity goes up. The most nonmetallic elements have the highest electronegativities. Fluorine is the element with the highest electronegativity; cesium has the lowest.

As you can see from Figure 7.11, electronegativity generally decreases as we move down in the Periodic Table and increases as we move across from left to right. You will recall that the same trends hold for "nonmetallic character." The nonmetals are clustered at the upper right of the Periodic Table. High electronegativity is characteristic of nonmetals, just as low ionization energy is typical of metals.

EXAMPLE 7.7

Using only the Periodic Table, arrange the following elements in order of increasing electronegativity: S, Cl, Se.

Solution

Locating these elements in the Periodic Table, we see that they form the pattern

S   Cl

Se

Sulfur should be more electronegative than selenium, since S lies above Se. Chlorine should be more electronegative than sulfur since Cl lies to the right of S. The predicted order is

Se < S < Cl

This agrees with the data in Figure 7.11 (2.4 < 2.5 < 3.0) and is also the order of increasing nonmetallic character.

Exercise

Of the three elements Se, S, Cl, which would you expect to have the highest ionization energy? Which would be the most metallic? Answer: Cl; Se.

Key Words

actinide
alkali metal
alkaline earth metal
atomic spectrum
Bohr model
electron cloud
electron configuration
electronegativity
excited state
ground state
ionization energy
lanthanide

main-group element
metallic character
orbital
orbital diagram
Periodic Table
principal energy level
quantum mechanics
quantum number
spin (electron)
sublevel
transition metal

Questions

7.1 What is meant by the ground state of an electron in an atom? an excited state?

7.2 Explain in your own words how the energy of an electron in an atom is "quantized."

7.3 Describe two different types of experiments that can be done to check the Bohr model of the hydrogen atom.

7.4 How does the quantum mechanical model of the hydrogen atom differ from the Bohr model?

7.5 Give the total capacity for electrons of the principal level for which n =
  a. 1   b. 2   c. 3   d. 4

7.6 How many electrons can enter each of the following sublevels?
  a. s   b. p   c. d   d. f

7.7 Arrange the following sublevels in order of increasing energy
  a. 4f   b. 4p   c. 4s   d. 4d

7.8 How many different sublevels are there in the principal level for which n is
  a. 1   b. 2   c. 3   d. 4

7.9 Explain what the notation $1s^2 2s^2 2p^3$ means.

7.10 What is the maximum number of electrons that can fit into a single orbital?

7.11 Describe the geometry of an s orbital; a p orbital.

7.12 What is the angle between any two of the three 2p orbitals?

7.13 When two electrons enter a single orbital, they have _____ spins.

7.14 When two electrons enter two different orbitals in the same sublevel, they are expected to have _____ spins.

7.15 How many electrons are there in the outermost principal energy level of a Group 1 element? a Group 2 element?

7.16 The elements in Groups 1 and 2 of the Periodic Table fill _____ sublevels.

7.17 The transition metals fill _____ sublevels.

7.18 The lanthanides (rare earths) fill the _____ sublevel.

7.19 The actinides fill the _____ sublevel.

7.20 Metallic character _____ as we go across in the Periodic Table and _____ as we go down.

7.21 Ionization energy _____ as we go across in the Periodic Table and _____ as we go down.

7.22 Electronegativity _____ as we go down in the Periodic Table and _____ as we go across.

Problems

Electron Configurations

7.23 Give the electron configuration of
  a. $_4$Be   b. $_{21}$Sc   c. $_{16}$S

7.24 Give the electron configuration of
  a. $_{28}$Ni   b. $_{10}$Ne   c. $_7$N

7.25 Identify the atom with the electron configuration
  a. $1s^2 2s^2 2p^5$
  b. $1s^2 2s^2 2p^6 3s^2 3p^6 4s^2 3d^6$
  c. $[_{18}Ar] 4s^2 3d^2$

7.26 Identify the atom that has the electron configuration
  a. $1s^2 2s^2 2p^4$
  b. $1s^2 2s^2 2p^6 3s^2 3p^1$
  c. $[_{18}Ar] 4s^2 3d^{10} 4p^3$

7.27 Identify the atom that has a single
  a. 2s electron   b. 3p electron   c. 3d electron

7.28 In the phosphorus atom (at. no. = 15), what percent of the electrons are located in the first principal energy level? the second principal level? the third principal level?

7.29 Of the first 36 elements, how many have at least one
 a. 1s electron   b. 2p electron   c. 3d electron

7.30 In the manganese atom (at. no. = 25), what is the total number of electrons in s sublevels? p sublevels? d sublevels?

Orbital Diagrams

7.31 Give the orbital diagrams of the following atoms
 a. $_{12}$Mg   b. $_{29}$Cu   c. $_{17}$Cl

7.32 Give the orbital diagrams, beyond $_{18}$Ar, for each of the following atoms
 a. $_{23}$V   b. $_{35}$Br   c. $_{27}$Co

7.33 Identify the atoms with the following orbital diagrams

	3s	3p	4s
a. [$_{10}$Ne]	(↑↓)	(↑↓)(↑)(↑)	( )
b. [$_{10}$Ne]	(↑↓)	( )( )( )	( )
c. [$_{10}$Ne]	(↑↓)	(↑↓)(↑↓)(↑↓)	(↑)

7.34 Identify the atoms with the following orbital diagrams

	4s	3d	4p
a. [$_{18}$Ar](↑↓)	(↑↓)(↑↓)(↑↓)(↑↓)(↑↓)	(↑)( )( )
b. [$_{18}$Ar](↑↓)	(↑)(↑)(↑)( )( )	( )( )( )
c. [$_{18}$Ar](↑↓)	(↑↓)(↑↓)(↑↓)(↑↓)(↑↓)	( )( )( )

7.35 State the number of half-filled orbitals in
 a. $_{11}$Na   b. $_{14}$Si   c. $_{22}$Ti

7.36 Which of the following could not represent orbital diagrams of atoms in their ground state?

	1s	2s	2p
a.	(↑↓)	(↑)	( )( )( )
b.	(↑)	(↑)	( )( )( )
c.	(↑↓)	(↑↓)	(↑↓)( )( )
d.	(↑↓)	(↑↓)	(↑)(↓)( )

Electron Configuration and the Periodic Table

(You may use the Periodic Table.)

7.37 Give the outer electron configuration of
 a. $_{49}$In   b. $_{51}$Sb   c. $_{53}$I

7.38 Identify the atoms with the following outer electron configurations
 a. $3s^23p^3$ b. $6s^26p^3$ c. $5s^25p^6$

7.39 What sublevel is partially filled in
 a. $_{42}$Mo   b. $_{55}$Cs   c. $_{74}$W

7.40 What sublevel is partially filled in
 a. $_{62}$Sm   b. $_{92}$U   c. $_{75}$Re

7.41 Give the outer electron configuration of
 a. $_{38}$Sr   b. $_{87}$Fr   c. $_{86}$Rn

7.42 What atom has the outer electron configuration
 a. $6s^1$ b. $6s^26p^5$ c. $4s^24p^3$

Trends in the Periodic Table

(You may use the Periodic Table to work out these problems.)

7.43 Of the elements in Group 2, which one has the highest
 a. metallic character
 b. ionization energy
 c. electronegativity

7.44 Of the elements in Group 7, which one has the highest
 a. metallic character
 b. ionization energy
 c. electronegativity

7.45 In which group in the Periodic Table would you expect to find the elements of lowest ionization energy?

7.46 Arrange the following elements in order of increasing metallic character
 a. Si   b. Al   c. Ga

7.47 Arrange the following elements in order of decreasing ionization energy
 a. Al   b. Na   c. Mg   d. K

7.48 Arrange the following elements in order of increasing electronegativity
 a. O   b. F   c. S   d. Se

Multiple Choice

7.49 The energy of the hydrogen electron in the fifth principal energy level, as calculated from Equation 7.1 is, in kilojoules per mole
 a. −1312   b. −262.4   c. −131.2
 d. −52.48   e. 0

7.50 The total number of electrons that can enter the fourth principal energy level is calculated to be
 a. 2   b. 8   c. 18   d. 32   e. some other number

7.51 A 2s orbital differs from a 1s orbital in
 a. geometric form
 b. capacity for electrons
 c. radius
 d. none of these

7.52 The first element to put an electron in a 4p orbital is
 a. $_{19}$K   b. $_{21}$Sc   c. $_{30}$Zn   d. $_{31}$Ga
 e. $_{36}$Kr

7.53 Which one of the following sublevels has the highest energy?
 a. 3s   b. 4f   c. 3d   d. 4d   e. 2p

7.54 Which one of the following atoms has the largest number of electrons with parallel spins?
 a. $_{14}$Si   b. $_{15}$P   c. $_{16}$S   d. $_{17}$Cl
 e. $_{18}$Ar

7.55 Which one of the following does not represent a ground-state electron configuration?
 a. $1s^22s^22p^4$ b. $1s^22s^12p^3$ c. $1s^22s^2$
 d. $1s^22s^1$

7.56 A strongly metallic element should have
 a. high ionization energy and low electronegativity
 b. high ionization energy and high electronegativity
 c. low ionization energy and low electronegativity
 d. low ionization energy and high electronegativity

7.57 Which two of the following elements would you expect to resemble each other most closely in metallic character?
 a. $_5$B and $_6$C   b. $_5$B and $_{13}$Al   c. $_5$B and $_{14}$Si   d. $_6$C and $_{14}$Si

7.58 In the calcium atom (at. no. = 20), how many sublevels contain electrons?
 a. 4   b. 5   c. 6   d. 7   e. 8

7.59 In $_{88}$Ra, all orbitals are filled. The total number of orbitals occupied is
 a. 22   b. 44   c. 88   d. 176
 e. some other number

7.60 The total number of possible orbitals in the third principal energy level is
 a. 1   b. 4   c. 9   d. 16   e. some other number

Chemical Bonding

Learning Objectives

After studying this chapter, you should be able to:

1. Explain the source of the energy evolved when an ionic compound is formed from the elements.
2. Write the electron configuration of a monatomic ion, given the atomic number of the element from which it is formed (Examples 8.1, 8.2).
3. Explain the source of the energy evolved when a covalent bond such as that in H_2 is formed.
4. Distinguish between single, double, and triple bonds.
5. Draw the Lewis structure of an atom of any main-group element (Example 8.3).
6. Draw the Lewis structure for a simple molecule (Examples 8.4, 8.5).
7. Describe the Lewis structures of several molecules that do not follow the octet rule, including BeF_2 and BF_3.
8. Predict the geometries of singly bonded molecules in which the central atom is surrounded by two, three, or four electron pairs (Example 8.6).
9. Knowing the geometry of a molecule, predict whether it is polar or nonpolar (Examples 8.7, 8.8).

I have yet to see any problem, however complicated, which, when you looked at it in the right way, did not become still more complicated.

PAUL ANDERSON

Photograph by Peter L. Kresan.

CHAPTER 8

In Chapter 3 we pointed out that isolated atoms are seldom found as such in nature. For the most part, they are too reactive to exist by themselves. In most substances, atoms are joined to one another by strong forces called chemical bonds. In this chapter we will look at the two most common types of bonds:

1. **Ionic bonds,** between oppositely charged ions in such compounds as NaCl (Na$^+$, Cl$^-$ ions) and KOH (K$^+$, OH$^-$ ions). The formulas and names of ionic compounds were discussed in Chapter 4. In this chapter (Section 8.1), we will look for answers to two questions related to ionic bonding. What is the driving force for the formation of ionic compounds from the elements? What are the electron configurations of the ions found in these compounds?
2. **Covalent bonds,** which consist of electron pairs shared between nonmetal atoms in certain elements (C, H$_2$, N$_2$, – –) and compounds (CH$_4$, NH$_3$, H$_2$O, – –). In Section 8.2, we will introduce covalent bonding, looking for the source of the energy evolved when such a bond is formed from isolated atoms.

Covalent bonds are most often found in molecules (H$_2$, N$_2$, CH$_4$, NH$_3$, H$_2$O, – –). The last three sections of this chapter are devoted to the properties of molecules. In Section 8.3, we will see how to draw Lewis structures to show the distribution of electrons between atoms in molecules. With that background, we will go on to consider the geometries (Section 8.4) and polarities (Section 8.5) of some simple molecules.

8.1 Ionic Bonding

The electrical attractive forces between oppositely charged ions are referred to as *ionic bonds*. These forces are very strong, as indicated by the fact that all ionic compounds are high-melting solids. Large amounts of energy and high temperatures are required to break ionic bonds, converting the rigid solid lattice to a liquid.

The formation of a binary ionic compound such as NaCl or CsF involves the transfer of electrons from metal to nonmetal atoms. As a result of this transfer, a metal atom is converted to a positively charged ion (cation). The nonmetal atom acquires one or more electrons to become a negatively charged ion (anion). Three points concerning the electron transfer deserve particular mention.

1. *Electron transfer is most likely to occur between a metal of low ionization energy and a nonmetal of high electronegativity.* We would expect cesium, with the lowest ionization energy of any metal, to give up electrons readily to fluorine, the nonmetal atom with the highest electronegativity. This does indeed happen. The product formed is CsF, which contains Cs^+ and F^- ions. In practice, ion formation usually occurs when any metal reacts with any nonmetal. Indeed, we can go a step further and say that *a compound containing a metal is almost always ionic.*
2. *The electrons transferred come from the outermost energy level of the metal atom and enter the outermost energy level of the nonmetal atom.* To illustrate what this statement means, consider the reaction between a sodium atom and a chlorine atom. In terms of electron configurations, we can write

$$_{11}Na\ (1s^22s^22p^63s^1) \rightarrow\ _{11}Na^+(1s^22s^22p^6)\ +\ e^-$$

$$_{17}Cl\ (1s^22s^22p^63s^23p^5)\ +\ e^- \rightarrow\ _{17}Cl^-(1s^22s^22p^63s^23p^6)$$

The electron that leaves the sodium atom is a 3s electron; that electron enters the 3p sublevel of chlorine. This process illustrates the general statement made in Chapter 7, that it is ordinarily the outer electrons of atoms that are involved in chemical reactions.
3. *The process is strongly exothermic.* Consider, for example, the reaction

$$Na(s)\ +\ \frac{1}{2}Cl_2(g) \rightarrow NaCl(s);\ \Delta H\ =\ -411\ kJ \qquad (8.1)$$

This thermochemical equation tells us that 411 kilojoules of heat are evolved when one mole of sodium chloride is formed from the elements. This energy is given off as a direct result of the formation of ionic bonds between Na^+ and Cl^- ions. These oppositely charged ions attract each other strongly, so a considerably amount of energy is released when they combine to form NaCl. In the solid, Na^+ ions are surrounded by Cl^- ions and vice versa (Figure 8.1). This is the most stable arrangement since it puts ions of opposite charge as close to one another as possible.

Figure 8.1 Ball-and-stick model of a crystal of NaCl (the Na⁺ ions actually touch the Cl⁻ ions). Each Cl⁻ ion is surrounded by six Na⁺ ions, and each Na⁺ ion by six Cl⁻ ions. A single crystal of NaCl contains many billions of ions arranged in this pattern.

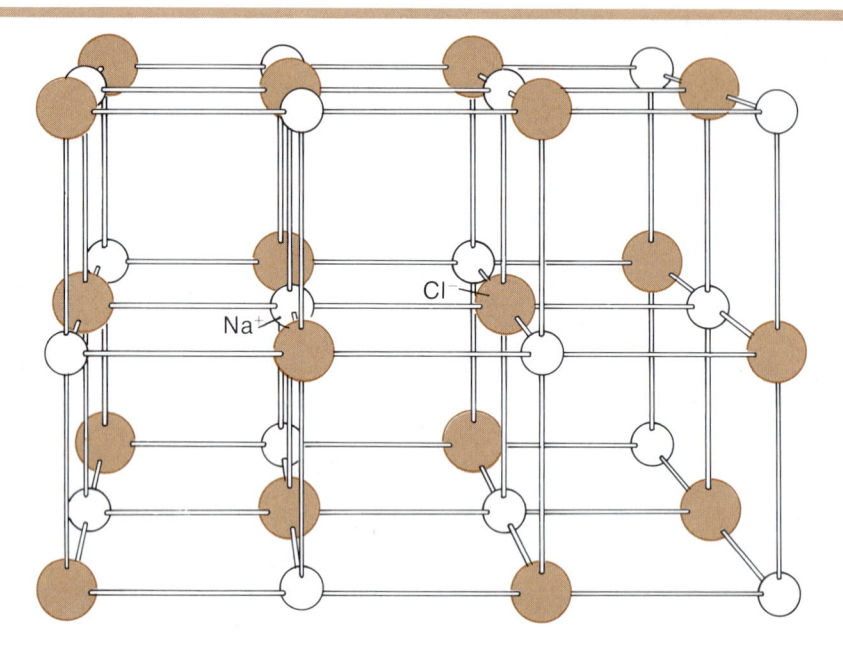

Electron Configurations of Ions

In Chapter 7, we derived the electron configurations of atoms. You will recall that this involves specifying the number of electrons in each sublevel. For example, the electron configuration of the noble gas neon (at. no. = 10) is

$$_{10}Ne \quad 1s^2 2s^2 2p^6$$

We will now look at the electron configurations of simple cations and anions.

You may remember from Chapter 4 that elements close to a noble gas in the Periodic Table form ions with the same number of electrons as the noble gas atom. This means that

Ions formed by elements in Groups 1, 2, 6, and 7 have the same electron configuration as that of the neighboring noble gas atom.

To understand what this statement means, consider the noble gas neon.

TABLE 8.1 Species with the Neon Structure

Atom or Ion	Number of Electrons	Electron Configuration
$_8O^{2-}$	8 + 2 = 10	$1s^2 2s^2 2p^6$
$_9F^-$	9 + 1 = 10	$1s^2 2s^2 2p^6$
$_{10}Ne$	10	$1s^2 2s^2 2p^6$
$_{11}Na^+$	11 − 1 = 10	$1s^2 2s^2 2p^6$
$_{12}Mg^{2+}$	12 − 2 = 10	$1s^2 2s^2 2p^6$

It is preceded in the Periodic Table by two nonmetals, oxygen in Group 6 (at. no. = 8) and fluorine in Group 7 (at. no. = 9). Following neon are two metals, sodium in Group 1 (at. no. = 11) and magnesium in Group 2 (at. no. = 12). Each of these elements forms a stable ion (O^{2-}, F^-, Na^+, Mg^{2+}) which has the same electron configuration as the neon atom (Table 8.1).

EXAMPLE 8.1
Give the electron configuration of
a. Cl^- b. K^+

Solution
a. Chlorine has an atomic number of 17. The Cl^- ion contains 17 + 1 = 18 electrons, the same number as the noble gas argon. The Cl^- ion has the electron configuration of argon, which is $1s^2 2s^2 2p^6 3s^2 3p^6$.
b. The atomic number of K is 19. There are 19 − 1 = 18 electrons in the K^+ ion. The electron configuration of K^+, like that of Cl^- and that of the Ar atom, is $1s^2 2s^2 2p^6 3s^2 3p^6$.

Exercise
There is an anion with a −2 charge and a cation with a +2 charge, both of which have the argon structure. What are the formulas of these ions? Answer: S^{2-}, Ca^{2+}.

The fact that so many ions have a noble gas structure is significant. It suggests that **a noble gas electron configuration is particularly stable.** Other atoms tend to acquire that structure by reacting to form ions. For example, when the elements potassium and chlorine react with one another, they form the ions K^+ and Cl^-. Both of these ions have the argon structure.

The noble gases themselves are extremely unreactive. The first three members of the group, He, Ne, and Ar, form no compounds with other elements. This again implies that the noble gas configuration is unusually stable. Atoms that have it tend to keep it.

So far, all the ions discussed in this section have been derived from main-group elements. However, as pointed out in Chapter 4, transition metals also form ions, typically with charges of +1, +2, or +3. In the first series of transition metals, we have the following ions

Cr^{2+}, Cr^{3+}; Mn^{2+}; Fe^{2+}, Fe^{3+}; Co^{2+}, Co^{3+}; Ni^{2+}; Cu^+, Cu^{2+}; Zn^{2+}

None of these ions have noble gas structures. They all have several more electrons than the preceding noble gas, argon (at. no. = 18). For example, for the ions of iron (at. no. = 26)

Fe^{2+} : no. of electrons = 26 − 2 = 24

Fe^{3+} : no. of electrons = 26 − 3 = 23

Chapter 8

The electron configurations of transition metal ions can be obtained by following a simple principle. When a transition metal atom is converted to a cation, *it is the outer s electrons that are lost first.* Consider, for example, the Fe^{2+} and Fe^{3+} ions

Atom	Electron Configuration	Ion	Electron Configuration
$_{26}Fe$	$[_{18}Ar]\ 4s^2 3d^6$	$_{26}Fe^{2+}$	$[_{18}Ar]\ 3d^6$
		$_{26}Fe^{3+}$	$[_{18}Ar]\ 3d^5$

Putting it another way, **in transition metal ions, electrons beyond the preceding noble gas are located in inner d orbitals** (e.g., 3d) **rather than outer s orbitals** (e.g., 4s).

EXAMPLE 8.2
What is the electron configuration of
a. Cu^+ b. Cu^{2+}

Solution

a. Copper has an atomic number of 29. The copper atom has 29 electrons. One of these is lost in forming Cu^+, leaving 28. There are ten electrons beyond the preceding noble gas, argon (at. no. = 18). This is just enough electrons to fill the 3d sublevel (recall that d sublevels have a capacity of ten electrons). Hence the electron configuration of the Cu^+ ion is

$Cu^+\ [_{18}Ar]\ 3d^{10}$

b. In Cu^{2+}, we have 29 − 2 = 27 electrons. There are nine electrons beyond argon. The electron configuration is

$Cu^{2+}\ [_{18}Ar]\ 3d^9$

Exercise
What is the electron configuration of Cr^{3+}? Answer: $[_{18}Ar]\ 3d^3$.

The electron configurations we have written for transition metal ions of the fourth period indicate that in these species the 3d sublevel is lower in energy than the 4s. You will recall (Chapter 7) that the opposite is true for the atoms in this period, where the 4s sublevel fills before the 3d. Apparently, there is a reversal in the relative energies of 3d and 4s when atoms are converted to ions.

8.2 Covalent Bonding

A covalent bond consists of an electron pair shared between two atoms. The simplest covalent bond is that joining hydrogen atoms in the H_2 molecule. When two hydrogen atoms, each with a single electron, come together, they form a bond. This may be shown as

H : H

where the dots represent electrons. More commonly, the covalent bond is shown as a dash written between the two atoms

H—H

The formation of a covalent bond between two hydrogen atoms is a strongly exothermic process

$H(g) + H(g) \rightarrow H_2(g); \Delta H = -436$ kJ

The process of bond formation is shown graphically in Figure 8.2. We start at the far right, with two isolated hydrogen atoms far apart from each other. At this point, there is no interaction between the atoms and the energy of the system is zero. As the atoms come closer together (moving to the left in Figure 8.2), they experience an attraction and energy is given off. This leads

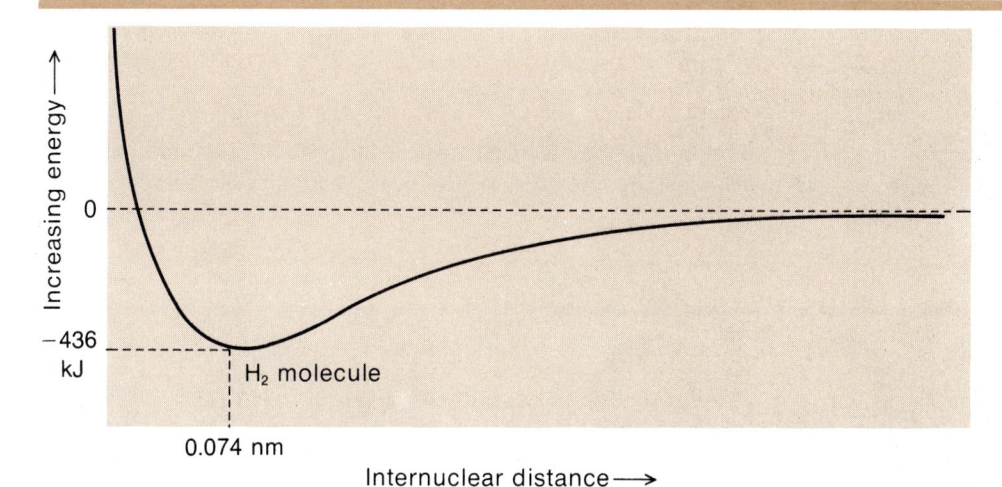

Figure 8.2 Energy of the H_2 molecule as a function of the distance between the two nuclei. The minimum in the curve occurs at the observed internuclear distance. Energy is compared to that of two separated hydrogen atoms.

gradually to an energy minimum, when the nuclei of the two atoms are 0.074 nm apart. The attractive energy at this point is 436 kJ/mol and the molecule is in its most stable state. If we try to squeeze the atoms closer together, repulsive forces become important and the energy curve rises steeply.

The question remains as to what causes the energy minimum in Figure 8.2. What is the source of the attractive energy between hydrogen atoms? Putting it another way, why does the sharing of a pair of electrons produce a stable molecule? One way to explain this involves the effect of bond formation on the electron clouds of the atoms. In a hydrogen atom, there is a single electron in a 1s orbital. When two hydrogen atoms approach one another closely, their 1s orbitals *overlap* (Figure 8.3). The two electrons, now attracted by both nuclei, spend most of their time between the nuclei. With that arrangement, attractive forces between particles of opposite charge (electron–proton) dominate. They are stronger than the repulsive forces between particles of like charge (electron–electron and proton–proton). Largely because of this factor, the hydrogen molecule is stable, to the extent of 436 kJ/mol.

A process similar to that just described for hydrogen takes place whenever two nonmetal atoms approach each other closely. Covalent bond formation occurs by the overlap of half-filled orbitals, one from each atom. The atoms joined may be the same as in H_2 and F_2

H—H   F—F

or different as in HF

H—F

The molecule formed may contain two atoms (one bond) as in H_2, F_2, and HF. In more complex molecules, there is more than one bond. In H_2O, there are two bonds, in NH_3, there are three, and in CH_4 there are four bonds.

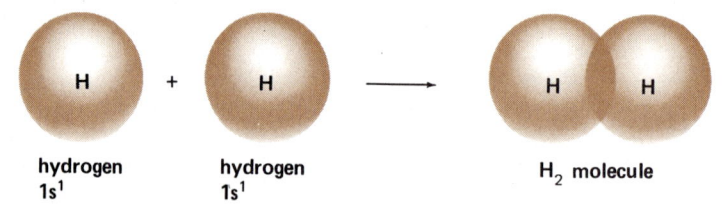

Figure 8.3 When two H atoms combine to form an H_2 molecule, the 1s orbitals on the two H atoms overlap, forming a new orbital that contains the two H atom electrons.

```
     O                N               H
    / \              /|\              |
   H   H            H | H          H—C—H
                      H               |
                                      H
  water, H₂O      ammonia, NH₃    methane, CH₄
```

Single, Double, and Triple Bonds

In all of the examples we have considered so far, two atoms were joined by a single pair of electrons. The bond formed this way is called a **single bond.** It is possible for two atoms to share more than one pair of electrons. If two pairs of electrons are shared, we say there is a **double bond** between the atoms. When three pairs of electrons are shared, there is a **triple bond.**

To illustrate the distinction between these three types of bonds consider the three hydrocarbons C_2H_6 (ethane), C_2H_4 (ethylene), and C_2H_2 (acetylene). The structures of these three molecules can be shown as

```
    H  H              H       H
    |  |               \     /
  H—C—C—H              C = C            H—C≡C—H
    |  |               /     \
    H  H              H       H
  ethane, C₂H₆      ethylene, C₂H₄     acetylene, C₂H₂
```

There is a *single* bond between the two carbon atoms in C_2H_6, represented by a single dash. In C_2H_4, the two dashes represent a *double* bond between the carbon atoms. Finally, there is a *triple* bond in C_2H_2, shown as three dashes. In all cases, a dash represents a pair of electrons.

Molecules, Polyatomic Ions, and Macromolecules

Covalent bonding is most often discussed in connection with molecular substances. Indeed, throughout the rest of this chapter, we will deal entirely with the properties of molecules. It is important to realize, however, that covalent bonding can occur in substances that are *not* molecular. In particular, we find covalent bonds in

1. *Polyatomic ions* The atoms that make up these ions are held to one another by covalent bonds. Consider, for example, three of the polyatomic ions considered in Chapter 4: the NH_4^+ ion, OH^- ion, and SO_4^{2-} ion. The structures of these ions may be shown as

```
      H   +                          O      2-
      |                              ‖
  [ H—N—H ]         [O—H]⁻      [ O—S—O ]
      |                              |
      H                              O
  ammonium ion     hydroxide ion    sulfate ion
```

Figure 8.4 Examples of network bonding in diamond and quartz. The bonds extend to the edges of the crystals; there are no small discrete molecules.

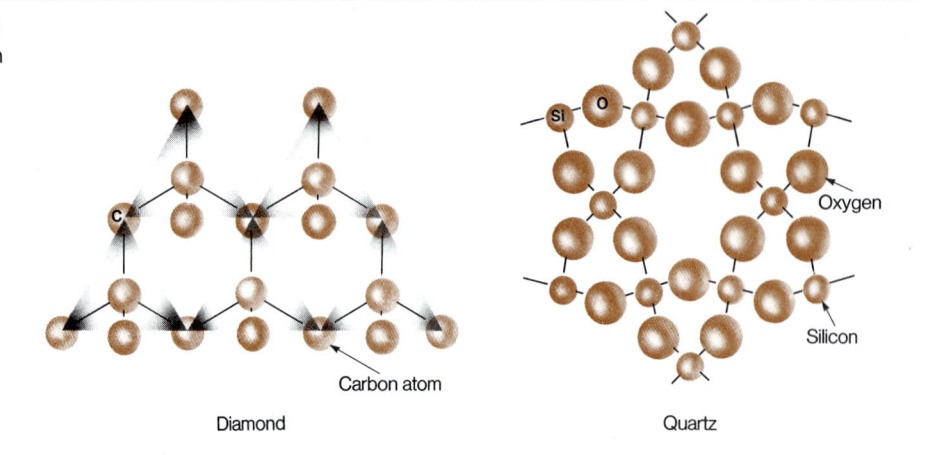

Diamond

Quartz

The dashes, as usual, represent covalent bonds, shared electron pairs. In the compound NH₄Cl, we have both ionic and covalent bonds. The NH_4^+ ion is held to the Cl^- ion by ionic bonds. Within the NH_4^+ ion, there are four covalent bonds joining nitrogen to hydrogen atoms.

2. *Macromolecular substances* In some cases, covalent bond formation can lead to structures of the type shown in Figure 8.4. Here, all the atoms are held together by a network of electron–pair bonds. There are no small discrete molecules. Substances that have this type of structure are referred to as being **macromolecular.** In effect, an entire crystal consists of one huge molecule.

Macromolecular substances are invariably high-melting solids. In this sense, they resemble ionic compounds. In both cases strong bonds (ionic or covalent) must be broken to melt the solid. Among the most familiar macromolecular substances are diamond (pure carbon) with a melting point of 3570°C and quartz (SiO_2) with a melting point of 1700°C.

8.3 Lewis Structures

As pointed out in Section 8.1, anions formed from nonmetal atoms, such as H^-, F^-, and O^{2-}, have noble gas electronic structures. In 1916, G. N. Lewis, a chemist at the University of California at Berkeley, pointed out that there is another, quite different way in which nonmetal atoms can acquire noble gas structures. They do this by sharing electrons with other atoms. Consider, for example, the hydrogen atom with its single electron. By forming a covalent bond, perhaps with another hydrogen atom

H· + H· → H : H

Gilbert Lewis

it gains a share in a second electron. We would say that, in the H$_2$ molecule, each hydrogen atom is surrounded by two electrons. In that sense, both hydrogen atoms have the electron configuration of the noble gas helium.

Lewis extended this idea to a wide variety of nonmetal atoms. In doing so he developed a symbolism that we still use today. This involves what are known as Lewis structures. In this section, we will first examine the Lewis structures of individual atoms. Then we will see how to draw Lewis structures for molecules in which nonmetal atoms are joined by covalent bonds.

Lewis Structures of Atoms

As we have noted already, it is the electrons in the outermost principal energy level that are important so far as the chemical properties of atoms are concerned. These are referred to as **valence electrons.** In drawing the Lewis structure of an atom, we show the valence electrons as dots distributed around the symbol of the atom.

Table 8.2 shows the Lewis structures of the atoms in the second period of the Periodic Table. Notice that

1. Only the valence electrons appear in the Lewis structure. For elements in this period, there are in each case two 1s electrons. These inner electrons are not shown.

Chapter 8

TABLE 8.2 Lewis Structures of the Atoms of the Second Period

Element	Group	Electron Configuration	Number of Valence Electrons	Lewis Structure
Li	1	$1s^2 2s^1$	1	Li·
Be	2	$1s^2 2s^2$	2	·Be·
B	3	$1s^2 2s^2 2p^1$	3	·Ḃ·
C	4	$1s^2 2s^2 2p^2$	4	·C̈·
N	5	$1s^2 2s^2 2p^3$	5	·N̈·
O	6	$1s^2 2s^2 2p^4$	6	·Ö·
F	7	$1s^2 2s^2 2p^5$	7	:F̈·
Ne	8	$1s^2 2s^2 2p^6$	8	:N̈e:

2. In the Lewis structure, we do not distinguish between s and p electrons. The number of valence electrons is the sum of the outer s and p electrons.
3. The valence electrons are arranged in a symmetrical way about the symbol of the atom (above, below, left, right).
4. **The number of valence electrons is equal to the group number in the Periodic Table.** This general principle applies to all the main-group elements beyond the first period. (In the first period, hydrogen has one valence electron; helium has two.)

EXAMPLE 8.3

Write Lewis structures for atoms of
a. Cl   b. S   c. Ar

Solution

a. Chlorine is in Group 7, so must have seven valence electrons. Its Lewis structure is similar to that of fluorine

:C̈l·

b. Sulfur is in Group 6, with six valence electrons

·S̈·

c. Argon, like all the noble gases except helium, has eight valence electrons.

:Är:

Exercise
How many valence electrons do the following atoms have: P, Si, Br? Answer: 5; 4; 7.

Lewis Structures of Molecules and the Octet Rule

The Lewis structures of atoms are readily combined to give Lewis structures for simple molecules. To illustrate the process involved, consider the hydrogen fluoride molecule, HF. The "starting materials" are a hydrogen atom (one valence electron) and a fluorine atom (seven valence electrons). The "product" is the HF molecule, in which there are eight valence electrons.

$$H\cdot + \cdot \ddot{\underset{..}{F}}: \qquad H—\ddot{\underset{..}{F}}:$$

Notice that in the Lewis structure of HF

1. The covalent bond between the atoms is shown as a dash, representing a pair of electrons.
2. The three *unshared pairs* of electrons around the fluorine atom are shown as dots.

By a similar process, we arrive at the following Lewis structures for molecules of water (H_2O), ammonia (NH_3), and methane (CH_4).

$$\underset{H \qquad H}{\ddot{O}} \qquad \underset{H \quad | \quad H}{\underset{H}{\ddot{N}}} \qquad H—\underset{H}{\overset{H}{\underset{|}{\overset{|}{C}}}}—H$$

Notice that in each of these molecules, *all the atoms have noble gas electronic structures*. Each hydrogen atom is surrounded by two valence electrons, the number found in helium. The other nonmetal atom (F, O, N, C) is surrounded by eight valence electrons. That is

F in HF: $2e^-$ in the H—F bond + $6e^-$ unshared = $8\ e^-$

O in H_2O: $4e^-$ in the two H—O bonds + $4e^-$ unshared = $8\ e^-$

N in NH_3: $6e^-$ in the three H—N bonds + $2e^-$ unshared = $8\ e^-$

C in CH_4: $8e^-$ in the four H—C bonds

Note that, in counting electrons around any atom, we include all the shared and unshared pairs. The number of valence electrons we usually get in that way is eight, the same number found in neon and indeed in all the noble gases except helium.

The general rule that atoms, through covalent bond formation, tend to acquire noble-gas structures is referred to as the **octet rule.** In most stable molecules, all the atoms except hydrogen are surrounded by eight valence electrons, an octet. This rule forms the basis for a systematic approach to writing Lewis structures for molecules.

Writing Lewis Structures for Molecules

For very simple molecules such as those just discussed, Lewis structures can often be written by inspection. Usually, though, you will save time and avoid confusion by using the following rules.

1. **Count the number of valence electrons available.** To do this, simply take the sum of those supplied by each atom. Remember that

   —a H atom has one valence electron.
   —a Group 4 atom (C, Si, Ge) has four valence electrons.
   —a Group 5 atom (N, P, As, Sb) has five valence electrons.
   —a Group 6 atom (O, S, Se, Te) has six valence electrons.
   —a Group 7 atom (F, Cl, Br, I) has seven valence electrons

2. **Draw a "skeleton" structure for the molecule, joining atoms by single bonds.** In a few very simple molecules, such as H_2 and HF, only one skeleton is possible. Usually, though, more than one skeleton is possible. That is, the atoms can be arranged in more than one order. Ultimately, experimental evidence must be used to find which arrangement of atoms is correct. In this course, you will either be given the correct skeleton or general rules that should enable you to find it.

3. **From the total number of valence electrons in (1), subtract two for each single bond in the skeleton.** This tells you how many valence electrons are left to work with.

4. **Distribute the remaining valence electrons as unshared pairs about the atoms in the molecule.** Try to do this in such a way that you get eight electrons around each atom, except hydrogen, which should have two.

   To illustrate this process, let's apply it to the phosphorus trichloride molecule, PCl_3.

   Step 1. Note that phosphorus is in Group 5 and chlorine is in Group 7.
   no. of valence e$^-$ = 5 + 3(7) = 26

   Step 2. Several skeletons are possible. The correct one can be obtained by applying the general rule that in molecules such as this, *where there are two or more atoms of the same type* (three Cl atoms), *these are usually bonded to a central atom* (P) *rather than to each other.* This leads to the skeleton

   $$\begin{array}{c} P \\ \diagup \mid \diagdown \\ Cl \quad \mid \quad Cl \\ Cl \end{array}$$

   Step 3. In the skeleton, there are three single bonds, consuming six valence electrons

   no. of valence e$^-$ left = 26 − 6 = 20

Step 4. In the skeleton, the phosphorus atom is surrounded by six valence electrons. It needs two more. Each chlorine atom in the skeleton is surrounded by two valence electrons and needs six more.

valence electrons needed = 2 + 3(6) = 20

Fortunately, as we found in Step 3, this is precisely the number of valence electrons that we have left. We distribute the 20 valence electrons as unshared pairs, giving a final Lewis structure

$$\ddot{\text{P}}$$
$$:\!\ddot{\text{Cl}}\quad\quad\ddot{\text{Cl}}\!:$$
$$:\!\ddot{\text{Cl}}\!:$$

EXAMPLE 8.4
Draw Lewis structures for
a. CCl_4 b. CH_4O

Solution
In each case, we follow the four-step procedure outlined above

a. (1) Carbon is in Group 4, Cl in Group 7. Hence, no. of valence e⁻ in CCl_4 = 4 + 4(7) = 32
(2) In CCl_4, the four Cl atoms are bonded to a central C atom.

$$\begin{array}{c} \text{Cl} \\ | \\ \text{Cl}-\text{C}-\text{Cl} \\ | \\ \text{Cl} \end{array}$$

(3) The four bonds in the skeleton consume eight valence electrons
no. of valence e⁻ left = 32 − 8 = 24
(4) The C atom already has its octet. Each of the four Cl atoms needs six more electrons (three unshared pairs). We have just enough electrons left to satisfy the Cl atoms. The Lewis structure of the CCl_4 molecule is

$$\begin{array}{c} :\!\ddot{\text{Cl}}\!: \\ | \\ :\!\ddot{\text{Cl}}-\text{C}-\ddot{\text{Cl}}\!: \\ | \\ :\!\ddot{\text{Cl}}\!: \end{array}$$

b. (1) no. of valence e⁻ in CH_4O = 4 + 4(1) + 6 = 14
(2) There are several possible skeletons. To arrive at the correct one, note that

—a hydrogen atom can form only one covalent bond (It needs only two electrons to have a noble-gas structure.)
—a carbon atom virtually always forms four bonds, with no unshared pairs.

Continued

The only skeleton consistent with these rules is

$$\begin{array}{c} H \\ | \\ H-C-O-H \\ | \\ H \end{array}$$

(3) There are five bonds in the skeleton
no. of valence e⁻ left = 14 − 10 = 4
(4) There are just enough valence electrons left to complete the oxygen octet. The Lewis structure is

$$\begin{array}{c} H \\ | \\ H-C-\ddot{O}-H \\ | \\ H \end{array}$$

Exercise
What is the Lewis structure of SCl_2? Answer

$$:\ddot{C}l-\ddot{S}-\ddot{C}l:$$

Lewis Structures Involving Multiple Bonds

Sometimes, when you reach the last step in drawing the Lewis structure of a molecule, you find there are too few electrons to go around. That is, the number of valence electrons left is less than that needed to satisfy the octet rule. If this happens, you economize by making one or more electron pairs do "double duty." Convert an unshared pair on one atom to a shared pair (bond) joining two atoms. This way, the electron pair becomes part of the octet of both atoms and is counted twice. The overall effect is to convert a single bond to a double bond

unshared pair + single bond → double bond

The counting rules here are simple. If there are too few electrons to satisfy the octet rule with a completely single-bonded structure

1. *Form a double bond to "save" two electrons.*
2. *Form a triple bond (or two double bonds) to save four electrons.*

To illustrate this process, consider the SO_2 molecule, which has 18 valence electrons. With a single-bonded structure, the best you can do is to write

$$:\ddot{O}-\ddot{S}-\ddot{O}:$$

This consumes all 18 electrons, but leaves only six around the sulfur atom, two short of an octet. In effect, you are two electrons short of what is needed. To solve this problem, move an unshared pair on one of the oxygen atoms to a position between oxygen and sulfur

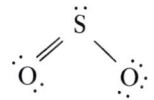

$$\ddot{O} \overset{\ddot{S}}{\underset{}{\diagup\!\!\diagdown}} \ddot{O}{:}$$

Now, without changing the total number of electrons, you have given each atom an octet. The structure just written is indeed the Lewis structure of the SO_2 molecule.

EXAMPLE 8.5
Write the Lewis structure of the N_2 molecule.

Solution
Since nitrogen is in Group 5

no. of valence e^- = 2(5) = 10

There is only one possible skeleton

N—N

Since the bond between the nitrogen atoms uses two electrons

no. of valence e^- left = 10 − 2 = 8

There is no way to give each atom an octet by distributing eight electrons as unshared pairs. The best you could do would be to write

$:\ddot{N}{-}\ddot{N}:$

Each nitrogen atom is surrounded by only six valence electrons. Each atom is two electrons short of an octet. There is a "deficiency" of four electrons. To remedy this, move an unshared pair from each nitrogen into the space between the atoms. This creates a triple bond, giving the final Lewis structure

$:N{\equiv}N:$

Exercise
What is the Lewis structure of formaldehyde, CH_2O? Answer

$$\text{H}-\underset{\underset{\text{H}}{|}}{\text{C}}=\ddot{\text{O}}$$

Exceptions to the Octet Rule

The octet rule is a very useful and quite general one. In drawing Lewis structures, you should always assume the octet rule applies unless specifically told otherwise. However, some molecules "violate" the octet rule. The only molecules of this type that need concern us here are those formed by the elements beryllium and boron. In BeF_2, the central beryllium atom is surrounded by only two electron pairs

$:\ddot{F}—Be—\ddot{F}:$

In the BF_3 molecule, the boron atom is surrounded by three electron pairs.

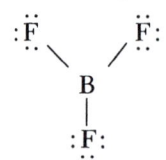

```
  :F̈           F̈:
    \         /
     \       /
      B
      |
     :F̈:
```

8.4 Molecular Geometry

Diatomic molecules such as Cl_2 or HF have a very simple geometry. These molecules are linear; the two atoms define a straight line. When a molecule contains more than two atoms, different geometries are possible. Consider a molecule XY_2, where two atoms of Y are bonded to a central atom X. If the two bonds are directed at an angle of 180° to each other, all three atoms will be on a straight line. Such a molecule is *linear*

Y—X—Y

Another, quite different geometry is possible. Suppose the angle between the two bonds is less than 180°. In that case, the three atoms will not be in a straight line. Instead, the molecule will be *bent*

```
       X
      / \
     Y   Y
```

Both geometries are known. Beryllium fluoride, BeF_2, is linear. Water, H_2O, is bent.

The geometries of simple molecules can be predicted using the principle: **The electron pairs around an atom in a molecule are directed so as to be as far apart as possible.**

In the rest of this section, we will apply this rule to predict the geometries of molecules in which a central atom is surrounded by two, three, or four pairs of electrons. These electron pairs may be single bonds or unshared pairs. (We will not consider here the geometries of molecules containing multiple bonds.)

Chemical Bonding

Two Electron Pairs

To orient two electron pairs so they will be as far apart as possible, we direct them at an angle of 180° to one another. This situation applies in BeF_2. As pointed out earlier, the Be atom in this molecule is surrounded by only two electron pairs, those used to form bonds with fluorine atoms. The geometry of the BeF_2 molecule is shown at the top of Figure 8.5. The three atoms are in a straight line; the molecule is **linear**.

Three Electron Pairs

It is not obvious how three electron pairs should be directed to get them as far apart as possible. However, a moment's thought should convince you that the way to do it is to direct them at 120° angles to each other. This means that the three pairs are directed from a point toward the corners of an equilateral triangle.

In BF_3, the central boron atom is surrounded by three pairs of elec-

Number of electron pairs	Orientation of electron pairs	Predicted bond angles	Examples
2	Straight line	180°	BeF_2 F——Be——F
3	Equilateral triangle	120°	BF_3
4	Tetrahedron	109.5°	CH_4

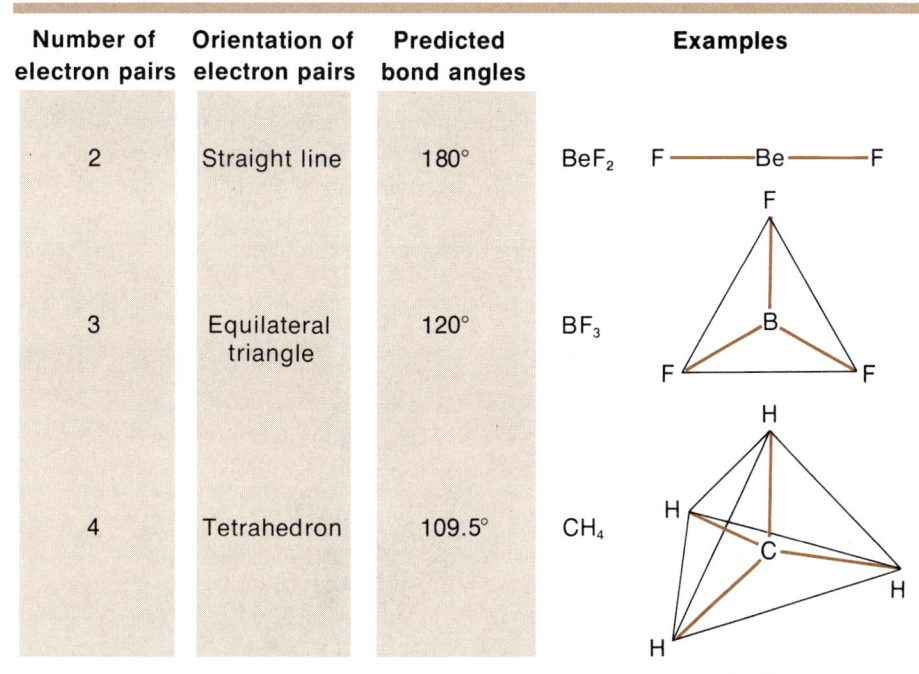

Figure 8.5 Geometry of electron pairs around a central atom (Be, B, C). The electron pairs orient themselves so as to be as far apart as possible. The most common situation involves four electron pairs directed toward the corners of a regular tetrahedron, as in CH_4.

Chapter 8

trons. These are the bonds to the three fluorine atoms. The geometry of the boron trifluoride molecule is shown in Figure 8.5. The boron atom is at the center of an **equilateral triangle** with the three fluorine atoms at the corners.

Four Electron Pairs

To get four electron pairs as far apart as possible, they should be directed toward the corners of a regular **tetrahedron**. This three-dimensional structure is outlined at the bottom of Figure 8.5. A tetrahedron has four sides and four corners; each side is an equilateral triangle. The interior angle of a regular tetrahedron is 109.5°. This means that the angle between any two of the four electron pairs is 109.5°.

As we have seen, most molecules obey the octet rule. This means that a central atom is ordinarily surrounded by four electron pairs. As a result, the tetrahedral arrangement of electron pairs is very common. We will now look at the geometries of three simple molecules of this type: CH_4, NH_3, and H_2O.

CH_4 In methane, CH_4, the carbon atom is located at the center of a tetrahedron with hydrogen atoms at each corner. We describe the CH_4 molecule as **tetrahedral**. Figure 8.6 shows "ball-and-stick" and space-filling models of the CH_4 molecule.

NH_3 In ammonia, NH_3, the central nitrogen atom is surrounded by four pairs of electrons. As in CH_4, these electron pairs are directed toward the corners of a regular tetrahedron. The nitrogen atom is located at the center. Three of the four corners of the tetrahedron are occupied by hydrogen atoms. There is an unshared pair of electrons at the fourth corner. This structure is shown at the left of Figure 8.7.

Ball and stick   Space filling

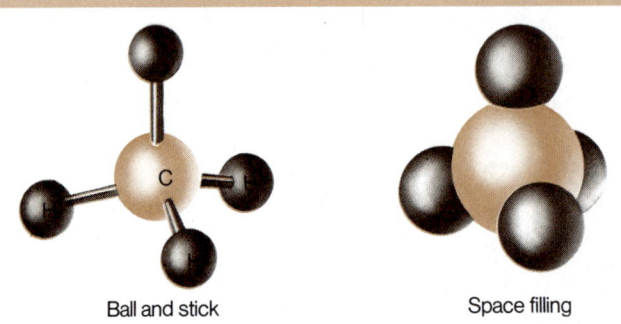

Figure 8.6 The CH_4 molecule. In methane, the carbon atom is at the center of a tetrahedron and the H atoms are at the corners. The molecule is tetrahedral and has a great deal of symmetry. Its geometry follows from the fact that the four pairs of electrons around the C atom are directed toward the corners of a tetrahedron.

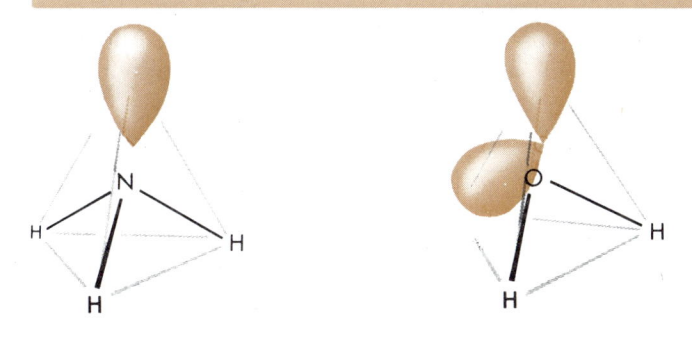

Ammonia   Water

Figure 8.7 In the NH₃ molecule *(left)* the four electron pairs around the N atom are directed toward the corners of a tetrahedron. Three of these pairs are involved in N—H bonds. The fourth pair is unshared. In H₂O *(right)*, two of the electron pairs around the O atom form O—H bonds, and two pairs are unshared.

In describing the geometry of the NH₃ molecule, or any other molecule, we *consider only the positions of the bonded atoms, not unshared electron pairs*. Note from Figure 8.7, or perhaps more obviously from Figure 8.8, that the nitrogen atom in NH₃ is located above the center of the equilateral triangle formed by the three hydrogen atoms. The four atoms form a pyramid. The nitrogen atom is at the top while the hydrogen atoms form the base of the pyramid. We would say that the NH₃ molecule is **pyramidal.**

H₂O   In the water molecule, H₂O, there are two bonds and two unshared pairs of electrons around the central oxygen atom. The four electron pairs, as in CH₄ and NH₃, are directed toward the corners of a regular tetra-

The ammonia molecule

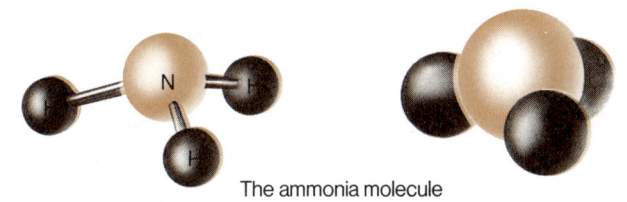

Figure 8.8 In describing the geometry of a molecule, we indicate the positions of the atoms and *not* of the electron pairs. In the ammonia molecule, the N atom lies above an equilateral triangle formed by the three H atoms, forming a pyramid. The molecule is a trigonal pyramid, or pyramidal in shape.

hedron (Figure 8.7, right). The geometry of the molecule is shown in Figure 8.9. We describe the geometry of the H$_2$O molecule as being **bent.** The three atoms are not in a straight line. Instead, the two bonds are directed at an angle slightly less than 109°.

EXAMPLE 8.6
Consider the PCl$_3$ molecule, whose Lewis structure is shown on p. 211. Describe its geometry.

Solution
The PCl$_3$ molecule resembles NH$_3$ in that the central atom, P, is surrounded by three bonded atoms (Cl atoms) and an unshared pair of electrons. The four electron pairs are directed toward the corners of a regular tetrahedron. However, we would describe the molecule as pyramidal. The phosphorus atom is at the top of the pyramid with the three chlorine atoms forming the base.

Exercise
What is the geometry of SCl$_2$, whose Lewis structure was shown in Example 8.4?
Answer: Bent.

We should emphasize that, **to predict the geometry of a molecule, you must know its Lewis structure.** The SCl$_2$ molecule is expected to be bent because there are two unshared pairs around the sulfur atom as well as two bonds. In BeF$_2$, where there are no unshared pairs on the Be atom, the molecule is linear. Again, even though BF$_3$ and PCl$_3$ have similar formulas, they have different geometries (equilateral triangle versus pyramid). These geometries can be predicted only if you realize that there are no unshared pairs on the central atom in BF$_3$ and one unshared pair in the case of PCl$_3$.

8.5 Polarity in Molecules

An important property of a molecule is its polarity, which tells us whether there are + and − poles present. Polarity depends upon the way elec-

Figure 8.9 In the water molecule the H—O—H bond angle is near 109°. The H$_2$O molecule is said to be bent.

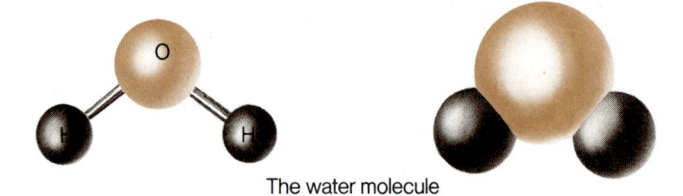

The water molecule

trons are distributed in the molecule. On this basis, we can distinguish between

1. **Nonpolar molecules,** in which the bonding electrons are distributed symmetrically through the molecule. There are no centers of negative and positive charge in such a molecule.
2. **Polar molecules,** in which the distribution of bonding electrons is unsymmetrical. There is a negative pole at one point in the molecule where the electron density is high. At another point, where the electron density is low, there is a positive pole. A polar molecule acts as a "dipole;" that is, it shows + and − poles.

Polar and nonpolar molecules behave quite differently in an electric field (Figure 8.10). Polar molecules line up in the field, with their positive poles directed toward the negative electrode. In contrast, nonpolar molecules are not affected by the field. They are arranged in a totally random way, whether the field is on or off.

Diatomic Molecules

We can readily decide whether a molecule containing only two atoms is nonpolar or polar. If the two atoms are the same as in H_2 and F_2

H—H   F—F

the molecule must be nonpolar. Since the atoms are identical, the bonding electrons must be equally shared between them. Hence, there can be no positive or negative poles in the molecule.

The situation changes when the two atoms differ, as in HF

H—F

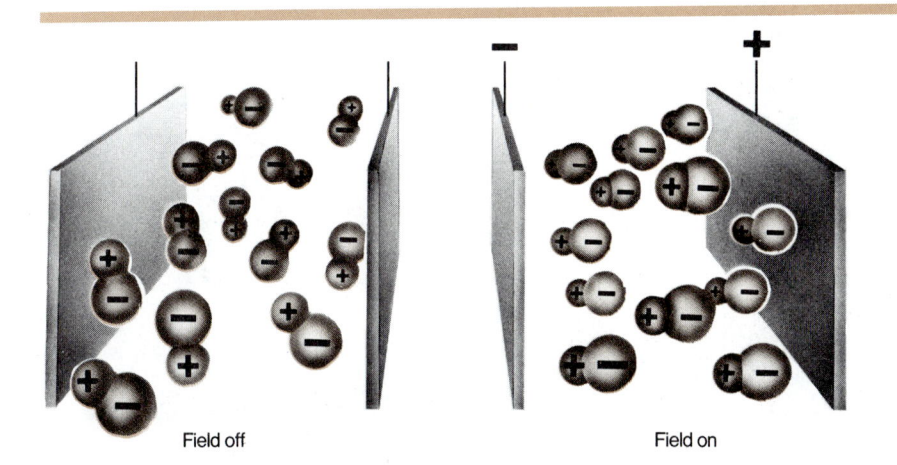

Figure 8.10 If polar molecules are subjected to an electric field, they tend to line up in a direction opposite to that of the field. This minimizes the electrostatic energy of the molecules. Nonpolar molecules are not oriented by an electric field.

Field off

Field on

You will recall (Chapter 7) that hydrogen and fluorine differ in electronegativity (2.1 for H, 4.0 for F). As a result, the bonding electrons are shifted toward the fluorine atom (Figure 8.11). This creates a negative pole at the F atom, where the electron density is relatively high. There is a positive pole at the H atom, where the electron density is relatively low.

EXAMPLE 8.7
Classify the following molecules as polar or nonpolar
a. Cl_2 b. ICl

Solution
a. nonpolar; the two atoms are identical.
b. polar; since chlorine is more electronegative than iodine (3.0 versus 2.5), the bonding electrons are shifted slightly toward chlorine. The Cl atom carries a partial negative charge, the iodine atom a partial positive charge.

Exercise
Give the formulas of two diatomic molecules containing oxygen, one of which is nonpolar, the other polar. Answer: O_2; NO, CO, – –.

All diatomic molecules where the two atoms differ are polar. The extent of polarity depends upon the difference in electronegativity between the two bonded atoms. The smaller this difference, the less strongly polar is the molecule. In HI (E.N. H = 2.1, I = 2.5), the difference in electronegativity is so small that the molecule is only slightly polar. In HF (E.N. H = 2.1, F = 4.0), polarity is much more pronounced.

Figure 8.11 In the HF molecule the F atom (shown at the right) is more electronegative than H. It will tend to attract electrons and so take on a negative charge. This leaves a positive charge on the H atom and makes the HF molecule polar. Any diatomic molecule containing different kinds of atoms will be polar.

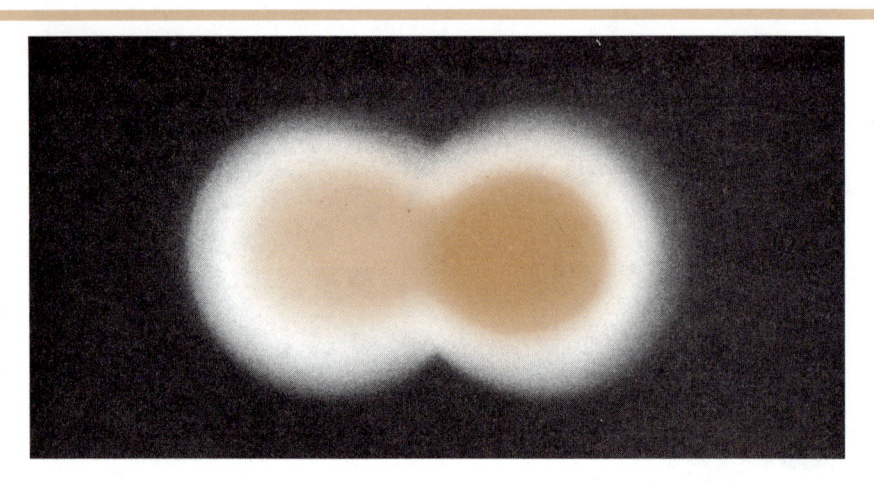

Chemical Bonding

Expanding upon this idea, we might think of a polar covalent bond such as that in HI or HF as being intermediate between two extremes

1. A nonpolar covalent bond in which the bonding electrons are equally shared between the two atoms.
2. An ionic bond in which there has been a complete transfer of electrons from the less electronegative to the more electronegative atom

This concept is shown schematically in Figure 8.12. The δ^+ and δ^- signs represent partial charges associated with a polar covalent bond. The + and − signs represent ionic charges such as Na^+ and Cl^-. In the nonpolar molecule at the bottom of the figure, there are no charges, partial or otherwise.

Polyatomic Molecules

When a molecule contains more than two atoms, we must consider its geometry to decide whether it is polar. Consider, as a simple example, a molecule AX_2. Suppose the central atom, A, is more electronegative than X. Two geometries are possible, bent and linear.

If the molecule is bent, AX_2 will be polar. A negative pole is located at the central atom, A. The positive pole is midway between the two X atoms.

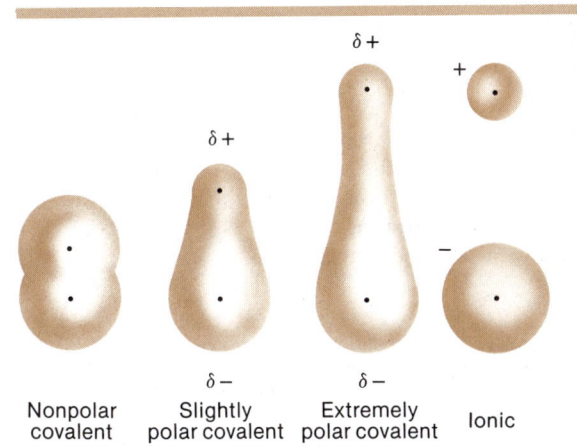

Figure 8.12 Models illustrating structural difference, proceeding from ionic structure to the nonpolar covalent molecule.

A molecule of this type will line up in the electrical field shown in Figure 8.10. The A atom will be oriented toward the + electrode, the X atoms toward the − electrode.

The situation is quite different if AX_2 is linear. Even though the bonds are polar, the molecule itself is nonpolar. The two polar bonds are in exactly opposite directions, at a 180° angle to each other. In effect, they cancel each other. There is no way a linear AX_2 molecule can line up in an electric field.

Both types of triatomic molecules are known. The most common example of a bent molecule is the one we discussed earlier, H_2O.

$$\underset{H \qquad H}{\overset{O}{\diagup\diagdown}}$$

The water molecule is polar. There is a negative pole at the oxygen atom (E.N. = 3.5) and a positive pole midway between the two hydrogen atoms (E.N. = 2.1). In contrast, the linear BeF_2 molecule is nonpolar.

$$\underset{180°}{F\!-\!Be\!-\!F}$$

The two bonds are polar (E.N. Be = 1.5, F = 4.0). However, since they are oriented at a 180° angle, they cancel each other. Putting it another way, in the BeF_2 molecule, the center of positive charge (Be atom) coincides with the center of negative charge (midway between the two F atoms). Anyway you look at it, BeF_2 should be, and is, nonpolar.

The argument we have used for triatomic molecules can be extended to more complex molecules. The general principle is a simple one: **A completely symmetrical molecule is nonpolar, even if the bonds are polar.** This is illustrated in Example 8.8.

EXAMPLE 8.8
Is the CH_4 molecule polar? the NH_3 molecule?

Solution
The CH_4 molecule is symmetrical (recall Figure 8.6). The four polar bonds cancel one another and the molecule is nonpolar. On the other hand, the NH_3 molecule is unsymmetrical. The highly electronegative nitrogen atom (E.N.=3.0) is sitting above the triangle formed by the three hydrogen atoms (E.N.=2.1). The NH_3 molecule is polar, with the negative pole at the nitrogen atom.

Exercise
Is the BF_3 molecule polar or nonpolar? Answer: Nonpolar.

Chemical Bonding

Key Words

bent molecule
covalent bond
double bond
electron configuration
electron pair repulsion
electronegativity
equilateral triangle molecule
ionic bond
Lewis structure
linear molecule
macromolecule
noble gas
nonpolar
octet rule
polar
polyatomic ion
pyramidal molecule
single bond
tetrahedral molecule
triple bond
unshared pair
valence electron

Questions

8.1 How does an ionic bond differ from a covalent bond?

8.2 Why do two nonmetals react to form a covalent rather than an ionic compound?

8.3 A monatomic ion differs from the corresponding atom only in the number of electrons in the _____ level.

8.4 What is the principal source of the energy evolved when an ionic compound is formed from the elements?

8.5 What is meant by the phrase "noble gas electron configuration"?

8.6 Explain why the formation of a covalent bond is an exothermic process.

8.7 How does a triple bond differ from a double bond? from a single bond?

8.8 What kind of bond joins N to H atoms in NH_4Cl? What kind of bond joins NH_4^+ to Cl^-?

8.9 What is meant by a "macromolecular" substance?

8.10 How do you determine the number of valence electrons in an atom of a main-group element?

8.11 State the octet rule. Does hydrogen follow this rule?

8.12 What is meant by the "skeleton structure" of a molecule?

8.13 How many covalent bonds does a hydrogen atom form? a carbon atom?

8.14 What do you do when you find there are too few electrons to "go around" in a Lewis structure?

8.15 Give the formulas of two molecules in which the central atom does not have a noble gas electron configuration.

8.16 To predict the geometry of a molecule, it is necessary to know how the electron pairs are oriented. What principle do you use to predict this orientation?

8.17 Two electron pairs around a central atom are directed at a _____ angle to each other.

8.18 Three electron pairs around an atom are directed towards the corners of a _____.

8.19 Four electron pairs around an atom are directed toward the corners of a _____.

8.20 What is meant by a "polar molecule"; a "nonpolar molecule"?

Problems

(You may use the Periodic Table in working out these problems or the multiple choice questions that follow.)

Electron Configurations of Ions

8.21 Give the electron configuration of
  a. Na^+ b. Ca^{2+} c. Rb^+

8.22 Give the electron configuration of
  a. S^{2-} b. Br^- c. N^{3-}

8.23 Give the formulas of four monatomic ions that have the electron configuration of argon.

8.24 Give the electron configuration, beyond the $_{18}Ar$ core, of
  a. Cr^{2+} b. Cr^{3+} c. Co^{2+} d. Co^{3+}

8.25 Give the electron configuration, beyond the $_{18}Ar$ core, of
  a. Fe^{3+} b. Ni^{2+} c. Cu^+ d. Zn^{2+}

8.26 Give the electron configuration, beyond the $_{36}Kr$ core, of
  a. Zr^{2+} b. Tc^{3+} c. Pd^{4+}

8.27 Which of the following have noble gas electron configurations?
  a. Sc^{3+} b. Ti^{2+} c. Ti^{4+} d. V^{3+} e. Cr^{3+}

8.28 List the noble gas that has the same electron configuration as
  a. K^+ b. I^- c. Mg^{2+} d. S^{2-} e. H^-

Lewis Structures

8.29 Give the Lewis structures of the following atoms
  a. Si   b. S   c. Cl   d. H

8.30 Give the Lewis structures of the following atoms
  a. P   b. Ar   c. O   d. F

8.31 How many valence electrons are there in each of the following molecules?
  a. CH_3Cl b. SCl_2 c. AsH_3 d. $SiCl_4$

8.32 Draw Lewis structures for each of the molecules in Problem 8.31.

8.33 Draw Lewis structures for
  a. CCl_2F_2 b. I_2 c. NH_3 d. NCl_3

8.34 Draw Lewis structures for
  a. N_2 b. CS_2 c. SeO_2

8.35 Draw Lewis structures for
  a. PCl_3 b. CO c. SO_3

8.36 None of the following molecules follows the octet rule; in each case there are less than eight electrons around one atom. Draw the Lewis structures
  a. BeF_2 b. BF_3 c. NO

8.37 The Lewis structure of a polyatomic anion is drawn in much the same way as for a molecule. The only difference is that, in counting valence electrons, you add the charge of the anion (1, 2, or 3) to the total contributed by the several atoms. State the total number of valence electrons in
  a. OH^- b. ClO_2^- c. CO_3^{2-} d. NO_3^-

8.38 Draw the Lewis structures of the polyatomic anions in Problem 8.37.

Geometry of Molecules

8.39 Describe the geometry of each of the following molecules
  a. Cl_2 b. BeF_2 c. BF_3

8.40 Describe the geometry of each of the following molecules
  a. CH_4 b. NH_3 c. H_2O d. HF

8.41 Describe the geometry of each of the molecules in Problem 8.31.

8.42 Describe the geometries of
  a. NCl_3 b. CH_2Cl_2 c. Cl_2O

Polarity of Molecules

8.43 Classify each of the following molecules as polar or nonpolar
  a. H_2 b. CO   c. N_2 d. HF

8.44 Which of the following molecules are polar?
 a. CH$_4$ b. NH$_3$ c. H$_2$O
 d. CH$_3$Cl
8.45 Which of the molecules in Problem 8.42 are polar?

8.46 Which of the following molecules are nonpolar?
 a. CCl$_4$ b. BeCl$_2$ c. BCl$_3$
 d. PCl$_3$

Multiple Choice

8.47 How many of the following species contain 4s electrons?

Sc, Sc^{3+}, Cr, Cr^{2+}, Cr^{3+}

 a. 1   b. 2   c. 3   d. 4   e. 5

8.48 Of the species in Question 8.47, how many contain 3d electrons?
 a. 1   b. 2   c. 3   d. 4   e. 5

8.49 In which one of the following compounds do both ions have the same electron configuration?
 a. NaCl   b. CaBr$_2$ c. RbBr
 d. NaH   e. K$_3$N

8.50 The bond between N and H in NH$_3$ could be described as
 a. an ionic bond
 b. a nonpolar covalent bond
 c. a polar covalent bond
 d. a double bond

8.51 The bond in the N$_2$ molecule is
 a. an ionic bond
 b. a polar covalent bond
 c. a single bond
 d. none of these

8.52 The number of valence electrons in the CO$_2$ molecule is
 a. 12   b. 16   c. 18   d. 44
 e. some other number

8.53 Of the following molecules, how many are polar?

CH$_4$, CH$_3$Cl, CH$_2$Cl$_2$, CHCl$_3$, CCl$_4$

 a. 1   b. 2   c. 3   d. 4   e. 5

8.54 In how many of the following molecules are all the atoms in the same plane?

HF, BeF$_2$, BF$_3$, CF$_4$

 a. 0   b. 1   c. 2   d. 3   e. 4

8.55 The angle between the bonds in a tetrahedral molecule is
 a. 90°   b. 109.5°   c. 120°   d. 180°
 e. 360°

8.56 How many of the following compounds would you expect to be ionic?

HF, NaF, CaF$_2$, PF$_3$, SF$_6$

 a. 1   b. 2   c. 3   d. 4   e. 5

8.57 How many of the following molecules are nonpolar?

N$_2$, O$_2$, NO, IBr, Br$_2$

 a. 1   b. 2   c. 3   d. 4   e. 5

8.58 In a molecule XY$_2$ with the Lewis structure

$:\ddot{Y}-\ddot{X}-\ddot{Y}:$

the angle between the bonds is expected to be
 a. 90°   b. 109.5°   c. 120°   d. 180°
 e. 360°

8.59 A reasonable Lewis structure for the COS molecule is
 a. $:\ddot{S}-\ddot{C}-\ddot{O}:$
 b. $:\ddot{S}=\ddot{C}-\ddot{O}:$
 c. $:\ddot{C}=\ddot{S}-\ddot{O}:$
 d. none of these

8.60 The total number of valence electrons in the acetic acid molecule, CH$_3$COOH, is
 a. 20   b. 22   c. 24   d. 26
 e. some other number

Solutions

Learning Objectives
After studying this chapter, you should be able to:

1. Distinguish between electrolytes and nonelectrolytes.
2. Distinguish between saturated, unsaturated, and supersaturated solutions.
3. Discuss the effect of temperature upon the solubilities of solids and gases; of pressure upon the solubilities of gases.
4. Apply the solubility rules (Table 9.2) to predict whether or not a precipitation reaction will occur when two solutions of ionic compounds are mixed (Example 9.1).
5. Write net ionic equations to describe precipitation reactions (Examples 9.2, 9.3).
6. Using Equation 9.7, calculate one of the three quantities: molarity, moles solute, liters solution, knowing the values of the other two quantities (Examples 9.4, 9.5).
7. Determine the volume of a solution required to prepare a given (larger) volume of a more dilute solution (Example 9.6).
8. Using Equation 9.10, calculate one of the three quantities: mass %, mass solute, mass solution, knowing the values of the other two quantities (Example 9.7).
9. Using Equation 9.11, calculate one of the three quantities: molality, moles solute, kilograms solvent, knowing the values of the other two quantities (Example 9.8).
10. Using Equations 9.14 and 9.15
    —calculate freezing points and boiling points of aqueous solutions of nonelectrolytes (Example 9.9).
    —determine the molar mass of a nonelectrolyte (Example 9.10).

> Every good laboratory consists of first-rate men working in great harmony to insure the progress of science; but down at the end of the hall is an unsociable, wrong-headed fellow working on unprofitable lines, and in his hands lies the hope of discovery.
>
> LORD RUTHERFORD

Photograph by Peter L. Kresan.

CHAPTER 9

In Chapter 2, we defined a solution as a homogeneous mixture of two or more substances. Typically, a solution is prepared by dissolving one substance called a **solute** in another substance called the **solvent.** Throughout this chapter, we will emphasize aqueous solutions, i.e., solutions in which the solvent is water. We will consider a variety of solutes. These may be gases (air, hydrogen chloride), liquids (methyl alcohol), or solids (sugar, sodium chloride).

Our main interest in this chapter will be in the structure and physical properties of water solutions. We will seek answers to the following general questions:

What solute particles (ions or molecules) are present in water solutions of such compounds as sodium chloride or methyl alcohol (Section 9.1)?

What factors affect the extent to which a particular solute dissolves in water? How do changes in temperature and pressure affect solubility (Section 9.2)?

How do we express the relative amounts of solute and solvent present in a solution? What units are used for solute concentrations (Section 9.4)?

How do the freezing point and boiling point of a solution compare with those of the pure solvent (Section 9.5)?

We will also consider what is perhaps the simplest type of chemical reaction that can occur in water solution. This is a precipitation reaction (Section 9.3), in which ions in solution combine to form an insoluble solid. We will discuss how to

—predict whether such a reaction will occur when two solutions are mixed.
—represent such reactions by chemical equations.

9.1 Solutes in Water

Perhaps the most useful way to classify solutes in water relates to the type of solute particle present. On this basis, we can distinguish between

1. **Nonelectrolytes,** which dissolve as molecules. Solutions of nonelectrolytes do not carry an electric current (Figure 9.1); molecules are not charged and so do not move in an electric field. One of the simplest nonelectrolytes is methyl alcohol, CH_3OH. The equation for the process by which it dissolves in water can be written

   $$CH_3OH(l) \rightarrow CH_3OH(aq) \tag{9.1}$$

2. **Electrolytes,** which go into solution as ions. Charged ions can move in an electric field, so solutions of electrolytes conduct an electric current.

   Compounds that are ionic in the solid state act as electrolytes in water. A simple example is sodium chloride, NaCl. As pointed out in Chapter 3, the solid consists of Na^+ and Cl^- ions. When sodium chloride dissolves in water, these ions separate. This process is shown graphically in Figure 9.2, p. 230. Solution occurs because of the attraction between polar water molecules on the one hand and Na^+ or Cl^- ions on the other. The solution process can be represented by the equation

   $$NaCl(s) \rightarrow Na^+(aq) + Cl^-(aq) \tag{9.2}$$

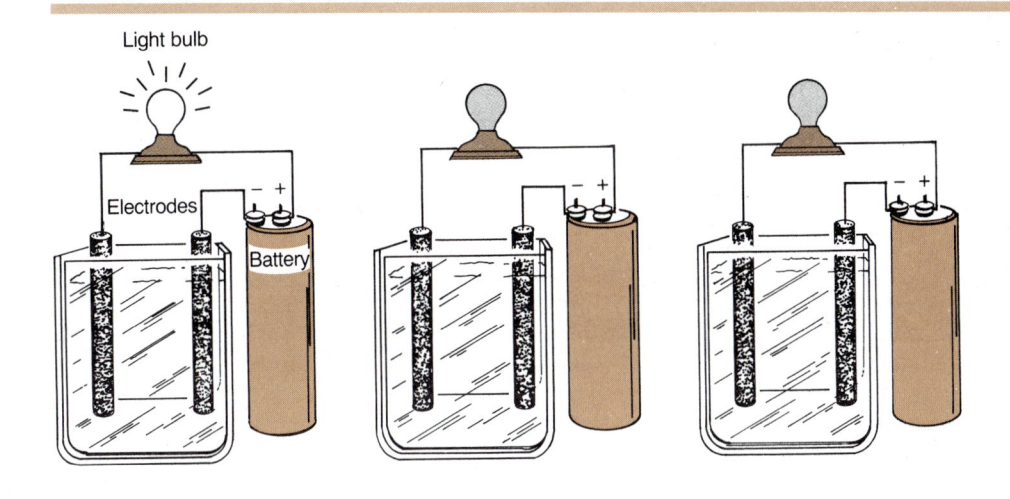

a. Solution of table salt
(an electrolytic solution)

b. Solution of methyl alcohol
(a nonelectrolytic solution)

c. Pure water
(a nonelectrolyte)

Figure 9.1 An apparatus for testing electrical conductivity. For an electric current to flow, the solution must contain ions, which serve as carriers of electric charge. Pure water contains very few ions, so it is not a conductor. If NaCl is dissolved in water, it forms ions. These conduct the current and the light bulb glows. Dissolving methyl alcohol in water produces only CH_3OH molecules in the solution, so it does not become a conductor.

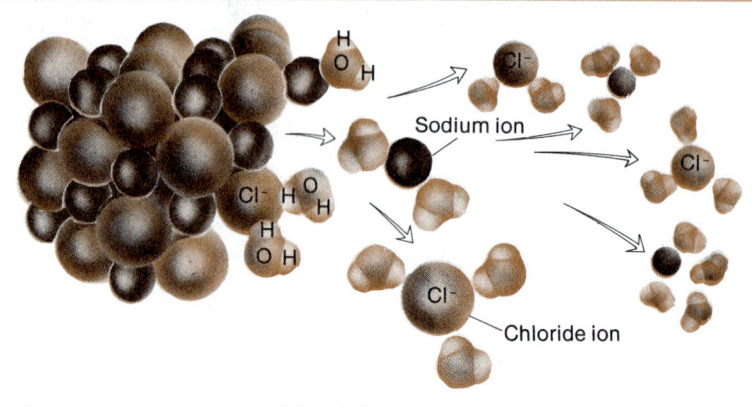

Figure 9.2 Sodium chloride, NaCl, dissolves readily in the polar solvent water. Chloride ions, Cl⁻, are attracted to the hydrogen atoms of the H$_2$O molecule, which carry a partial positive charge. At the same time, Na$^+$ ions are attracted to the oxygen atom of the H$_2$O molecule, which carries a partial negative charge.

Similar equations can be written for the dissolving of other ionic compounds

$$CaCl_2(s) \rightarrow Ca^{2+}(aq) + 2\ Cl^-(aq) \tag{9.3}$$

$$KNO_3(s) \rightarrow K^+(aq) + NO_3^-(aq) \tag{9.4}$$

9.2 Factors Affecting Solubility

The solubility of a substance in water can be expressed in many different ways. We may report the number of grams of solute per hundred grams of water (Table 9.1). Another approach is to quote the number of moles of solute per liter of solution. Solubilities such as those listed in Table 9.1 tell us how much solute there is in a *saturated* solution. A saturated solution is one that

TABLE 9.1 Solubilities of Compounds in Water

	Solubility at 25°C, 1 atm	
	grams solute/100 g water	moles solute/liter solution
CaCO$_3$(s)	0.0007	0.00007
CH$_4$(g)	0.0022	0.0014
C$_6$H$_6$(l)	0.13	0.017
PbCl$_2$(s)	0.44	0.016
NaCl(s)	36	5.4
NaOH(s)	79	16
CH$_3$OH(l)	∞	∞

is prepared by stirring or shaking with an excess of solute (Figure 9.3). At 25°C and 1 atm, a saturated solution of sodium chloride contains 36 g of NaCl per 100 g of water.

A solution that contains less solute than there is in the saturated solution is said to be *unsaturated*. A solution containing 30 g NaCl/100 g water at 25°C and 1 atm is unsaturated. Under certain conditions, it is possible to prepare a *supersaturated* solution, containing more solute than the saturated solution. A solution containing 40 g NaCl/100 g water at 25°C and 1 atm would be supersaturated.

Solubility in water depends strongly upon the nature of the solute. Many ionic solutes, including NaCl and NaOH (Table 9.1) are very soluble in water. A few ionic solutes are quite insoluble in water (note $CaCO_3$ in Table 9.1). We will have more to say on this topic in Section 9.3. Most molecular solutes, including methane, CH_4, and benzene, C_6H_6, have very low water solubilities (Table 9.1). Gasoline, a blend of molecular hydrocarbons, does not "mix" with water. That is, it does not dissolve appreciably in water.

A few molecular solutes, including methyl alcohol, CH_3OH, are soluble in water in all proportions. For this to happen, there must be a strong attraction between the solute molecule and an H_2O molecule. The intermolecular attractive force involved is called a *hydrogen bond*. Hydrogen bonding can occur with any molecule in which a hydrogen atom is bonded to oxygen,

(A) 50 g NaCl(s) + 100 g water
No stirring

(B) 30 g NaCl(s) + solution
Stirring 5 minutes

(C) 14 g NaCl(s) + solution
Stirring 10 minutes

(D) 14 g NaCl(s) + solution
Stirring 1 hour

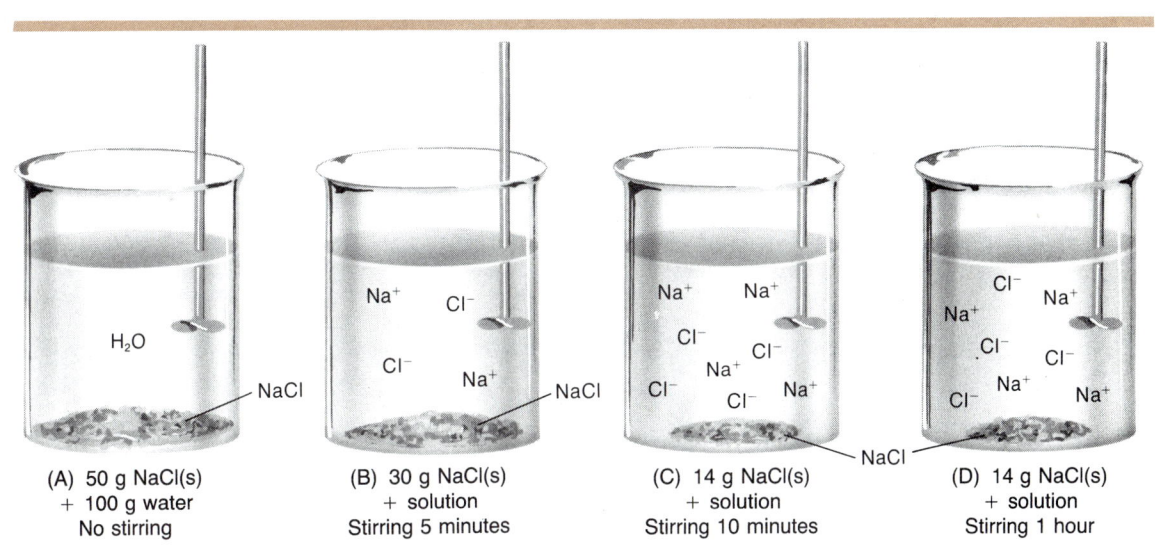

Figure 9.3 Formation of a saturated solution of NaCl. When NaCl is added to water, most of the solid goes to the bottom. On stirring, the solid slowly dissolves. After a period of time, the amount of solid in solution becomes constant, and the solution is said to be saturated.

nitrogen, or fluorine. The hydrogen bonding between CH_3OH and H_2O molecules is shown in Figure 9.4. Other molecular solutes that are very soluble in water because of hydrogen bonding include ammonia, NH_3, and hydrogen fluoride, HF.

For a given solute, water solubility depends upon two external factors. These are the temperature and pressure at which the solution is formed.

Temperature

The solubility of a solid usually increases with temperature (Figure 9.5). The magnitude of the increase varies greatly from one solid to another. The water solubility of potassium nitrate, KNO_3, at 100°C is about six times that at 25°C. In contrast, the solubility of sodium chloride, NaCl, increases only slightly with temperature (36 g/100 g water at 25°C, 40 g/100 g water at 100°C).

When a hot, saturated solution of a solid is cooled, we ordinarily find that some of the solid crystallizes out. Consider, for example, a solution prepared by saturating 100 g of water at 100°C with potassium nitrate, KNO_3. From Figure 9.5, we see that this solution should contain 240 g of KNO_3 (solubility at 100°C: 240 g KNO_3/100 g water). Suppose now that this solution is cooled to 25°C. Here, the solubility of potassium nitrate is much lower, only 40 g/100 g of water. At this point, we expect

240 g − 40 g = 200 g

of KNO_3 crystals to separate from solution.

The process just described is typical of what ordinarily happens when a hot solution of a solid is cooled. Excess solute crystallizes out, leaving a saturated solution. Sometimes, though, this doesn't happen. When the solution is cooled, no solid appears. Instead, the solution becomes supersaturated with solid. That is, there is more solid present than you would predict on the basis of its solubility. If a supersaturated solution is stirred or "seeded" with a so-

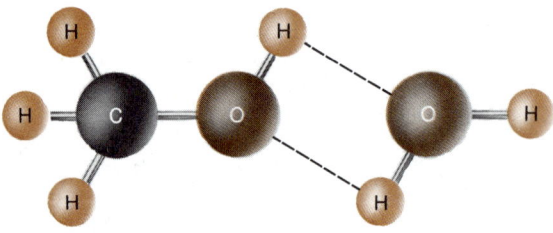

Figure 9.4 Hydrogen bonding in a solution containing CH_3OH molecules *(left)* and H_2O molecules *(right)*. The hydrogen bonds are shown as dotted lines. They bridge oxygen atoms in the two molecules. H_2O molecules also form hydrogen bonds with each other.

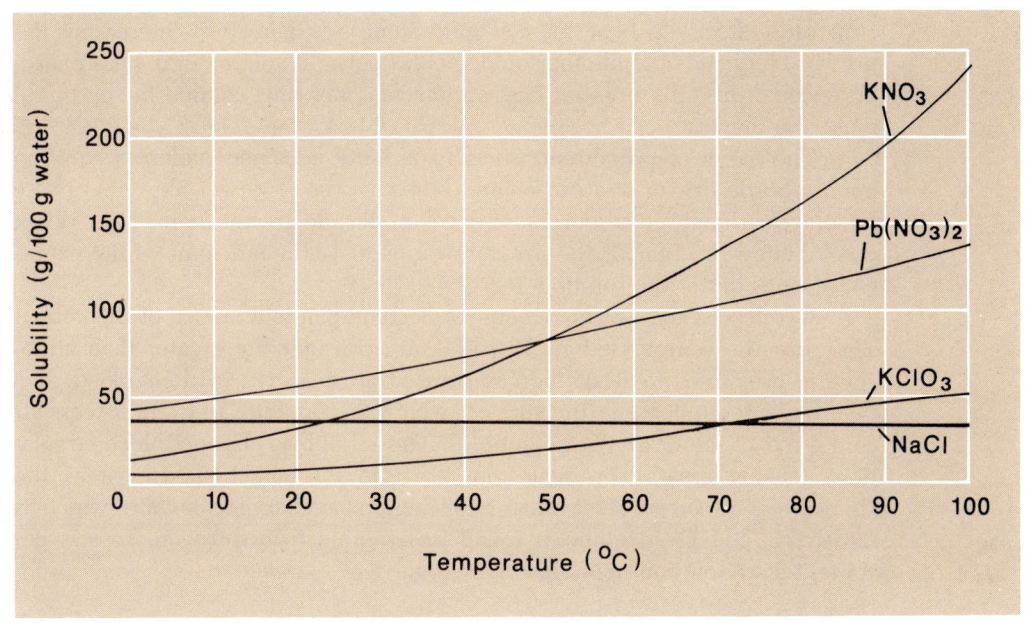

Figure 9.5 The water solubility of these four ionic compounds, and indeed that of most solids, increases with temperature.

lute crystal, the excess solid separates. The solution that remains at that point is saturated.

In contrast to solids, gases become less soluble in water when the temperature rises. You have probably observed this effect when heating water in a beaker or saucepan. Bubbles of air form as the temperature increases. This happens because the solubility of air in water decreases with temperature. The solubility at 100°C is only about half that at 25°C. The excess air comes out of solution as the water is heated.

An increase in water temperature can have an adverse effect on marine life. Fish obtain the oxygen they need from air dissolved in water. When the temperature rises, the amount of air in the water decreases. If the effect is too large, fish, particularly trout and salmon, may die. Increases in temperature come about when power plants or factories take in water from a stream or lake and use it in a "heat exchanger." When the hot water is discharged back into the stream or lake, the temperature of the body of water rises.

Pressure

Pressure affects solubility appreciably only with gaseous solutes. At a given temperature, the solubility of a gas is directly proportional to its pressure.

For example, compressed air at 5 atm is five times as soluble as is air at 1 atm. This effect can be explained quite simply (Figure 9.6). Increasing the pressure increases the number of molecules per unit volume in the gas phase. To balance this, the number of gas molecules per unit volume in the liquid must also increase.

The effect of pressure upon gas solubility becomes evident when you open a bottle or can of a carbonated beverage (beer, ginger ale, etc). These beverages are bottled at a pressure of cabon dioxide greater than 1 atm. When the container is opened, the pressure drops to 1 atm and some of the excess CO_2 comes out of the solution.

Careless scuba divers sometimes become painfully aware of this effect. They breathe compressed air at a pressure considerably greater than atmospheric pressure. Some of the compressed air dissolves in their blood. If a diver rises too quickly to the surface, where the pressure is 1 atm, excess air comes out of solution as large bubbles. These bubbles obstruct blood circulation, causing great pain. One way to avoid this affliction, known as the "bends," is to substitute a mixture of helium and oxygen for air, which is mostly N_2 and O_2. Helium is much less soluble than nitrogen, so less gas comes out of solution upon decompression.

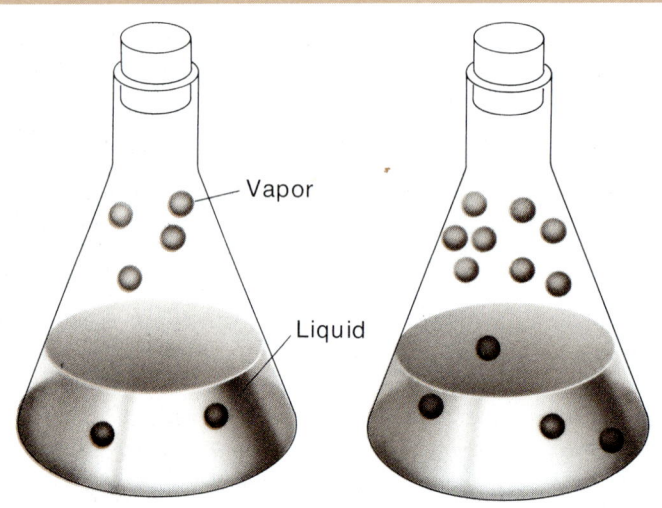

Figure 9.6 Effect of pressure on gas solubility. Doubling the pressure of the gas doubles the concentration in the vapor phase (eight molecules instead of four). To compensate for this, the concentration of gas in the liquid phase also doubles (four molecules instead of two). More generally, solubility is directly proportional to the pressure of the gas.

9.3 Solubility of Ionic Compounds; Precipitation Reactions

Ionic compounds vary greatly in their water solubility. Consider, for example, the four ionic solids listed in Table 9.1. We would call $CaCO_3$ (solubility = 7×10^{-5} mol/L) "insoluble." The two compounds NaCl (solubility = 5.4 mol/L) and NaOH (solubility = 16 mol/L) are clearly "soluble." Lead chloride, $PbCl_2$ (solubility = 0.016 mol/L) falls somewhere in between.

In practice, it is convenient to establish an arbitrary cut-off point for electrolyte solubility at about 0.1 mol/L. We will call any compound whose solubility at 25°C is greater than 0.1 mol/L "soluble." Compounds with solubilities less than 0.1 mol/L are "insoluble." On this basis, we arrive at the *solubility rules* given in Table 9.2.

The solubility rules are easy to interpret. From Table 9.2 it should be clear, for example, that

—$Cu(NO_3)_2$ must be soluble since all nitrates are soluble.
—$CuCl_2$ is soluble since it is not one of the chlorides listed as insoluble.
—CuS is insoluble since it is not one of the sulfides listed as soluble.

Precipitation Reactions

When water solutions of two different ionic compounds are mixed, it sometimes happens that an insoluble solid separates out of solution. This solid is referred to as a *precipitate*. It is formed when the positive ion of one solution combines with the negative ion of the other solution. The reaction that occurs is referred to as *precipitation*.

To illustrate a typical precipitation reaction, consider the experiment shown in Figure 9.7, p. 236. The first tube contains a water solution of $AgNO_3$, containing Ag^+ and NO_3^- ions. To this is added a water solution of NaCl (Na^+, Cl^- ions). As you can see from the second test tube, a solid forms when these two solutions are mixed. The precipitate forms at the surface, where

TABLE 9.2 Solubility Rules

NO_3^-	All nitrates are soluble.
Cl^-	All chlorides are soluble except AgCl, Hg_2Cl_2, and $PbCl_2$.
SO_4^{2-}	Most sulfates are soluble; exceptions include $SrSO_4$, $BaSO_4$, and $PbSO_4$.
CO_3^{2-}	All carbonates are insoluble except those of the Group 1 elements and NH_4^+.
OH^-	All hydroxides are insoluble except those of the Group 1 elements, $Sr(OH)_2$ and $Ba(OH)_2$. ($Ca(OH)_2$ is slightly soluble.)
S^{2-}	All sulfides except those of Groups 1 and 2 elements and NH_4^+ are insoluble.

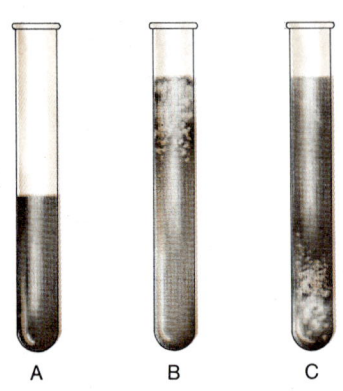

Figure 9.7 When a solution of NaCl is added to the solution of silver nitrate shown in (A) a white precipitate forms (B). This slowly settles out (C). It can be identified as AgCl, which is very insoluble in water. It forms when Ag^+ ions from the $AgNO_3$ solution come in contact with Cl^- ions from the NaCl solution.

the solutions first come in contact. As time passes (third test tube), the insoluble solid settles to the bottom.

Referring back to Table 9.2, we can deduce what happened in this experiment. One solution contains Ag^+ ions, the other Cl^- ions. When these solutions are mixed, these ions come in contact with each other. Notice from Table 9.2 that AgCl is insoluble in water. Hence, everytime a Ag^+ ion collides with a Cl^- ion, AgCl must precipitate out of solution. The solid shown in the second and third test tubes is AgCl, formed by the reaction between Ag^+ and Cl^- ions. The other ions present (NO_3^- ions in the $AgNO_3$ solution, Na^+ in the NaCl solution) take no part in the reaction. They are sometimes referred to as "spectator ions."

Using the solubility rules it is possible to predict whether a precipitation reaction will occur when two solutions are mixed. The reasoning involved is shown in Example 9.1.

EXAMPLE 9.1

Suppose solutions of the following ionic compounds are mixed
a. $Pb(NO_3)_2$ and KCl   b. K_2SO_4 and $CuCl_2$
Using Table 9.2 predict whether a precipitation reaction will occur in each case. If it does, give the formula of the precipitate.

Solution
a. The ions present in the two solutions are

   $Pb(NO_3)_2$ solution: Pb^{2+}, NO_3^-

   KCl solution: K^+, Cl^-

   The only way a precipitate can form is for a cation from one solution to combine

Continued

with an anion from the other solution to form an insoluble solid. There are two possibilities.
(1) Pb^{2+} ions from the Pb(NO$_3$)$_2$ solution could combine with Cl$^-$ ions from the KCl solution to form solid PbCl$_2$.
(2) K$^+$ ions from the KCl solution could combine with NO$_3^-$ ions from the Pb(NO$_3$)$_2$ solution to form solid KNO$_3$.

Referring to Table 9.2, we see that Reaction 1 should indeed occur, since PbCl$_2$ is insoluble. In contrast, Reaction 2 will not occur. Potassium nitrate, like all other nitrates, is soluble in water and will not precipitate out of solution.

We conclude that when water solutions of Pb(NO$_3$)$_2$ and KCl are mixed, a precipitation reaction will occur. The precipitate is lead chloride, PbCl$_2$.

b. We follow the same reasoning as in part a.

Ions Present

K$_2$SO$_4$ solution: K$^+$, SO$_4^{2-}$

CuCl$_2$ solution: Cu^{2+}, Cl$^-$

Possible Precipitates

KCl (K$^+$ from K$_2$SO$_4$ solution + Cl$^-$ from CuCl$_2$ solution)

CuSO$_4$ (Cu^{2+} from CuCl$_2$ solution, SO$_4^{2-}$ from K$_2$SO$_4$ solution)

Reference to Table 9.2 should convince you that both KCl and CuSO$_4$ are water-soluble. Neither of these compounds will form in a precipitation reaction. No such reaction occurs when solutions of K$_2$SO$_4$ and CuCl$_2$ are mixed.

Exercise
Four test tubes each contain solutions of MgCl$_2$. A solution of AgNO$_3$ is added to the first tube, NaOH to the second, CuSO$_4$ to the third, and Na$_2$CO$_3$ to the fourth. How many precipitates form? Answer: Three.

Net Ionic Equations

Precipitation reactions and other reactions involving ions in solutions are most often described by a certain type of chemical equation. This is known as a **net ionic equation.** One such equation is

$$Ag^+(aq) + Cl^-(aq) \rightarrow AgCl(s) \tag{9.5}$$

This net ionic equation is written to describe the formation of a precipitate of AgCl when a solution containing Ag$^+$ ions is mixed with another solution containing Cl$^-$ ions. In another case, the product of a precipitation reaction might be lead chloride, PbCl$_2$. This solid forms when solutions containing Pb^{2+} and Cl$^-$ ions are mixed. The net ionic equation for the reaction is

$$Pb^{2+}(aq) + 2\,Cl^-(aq) \rightarrow PbCl_2(s) \tag{9.6}$$

In a net ionic equation, as in all chemical equations;

— there is a balance of both atoms and charge on the two sides of the equation. In the net ionic equation for a precipitation reaction, such as 9.5 or 9.6, there is a net charge of zero on both sides.
— the states of different species are indicated by symbols (aq), (s), (l), (g), written after the formula. In a precipitation reaction, the reactants (left side) are ions in water solution. The product (right side) is an insoluble solid.
— only those species that take part in the reaction are included in the equation. We omit any "spectator ions" that are in solution before and after reaction.

To write a net ionic equation for a precipitation reaction, it is best to follow a systematic procedure. To illustrate that procedure, let us derive the net ionic equation for the reaction shown in Figure 9.7. We follow a four-step path.

1. The first step in writing a net ionic equation or any other chemical equation, is to **find out what happened in the reaction.** Following the reasoning outlined on p. 236, we decided that in this case the precipitate was silver chloride, AgCl(s).

*2. Represent the reaction by an "un-ionized" equation. Here, we use the formulas of the various compounds involved in the reaction

$$AgNO_3\,(aq) + NaCl\,(aq) \rightarrow AgCl\,(s) + NaNO_3\,(aq)$$

This equation says that water solutions of silver nitrate, $AgNO_3$, and sodium chloride, NaCl, react with each other. The products include an insoluble solid, AgCl, and a water solution of sodium nitrate, $NaNO_3$.

*3. Rewrite the equation, taking account of the fact that any ionic compound in water solution is present as individual ions. This involves substituting aqueous ions for each soluble ionic compound in the equation. In this case, note that $AgNO_3$ in solution consists of Ag^+ and NO_3^- ions. Similarly, NaCl(aq) is really Na^+(aq) and Cl^-(aq); $NaNO_3$(aq) consists of Na^+(aq) + NO_3^-(aq). Hence we write;

$$Ag^+(aq) + NO_3^-(aq) + Na^+(aq) + Cl^-(aq) \rightarrow AgCl(s) \\ + Na^+(aq) + NO_3^-(aq)$$

An equation such as the one just written is called a "total ionic" equation. It includes all the ions present in water solution, before and after reaction.

4. Cancel out "spectator ions" that appear on both sides of the equation.

---

*Steps 2 and 3 may be omitted as you gain more practice in writing net ionic equations.

Here, we cancel Na$^+$ and NO$_3^-$ ions from both sides to arrive at the final, net ionic equation

$$Ag^+(aq) + Cl^-(aq) \rightarrow AgCl(s)$$

EXAMPLE 9.2
Consider the reaction discussed in Example 9.1a. Write a net ionic equation for this reaction, following the procedure just described.

Solution
The first (and most important) step was accomplished in Example 9.1a, where we decided that the product was PbCl$_2$(s). The "un-ionized" equation is

$$Pb(NO_3)_2\ (aq) + 2\ NaCl\ (aq) \rightarrow PbCl_2(s) + 2\ NaNO_3\ (aq)$$

Note that 2 mol of NaCl are required to form 1 mol of PbCl$_2$; 2 mol of NaNO$_3$ are formed in solution. The total ionic equation is

$$Pb^{2+}\ (aq) + 2NO_3^-\ (aq) + 2Na^+\ (aq) + 2Cl^-\ (aq) \rightarrow PbCl_2(s) + 2Na^+\ (aq) + 2NO_3^-\ (aq)$$

We now cancel the spectator ions, Na$^+$ and NO$_3^-$, to obtain

$$Pb^{2+}\ (aq) + 2\ Cl^-\ (aq) \rightarrow PbCl_2(s)$$

Exercise
When solutions of lead perchlorate, Pb(ClO$_4$)$_2$, and hydrochoric acid, HCl, are mixed, lead chloride precipitates. Write a net ionic equation for the reaction. Answer: Pb^{2+} (aq) + 2 Cl$^-$ (aq) → PbCl$_2$(s).

EXAMPLE 9.3
When solutions of MgCl$_2$ and KOH are mixed, a white precipitate forms. Write a net ionic equation for the precipitation reaction involved.

Solution
The first step is to decide what the precipitate is. Following the reasoning described in Example 9.1, you should conclude that there are two possible precipitates, Mg(OH)$_2$ and KCl. Referring to Table 9.2, it should be clear that Mg(OH)$_2$ is insoluble while KCl is soluble. The product of the precipitation reaction is Mg(OH)$_2$.

The "un-ionized" equation is

$$MgCl_2(aq) + 2\ KOH(aq) \rightarrow Mg(OH)_2(s) + 2\ KCl(aq)$$

The total ionic equation is

Continued

Chapter 9

> $Mg^{2+}(aq) + 2Cl^-(aq) + 2K^+(aq) + 2\,OH^-(aq) \rightarrow Mg(OH)_2(s)$
> $\phantom{Mg^{2+}(aq) + 2Cl^-(aq) + 2K^+(aq) + 2\,OH^-(aq) \rightarrow} + 2K^+(aq) + 2Cl^-(aq)$
>
> The net ionic equation is
>
> $Mg^{2+}(aq) + 2\,OH^-(aq) \rightarrow Mg(OH)_2(s)$
>
> **Exercise**
> Solutions of $MgCl_2$ and $CuSO_4$ are mixed. Write a net ionic equation for any precipitation reaction that occurs. Answer: No reaction, no equation.

Net ionic equations are by no means restricted to precipitation reactions. Indeed, net ionic equations can be and are used to represent any type of reaction taking place in water solution. Included are acid–base reactions (Chapter 11) and oxidation–reduction reactions (Chapter 12). We will have more to say about net ionic equations in those two chapters.

9.4 Concentrations of Solutions

There are several different ways to describe the relative amounts of solute present in different solutions. Qualitatively, we can do this by using the words "dilute" and "concentrated." Quite simply, a dilute solution contains less solute, in a given amount of solution, than a concentrated solution does. For example, dilute hydrochloric acid contains less HCl per liter than concentrated hydrochloric acid does.

Usually, we will need to express quantitatively the concentration of solute in a solution. One way to do this is to specify the **molarity** of solute. This concentration unit was mentioned in Chapter 3; it will be discussed in some detail here. Two other concentration units, *mass percent* and *molality*, are introduced here.

Molarity

Concentrations of reagents in solution are most often given in terms of the molarity of solute. Molarity tells us how many moles of solute there are per liter of solution

$$\text{molarity (M)} = \frac{\text{moles solute}}{\text{liters solution}}$$

(9.7)

We can readily calculate molarity if we know the amount of solute in a given volume of solution (Example 9.4).

EXAMPLE 9.4

A solution is prepared by dissolving 25.0 g of NaCl in enough water to form 525 cm³ of solution. What is the molarity of NaCl in this solution?

Solution
To calculate the molarity of NaCl, we must find the number of moles of NaCl per liter of solution. We know the number of grams per cubic centimeter (25.0 g NaCl/525 cm³). We need to convert grams to moles and cubic centimeters to liters

1 mol NaCl = 58.4 g NaCl; 1 L = 1000 cm³

Carrying out these two conversions in succession

$$\text{M NaCl} = \frac{25.0 \text{ g NaCl}}{525 \text{ cm}^3} \times \frac{1 \text{ mol NaCl}}{58.4 \text{ g NaCl}} \times \frac{1000 \text{ cm}^3}{1 \text{ L}} = 0.815 \frac{\text{mol NaCl}}{\text{L}}$$

Exercise
Suppose we had dissolved 2.50 g of NaCl (instead of 25.0 g) in 52.5 cm³ of solution (instead of 525 cm³). What then would be the molarity of NaCl? Answer: 0.815 M.

Sometimes you will need to calculate the amount of solute in a given volume of solution. In other cases, you may want to know the volume of solution required to contain a specified amount of solute. Here, two approaches are possible

1. Solve Equation 9.7 for the required quantity (number of moles or number of liters) and substitute

   moles solute = molarity × liters solution   (9.8)

   $$\text{liters solution} = \frac{\text{moles solute}}{\text{molarity}} \quad (9.9)$$

2. Use molarity as a conversion factor to relate moles of solute to liters of solution. The use of these two approaches is illustrated in Example 9.5.

EXAMPLE 9.5

Consider a 0.125 M solution of KNO₃.
a. How many moles of KNO₃ are there in 0.352 L of this solution?
b. What volume of this solution contains 0.100 mol of KNO₃?

Solution
a. Method 1:

   moles solute = M × liters solution

Continued

$$= 0.125 \frac{\text{mol}}{\text{L}} \times 0.352 \text{ L} = 0.0440 \text{ mol}$$

Method 2:

Since the molarity is 0.125 mol/L: 0.125 mol = 1L

$$\text{moles solute} = 0.352 \text{ L} \times \frac{0.125 \text{ mol}}{1 \text{ L}} = 0.0440 \text{ mol}$$

b. Method 1

Liters solution = (moles solute)/M

$$= \frac{0.100 \text{ mol}}{0.125 \text{ mol/L}} = 0.800 \text{ L}$$

Method 2:

$$\text{liters solution} = 0.100 \text{ mol} \times \frac{1 \text{ L}}{0.125 \text{ mol}} = 0.800 \text{ L}$$

Exercise
What volume of this solution is required to contain 10.0 g of KNO_3 (molar mass = 101.1 g/mol)? Answer: 0.791 L.

To prepare a water solution to a given molarity, you usually start by weighing out a calculated amount of solute. Then you add enough water to dissolve the solute. Finally, you add more water to reach the desired volume of solution (Figure 9.8). Suppose, for example, you want to make 500 cm³ of 0.200 M NaOH. To do this, you need

$$0.500 \text{ L} \times 0.200 \frac{\text{mol}}{\text{L}} = 0.100 \text{ mol NaOH}$$

The molar mass of NaOH is 40.0 g/mol. Hence, you would weigh out

$$0.100 \text{ mol} \times \frac{40.0 \text{ g}}{1 \text{ mol}} = 4.00 \text{ g NaOH}$$

This would be dissolved in enough water to give 500 cm³ of solution. This way, you would obtain half a liter of 0.200 M NaOH.

There is another way to prepare 500 cm³ of 0.200 M NaOH. In this approach, you start with a more concentrated solution, perhaps 1.00 M, and dilute with water. Recall that you need 0.100 mol of NaOH. To obtain this, you could measure out 0.100 L of 1.00 M NaOH

$$\text{liters of 1.00-M solution} = \frac{0.100 \text{ mol}}{1.00 \text{ mol/L}} = 0.100 \text{ L} \quad (100 \text{ cm}^3)$$

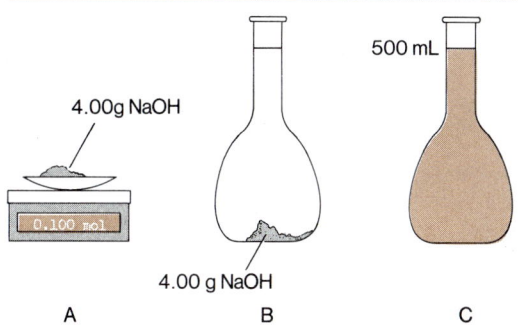

Figure 9.8 Preparation of 500 cm³ of 0.200 M NaOH solution. First, 0.100 mol (4.00 g) of NaOH is weighed out on an analytical balance (A). This sample of NaOH is then transferred to a 500 mL volumetric flask (B). Enough water is added to the flask to completely dissolve the sample. Water is then added, with swirling, until the level reaches the 500 mL mark on the neck of the flask.

To the 100 cm³ of 1.00 M NaOH, you would add enough water (about 400 cm³) to obtain 500 cm³ of 0.200 M solution.

The process just described is shown graphically in Figure 9.9. It is quite a useful way to prepare a dilute solution from one that is more concentrated. One advantage of this "dilution" approach is that only volume measurements

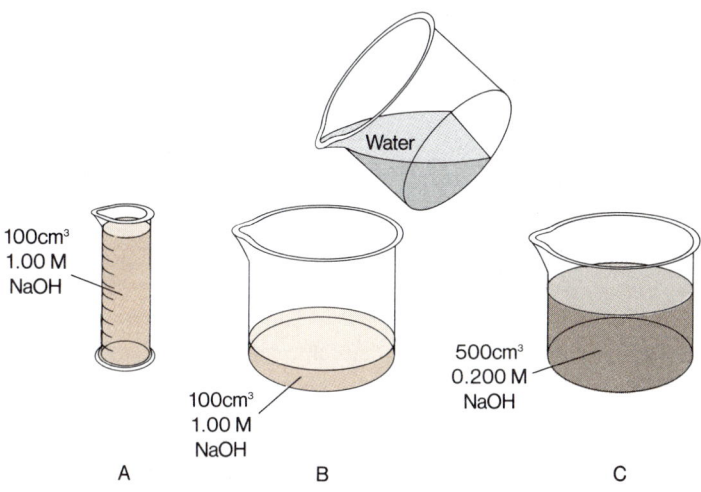

Figure 9.9 Preparation of 500 cm³ of 0.200 M NaOH by dilution. First, 100 cm³ of 1.00 M NaOH is measured out in a graduated cylinder (A). This is then transferred to a large beaker (B). Enough water (approximately 400 cm³) is added to give a final volume of 500 cm³ (C). The concentration of NaOH in the final solution is 0.200 M.

are needed. No weighings are required. To prepare a solution by dilution, you proceed as follows

1. *Calculate the number of moles of solution required*

   moles solute = (molarity dilute solution) × (liters dilute solution)

2. *Calculate the volume of the more concentrated solution required to give the number of moles of solute in step 1.*

   $$\text{liters concentrated solution} = \frac{\text{moles solute}}{\text{molarity concentrated solution}}$$

3. *Measure out the volume of the concentrated solution calculated in step 2. Add enough water to give the desired volume of the dilute solution.*

EXAMPLE 9.6
How would you prepare 2.00 L of 0.300 M HCl, starting with 6.00 M HCl?

Solution
Let's follow the general approach just described

(1) moles HCl required = $0.300 \, \frac{\text{mol}}{\text{L}} \times 2.00 \, \text{L} = 0.600 \, \text{mol}$

(2) liters of 6.00 M HCl required = $\frac{0.600 \, \text{mol}}{6.00 \, \text{mol/L}} = 0.100 \, \text{L}$

(3) Measure out 0.100 L (100 cm^3) of 6.00 M HCl. Dilute with water to a final volume of 2.00 L

Exercise
About how much water would be required in Step 3? *Answer:* 1.9 L.

Mass Percent

In the laboratory, concentrations are sometimes expressed in terms of mass percent. A "20%" glucose solution contains 20 percent by mass of glucose. A 100-g sample of such a solution would contain 20 g of glucose (and 80 g of water). In general, the mass percent of solute is defined by the equation:

$$\text{mass percent} = \frac{\text{mass solute}}{\text{total mass solution}} \times 100 \tag{9.10}$$

EXAMPLE 9.7
A solution is prepared by dissolving 12 g of NaCl in 52 g of water.
a. What is the mass percent of sodium chloride in the solution?

Continued

b. What is the mass percent of water in the solution?
c. How many grams of NaCl are there in 25 g of this solution?

Solution
a. The total mass of the solution is: 12 g + 52 g = 64 g
   Substituting in Equation 9.10;

$$\text{Mass percent NaCl} = \frac{\text{mass NaCl}}{\text{total mass solution}} \times 100 = \frac{12 \text{ g}}{64 \text{ g}} \times 100 = 19$$

b. Since there are only two components, the mass percents must add up to 100
   mass percent water = 100 − 19 = 81
c. Solving Equation 9.10 for mass of solute

$$\text{mass NaCl} = \text{total mass solution} \times \frac{\text{mass percent NaCl}}{100}$$

$$= 25 \text{ g} \times \frac{19}{100} = 4.8 \text{ g}$$

Exercise
What total mass of this solution would have to be taken to obtain 5.0 g of NaCl?
Answer: 27 g.

Percent by volume is sometimes used as a concentration unit when both solute and solvent are liquids. A solution prepared by diluting 10 cm³ of ethyl alcohol to 100 cm³ with water would contain 10% by volume of alcohol. The concentration of alcoholic beverages is most often given in terms of the "proof," which is twice the volume percent of ethyl alcohol. A "110 proof" whiskey contains 55% by volume of alcohol.

Molality

Still another way to specify the concentration of solute in a solution is to give the *molality*. This tells us how many moles of solute there are per kilogram of solvent

$$\text{molality (m)} = \frac{\text{moles solute}}{\text{kilograms solvent}} \quad (9.11)$$

EXAMPLE 9.8
What is the molality of a solution in which 6.20 g of NaCl is dissolved in 86.00 g of water?

Solution
To calculate the molality of NaCl, we must find the number of moles of NaCl per kilogram of water. We know the number of grams of NaCl per gram of water (6.20

Continued

> g NaCl/86.00 g water). We need to convert grams of NaCl to moles and grams of water to kilograms, using the relations
>
> 1 mol NaCl = 58.4 g NaCl ;   1 kg water = 1000 g water
>
> Carrying out these two conversions in succession
>
> $$m\ NaCl = \frac{6.20\ g\ NaCl}{86.00\ g\ water} \times \frac{1\ mol\ NaCl}{58.4\ g\ NaCl} \times \frac{1000\ g\ water}{1\ kg\ water}$$
> $$= 1.23\ mol\ NaCl/kg\ water$$
>
> **Exercise**
> What mass of NaCl is required to prepare a 0.100 m solution containing 200.0 g of water? Answer 1.17 g NaCl.

For dilute water solutions, the molality (m) of solute is nearly equal to its molarity (M). This reflects the fact that one liter of such a solution contains very close to one kilogram of water. Consider, for example, a 1.000 M solution of NaCl. It turns out that one liter of this solution contains 1001 g of water, in addition to one mole of NaCl. The molality is

$$\frac{1.000\ mol}{1.001\ kg} = 0.999\ m$$

We see that in this case molarity and molality differ by only one part per thousand (1.000 versus 0.999).

9.5 Freezing Points and Boiling Points of Solutions

The freezing points of aqueous solutions are *lower* than that of pure water (0°C). You take advantage of this effect when you add antifreeze to your car radiator. A 50% solution of ethylene glycol, $C_2H_6O_2$, does not freeze until the temperature drops to −34°C (−29°F). Highway crews use the same principle to keep roads ice free in winter. They spread a mixture of sand and either NaCl or $CaCl_2$ on the road. These two electrolytes form concentrated solutions that do not freeze at temperatures as low as −20°C.

The boiling points of aqueous solutions are ordinarily *higher* than that of pure water (100°C at 1 atm). For example, a concentrated water solution of ethylene glycol boils at a temperature well above 100°C. Hence, keeping antifreeze of this type in your car radiator during the summer helps to avoid "boilover." Saturated solutions of NaCl and $CaCl_2$ also boil considerably above 100°C.

Equations for Freezing Point Lowering, Boiling Point Elevation

Freezing point lowering and boiling point elevation are **colligative properties.** That is, they are

—directly proportional to the concentration of particles.
—essentially independent of the type of solute particle.

For nonelectrolytes, these relations lead to two quite simple equations

$$\Delta T_f = k_f \times m \tag{9.12}$$

$$\Delta T_b = k_b \times m \tag{9.13}$$

In these equations ΔT_f and ΔT_b are the freezing point lowering and boiling point elevation, respectively. Both are positive quantities; they are defined by the relations

ΔT_f = freezing point pure solvent − freezing point solution

ΔT_b = boiling point solution − boiling point pure solvent

m is the molality of solute, as defined by Equation 9.11, and k_f and k_b are constants for a particular solvent (Table 9.3). For water;

$k_f = 1.86°C$; $k_b = 0.52°C$

For water solutions, we can rewrite Equations 9.12 and 9.13 as

$$\Delta T_f = 1.86°C \times m \tag{9.14}$$

$$\Delta T_b = 0.52°C \times m \tag{9.15}$$

Using Equations 9.14 and 9.15, we can predict that

—a one molal solution of a nonelectrolyte will have a freezing point lowering of about 1.86°C; it will start to freeze at about −1.86°C.
—a one molal solution of a nonelectrolyte will have a boiling point elevation of about 0.52°C; its normal boiling point will be about 100.52°C.

TABLE 9.3 Molal Freezing Point and Boiling Point Constants

Solvent	Freezing Point	k_f	Boiling Point	k_b
Water	0°C	1.86	100°C	0.52
Acetic acid	17	3.90	118	2.93
Benzene	5.50	5.10	80.0	2.53
Cyclohexane	6.5	20.2	81	2.79
Camphor	178	40.0	208	5.95
p-Dichlorobenzene	53	7.1	—	—

More generally, we can use these equations to estimate the freezing and boiling points of solutions of nonelectrolytes of known molality (Example 9.9).

EXAMPLE 9.9
Antifreeze is prepared by adding 20.0 g of ethylene glycol, $C_2H_6O_2$, to 100.0 g of water. Determine, using Equations 9.14 and 9.15, the freezing point and normal boiling point of this solution.

Solution
Let's first calculate the molality of the solute. The molar mass of $C_2H_6O_2$ is

$(24.0 + 6.0 + 32.0)$ g/mol $= 62.0$ g/mol

We now proceed as in Example 9.8

$$m\ C_2H_6O_2 = \frac{20.0\ g\ C_2H_6O_2}{100.0\ g\ water} \times \frac{1\ mol\ C_2H_6O_2}{62.0\ g\ C_2H_6O_2} \times \frac{1000\ g\ water}{1\ kg\ water}$$

$= 3.23$ m

Now we can calculate ΔT_f and ΔT_b

Equation 9.14: $\Delta T_f = 3.23 \times 1.86°C = 6.01°C$

Equation 9.15: $\Delta T_b = 3.23 \times 0.52°C = 1.7°C$

The freezing point is 6.01°C below that of pure water, 0.00°C

fp $= 0.00°C - 6.01°C = -6.01°C$

The boiling point is 1.7°C above that of pure water, 100.0°C

bp $= 100.0°C + 1.7°C = 101.7°C$

(This is not a very effective antifreeze; the solution freezes at $-6°C$, about 21°F. Better add more ethylene glycol!)

Exercise
Suppose you added 20.0 g of ethyl alcohol, C_2H_6O, to 100.0 g of water. Would the freezing point of the solution be higher or lower than that calculated in this example? Answer: Lower.

A common method of determining the molar mass of a nonelectrolyte involves freezing point lowering. A sample of the nonelectrolyte is weighed out and dissolved in a known mass of solvent. The freezing point of the solution is measured. The molar mass of the solute is calculated as shown in Example 9.10.

EXAMPLE 9.10
A solution containing 1.00 g of a nonelectrolyte X in 10.0 g of water freezes at $-1.42°C$. Calculate the molar mass of X.

Continued

Solution
To obtain the molar mass, we follow a three-step path. We first calculate the molality of the solution, using Equation 9.14. Then, since we know the mass of solvent, we can use Equation 9.11 to obtain the number of moles of solute. Finally knowing both the number of moles of solute and the number of grams (1.00 g), we can obtain the molar mass.

(1) The freezing point lowering, ΔT_f, is

$$0 - (1.42°C) = 1.42°C$$

From Equation 9.14

$$m = \frac{\Delta T_f}{1.86°C} = \frac{1.42°C}{1.86°C} = 0.763$$

(2) From Equation 9.11:

moles solute = molality × kg solvent

$$= 0.763 \times \frac{10.0}{1000} = 0.00763 \text{ mol}$$

(3) 0.00763 mol = 1.00 g

$$\text{molar mass} = \frac{1.00 \text{ g}}{0.00763 \text{ mol}} = 131 \text{ g/mol}$$

Exercise
What is the boiling point of this solution at 1 atm pressure? Answer: 100.40°C.

Key Words

boiling point elevation
colligative property
concentrated solution
dilute solution
electrolyte
freezing point lowering
hydrogen bond
mass percent
molality
molar mass
molarity

mole
net ionic equation
nonelectrolyte
precipitate
saturated solution
solubility rules
solute
solvent
spectator ion
supersaturated solution
unsaturated solution

Questions

9.1 How does an electrolyte differ from a nonelectrolyte?

9.2 Explain what is meant by a saturated solution; an unsaturated solution; a supersaturated solution.

9.3 A crystal of KNO_3 is added to a solution of that solid in water. The crystal does not dissolve, nor does more solute crystallize. Is the solution saturated, unsaturated, or supersaturated?

9.4 What is a hydrogen bond? In what kinds of molecules can hydrogen bonding occur?

9.5 An increase in temperature ordinarily _____ the water solubility of solids and _____ the water solubility of gases.

9.6 An increase in pressure _____ the solubility of _____ in water.

9.7 All chlorides are soluble in water except _____, _____, and _____.

9.8 Most sulfates are soluble; exceptions include _____ and _____.

9.9 Most carbonates are _____; however, Group 1 carbonates are _____.

9.10 Of the Group 2 hydroxides, the most soluble is _____.

9.11 All transition metal sulfides are _____ in water.

9.12 Define the concentration terms molarity and molality.

9.13 Explain how a solution is prepared to a desired molarity, starting with
 a. pure solute
 b. a more concentrated solution of the solute

9.14 Water solutions freeze _____ 0°C and usually boil _____ 100°C.

9.15 What quantities must be measured to determine the molar mass of a nonelectrolyte by freezing point lowering?

Problems

Precipitation Reactions
(Use the Periodic Table where needed for atomic masses.)

9.16 Identify the precipitate formed when the following solutions are mixed
 a. $Pb(NO_3)_2$ and HCl
 b. $Pb(NO_3)_2$ and H_2SO_4
 c. $Pb(NO_3)_2$ and Na_2S

9.17 Identify the precipitates formed, if any, when the following solutions are mixed
 a. $AgNO_3$ and Na_2CO_3
 b. $AsCl_3$ and K_2S
 c. $NaCl$ and KOH
 d. $CuSO_4$ and $NaOH$

9.18 Write balanced net ionic equations for the precipitation reactions in Problem 9.16.

9.19 Write balanced net ionic equations for the precipitation reactions in Problem 9.17.

9.20 Write balanced net ionic equations for the precipitation reactions that occur when the following solutions are mixed
 a. $Ba(NO_3)_2$ and $Fe_2(SO_4)_3$
 b. $AgNO_3$ and Na_2S
 c. $NiCl_2$ and $NaOH$
 d. $Al(NO_3)_3$ and KOH

9.21 Write balanced net ionic equations for the precipitation reactions, if any, that occur when the following solutions are mixed
 a. $NiSO_4$ and KNO_3
 b. $NiSO_4$ and $LiOH$
 c. $NiSO_4$ and $Ba(OH)_2$
 d. $Zn(NO_3)_2$ and Na_2CO_3

9.22 When solutions of sodium chromate and silver nitrate are mixed, a red precipitate forms
 a. Give the formulas of the two possible precipitates
 b. Identify the actual precipitate
 c. Write a net ionic equation for the precipitation reaction involved

9.23 When solutions of iron(III) nitrate and sodium phosphate are mixed, a tan precipitate forms.

a. Give the formulas of the two possible precipitates
b. Identify the precipitate and write a net ionic equation for its formation

9.24 Convert the following to net ionic equations
  a. $(NH_4)_2S(aq) + Pb(NO_3)_2(aq) \rightarrow PbS(s) + 2\,NH_4NO_3(aq)$
  b. $FeCl_3(aq) + 3\,LiOH(aq) \rightarrow Fe(OH)_3(s) + 3\,LiCl(aq)$
  c. $Hg_2(NO_3)_2(aq) + CaCl_2(aq) \rightarrow Ca(NO_3)_2(aq) + Hg_2Cl_2(s)$

9.25 Convert the following to net ionic equations
  a. $2Ag^+(aq) + 2NO_3^-(aq) + 2K^+(aq) + CrO_4^{2-}(aq) \rightarrow Ag_2CrO_4(s) + 2K^+(aq) + 2NO_3^-(aq)$
  b. $Ca^{2+}(aq) + SO_4^{2-}(aq) + 2K^+(aq) + 2F^-(aq) \rightarrow CaF_2(s) + 2K^+(aq) + SO_4^{2-}(aq)$
  c. $Ba^{2+}(aq) + 2Cl^-(aq) + 2Na^+(aq) + SO_4^{2-}(aq) \rightarrow BaSO_4(s) + 2Na^+(aq) + 2Cl^-(aq)$

Molarity

9.26 Calculate the molarity of a solution containing
  a. 1.20 mol of Na_3PO_4 in 656 cm^3 of solution
  b. 15.0 g of KOH in 1.24 L of solution
  c. 32.0 g of C_2H_5OH in 94.6 cm^3 of solution

9.27 Determine the number of moles of solute in
  a. 12.0 L of 0.320 M NaCl solution
  b. 159 cm^3 of 0.105 M $FeCl_3$ solution
  c. 1.69 L of a solution containing 3.45 g of $CaCl_2$ per liter

9.28 Find the volume of 1.00 M K_2CrO_4 solution that contains
  a. 1.00 mol of K_2CrO_4
  b. 0.00600 mol of K_2CrO_4
  c. 1.00 g of K_2CrO_4

9.29 A solution is prepared by dissolving 12.0 g of ethylene glycol, $C_2H_6O_2$, in water and diluting to 60.0 cm^3.
  a. What is the molarity of ethylene glycol?
  b. What volume of this solution contains 0.100 mol of $C_2H_6O_2$?
  c. How many moles of ethylene glycol are there in 1.00 cm^3 of this solution?

9.30 You are given 20.0 L of a 0.104 M solution of $AgNO_3$
  a. What is the total mass in grams of $AgNO_3$ in this solution?
  b. If you want 1.00 mol of silver nitrate, what volume of solution should you take?

9.31 Describe in some detail how you would prepare 500.0 cm^3 of 0.1030 M KCl, starting with the solid.

9.32 What volume of 6.0 M HCl is required to prepare 2.5×10^2 cm^3 of 0.10 M HCl?

9.33 What volume of 0.200 M KCN solution is required to prepare 3.00×10^2 cm^3 of a KCN solution with a molarity of 2.50×10^{-3}?

9.34 Concentrated nitric acid contains 72.0 mass percent of HNO_3 and has a density of 1.42 g/cm^3.
  a. What is the mass of HNO_3 in one liter of the concentrated acid?
  b. What is the molarity of HNO_3?

Mass Percent

9.35 What mass of glucose, $C_6H_{12}O_6$, is needed to prepare 5.00×10^2 g of a solution in which the mass percent of glucose is 12.0? What mass of water is required?

9.36 What is the mass percent of NaOH in a solution prepared by adding 80.0 g of NaOH to 126 g of water?

9.37 Concentrated sulfuric acid contains 96 mass percent H_2SO_4. How many grams of H_2SO_4 are there in 58 g of this solution? How many grams of water?

9.38 For a 12.0% solution of HCl in water, calculate
  a. the mass of HCl in 25.0 g of solution
  b. the mass of water in 75.0 g of solution
  c. the mass of solution that contains 6.00 g of HCl

Molality

9.39 What is the molality of a solution prepared by dissolving

Continued

a. 15.0 g of glucose, $C_6H_{12}O_6$, in 100.0 g of water?
b. 15.0 g of ethylene glycol, $C_2H_6O_2$, in 100.0 g of water?

9.40 How many grams of sucrose, $C_{12}H_{22}O_{11}$, must be added to 80.0 g of water to prepare a 0.250 molal solution?

9.41 A solution is prepared by dissolving 20.0 g of ethyl alcohol, C_2H_5OH, in 60.0 g of water. Calculate
a. the molality of C_2H_5OH
b. the mass percent of C_2H_5OH

9.42 Dilute HCl contains 20.0 mass percent of HCl. What is the molality of HCl in this solution?

Freezing Point, Boiling Point

9.43 Calculate the freezing point and normal boiling point of a water solution containing 10.0 g of citric acid, $C_6H_8O_7$, in 60.0 g of water.

9.44 What are the freezing point and normal boiling point of a water solution containing 208 g of sugar, $C_{12}H_{22}O_{11}$, in 762 g of water?

9.45 How many grams of urea, $CO(NH_2)_2$, must be dissolved in 100.0 g of water to obtain a solution
a. freezing at $-1.00°C$?
b. boiling at $100.50°C$?

9.46 What mass of C_2H_5OH must be added to a kilogram of water to yield a solution freezing at $-10.0°F$?

9.47 A water solution contains 20.0 mass percent ethylene glycol, $C_2H_6O_2$
a. What is the molality of ethylene glycol?
b. What is the freezing point of the solution?

9.48 A solution containing 2.00 g of a certain nonelectrolyte dissolved in 30.0 g of water freezes at $-0.620°C$. What is the molar mass of the nonelectrolyte?

9.49 Calculate the molar mass of a nonelectrolyte if a solution of 1.110 g dissolved in 9.858 g of water freezes at $-2.64°C$.

9.50 When 6.84 g of a nonelectrolyte is dissolved in 60.0 g of benzene, the solution formed freezes at 2.15°C. What is the molar mass of the nonelectrolyte? (Refer to Table 9.3 for necessary data.)

Multiple Choice

9.51 How many moles of ions are formed when 0.120 mol of $Fe(NO_3)_3$ dissolves in water?
a. 0.120   b. 0.240   c. 0.360
d. 0.480   e. 0.600

9.52 To form a supersaturated solution of a gas in a liquid, you could
a. increase the temperature
b. reduce the temperature
c. increase gas pressure
d. do any of the above

9.53 Which one of the following would *not* form hydrogen bonds with water?
a. CH_4 b. CH_3OH c. NH_3
d. HF

9.54 How many of the following compounds are soluble in water?

$CaCO_3$, CaS, $CaCl_2$, $Ca(NO_3)_2$

a. 0   b. 1   c. 2   d. 3   e. 4

9.55 What volume (liters) of 0.200 M NaOH can be prepared from 10.0 g of NaOH (molar mass = 40.0 g/mol)?
a. 0.0500   b. 0.800   c. 1.00
d. 1.25   d. 20.0

9.56 What volume of 0.400 M NaOH can be prepared from 525 cm^3 of 0.500 M NaOH?
a. 420 cm^3 b. 500 cm^3 c. 525 cm^3
d. 656 cm^3 e. 1000 cm^3

9.57 Which of the following water solutions contains the largest number of moles of ions per liter?
a. 0.200 M K_2CrO_4 b. 0.400 M NaCl
c. 0.100 M $Al(NO_3)_3$ d. 0.100 M $Al_2(SO_4)_3$

9.58 A solution contains 12.0 g of sugar in 200.0 cm^3; the density of the solution is 1.02 g/cm^3. The mass percent of sugar is
a. 5.66   b. 5.88   c. 6.00   d. 6.12
e. 6.38

9.59 Five solutions each contain 1.00 g of solute in 10.0 g of water. The solutes are those listed below. Which has the lowest freezing point?
a. CH_3OH b. C_2H_5OH c. $C_2H_6O_2$
d. $C_6H_{12}O_6$

9.60 A water solution freezes at $-0.93°C$. Its normal boiling point is approximately
a. 99.74°C   b. 100.00°C   c. 100.26°C
d. 100.52°C   e. 100.93°C

Reaction Rate and Chemical Equilibrium

Learning Objectives
After studying this chapter, you should be able to:

1. Explain what is meant by activation energy and how it affects reaction rate.
2. Predict the effect upon reaction rate (Example 10.1) of:
   — changing the concentration (or pressure) of reactant.
   — changing the temperature.
   — adding a catalyst.
3. Given the chemical equation for an equilibrium system, write the corresponding expression for K_c (Example 10.2).
4. Given equilibrium concentrations, or data from which these concentrations can be calculated, determine K_c (Examples 10.3, 10.4).
5. Correlate extent of reaction (large or small) with the value of the equilibrium constant for the reaction (large or small).
6. Given K_c and all but one of the concentrations in an equilibrium system, calculate that concentration (Examples 10.5, 10.6).
7. Predict the effect upon an equilibrium system of:
   — adding or removing a reactant or product (Example 10.7).
   — increasing or decreasing the pressure (Example 10.8).
   — increasing or decreasing the temperature, given the value of ΔH (Example 10.9).
8. Knowing the sign of ΔH for a reaction, predict whether K_c will increase or decrease when the temperature increases.

> It ain't the things we don't know that hurt us. It's the things we do know that ain't so.
>
> ARTEMUS WARD

Photograph by Peter L. Kresan.

CHAPTER 10

When we carry out a reaction in the laboratory, we must be concerned with two features of the reaction.

1. *The rate of reaction.* Ordinarily, we want to make a reaction occur rather quickly, perhaps within a few minutes or a few hours. In this chapter, we will look at what is meant by reaction rate and how rate is related to a quantity called activation energy (Section 10.1). With that background, we will see how reaction rate is affected by changes in reactant concentration, changes in temperature, or addition of a catalyst (Section 10.2).
2. *The extent of reaction.* Ordinarily, we want as high a yield of product as possible. Ideally, we would like a 100% conversion of reactants to products. In practice, this seldom occurs. Instead, we may obtain a mixture containing appreciable amounts of reactants as well as products. To understand why this happens, we need to look at the principles of chemical equilibrium (Section 10.3). The ratio of products to reactants at equilibrium can be described using a quantity called an equilibrium constant. The form and meaning of the equilibrium constant for gaseous systems, K_c, will be examined in Sections 10.4 and 10.5. The position of an equilibrium, like the rate of a reaction, is sensitive to external conditions. In Section 10.6, we will see how changes in concentration, pressure, or temperature affect the relative amounts of products and reactants at equilibrium.

10.1 Rate of Reaction; Activation Energy

"Rate" always describes how a quantity changes with time. That quantity may be distance, as is the case when you drive an automobile at a rate of 40 miles per hour. In a chemical reaction, the quantity that is changing is concentration. *The rate of reaction is a positive ratio that tells us how the concentration of a reactant or product changes with time.* To understand what this statement means, consider the reaction

$$CO(g) + NO_2(g) \rightarrow CO_2(g) + NO(g) \qquad (10.1)$$

We might take the rate of this reaction to be the change of concentration of CO_2 with time

$$\text{rate} = \frac{\Delta \text{conc. } CO_2}{\Delta t}$$

Here the symbol Δ means, as always, "final minus initial." If the concentration of CO_2 increases from 0 to 0.10 mol/L in one minute

$$\text{rate} = \frac{(0.10 - 0) \text{ mol/L}}{(1 - 0) \text{ min}} = 0.10 \frac{\text{mol/L}}{\text{min}}$$

In a general qualitative way, we distinguish between "fast" and "slow" reactions. A fast reaction is one in which the concentration of product builds up quickly. An example would be a reaction with a rate of, let's say

$$1.0 \frac{\text{mol/L}}{\text{second}}$$

At this rate, product would appear rapidly. Within a few seconds at most, reactants would be consumed. The reaction might well be over before you realized it had started. In contrast, consider a reaction where the rate is

$$0.010 \frac{\text{mol/L}}{\text{day}}$$

This, by any standard, is a slow reaction. It will take more than a week for the concentration of product to reach 0.1 mol/L. This is not the kind of reaction you like to carry out in a three-hour lab period.

Reaction Path; Activation Energy

To understand why some reactions are fast and others slow, it is helpful to consider the path by which reactions take place. Let's focus on a particular reaction, that between CO and NO_2

$$CO(g) + NO_2(g) \rightarrow CO_2(g) + NO(g) \qquad (10.1)$$

We believe that this reaction occurs in a single step: the collision of a CO molecule with an NO_2 molecule. However, we also know that *only a small fraction of such collisions result in reaction*. The rate of reaction that we observe is much, much, smaller than the rate of collision calculated from the kinetic theory of gases.

We can readily see why collisions between CO and NO_2 molecules are seldom effective. For reaction to occur, an oxygen atom must split off from an NO_2 molecule. In other words, a chemical bond must be broken. This requires the absorption of energy, which can only come from one source: the kinetic energy of the colliding molecules. Only if the CO and NO_2 molecules are moving very rapidly will they collide with enough energy to break bonds.

Chapter 10

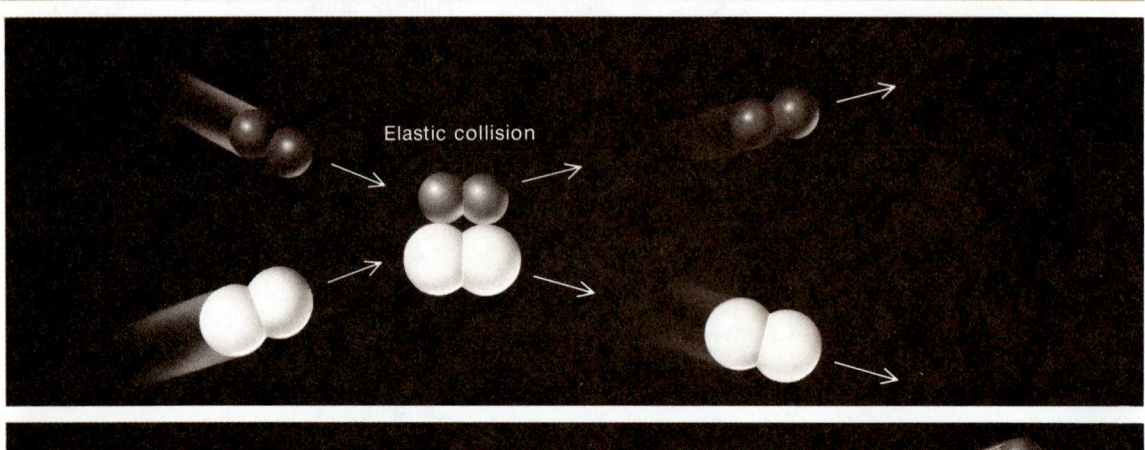

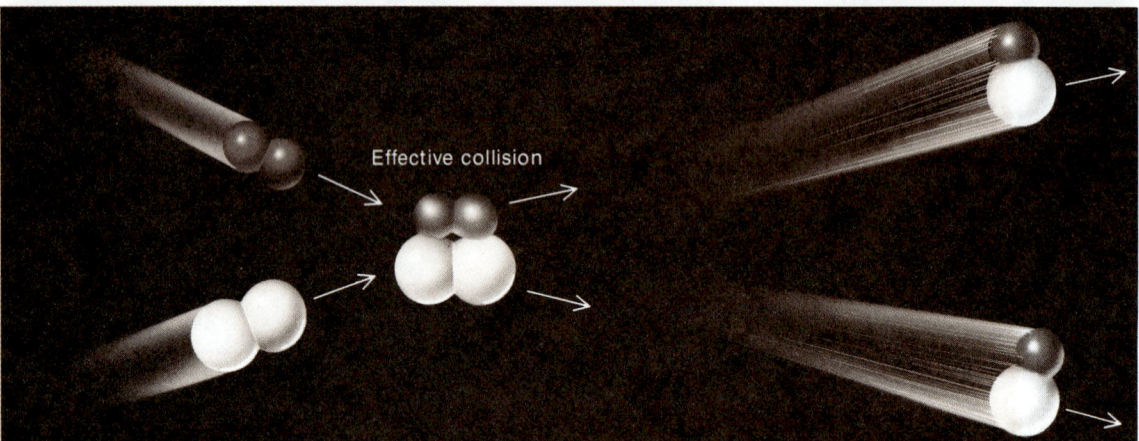

Figure 10.1 If two molecules collide, they usually just bounce off each other unchanged, in a so-called elastic collision. If the energy of the collision is high enough, E_a, then the atoms in the colliding molecules can interact to produce products. Such a collision would be called effective.

At ordinary temperatures, most molecules are moving much too slowly for this to happen. They simply bounce off one another without reacting (Figure 10.1).

The energy that colliding molecules must possess for reaction to occur is called the **activation energy**. This quantity is given the symbol E_a and is commonly expressed in kilojoules per mole of reactant. For the CO–NO$_2$ reaction

$$CO(g) + NO_2(g) \rightarrow CO_2(g) + NO(g) \; ; \; E_a = 134 \text{ kJ}$$

This means that CO and NO$_2$ molecules must have a combined energy of 134

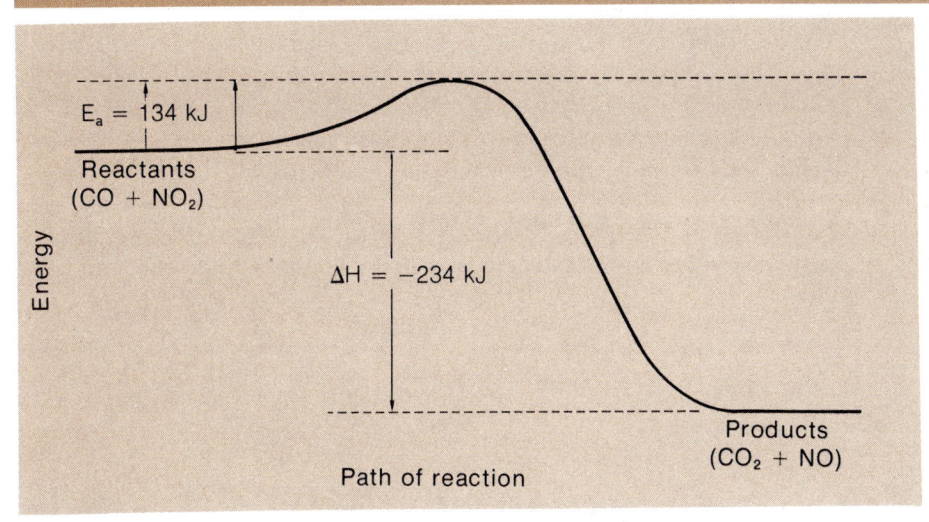

Figure 10.2 Concept of activation energy. During the reaction step, the activation energy E_a, about 134 kJ, must be furnished to the reactants for each mole of CO that reacts. This energy activates each CO–NO$_2$ complex to the point where reaction can proceed.

kJ (per mole) if they are to react upon collision (Figure 10.2). It turns out that 134 kJ is roughly the amount of energy required to break or markedly weaken one of the nitrogen-to-oxygen bonds in NO_2. This is not surprising, since for reaction to occur, an oxygen atom must be transferred from an NO_2 molecule to a CO molecule.

Many reactions occur by a more complex path than the one just described. Several steps may be required, perhaps involving a series of collisions. However, every reaction has a characteristic activation energy, similar in nature to that for the CO–NO$_2$ reaction. We can think of E_a as an energy barrier which the reactants must surmount before being converted to products. Even though the overall energy change for the reaction may be negative, the activation energy is always positive. The situation is analogous to that of a bicyclist traveling from Denver (elevation = 1.6 km) to San Francisco (elevation = 0.0 km). Overall, he is going downhill, but it's hard to convince him of that while he's struggling to get over the Continental Divide (elevation = 3.4 km).

The concept of activation energy helps us understand why some reactions are very fast while others tend to be slow. In general, we expect an inverse relationship between activation energy and reaction rate.

1. Reactions that have a large activation energy, perhaps 200 kJ or greater, ordinarily have a slow rate. Colliding molecules rarely have enough energy

to bring about reaction. Of the total number of collisions, only a tiny fraction are effective. High activation energy is characteristic of reactions in which the bonds in the reactant molecules are very strong.

2. Reactions with a small activation energy tend to have a fast rate. In the extreme, if the activation energy is zero, every collision results in reaction. Such is the case in a simple precipitation reaction such as

$$Ag^+(aq) + Cl^-(aq) \rightarrow AgCl(s) \; ; \; E_a \approx 0$$

This reaction occurs instantaneously when the ions come in contact.

10.2 Effect of Changes in Conditions upon Reaction Rate

For most reactions, it is possible to change the rate over a wide range by adjusting the conditions of reaction. To do this, we may

— change the concentration of a reactant.
— change the temperature.
— add a catalyst.

We will now consider how these three factors affect reaction rate.

Concentration of Reactant

Ordinarily, we find that **an increase in concentration of reactant increases the rate of reaction.** We can readily see why this would be the case (Figure 10.3). When the concentration of reactant is increased, there are more molecules per unit volume. This increases the number of collisions in a given time which in turn increases the reaction rate. For many reactions, doubling the concentration of a reactant approximately doubles the reaction rate.

Figure 10.3 Effect of concentration of a reactant, A, on collision rate. When the concentration of A is doubled, there are twice as many collisions with the other reactant.

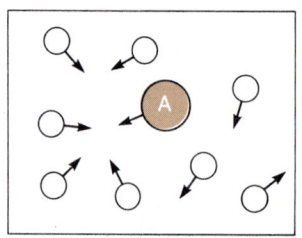

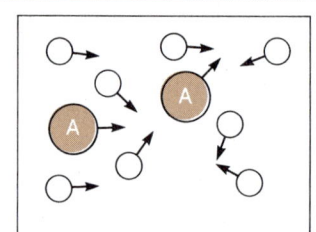

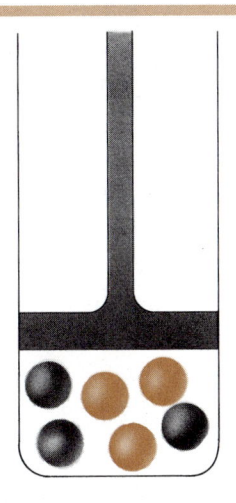

Figure 10.4 When a gas mixture is compressed (right), the same number of molecules occupy a smaller volume. Hence, the concentration of molecules increases. This in turn leads to more frequent collisions and hence a faster reaction. The effect of compression is always to increase reaction rate.

There are two different ways to increase the concentration of a gaseous reactant.

1. *Compress the reaction mixture, i.e., reduce the volume.* With the same number of molecules in a smaller volume, concentration increases (Figure 10.4). In order to compress a system, we must increase the pressure. We might say, then, than *an increase in pressure increases the rate of a gas–phase reaction*. It does this by increasing the concentration of reactant molecules.
2. *Add reactant to the system.* With more reactant molecules in the same volume, concentration increases. The rate of reaction also increases.

The effect of reactant concentration explains a common observation about rates. We ordinarily find that a reaction slows down as time passes. The rate, which may be rapid at first, decreases steadily. Eventually, it drops to zero. This pattern makes sense when we realize that reactant concentration also decreases with time. As more and more reactant molecules are consumed, their concentration drops. Collisions become less frequent and rate decreases.

Temperature

An increase in temperature almost always increases reaction rate. Potatoes cook a lot faster in boiling water, at 100°C, than they do in hot water at 90°C. To prevent foods from spoiling, they are kept cold in a refrigerator at 5°C. To slow down the process even further, you can put them in a freezer at −15°C.

To estimate the effect of temperature upon reaction rate, we often use a simple and very approximate rule. The rate of a reaction is about doubled when the temperature is increased by 10°C. Using the rule, we can predict that

— the time required to cook potatoes should be cut in half by raising the temperature from 90 to 100°C.
— milk in a refrigerator at 5°C will stay fresh four times as long as at room temperature, 25°C. (Two 10° intervals are involved here: 2 × 2 = 4).

The effect of temperature upon reaction rate can be explained in terms of kinetic theory. Recall from Chapter 6 that raising the temperature greatly increases the fraction of molecules with very high kinetic energies. As we pointed out earlier in this chapter, these are the molecules that are most likely to react when they collide. They can supply the activation energy required to break bonds and cause reaction to take place.

This effect is shown graphically in Figure 10.5. Here we show the distribution of kinetic energies among molecules at two different temperatures. One of these temperatures (T + 10) is 10°C higher than the other (T). Molecules with energies equal to or greater than the activation energy, E_a, fall in the shaded area at the right. Notice that this area is about twice as large at the higher temperature. Hence, the number of effective collisions should increase by a factor of two in going from T to T + 10. If this is the case, the rate must also double.

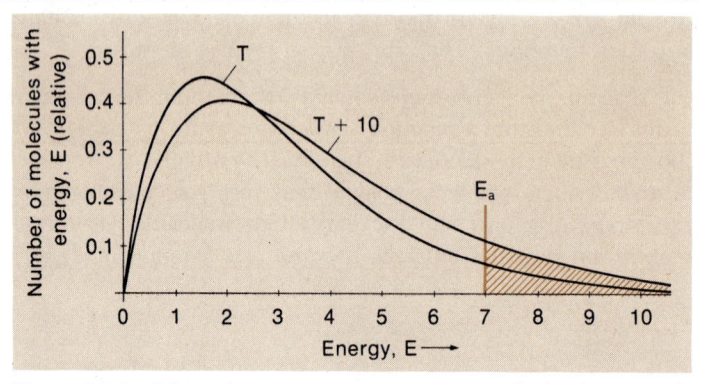

Figure 10.5 When the temperature is increased, the fraction of molecules with very high energies increases sharply. Hence, many more molecules possess the activation energy, E_a, and reaction occurs more rapidly. If E_a is of the order of 50 kJ, an increase in temperature of 10°C approximately doubles the number of molecules having energy E_a or greater, and thus doubles the reaction rate.

Catalyst

Often it is possible to increase the rate of a reaction by adding a material called a catalyst. By definition, **a catalyst changes the rate of a reaction without being consumed by it.** Catalysts are often used in the general chemistry laboratory. Consider, for example, a common method of preparing small amounts of oxygen

$$2 \text{ H}_2\text{O}_2(\text{aq}) \rightarrow 2 \text{ H}_2\text{O} + \text{O}_2(\text{g})$$

Ordinarily this reaction occurs quite slowly. A 3% water solution of hydrogen peroxide, H_2O_2, can be kept for several months without decomposing. However, if you add a few crystals of MnO_2, reaction occurs almost instantly. Manganese(IV) oxide acts as a catalyst; it speeds up the reaction yet can be recovered intact when the reaction is over.

The catalytic converter in an automobile contains a mixture of two metals, Pt and Pd. On the surface of the mixture, carbon monoxide formed by the incomplete combustion of gasoline is converted to CO_2

$$2 \text{ CO}(\text{g}) + \text{O}_2(\text{g}) \rightarrow 2 \text{ CO}_2(\text{g})$$

In the absence of a catalyst, this reaction occurs quite slowly.

A catalyst changes the path by which a reaction occurs. In doing so, it lowers the activation energy (Figure 10.6). Notice that the catalyst does not change the energies of products or reactants. The overall change in energy for the reaction is the same as for the uncatalyzed reaction. Finding a catalyst

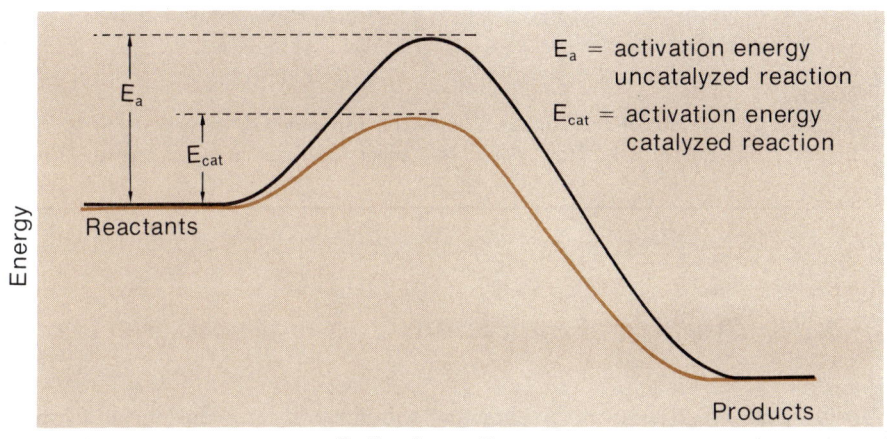

Figure 10.6 By changing the path by which a reaction occurs, a catalyst can lower the activation energy that is required, and so speed up the reaction.

is a little like finding a pass through a mountain range. The relative positions of the valleys on both sides are unchanged, but it's a lot easier to get from one to the other.

Catalysts are widely used in industrial processes to increase reaction rates. Often it is possible to find a catalyst that will increase the rate of the desired reaction without changing the rate of an undesired side reaction. By finding the proper catalyst, a chemist may obtain the best of both worlds, getting both a fast reaction and one that produces nearly pure product.

EXAMPLE 10.1
Assume that the reaction

$$N_2(g) + O_2(g) \rightarrow 2\ NO(g)$$

comes about as the result of a collision between N_2 and O_2 molecules. How will the rate of reaction be affected by
a. compressing the reaction mixture?
b. adding N_2 to the reaction mixture (constant volume)?
c. adding NO to the reaction mixture (constant volume)?
d. reducing the temperature?

Solution
a. Rate should increase. There are more molecules per unit volume, so the rate of collision increases.
b. Rate should increase. With more N_2 molecules in the same volume, the rate of collision increases.
c. No effect expected. The concentrations of N_2 and O_2 remain unchanged.
d. Rate should decrease. Fewer molecules have the kinetic energy required for collision to result in reaction.

Exercise
Suppose the volume of the container in which this reaction is carried out is increased from 1 L to 2 L. How would this affect the reaction rate? Answer: Decrease.

10.3 Chemical Equilibrium in a Gaseous System

To illustrate what is meant by chemical equilibrium, let's consider an experiment involving gases. We start by placing 1.00 mol of gaseous PCl_5 in a one-liter container at 250°C (Figure 10.7). As time passes, it is clear that a reaction is taking place. For one thing, the pressure inside the container increases, even though the temperature and volume remain constant. Moreover, a

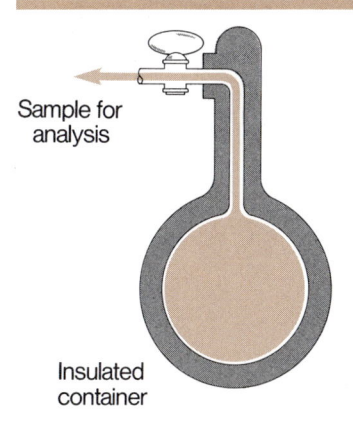

Figure 10.7 Apparatus that might be used to study the PCl_5–PCl_3–Cl_2 equilibrium. Samples of the gas present would be removed and analyzed to establish that equilibrium existed and the equilibrium concentrations of each of the species.

greenish-yellow color characteristic of $Cl_2(g)$ develops. We interpret this evidence to mean that the following reaction is taking place

$$PCl_5(g) \rightarrow PCl_3(g) + Cl_2(g) \tag{10.2}$$

To follow the course of this reaction, we might withdraw samples at ten-minute intervals and analyze them. When we do this, we discover that *Reaction 10.2 does not go to completion*. No matter how long we wait, there is some unreacted PCl_5 present along with the products, PCl_3 and Cl_2. Referring to Figure 10.8, we see that after about 40 minutes, the concentration of PCl_5 levels off at 0.80 mol/L. At the same time, the concentrations of PCl_3 and Cl_2 become constant at 0.20 mol/L. We refer to this mixture as an **equilibrium system** containing the three gases PCl_5, PCl_3, and Cl_2. The characteristics of an equilibrium system are

1. It contains all the substances involved in the reaction, both products and reactants.
2. Its composition does not change with time.

To further understand what chemical equilibrium involves, it helps to look at what is happening on a molecular level. When PCl_5 is placed in the container at 250°C, it begins to decompose by Reaction 10.2. A molecule of PCl_5 dissociates to form a molecule of PCl_3 and a molecule of Cl_2

$$PCl_5(g) \rightarrow PCl_3(g) + Cl_2(g) \tag{10.2}$$

At first, this is the only reaction taking place. However, as the concentrations of PCl_3 and Cl_2 build up, the reverse reaction starts to occur. That is, molecules of PCl_3 and Cl_2 combine to form a molecule of PCl_5

$$PCl_3(g) + Cl_2(g) \rightarrow PCl_5(g) \tag{10.3}$$

As time passes, this reaction speeds up while the forward reaction (Reaction

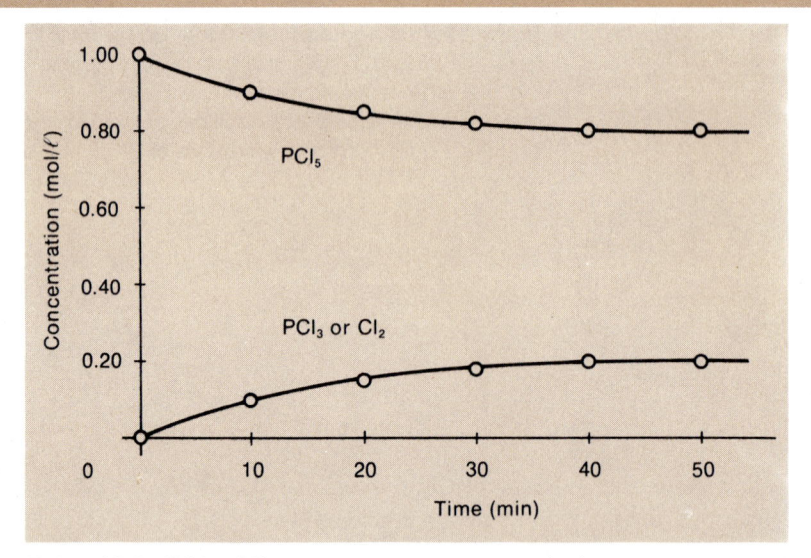

Figure 10.8 When PCl₅ is put into a container at high temperatures, some, but not all, of it decomposes to PCl₃ and Cl₂. The number of moles of PCl₅ that break down will equal the number of moles of PCl₃ formed or the number of moles of Cl₂ formed. After a certain period of time the concentrations of all three species will stop changing. At that point the system is in a state of chemical equilibrium.

10.2) slows down. After about 40 minutes, the rates of the two processes become equal. When this happens, the concentrations of PCl₅, PCl₃, and Cl₂ no longer change with time. The equilibrium reached in this way is described by the equation

$$PCl_5(g) \rightleftharpoons PCl_3(g) + Cl_2(g) \tag{10.4}$$

The double arrow implies that the equilibrium is *dynamic*. Forward and reverse reactions are taking place at the same rate.

Relation Between [PCl₅], [PCl₃], and [Cl₂]

There are many different ways to establish the equilibrium indicated by Equation 10.4. Table 10.1 shows the results of four different experiments relating to this system. Experiment 1 is that described above. We start with pure PCl₅ at a concentration of 1.00 mol/L and allow the system to come to equilibrium. Experiment 2 is similar to Experiment 1, except that the original concentration of PCl₅ is smaller, 0.60 mol/L. In Experiments 3 and 4, we approach equilibrium from the other side of the system. That is, we start with a mixture of PCl₃ and Cl₂. They react, by Reaction 10.3, to form some PCl₅.

For each experiment in Table 10.1, we list the final equilibrium concentrations of PCl₅, PCl₃, and Cl₂. You might wonder whether there is any rela-

TABLE 10.1 Original and Equilibrium Concentrations (mol/L) in Experiments 1–4

	Expt. 1		Expt. 2		Expt. 3		Expt. 4	
	Orig.	Equil.	Orig.	Equil.	Orig.	Equil.	Orig.	Equil.
Conc. PCl_5	1.00	0.80	0.60	0.45	0.00	0.80	0.00	0.40
Conc. PCl_3	0.00	0.20	0.00	0.15	1.00	0.20	0.60	0.20
Conc. Cl_2	0.00	0.20	0.00	0.15	1.00	0.20	0.50	0.10

tionship between these quantities, valid for all four experiments. It turns out that there is. In each case

$$\frac{(\text{equil. conc. } PCl_3) \times (\text{equil. conc. } Cl_2)}{(\text{equil. conc. } PCl_5)} = 0.050$$

To check that this is the case, substitute numbers

Expt. 1 $\quad \dfrac{0.20 \times 0.20}{0.80} = \dfrac{0.040}{0.80} = 0.050$

Expt. 2 $\quad \dfrac{0.15 \times 0.15}{0.45} = \dfrac{0.0225}{0.45} = 0.050$

Expt. 3 $\quad \dfrac{0.20 \times 0.20}{0.80} = \dfrac{0.040}{0.80} = 0.050$

Expt. 4 $\quad \dfrac{0.20 \times 0.10}{0.40} = \dfrac{0.020}{0.40} = 0.050$

Notice that this relationship is valid only for *equilibrium* concentrations. It does not hold for the original concentrations listed in Table 10.1. Neither would it apply to data obtained on the way to equilibrium. To emphasize this, we write equilibrium concentrations in a special way. We enclose them in square brackets. For this system at 250°C

$$\frac{[PCl_3] \times [Cl_2]}{[PCl_5]} = 0.050$$

where **the symbol [] stands for equilibrium concentration in moles per liter.**

10.4 Equilibrium Constants for Gaseous Systems

The number just obtained for the PCl_5–PCl_3–Cl_2 equilibrium system at 250°C, 0.050, is called an **equilibrium constant** and is given the symbol K_c. Thus for the system

$$PCl_5(g) \rightleftharpoons PCl_3(g) + Cl_2(g)$$

$$K_c = \frac{[PCl_3] \times [Cl_2]}{[PCl_5]} = 0.050 \text{ at } 250°C$$

In words, this relation tells us that

1. At equilibrium, the product of the concentration of PCl_3 times that of Cl_2 divided by that of PCl_5 has a constant value.
2. At 250°C, this constant, K_c, has the value 0.050.
3. There is no way, at 250°C, that we can have a system containing PCl_5, PCl_3, and Cl_2 at equilibrium unless, in that system, the quotient

$$\frac{[PCl_3] \times [Cl_2]}{[PCl_5]}$$

equals 0.050.

Throughout the rest of this chapter, we will deal with equilibrium constants for many different gaseous systems. The expressions we write for K_c for different systems may look quite different. The numerical value of K_c will certainly differ from one system to another. However, in all cases, the value of K_c is independent of

— the original concentrations of reactants or products.
— container volume.
— the pressure of the system.

It is for these reasons that we refer to K_c as an equilibrium "constant." Its value does not change when we change original concentrations, volume, or pressure. (On the other hand, K_c does change with temperature, as we will see in Section 10.6.)

Form of K_c Expression

We have just seen that for the equilibrium system

$$PCl_5(g) \rightleftharpoons PCl_3(g) + Cl_2(g)$$

the form of the equilibrium constant expression is

$$K_c = \frac{[PCl_3] \times [Cl_2]}{[PCl_5]}$$

An analogous expression can be written for every gaseous system at equilibrium. To deduce the form of that expression, it may be helpful to look at two more systems

System	Expression for K_c
$2HI(g) \rightleftharpoons H_2(g) + I_2(g)$	$K_c = \dfrac{[H_2] \times [I_2]}{[HI]^2}$
$2SO_2(g) + O_2(g) \rightleftharpoons 2SO_3(g)$	$K_c = \dfrac{[SO_3]^2}{[SO_2]^2 \times [O_2]}$

From these examples it should be clear that, in the expression for K_c

1. The equilibrium concentrations of products (right side of equation) appear in the numerator. The equilibrium concentrations of reactants (left side of equation) appear in the denominator.

2. Each equilibrium concentration is raised to a power equal to the coefficient of the substance in the balanced equation (e.g., 1 for H_2, 1 for I_2, 2 for HI).

3. If there is more than one product, the concentration terms are multiplied by one another (e.g., $[H_2] \times [I_2]$). Similarly, if there is more than one reactant, their concentration terms are multiplied (e.g., $[SO_2]^2 \times [O_2]$).

EXAMPLE 10.2

Write the expression for K_c for

$N_2(g) + 3 H_2(g) \rightleftharpoons 2 NH_3(g)$

Solution
Following the rules just cited, we have

$$K_c = \frac{[NH_3]^2}{[N_2] \times [H_2]^3}$$

In other words, for this system, the ratio $[NH_3]^2/[N_2] \times [H_2]^3$ has a fixed value at a given temperature, independent of volume, pressure, or original concentration.

Exercise
What is the expression for K_c for the system $N_2O_4(g) \rightleftharpoons 2 NO_2(g)$? Answer: $K_c = [NO_2]^2/[N_2O_4]$.

Evaluation of K_c

To determine the magnitude of the equilibrium constant for a system, we must know the equilibrium concentrations of all the reactants and products. If these concentrations are given directly, K_c is readily calculated.

EXAMPLE 10.3

For the equilibrium: $2\ HI(g) \rightleftharpoons H_2(g) + I_2(g)$ at 520°C, it is found that in a certain equilibrium mixture

$[H_2]$ = 0.029 mol/L; $[I_2]$ = 0.020 mol/L; $[HI]$ = 0.19 mol/L

Calculate K_c for this system at 520°C.

Solution
The expression for K_c is

$$K_c = \frac{[H_2] \times [I_2]}{[HI]^2}$$

Substituting numbers,

$$K_c = \frac{(0.029) \times (0.020)}{(0.19)^2} = 0.016$$

Exercise
For the system referred to in Example 10.2 at a certain temperature, $[NH_3]$ = $[N_2]$ = $[H_2]$ = 0.10 mol/L. Calculate K_c. Answer: 1.0×10^2.

Sometimes, calculating K_c is a bit more involved. You may have to first determine one or more of the equilibrium concentrations (Example 10.4).

EXAMPLE 10.4

Consider the equilibrium

$$2\ HI(g) \rightleftharpoons H_2(g) + I_2(g)$$

At 400°C, a student starts with pure HI at a concentration of 0.22 mol/L. Some of the HI decomposes, forming H_2 and I_2. The equilibrium concentration of H_2 is found to be 0.020 mol/L. What is
a. $[I_2]$ b. $[HI]$ c. K_c at 400°C?

Solution
a. Note from the balanced equation that, for every mole of H_2 formed, one mole of I_2 is formed at the same time. In this case, 0.020 mol/L of H_2 was formed. An equal number of moles per liter of I_2 must have formed at the same time. Hence, the equilibrium concentration of I_2, like that of H_2, must be 0.020 mol/L.
b. The balanced equation tells us that *two* moles of HI are required to form *one* mole of H_2. In order to form 0.020 mol/L of H_2, twice as much HI, 0.040 mol/L, must have decomposed. Since there was 0.22 mol/L of HI to begin with

$[HI]$ = 0.22 mol/L − 0.040 mol/L = 0.18 mol/L

Continued

It may help to summarize the reasoning in (a) and (b) by a table

	Original Concentration (mol/L)	Change in Concentration (mol/L)	Equilibrium Concentration (mol/L)
HI	0.22	−0.040	0.18
H_2	0.00	+0.020	0.020
I_2	0.00	+0.020	0.020

c. $K_c = \dfrac{[H_2] \times [I_2]}{[HI]^2} = \dfrac{(0.020) \times (0.020)}{(0.18)^2} = 0.012$

Exercise
Consider the system $N_2O_4(g) \rightleftharpoons 2\,NO_2(g)$. At 100°C, a student starts with pure N_2O_4 at a concentration of 0.100 mol/L. It is found that $[NO_2]$ = 0.120 mol/L. Calculate $[N_2O_4]$ and K_c. Answer: 0.040 mol/L; 0.36.

10.5 Equilibrium Constant and Extent of Reaction

Depending upon the position of the equilibrium involved, we may get a yield of 100%, 90%, - - - 0% in a reaction. The extent of reaction is related to the magnitude of the equilibrium constant, K_c. In this section, we will consider the nature of this relationship.

K_c Very Small

If the equilibrium constant for a reaction is very small, very little product will be formed. Consider, for example, the reaction

$$Cl_2(g) \rightleftharpoons 2\,Cl(g) \qquad (10.5)$$

The equilibrium constant for this reaction at 25°C is 1×10^{-38}, a very small number indeed

$$K_c = 1 \times 10^{-38} = \frac{[Cl]^2}{[Cl_2]}$$

From the expression for K_c, we see that [Cl] appears in the numerator while $[Cl_2]$ is in the denominator. In order for the ratio $[Cl]^2/[Cl_2]$ to be 1×10^{-38}, the concentration of Cl must be very small relative to that of Cl_2. In other words, Reaction 10.5 proceeds to only a tiny extent. At room temperature, virtually all of a sample of elemental chlorine is in the form of Cl_2 rather than Cl atoms.

EXAMPLE 10.5

In an equilibrium system at 25°C, $[Cl_2] = 1 \times 10^{-2}$ mol/L. Taking K_c to be 1×10^{-38}, calculate the equilibrium concentration of Cl atoms.

Solution

We start with the equation

$$\frac{[Cl]^2}{[Cl_2]} = K_c$$

Solving for $[Cl]^2$,

$$[Cl]^2 = K_c \times [Cl_2]$$

Substituting numbers,

$$[Cl]^2 = (1 \times 10^{-38})(1 \times 10^{-2}) = 1 \times 10^{-40}$$

Extracting square roots

$$[Cl] = 1 \times 10^{-20} \text{ mol/L}$$

We see that the concentration of chlorine atoms is very small indeed. For every Cl atom, there are

$$\frac{1 \times 10^{-2}}{1 \times 10^{-20}} = 1 \times 10^{18} \text{ Cl}_2 \text{ molecules}$$

Exercise

When $[Cl] = 1 \times 10^{-22}$ mol/L, what is $[Cl_2]$? Answer: 1×10^{-6} M.

K_c Very Large

If the equilibrium constant for a system is very large, we expect the forward reaction to go virtually to completion. As an example, consider the reaction

$$2 \text{ CO(g)} + \text{O}_2(g) \rightleftharpoons 2 \text{ CO}_2(g) \tag{10.6}$$

At 500°C

$$K_c = \frac{[CO_2]^2}{[CO]^2 \times [O_2]} = 4 \times 10^{30} \text{ (at 500°C)}$$

Since K_c is so large, the concentration of CO_2 at equilibrium must be much, much larger than those of CO and O_2. This means that Reaction 10.6 must go nearly to completion. Virtually all of the carbon monoxide and oxygen are converted to carbon dioxide at equilibrium.

Experimentally, we find that the prediction just made is confirmed. In the catalytic converter of an automobile, Reaction 10.6 goes essentially to completion. Any carbon monoxide in the exhaust is converted to carbon dioxide in the presence of the Pt–Pd catalyst at 500°C.

There is one restriction on this type of prediction. *The magnitude of K_c tells us nothing about the rate of reaction.* Often, reactions for which K_c is very large take place slowly. As an example, consider Reaction 10.6, this time at 25°C. Here, K_c is even larger than before

$$K_c = \frac{[CO_2]^2}{[CO]^2 \times [O_2]} = 2 \times 10^{91} \text{ at } 25°C$$

From the magnitude of K_c, you would expect Reaction 10.6 to go virtually to completion at 25°C. In principle, it does; at *equilibrium*, the concentration of CO_2 is much, much greater than those of CO and O_2. However, at 25°C, the rate is so small that the reaction takes almost forever to reach equilibrium. Carbon monoxide exposed to oxygen at room temperature in the absence of a catalyst stays around for a long, long time. In general, to be sure that a reaction will give a high yield of product in a reasonable time, two conditions must be met

1. K_c must be large.
2. The rate must be large enough so that equilibrium is reached quickly.

K_c Neither Very Large nor Very Small

We have seen what happens at the extremes, where K_c is very large or very small. Suppose, though, that K_c has an "in-between" value. Perhaps it lies between 10^{-5} and 10^5, to pick a couple of numbers. In that case, we cannot say that the extent of reaction will be virtually zero (K_c very small) or nearly 100% (K_c very large). Instead, we expect to get an equilibrium mixture containing appreciable amounts of both products and reactants.

To illustrate what happens when K_c is neither very large nor very small, consider the system

$$2 \text{ HI}(g) \rightleftharpoons H_2(g) + I_2(g)$$

As pointed out earlier, K_c at 520°C is 0.016

$$K_c = \frac{[H_2] \times [I_2]}{[HI]^2} = 0.016$$

Suppose now that, in an equilibrium mixture at 520°C, $[H_2] = [I_2] = 0.10$ mol/L. Then

$[HI]^2 \times 0.016 = (0.10)^2$

$[HI]^2 = 0.010/0.016 = 0.63; [HI] = 0.80$ mol/L

The equilibrium mixture contains about 80% HI (conc. = 0.80 mol/L), 10%

H$_2$ (conc. = 0.10 mol/L) and 10% I$_2$ (conc. = 0.10 mol/L). In other words, all three substances are present in appreciable amounts in the equilibrium mixture.

EXAMPLE 10.6

Industrially, ammonia is made by the reaction

N$_2$(g) + 3 H$_2$(g) ⇌ 2 NH$_3$(g)

For this system at 400°C, K$_c$ = 0.50. Suppose [N$_2$] = 1.2 mol/L, [H$_2$] = 3.6 mol/L. What is [NH$_3$]?

Solution

The equilibrium constant expression is

$$\frac{[NH_3]^2}{[N_2] \times [H_2]^3} = K_c$$

Solving for [NH$_3$]^2

$$[NH_3]^2 = K_c \times [N_2] \times [H_2]^3$$

Substituting numbers

$$[NH_3]^2 = 0.50 \times 1.2 \times (3.6)^3 = 28$$

Solving for [NH$_3$]

[NH$_3$] = 5.3 mol/L

We see that the concentration of ammonia at equilibrium is roughly comparable to those of N$_2$ and H$_2$. The equilibrium mixture would contain about as much product (5.3 mol/L) as reactants (4.8 mol/L for both N$_2$ and H$_2$).

Exercise

Suppose that at 400°C, [N$_2$] = [NH$_3$] = 0.10 mol/L. What is [H$_2$]? Answer: [H$_2$]^3 = 0.20; [H$_2$] = 0.58 mol/L.

When K$_c$ is neither very small nor very large, the position of the equilibrium is very sensitive to external conditions. By varying these conditions, we can adjust the yield of products within rather wide limits. In Section 10.6, we will consider how this is done. All the systems we will work with will be ones where K$_c$ has an "in-between" value.

10.6 Changes in Gaseous Equilibrium Systems

After a system has reached equilibrium, it is still possible to change the relative amounts of products and reactants. To do this, we can

1. Add or remove a gaseous reactant or product.
2. Compress or expand a gaseous system, thereby changing the total pressure.
3. Change the temperature by heating or cooling the system.

(Incidentally, it is *not* possible to change the position of an equilibrium by adding a catalyst. A catalyst can only change the rate of a reaction; it has no effect upon the equilibrium.)

We can predict the direction in which an equilibrium will shift when we change the external conditions. To do this, we apply a principle first suggested by the French chemist Henri Le Chatelier in 1884: **When a system at equilibrium is disturbed by a change in conditions, the system shifts so as to partially cancel the effect of the change.**

This means that the final position of the equilibrium will be somewhere between

—the original equilibrium position.
—the unstable position immediately after the change.

Adding or Removing a Reactant or Product

This situation offers perhaps the simplest illustration of Le Chatelier's principle. **If a system at equilibrium is disturbed by adding a gaseous reactant or product, reaction will occur in such a direction (forward or reverse) as to consume part of the added gas. Conversely, if a gaseous species is removed, the system will shift so as to restore part of that species.**

To illustrate this rule, let's consider the equilibrium

$$PCl_5(g) \rightleftharpoons PCl_3(g) + Cl_2(g)$$

Suppose that, to the equilibrium system, we add some PCl_5. According to Le Chatelier's principle, the forward reaction ($PCl_5 \rightarrow PCl_3 + Cl_2$) should occur. Part of the added PCl_5 will decompose, forming PCl_3 and Cl_2. The final equilibrium concentration of PCl_5 will be

—somewhat larger than it was originally.
—somewhat smaller than it was immediately after the PCl_5 was added.

Table 10.2 illustrates this reasoning with some numbers. In Experiment 1, we first establish equilibrium at 250°C in a one-liter container. At this point, we have 0.80 mol PCl_5, 0.20 mol PCl_3, and 0.20 mol Cl_2. We then disturb this equilibrium by adding 0.50 mol PCl_5. Temporarily, the amount of PCl_5 increases to 1.30 mol. Then the forward reaction

$$PCl_5(g) \rightarrow PCl_3(g) + Cl_2(g)$$

occurs; 0.05 mol PCl_5 decomposes. This partially cancels the change. The final equilibrium concentration of PCl_5, 1.25 mol/L, is intermediate between the original equilibrium concentration, 0.80 mol/L, and the temporary value, 1.30

mol/L. The final concentrations of PCl$_3$ and Cl$_2$ are larger, by 0.05 mol/L, than they were originally.

At the bottom of Table 10.2, we show the results of a second experiment with the PCl$_5$–PCl$_3$–Cl$_2$ system. This time, we *remove* 0.50 mol of PCl$_5$. The reverse reaction:

$$PCl_5(g) \leftarrow PCl_3(g) + Cl_2(g)$$

occurs to restore part of the PCl$_5$. The concentration of PCl$_5$, which dropped from 0.80 to 0.30 mol/L when PCl$_5$ was removed, bounces part way back. It goes up to 0.36 mol/L. Since PCl$_3$ and Cl$_2$ are consumed to form PCl$_5$, their final equilibrium concentrations are lower than they were originally.

EXAMPLE 10.7

Consider the equilibrium: $2\ CO(g) + O_2(g) \rightleftharpoons 2\ CO_2(g)$
In which direction does reaction occur if
a. O$_2$ is added to the equilibrium system?
b. O$_2$ is withdrawn from the equilibrium system?
c. CO$_2$ is withdrawn from the equilibrium system?

Solution
a. $2\ CO(g) + O_2(g) \rightarrow 2\ CO_2(g)$; part of added O$_2$ is consumed.
b. $2\ CO(g) + O_2(g) \leftarrow 2\ CO_2(g)$; part of withdrawn O$_2$ is restored.
c. $2\ CO(g) + O_2(g) \rightarrow 2\ CO_2(g)$; part of withdrawn CO$_2$ is restored.

Exercise
Suppose that in (a), the original equilibrium concentration of O$_2$ is 0.10 mol/L. Enough O$_2$ is added to temporarily double its concentration. When equilibrium is restored [O$_2$] could be which one of the following: 0.05 M, 0.10 M, 0.15 M, 0.20 M, 0.30 M? Answer: 0.15 M.

TABLE 10.2 Effect of Adding or Removing PCl$_5$ on the System: PCl$_5$ (g) $\rightleftharpoons$ PCl$_3$(g) + Cl$_2$(g); (250°C in a 1.00-L container)

Experiment 1: Adding 0.50 mol PCl$_5$

	Original Concentration (Equilibrium)	Temporary Concentration (Nonequilibrium)	Final Concentration (Equilibrium)
PCl$_5$	0.80 mol/L	1.30 mol/L	1.25 mol/L
PCl$_3$	0.20	0.20	0.25
Cl$_2$	0.20	0.20	0.25

Experiment 2: Removing 0.50 mol PCl$_5$

	Original Concentration (Equilibrium)	Temporary Concentration (Nonequilibrium)	Final Concentration (Equilibrium)
PCl$_5$	0.80 mol/L	0.30 mol/L	0.36 mol/L
PCl$_3$	0.20	0.20	0.14
Cl$_2$	0.20	0.20	0.14

Expansion or Compression

There is another way in which we might disturb the equilibrium system

$$PCl_5(g) \rightleftharpoons PCl_3(g) + Cl_2(g)$$

Suppose we were to increase the volume, as shown in Figure 10.9. The immediate effect would be to reduce the concentration of gas molecules. The number of molecules per unit volume decreases because we have the same number of molecules in a larger volume.

According to Le Chatelier's principle, the system will shift to partially cancel the decrease in concentration of molecules. To accomplish this, some PCl_5 decomposes

$$PCl_5(g) \rightarrow PCl_3(g) + Cl_2(g)$$

This reaction increases the overall concentration of molecules, because we get two molecules (PCl_3 and Cl_2) from one molecule (PCl_5). So, this reaction occurs when the volume of the system increases.

Suppose now that we compress the PCl_5–PCl_3–Cl_2 system, decreasing the volume. The immediate effect is to increase the concentration of gas molecules; the same number of molecules occupies a smaller volume. To partially offset this effect, the reverse reaction

$$PCl_5(g) \leftarrow PCl_3(g) + Cl_2(g)$$

occurs. This decreases the total number of gas molecules (2 molecules → 1

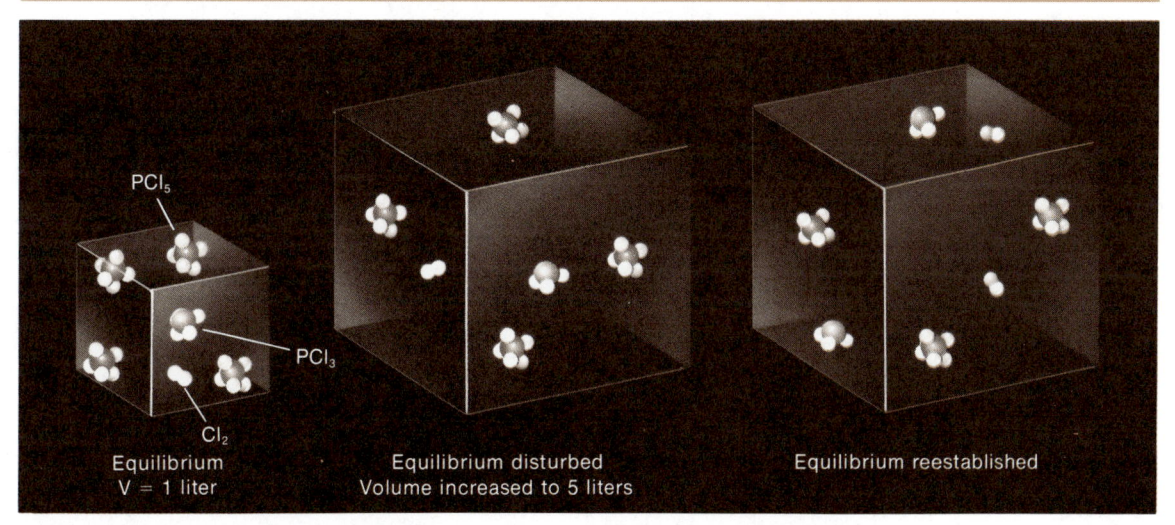

Figure 10.9 Diagram showing the effect of an expansion on an equilibrium mixture of PCl_5, PCl_3, and Cl_2. The equilibrium system will tend to shift toward PCl_3 and Cl_2 to counteract the decrease in concentrations caused by the expansion.

molecule). The concentration of gas molecules decreases; there are fewer molecules in the same volume.

The argument just made can be applied to all gaseous systems. Simple rules can be used to predict the effect of expansion or compression upon an equilibrium system. In stating these rules, we usually refer to expansion as a "decrease in pressure" and compression as an "increase in pressure." The rules are:

1. **When the pressure on an equilibrium system is decreased, a reaction occurs that increases the number of moles of gas.** Examples include

    $PCl_5(g) \rightarrow PCl_3(g) + Cl_2(g)$; 1 mol gas → 2 mol gas

    $2 CO_2(g) \rightarrow 2 CO(g) + O_2(g)$; 2 mol gas → 3 mol gas

2. **When the pressure on an equilibrium system is increased, a reaction occurs that decreases the number of moles of gas.** Examples include

    $PCl_3(g) + Cl_2(g) \rightarrow PCl_5(g)$; 2 mol gas → 1 mol gas

    $2 CO(g) + O_2(g) \rightarrow 2 CO_2(g)$; 3 mol gas → 2 mol gas

---

EXAMPLE 10.8
Predict what reaction will occur, if any, when the pressure is increased on each of the following equilibrium systems
a. $2 NO_2(g) \rightleftharpoons 2 NO(g) + O_2(g)$
b. $2 HI(g) \rightleftharpoons H_2(g) + I_2(g)$
c. $2 HI(g) \rightleftharpoons H_2(g) + I_2(s)$

Solution
a. $2 NO(g) + O_2(g) \rightarrow 2 NO_2(g)$; 3 mol gas → 2 mol gas
b. No reaction, because there are the same number of moles of gas on each side of the equation.
c. $2 HI(g) \rightarrow H_2(g) + I_2(s)$; 2 mol gas → 1 mol gas
    Note that it is the number of moles of *gas* that is important.

Exercise
To increase the yield of SO_3 in the reaction: $2 SO_2(g) + O_2(g) \rightleftharpoons 2 SO_3(g)$, would you increase or decrease the pressure? Answer: Increase.

Changing the Temperature

Let's consider once again the equilibrium system

$PCl_5(g) \rightleftharpoons PCl_3(g) + Cl_2(g)$

Suppose now that we supply heat to the system, perhaps using a Bunsen burner. The immediate effect is to raise the temperature. According to Le Chatelier's principle, the system will shift to counteract this effect. It can do this in a simple way. The forward reaction is endothermic, i.e., it absorbs heat

$$PCl_5(g) \rightarrow PCl_3(g) + Cl_2(g); \Delta H = +92.5 \text{ kJ}$$

Hence this reaction occurs, absorbing part of the heat supplied. This makes the temperature increase smaller than it would otherwise be. When equilibrium is finally restored, there will be more products (PCl_3, Cl_2) and less reactant (PCl_5).

We can readily extend this argument to deduce what will happen to the PCl_5–PCl_3–Cl_2 equilibrium system when the temperature is reduced. We start by noting that *ΔH for the reverse reaction is equal in magnitude but opposite in sign to that for the forward reaction.* Thus we have

$$PCl_3(g) + Cl_2(g) \rightarrow PCl_5(g); \Delta H = -92.5 \text{ kJ}$$

If the system is cooled, this exothermic reaction occurs, giving off heat. The evolution of heat makes the temperature decrease smaller than it would otherwise be. When equilibrium is restored, there is more PCl_5, less $PCl_3 + Cl_2$.

In general, we can say that

1. **When the temperature of an equilibrium system is increased, a reaction that is endothermic ($\Delta H > 0$) takes place.**

    $PCl_5(g) \rightarrow PCl_3(g) + Cl_2(g); \Delta H = +92.5 \text{ kJ}$

    $2 CO_2(g) \rightarrow 2 CO(g) + O_2(g); \Delta H = +566.0 \text{ kJ}$

2. **When the temperature of an equilibrium system is decreased, a reaction that is exothermic ($\Delta H < 0$) takes place.**

    $PCl_3(g) + Cl_2(g) \rightarrow PCl_5(g); \Delta H = -92.5 \text{ kJ}$

    $2 CO(g) + O_2(g) \rightarrow 2 CO_2(g); \Delta H = -566.0 \text{ kJ}$

EXAMPLE 10.9

Consider the system

$2 SO_2(g) + O_2(g) \rightleftharpoons 2 SO_3(g); \Delta H = -198.2 \text{ kJ}$

What will happen if the temperature is increased? decreased?

Solution

The forward reaction is exothermic. (The quoted value of ΔH always applies to the reaction from left to right.)

Continued

> $2 SO_2(g) + O_2(g) \rightarrow 2 SO_3(g); \Delta H = -198.2$ kJ
>
> The reverse reaction is endothermic; ΔH has the same magnitude but the opposite sign.
>
> $2 SO_3(g) \rightarrow 2 SO_2(g) + O_2(g); \Delta H = +198.2$ kJ
>
> Hence an increase in temperature will cause SO_3 to decompose to SO_2 and O_2. Conversely, a decrease in temperature will cause the reaction: $2 SO_2(g) + O_2(g) \rightarrow 2 SO_3(g)$ to occur; SO_3 will form at the expense of SO_2 and O_2.
>
> **Exercise**
> When the system $N_2O_4(g) \rightleftharpoons 2 NO_2(g)$ is heated, more NO_2 is formed. What is the sign of ΔH for the forward reaction? Answer: $+$.

As we mentioned earlier, a change in temperature is the one factor that can change the equilibrium constant for a reaction. We sometimes describe the effect of temperature upon the position of an equilibrium in terms of its effect upon K_c. The general rule is that an *increase in temperature*

—*increases K_c if the forward reaction is endothermic ($\Delta H > 0$).*
—*decreases K_c if the forward reaction is exothermic ($\Delta H < 0$).*

Thus we would predict that for the system

$$PCl_5(g) \rightleftharpoons PCl_3(g) + Cl_2(g); \Delta H = +92.5 \text{ kJ}$$

the equilibrium constant, K_c, should increase as the temperature rises. Conversely, for the system

$$2 SO_2(g) + O_2(g) \rightleftharpoons 2 SO_3(g); \Delta H = -198.2 \text{ kJ}$$

K_c should decrease as the temperature increases. As you can see from Table 10.3, these predictions are confirmed.

TABLE 10.3 Effect of Temperature upon the Equilibrium Constant

	$PCl_5(g) \rightleftharpoons PCl_3(g) + Cl_2(g); \Delta H = +92.5$ kJ				
t (°C)	200	250	300	400	500
K_c	0.0049	0.050	0.25	3.7	28

	$2 SO_2(g) + O_2(g) \rightleftharpoons 2 SO_3(g); \Delta H = -198.2$ kJ				
t (°C)	500	600	700	800	900	1000
K_c	150,000	5000	350	40	6.6	1.5

Key Words

activation energy   equilibrium constant, K_c

catalyst
equilibrium
Le Chatelier's principle
rate of reaction

Questions

10.1 Explain in your own words what is meant by reaction rate.
10.2 Why do very few collisions result in reaction?
10.3 Define activation energy.
10.4 A reaction with a large activation energy is expected to have a _____ rate. A reaction with a small activation energy usually has a _____ rate.
10.5 List three factors that can affect the rate of reaction.
10.6 What happens to reaction rate when concentration is increased? When a gaseous reaction mixture is compressed?
10.7 An increase in temperature of about _____ °C doubles reaction rate.
10.8 Explain in terms of kinetic theory why an increase in temperature increases reaction rate.
10.9 What is a catalyst? How does a catalyst increase the rate of reaction?
10.10 Why is a chemical system at equilibrium referred to as "dynamic"?
10.11 If the equilibrium constant is of the order of 10^{10}, the forward reaction proceeds to a _____ extent.
10.12 If the equilibrium constant is of the order of 10^{-10}, the forward reaction proceeds to a _____ extent.
10.13 If an equilibrium mixture contains appreciable amounts of both reactants and products, the value of K_c is neither _____ nor _____.
10.14 List three ways in which the position of an equilibrium can be changed.
10.15 What effect does a catalyst have on the position of a chemical equilibrium?
10.16 State Le Chatelier's principle.
10.17 Describe what happens to an equilibrium system if a gaseous reactant is added.
10.18 How do you decide what effect a change in pressure will have upon the position of a chemical equilibrium?
10.19 How do you decide upon the effect of a change in temperature upon the position of a chemical equilibrium?
10.20 If a reaction is endothermic, K_c _____ as the temperature increases; if the reaction is exothermic, K_c _____ as the temperature increases.

Problems

Factors Affecting Reaction Rates

10.21 Consider the reaction $2\ NO(g) + O_2(g) \rightarrow 2\ NO_2(g)$. What effect would you expect each of the following changes to have on the rate of this reaction?

a. increase in NO concentration
b. increase in NO_2 concentration
c. decrease in temperature
d. compression
e. expansion

Chapter 10

10.22 Consider the reaction $CO(g) + NO_2(g) \rightarrow CO_2(g) + NO(g)$. Describe at least four different ways in which the rate of this reaction could be increased.

10.23 Consider the reaction $H_2(g) + I_2(g) \rightarrow 2\ HI(g)$. How would you adjust the temperature and pressure to slow down this reaction?

Expression for K_c

10.24 Write expressions for K_c for the following systems
a. $2\ H_2(g) + O_2(g) \rightleftharpoons 2\ H_2O(g)$
b. $SO_3(g) \rightleftharpoons SO_2(g) + \frac{1}{2} O_2(g)$
c. $N_2(g) + 3\ H_2(g) \rightleftharpoons 2\ NH_3(g)$

10.25 Write expressions for K_c for the following systems
a. $2\ HI(g) \rightleftharpoons H_2(g) + I_2(g)$
b. $H_2(g) + I_2(g) \rightleftharpoons 2\ HI(g)$
c. $\frac{1}{2} H_2(g) + \frac{1}{2} I_2(g) \rightleftharpoons HI(g)$

10.26 Write chemical equations corresponding to the following K_c expressions
a. $K_c = \dfrac{[NO]^4 \times [H_2O]^6}{[NH_3]^4 \times [O_2]^5}$
b. $K_c = \dfrac{[H_2O]^3 \times [NO]^2}{[O_2]^{\frac{5}{2}} \times [NH_3]^2}$
c. $K_c = \dfrac{[NH_3]^4 \times [O_2]^5}{[NO]^4 \times [H_2O]^6}$

10.27 Write chemical equations corresponding to the following expressions for K_c
a. $K_c = \dfrac{[NO]^2}{[N_2] \times [O_2]}$
b. $K_c = \dfrac{[NO_2]^2}{[N_2O_4]}$
c. $K_c = \dfrac{[SF_6]}{[S] \times [F_2]^3}$

Calculation of K_c

10.28 Consider the system $PCl_5(g) \rightleftharpoons PCl_3(g) + Cl_2(g)$. At 300°C, it is found that $[PCl_5] = 0.12$ M, $[PCl_3] = 0.18$ M, $[Cl_2] = 0.16$ M. Calculate K_c.

10.29 Calculate K_c for the system $H_2(g) + I_2(g) \rightleftharpoons 2\ HI(g)$ if the equilibrium concentrations are found to be $[H_2] = 3.8 \times 10^{-3}$ M, $[I_2] = 1.5 \times 10^{-3}$ M, and $[HI] = 1.7 \times 10^{-2}$ M.

10.30 When NO_2 and SO_2 are allowed to react and reach equilibrium, it is found that $[SO_3] = 0.50$ M, $[NO] = 0.40$ M, $[NO_2] = 1.4$ M, and $[SO_2] = 1.1$ M. Calculate the equilibrium constant for the reaction
$$SO_2(g) + NO_2(g) \rightleftharpoons NO(g) + SO_3(g)$$

10.31 For the system $N_2O_4(g) \rightleftharpoons 2\ NO_2(g)$ at 100°C, $[N_2O_4] = [NO_2] = 0.36$ M. Calculate K_c.

10.32 Consider the system $PCl_5(g) \rightleftharpoons PCl_3(g) + Cl_2(g)$. At 400°C, 1.00 mol of PCl_5 is put in a 1.00-L container. At equilibrium, 0.18 mol of PCl_5 is left. Calculate
a. $[PCl_5]$ b. $[PCl_3]$ c. $[Cl_2]$
d. K_c

10.33 Consider the system $2\ SO_3(g) \rightleftharpoons 2\ SO_2(g) + O_2(g)$. At 800°C, 0.100 mol of SO_3 is placed in a 1.00-L container. At equilibrium, 0.0500 mol of SO_3 is left. Calculate
a. $[SO_3]$ b. $[SO_2]$ c. $[O_2]$ d. K_c

Calculation of Equilibrium Concentrations

10.34 For the system $PCl_5(g) \rightleftharpoons PCl_3(g) + Cl_2(g)$ at 250°C, $K_c = 0.050$. Calculate $[PCl_5]$ if $[PCl_3] = 0.012$ M, $[Cl_2] = 0.34$ M.

10.35 For the system $2\ SO_3(g) \rightleftharpoons 2\ SO_2(g) + O_2(g)$ at 800°C, $K_c = 0.025$. Calculate $[SO_2]$ if $[SO_3] = [O_2] = 0.10$ M.

10.36 For the system $Cl_2(g) \rightleftharpoons 2\ Cl(g)$, $K_c = 1 \times 10^{-38}$ at 25°C. Calculate
a. $[Cl_2]$ when $[Cl] = 1 \times 10^{-20}$ M
b. $[Cl]$ when $[Cl_2] = 0.015$ M

10.37 The equilibrium constant for the reaction $N_2O_4(g) \rightleftharpoons 2 NO_2(g)$ is 0.36 at 100°C. Calculate
  a. $[NO_2]$ when $[N_2O_4] = 0.10$ M
  b. $[N_2O_4]$ when $[NO_2] = 0.10$ M

10.38 Consider the system $N_2(g) + 3 H_2(g) \rightleftharpoons 2 NH_3(g)$. K_c is 0.50 at 400°C. Calculate $[NH_3]$ when $[N_2] = [H_2] = 0.010$ M

Le Chatelier's Principle

10.39 Consider the system $N_2O_4(g) \rightleftharpoons 2 NO_2(g)$. Which way does the equilibrium shift if
  a. N_2O_4 is added?
  b. NO_2 is added?
  c. N_2O_4 is removed?
  d. NO_2 is removed?

10.40 Which way does the equilibrium in Problem 10.39 shift if the system is compressed? expanded?

10.41 Consider the system $N_2(g) + 3 H_2(g) \rightleftharpoons 2 NH_3(g)$. What happens to the position of this equilibrium if
  a. H_2 is added?
  b. the system is compressed?
  c. NH_3 is removed?

10.42 Predict the effect of compression on the positions of the following equilibria
  a. $2 NO_2(g) \rightleftharpoons 2 NO(g) + O_2(g)$
  b. $4 NH_3(g) + 5 O_2(g) \rightleftharpoons 4 NO(g) + 6 H_2O(g)$
  c. $2 CO(g) + O_2(g) \rightleftharpoons 2 CO_2(g)$

10.43 Predict the effect of an increase in volume upon the positions of the following equilibria
  a. $H_2(g) + Br_2(g) \rightleftharpoons 2 HBr(g)$
  b. $H_2(g) + Br_2(l) \rightleftharpoons 2 HBr(g)$
  c. $CaCO_3(s) \rightleftharpoons CaO(s) + CO_2(g)$
  d. $2 NH_3(g) \rightleftharpoons N_2(g) + 3 H_2(g)$

10.44 For the system

$$2 CO(g) + O_2(g) \rightleftharpoons 2 CO_2(g); \Delta H = -586 \text{ kJ}$$

what effect will an increase in temperature have upon
  a. the position of the equilibrium?
  b. the value of K_c?

10.45 For the system

$$2 NO_2(g) \rightleftharpoons 2 NO(g) + O_2(g); \Delta H = +113 \text{ kJ}$$

predict the effect of an increase in temperature upon
  a. the position of the equilibrium
  b. the value of K_c

10.46 Predict what effect increases in temperature and pressure will have upon the position of the equilibrium $N_2(g) + O_2(g) \rightleftharpoons 2 NO(g); \Delta H = +90 \text{ kJ}$

10.47 Consider the Haber process for the synthesis of ammonia

$$N_2(g) + 3 H_2(g) \rightleftharpoons 2 NH_3(g); \Delta H = -92 \text{ kJ}$$

If your objective is to obtain as high a yield as possible of ammonia, how would you adjust the temperature and pressure?

Multiple Choice

10.48 For the system $PCl_5(g) \rightleftharpoons PCl_3(g) + Cl_2(g)$, suppose you start with 1.00 mol of PCl_5 and 1.00 mol of Cl_2 in a 1.00-L container. At equilibrium, the concentration of chlorine will be
  a. less than 1.00 M
  b. 1.00 M
  c. greater than 1.00 M
  d. cannot answer without knowing the value of K_c

10.49 For the system $2\,HI(g) \rightleftharpoons H_2(g) + I_2(g)$, K_c = 0.010 at a certain temperature. The ratio $[H_2]^{\frac{1}{2}} \times [I_2]^{\frac{1}{2}}/[HI]$ is
  a. 1.0×10^{-4}
  b. 1.0×10^{-2}
  c. 1.0×10^{-1}
  d. 1.0

10.50 For the system $N_2(g) + 2\,O_2(g) \rightleftharpoons 2\,NO_2(g)$, the expression for K_c is
  a. $\dfrac{[N_2] \times [O_2]}{[NO_2]}$
  b. $\dfrac{[N_2] \times [O_2]^2}{[NO_2]^2}$
  c. $\dfrac{[NO_2]}{[N_2] \times [O_2]}$
  d. $\dfrac{[NO_2]^2}{[N_2] \times [O_2]^2}$

10.51 For the system $2\,HI(g) \rightleftharpoons H_2(g) + I_2(g)$, it is found that $[H_2] = [I_2] = 0.01$ M, $[HI] = 0.1$ M. This means that K_c is
  a. 1×10^{-4} b. 1×10^{-3} c. 1×10^{-2}
  d. 1×10^{-1} e. 1

10.52 For the system $2\,SO_3(g) \rightleftharpoons 2\,SO_2(g) + O_2(g)$, K_c = 400. In a certain equilibrium system, $[O_2] = [SO_3] = 1.0$ M. The concentration of SO_2 in moles per liter is
  a. 2.5×10^{-3} b. 2.0×10^{-1} c. 1.0
  d. 2.0×10^1 e. 4.0×10^2

10.53 Given the following data

	K_c
$N_2(g) + O_2(g) \rightleftharpoons 2\,NO(g)$;	1×10^{-30}
$2\,H_2(g) + O_2(g) \rightleftharpoons 2\,H_2O(g)$;	2×10^{81}
$2\,CO_2(g) \rightleftharpoons 2\,CO(g) + O_2(g)$;	4×10^{-92}

Which one of the following is most likely to dissociate to give O_2?
  a. H_2O b. CO_2 c. NO   d. none of these will dissociate

10.54 For the system $2\,SO_3(g) \rightleftharpoons 2\,SO_2(g) + O_2(g)$, $K_c = 3 \times 10^{-3}$ at 700°C and 2×10^{-2} at 800°C. This means that ΔH for the forward reaction is
  a. less than zero   b. zero   c. greater than zero

10.55 Given $N_2O_4(g) \rightleftharpoons 2\,NO_2(g)$; $\Delta H = +58$ kJ To get as high a yield as possible of NO_2 from N_2O_4, you should
  a. increase both the temperature and the pressure
  b. decrease both the temperature and the pressure
  c. increase the temperature, decrease the pressure
  d. decrease the temperature, increase the pressure

10.56 A certain reaction is exothermic. The activation energy, E_a, for this reaction
  a. is a positive quantity
  b. is a negative quantity
  c. may be either positive or negative

10.57 Increasing the temperature increases reaction rate principally because
  a. most reactions are endothermic
  b. the fraction of fast moving molecules is greatly increased
  c. collisions occur more often
  d. the activation energy decreases

10.58 To decrease the activation energy for a reaction, you could
  a. increase the reactant concentrations
  b. raise the temperature
  c. compress the system
  d. find an appropriate catalyst

10.59 For which one of the following reactions would increases in temperature and pressure increase both the rate of reaching equilibrium and the yield of product?
  a. $N_2(g) + O_2(g) \rightleftharpoons 2\,NO(g)$: $\Delta H > 0$
  b. $N_2(g) + 3H_2(g) \rightleftharpoons 2\,NH_3(g)$; $\Delta H < 0$
  c. $SO_2(g) + \dfrac{1}{2}O_2(g) \rightleftharpoons SO_3(g)$; $\Delta H < 0$
  d. $2\,CO_2(g) \rightleftharpoons 2\,CO(g) + O_2(g)$; $\Delta H > 0$
  e. $2\,HI(g) \rightleftharpoons H_2(g) + I_2(s)$; $\Delta H > 0$

10.60 The rate of a certain reaction is 3.6×10^{-1} mol/L · min. The rate expressed in mol/L · hr is
 a. 1.0×10^{-4}
 b. 6.0×10^{-3}
 c. 3.6×10^{-1}
 d. 22
 e. 1.3×10^{3}

Acids and Bases

Learning Objectives
After studying this chapter, you should be able to:

1. Calculate one of the two quantities [H$^+$] or [OH$^-$] given the value of the other quantity (Example 11.1).
2. Calculate one of the two quantities [H$^+$] or pH, given the value of the other quantity (Examples 11.3, 11.4).
3. Given either the [H$^+$] or the pH of a solution, classify the solution as acidic, basic, or neutral (Example 11.2).
4. Write the formulas of the six common strong acids and the equations for their dissociation in water.
5. Calculate K$_a$ for a weak acid HX, given [H$^+$] and the original concentration of HX (Example 11.5).
6. Calculate [H$^+$] in a solution of a weak acid HX, given K$_a$ and [HX] (Example 11.6).
7. Write the formulas of the common strong bases and the equations for their dissociation in water.
8. Distinguish between a strong acid and a weak acid; between a strong base and a weak base (Example 11.7).
9. Write net ionic equations (Example 11.8) for the reaction of:
   —a strong acid with a strong base (Equation 11.18).
   —a weak acid with a strong base (Equation 11.21).
   —a strong acid with the weak base ammonia (Equation 11.22).
10. Use acid–base titration data to calculate the concentration of an acid or base in water solution (Example 11.9).
11. Given the equation for a Brønsted–Lowry acid–base reaction, identify the acid and base (Example 11.10).
12. Given the equation for a Lewis acid–base reaction, identify the acid and base.

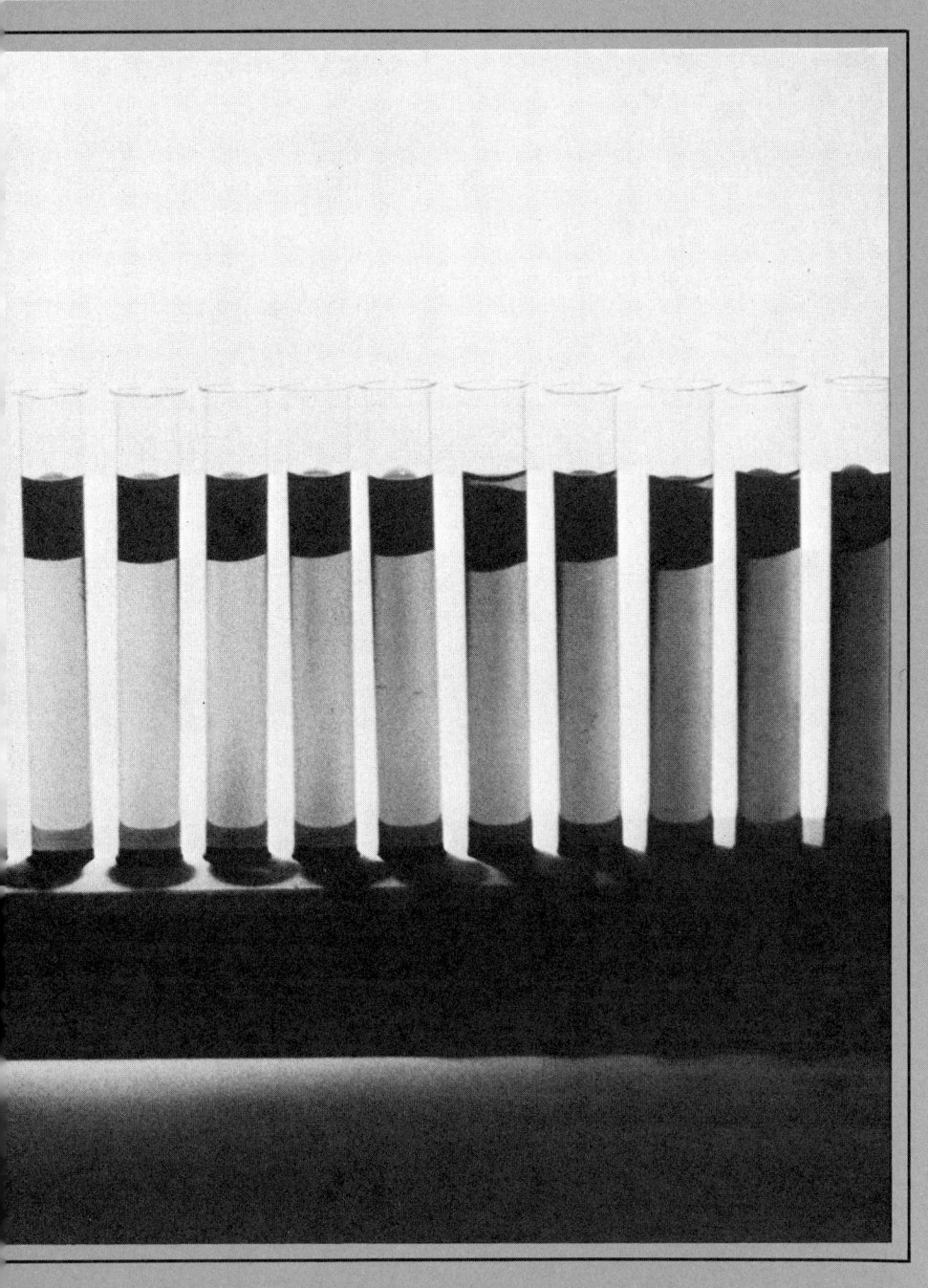

> Here is the world, sound as a nut, perfect, not the smallest piece of chaos left, never a stitch nor an end, not a mark of haste, or botching, or second thought; but the theory of the world is a thing of shreds and patches.
>
> RALPH WALDO EMERSON

From Harris, W. E., and B. Kratochvil: *An Introduction to Chemical Analysis*. Philadelphia, Saunders College Publishing, 1981.

CHAPTER 11

In earlier chapters, we occasionally mentioned acids and bases. We are now ready to look more closely at the properties and reactions of these compounds. Our emphasis will be on aqueous solutions of acids and bases. Such solutions are common both in the laboratory and the household. Vinegar and lemon juice are acidic; ammonia and milk of magnesia are basic.

All acidic water solutions have certain properties in common.

1. They have a sour taste. Acetic acid makes vinegar sour. The sharp taste of lemon juice is due to another organic acid, citric acid. Sulfuric acid probably has a sour taste too, but no one in his or her right mind would make that test.
2. They turn the color of an organic dye called litmus from blue to red.
3. They react with several metals, including zinc and aluminum, to generate hydrogen gas.
4. They react with basic solutions to neutralize their properties.

Basic water solutions also have certain common properties.

1. They have a slippery feeling. You may have noticed this property in using household ammonia, strong detergents, or laundry soap. All of these form basic water solutions.
2. They turn the color of litmus from red to blue.
3. They react with several cations, including Mg^{2+} and Fe^{3+}, to form insoluble precipitates.
4. They react with acidic solutions to neutralize their properties.

We begin this chapter by looking at the nature and concentrations of the ions found in acidic and basic solutions (Section 11.1). Then we consider what kinds of substances act as acids (Section 11.2) or bases (Section 11.3). Acid–base reactions are discussed in Section 11.4. Finally, in Section 11.5, we will look at some more general models of acids and bases that go beyond water solutions.

11.1 Acidic and Basic Water Solutions

The properties of acidic water solutions are due to the H^+ ion (proton). It is this ion that is responsible for the sour taste of vinegar and the color change (blue to red) with litmus. It is also H^+ ions in water solution that react with zinc or aluminum

$$Zn(s) + 2\ H^+(aq) \rightarrow Zn^{2+}(aq) + H_2(g) \tag{11.1}$$

$$Al(s) + 3\ H^+(aq) \rightarrow Al^{3+}(aq) + \frac{3}{2} H_2(g) \tag{11.2}$$

In a sense, the equations we have just written are oversimplified. The H^+ ion, a bare proton, does not exist by itself in water. Instead, it attaches itself to a water molecule to form a species called a *hydronium ion*

$$H^+ + H\!-\!\ddot{O}\!-\!H \rightarrow \left[H\!-\!\underset{\underset{H}{|}}{O}\!-\!H \right]^+$$

proton   hydronium ion

The properties of acidic water solutions are often ascribed to the H_3O^+ ion. However, we will keep things simple by using $H^+(aq)$ in all the equations we write. (Actually, most cations are bonded to H_2O molecules in solution, but we usually ignore this when we write equations involving those ions.)

The properties of basic water solutions are due to the hydroxide ion, OH^-. It is this ion that is responsible for the slippery feeling and the color change with litmus (red to blue). The OH^- ion also is involved in the precipitation reactions with Mg^{2+} and Fe^{3+}. The net ionic equations for these reactions are

$$Mg^{2+}(aq) + 2\ OH^-(aq) \rightarrow Mg(OH)_2(s) \tag{11.3}$$

$$Fe^{3+}(aq) + 3\ OH^-(aq) \rightarrow Fe(OH)_3(s) \tag{11.4}$$

Relation Between [H⁺] and [OH⁻]

As we have just seen, H^+ ions make a water solution acidic; OH^- ions make a solution basic. Could we just say then that a water solution containing H^+ ions is acidic while one containing OH^- ions is basic? Not quite, as it turns out. You may be surprised to learn that all water solutions contain both of these ions. This happens because water itself is slightly ionized

$$H_2O \rightleftharpoons H^+(aq) + OH^-(aq) \tag{11.5}$$

As a result of this equilibrium, there must be at least a few H^+ ions in every water solution, no matter how strongly basic it may be. By the same token, an acidic solution must contain at least a few OH^- ions.

To relate the concentrations of H^+ and OH^- ions in solution, we need to set up an equilibrium constant expression for the system shown in Equation 11.5. Following the rules discussed in Chapter 10, we have

$$K = \frac{[H^+] \times [OH^-]}{[H_2O]} \quad (11.6)$$

Here, K is the equilibrium constant. The terms $[H^+]$, $[OH^-]$, and $[H_2O]$ refer to the equilibrium concentrations in moles per liter of these three species.

The expression just written can be simplified considerably. The concentration of H_2O molecules in water solution is very large and virtually constant. Adding H^+ ions, OH^- ions, or any other ions you might think of, has almost no effect on the number of H_2O molecules per unit volume. In other words, we can say that

$[H_2O]$ = constant

Substituting in Equation 11.6, we have;

$$K = \frac{[H^+] \times [OH^-]}{\text{constant}}$$

or

$K \times \text{constant} = [H^+] \times [OH^-]$

The left side of this equation is the product of two constants. For convenience, we combine the constants, using the symbol K_w to stand for their product. By experiment, we find that K_w is 1.0×10^{-14} at 25°C. In other words

$$K_w = [H^+] \times [OH^-] = 1.0 \times 10^{-14} \quad (11.7)$$

Equation 11.7 is easier to work with than 11.6 because it involves only two concentration terms rather than three.

Equation 11.7 tells us that

1. *The extent of ionization of water (Equation 11.5) is very small.* Consider pure water, where there are equal numbers of H^+ and OH^- ions. Here

$[H^+] = [OH^-]$

So

$[H^+]^2 = 1.0 \times 10^{-14}$; $[H^+] = 1.0 \times 10^{-7} M = [OH^-]$

In other words, in pure water the concentrations of both H^+ and OH^- are very small, approximately 10^{-7} M. Only a tiny fraction (less than one in a hundred million) of the H₂O molecules are ionized. Small wonder that pure water does not conduct electricity!

2. *The concentrations of H^+ and OH^- in water solution are inversely proportional to one another* (Figure 11.1). The relation $[H^+] \times [OH^-] = K_w$ is analogous to Boyle's Law: $PV = k_1$. Just as gas pressure decreases when volume increases, so $[H^+]$ decreases when $[OH^-]$ increases. To illustrate what this means, suppose we start with pure water, in which, as we have pointed out

$$[H^+] = [OH^-] = 1.0 \times 10^{-7} \text{ M}$$

Suppose we now add enough NaOH to the water to increase the concentration of OH^- ions by a factor of ten, from 1.0×10^{-7} to 1.0×10^{-6} M

$$[OH^-] = 1.0 \times 10^{-6} \text{ M}$$

From Equation 11.7, we see that

$$[H^+] \times (1.0 \times 10^{-6}) = 1.0 \times 10^{-14}$$

$$[H^+] = \frac{1.0 \times 10^{-14}}{1.0 \times 10^{-6}} = 1.0 \times 10^{-8} \text{ M}$$

Notice that $[H^+]$ is only 1/10 of its value in pure water (i.e., $10^{-8}/10^{-7} =$

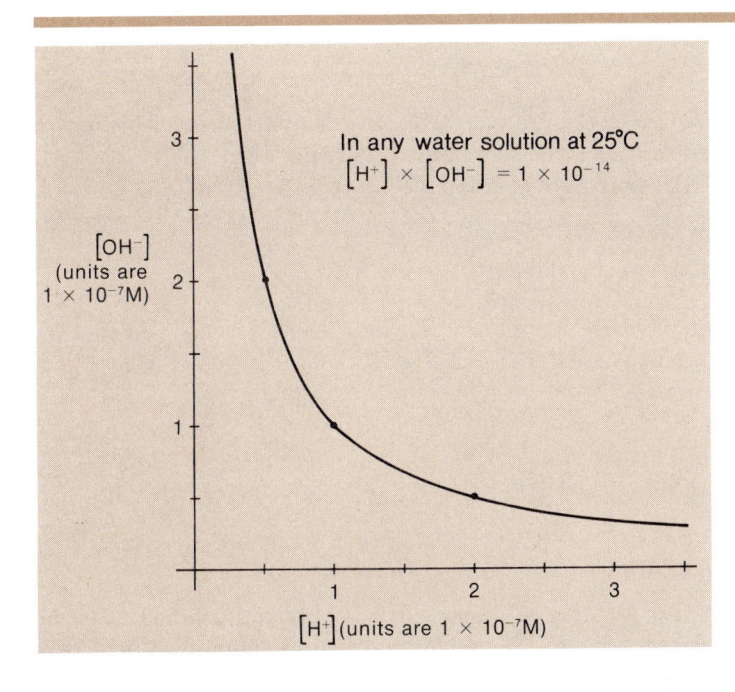

Figure 11.1 A graph of OH^- versus H^+ looks very much like a graph of gas volume versus pressure (compare Figure 6.7, p. 144). In both cases, the two variables are inversely proportional to one another. When H^+ gets larger, OH^- gets smaller.

0.1). In other words, when [OH$^-$] increases by a factor of ten, [H$^+$] decreases to 1/10 of its original value.

EXAMPLE 11.1
In a 0.020 M solution of NaOH, the concentration of OH$^-$ ions is 0.020 M. What is the concentration of H$^+$ ions?

Solution
Solving Equation 11.7 for [H$^+$]

$$[H^+] = \frac{1.0 \times 10^{-14}}{[OH^-]}$$

Substituting [OH$^-$] = 0.020 M = 2.0×10^{-2} M

$$[H^+] = \frac{1.0 \times 10^{-14}}{2.0 \times 10^{-2}} = 0.50 \times 10^{-12} \text{ M} = 5.0 \times 10^{-13} \text{ M}$$

Exercise
What is [OH$^-$] in a solution 0.0050 M in H$^+$? *Answer:* 2.0×10^{-12} M.

Equation 11.7 offers a simple way of distinguishing among neutral, acidic and basic solutions.

1. In a **neutral** solution, there are equal numbers of H$^+$ and OH$^-$ ions. The concentrations of both of these ions are those in pure water, 1.0×10^{-7} M

   [H$^+$] = [OH$^-$] = 1.0×10^{-7} M

2. In an **acidic** solution, there are more H$^+$ ions than OH$^-$ ions. This means that the concentration of H$^+$ ions must be greater than 1.0×10^{-7} M, while that of OH$^-$ must be less than 1.0×10^{-7} M.

   [H$^+$] > [OH$^-$]

   [H$^+$] > 1.0×10^{-7} M > [OH$^-$]

3. In a **basic** solution, there are more OH$^-$ ions than H$^+$ ions. The concentration of OH$^-$ ions is greater than 1.0×10^{-7} M; that of H$^+$ is less than this number.

   [OH$^-$] > [H$^+$]

   [OH$^-$] > 1.0×10^{-7} M > [H$^+$]

pH

As we have seen, it is possible to describe the acidity of a solution in terms of the concentration of H$^+$ ions. We might say that a solution in which [H$^+$]

= 10^{-1} M is "highly acidic." Another solution with [H$^+$] = 10^{-5} M would be "slightly acidic," and so on. There is one drawback to expressing acidity in this way. Many people are unfamiliar with the meaning of exponential numbers or uncomfortable with their use. With this in mind, Peder Sørensen in 1909 suggested another way to describe acidity. This involves the use of a quantity known as **pH**, defined by the relation

$$pH = -\log[H^+] \tag{11.8}$$

Using Equation 11.8, we can readily calculate pH when the concentration of H$^+$ is a simple exponential such as 10^{-1} M or 10^{-5} M. To do this, we use the fact that

$$\log(10^{-x}) = -x$$

For a solution in which [H$^+$] = 10^{-1} M

$$pH = -\log(10^{-1}) = -(-1) = 1$$

Again, if

$$[H^+] = 10^{-5} \text{ M}; \quad pH = -\log(10^{-5}) = -(-5) = 5$$

Table 11.1 gives the pH values for a series of solutions in which the concentration of H$^+$ varies from 1 to 1 × 10^{-14} M. Notice that

1. The pH values cover a range from 0 to 14. A 1 M solution of HCl, in which [H$^+$] = 1 M, has a pH of 0. A 1 M NaOH solution, where [OH$^-$] = 1 M and [H$^+$] = 10^{-14} M, has a pH of 14. Most of the solutions you work with in the laboratory, like those in the table, have pH's that fall between 0 and 14.

TABLE 11.1 Relation Between [H$^+$], [OH$^-$], and pH

	[H$^+$]	[OH$^-$]	pH	
1 M HCl	1 M	10^{-14} M	0	
0.1 M HCl	10^{-1} M	10^{-13} M	1	
0.01 M HCl	10^{-2} M	10^{-12} M	2	
0.001 M HCl	10^{-3} M	10^{-11} M	3	
0.000 1 M HCl	10^{-4} M	10^{-10} M	4	
0.000 01 M HCl	10^{-5} M	10^{-9} M	5	
0.000 001 M HCl	10^{-6} M	10^{-8} M	6	
pure water	10^{-7} M	10^{-7} M	7	Decreasing acidity
0.000 001 M NaOH	10^{-8} M	10^{-6} M	8	
0.000 01 M NaOH	10^{-9} M	10^{-5} M	9	
0.000 1 M NaOH	10^{-10} M	10^{-4} M	10	
0.001 M NaOH	10^{-11} M	10^{-3} M	11	
0.01 M NaOH	10^{-12} M	10^{-2} M	12	
0.1 M NaOH	10^{-13} M	10^{-1} M	13	
1 M NaOH	10^{-14} M	1 M	14	↓

Chapter 11

2. A solution that is neutral, with $[H^+] = [OH^-] = 10^{-7}$ M, has a pH of 7. Acidic solutions, where $[H^+] > 10^{-7}$ M, have a pH less than 7. Basic solutions, where $[H^+] < 10^{-7}$ M, have a pH greater than 7. That is

   acidic solution   pH < 7

   neutral solution   pH = 7

   basic solution   pH > 7

3. The pH of a solution increases as $[H^+]$ decreases. For example, when $[H^+]$ decreases from 10^{-1} M to 10^{-2} M, the pH increases from 1 to 2. In general, the higher the pH of a solution, the lower its H^+ ion concentration.

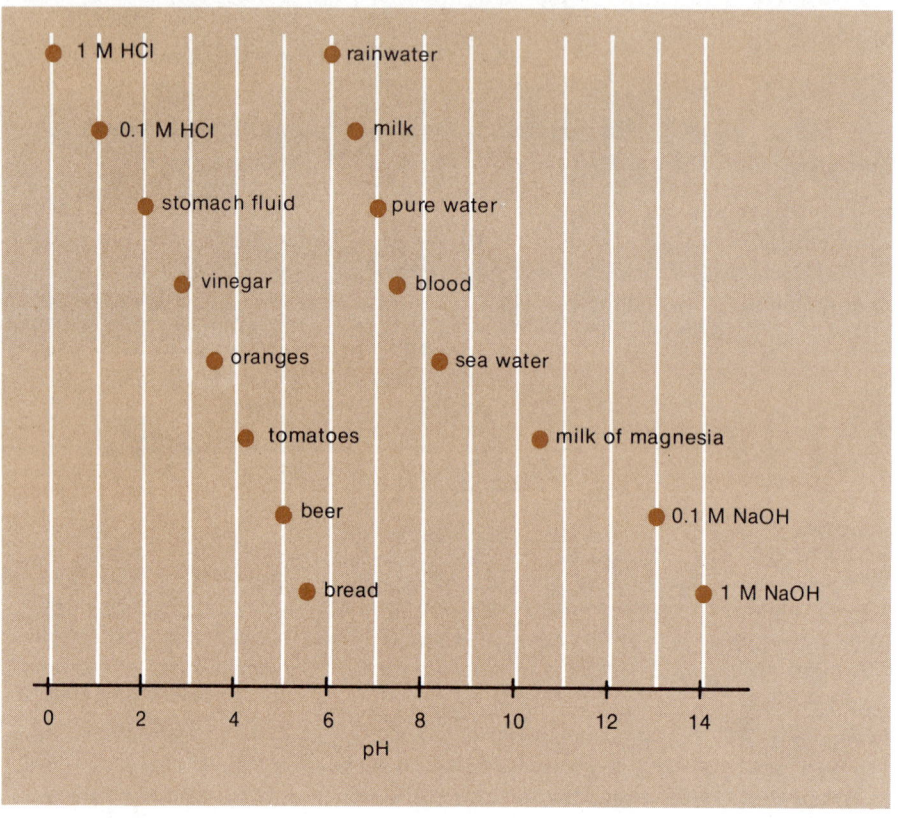

Figure 11.2 Most water solutions are not neutral because most solutes have acidic or basic properties.

EXAMPLE 11.2

Figure 11.2 gives the pH of several common solutions. Consider the four solutions: beer, blood, seawater, and vinegar.
a. Which of these solutions are acidic? basic?
b. Of the four solutions, which one has the highest concentration of H^+? the lowest?

Solution
a. Vinegar and beer, each with a pH less than 7, are acidic. Blood and seawater, with pH's greater than 7, are basic.
b. Vinegar has the lowest pH (≈ 3), so has the highest $[H^+]$. Seawater, with the highest pH (≈ 8.5), has the lowest $[H^+]$.

Exercise
"Acid rain" typically has a pH of 4. What is the ratio of the H^+ ion concentration in acid rain to that in ordinary rainwater (Figure 11.2)? Answer: 100 : 1.

As we have seen, Equation 11.8 can be used to find, by inspection, the pH of a solution when $[H^+]$ can be expressed as a simple exponent, such as 10^{-1} or 10^{-5}. In general

if $[H^+] = 10^{-x}$, then pH = x

Extracting the pH is a bit more complex if $[H^+]$ is an exponential number such as 4×10^{-5} M or 3.6×10^{-9} M. Here, it is simplest to use your calculator (Example 11.3).

EXAMPLE 11.3

What is the pH of a solution if $[H^+] = 4 \times 10^{-5}$ M?

Solution
There are a couple of different ways of setting up this calculation on your calculator.
(1) Enter 4×10^{-5} in exponential notation on your calculator (see Appendix 1). Now, press the "log" key. You should find that the logarithm of 4×10^{-5} is -4.4. Since pH $= -\log [H^+]$, it follows that pH = 4.4.
(2) Use the fact that the logarithm of the product of two numbers is the sum of their logarithms. Thus,

$$\log (4 \times 10^{-5}) = \log 4 + \log 10^{-5}$$

Find log 4 on your calculator; you should get log 4 = 0.6. The log of 10^{-5} is -5. Hence,

Continued

$$\log (4 \times 10^{-5}) = 0.6 - 5.0 = -4.4$$

Since pH = $-\log[H^+]$,

$$pH = -(-4.4) = 4.4$$

Exercise
What is the pH of a solution in which $[H^+] = 3.6 \times 10^{-9}$ M? Answer: pH = 8.44.

Sometimes you will need to convert a known pH to the corresponding concentration of H^+ ion. To do this, it is convenient to write

$$[H^+] = 10^{-pH} \tag{11.9}$$

If the pH is a simple integer, $[H^+]$ is readily found. For example, for a solution of pH 11, $[H^+] = 10^{-11}$ M. If the pH is not integral (e.g., pH = 11.3), you will need to use your calculator to find $[H^+]$ (Example 11.4).

EXAMPLE 11.4
What is $[H^+]$ in a solution of pH 11.3?

Solution
Applying Equation 11.9; $[H^+] = 10^{-11.3}$. First enter 11.3 on your calculator. Now change the sign to -11.3 by pushing the $+/-$ key. The next operation will depend upon the type of calculator you have. Either
(1) press the 10^x key, if you have one.
(2) press first the INV key and then the LOG key.
You should find that $10^{-11.3} = 5 \times 10^{-12}$. Hence; $[H^+] = 5 \times 10^{-12}$ M. If your calculator won't do this operation for you, you should get one that will (see Appendix 1).

Exercise
What is $[H^+]$ in a solution of pH 2.71? Answer: 1.9×10^{-3} M.

11.2 Strong and Weak Acids

For our purposes, an acid can be defined to be a species that reacts to form H^+ ions in water solution. Depending upon the extent to which this reaction occurs, an acid can be classified as strong or weak.

Strong Acids

A strong acid is one that dissociates completely to form H^+ ions in water solution. Hydrochloric acid is a typical strong acid. In dilute water solution, HCl is 100% dissociated to H^+ and Cl^- ions. In other words, the reaction

$$HCl(aq) \rightarrow H^+(aq) + Cl^-(aq) \tag{11.10}$$

goes to completion. If you add 0.10 mol of HCl to a liter of water, you will find that

$[H^+] = 0.10$ M; $[Cl^-] = 0.10$ M; pH $= 1.00$

The concentration of HCl molecules in this solution is zero.

There are only six common strong acids. They are

HCl hydrochloric acid
HBr hydrobromic acid
HI hydriodic acid
HClO$_4$ perchloric acid
HNO$_3$ nitric acid
H$_2$SO$_4$ sulfuric acid

In each case, the strong acid is completely dissociated to ions in water solution. The dissociation reactions are entirely analogous to that with HCl

$$HNO_3(aq) \rightarrow H^+(aq) + NO_3^-(aq) \tag{11.11}$$

$$H_2SO_4(aq) \rightarrow H^+(aq) + HSO_4^-(aq) \tag{11.12}$$

Of the six strong acids, you are likely to come across only three in the general chemistry laboratory. These are hydrochloric acid, nitric acid, and sulfuric acid. Each acid is available at two different concentrations. These are

dilute HCl 6 M   concentrated HCl 12 M
dilute HNO$_3$ 6 M   concentrated HNO$_3$ 16 M
dilute H$_2$SO$_4$ 3 M   concentrated H$_2$SO$_4$ 18 M

Concentrated acids are generally unpleasant to work with, for reasons that have little to do with their acid strength. A solution of 12 M HCl gives off fumes of HCl(g), which has a penetrating, acrid odor. Concentrated nitric acid often has a yellow or brown color due to NO$_2$, a gas with an odor at least as unpleasant as that of HCl. Concentrated sulfuric acid is the most dangerous of the lot. It is an oily liquid that gives off a great deal of heat when it reacts with water. If you spill any concentrated acid on your skin or clothing, flood it with plenty of water and call for your instructor.

Weak Acids

A weak acid is only partially dissociated in water. Hydrofluoric acid is a typical weak acid. When added to water, hydrogen fluoride forms an equilibrium mixture containing HF molecules, H^+ ions, and F^- ions

$$HF(aq) \rightleftharpoons H^+(aq) + F^-(aq) \tag{11.13}$$

If 0.100 mol of HF is added to a liter of water, only about 8% of its ionizes. That is:

$[H^+] = 0.0080$ M; $[F^-] = 0.0080$ M

Most of the hydrogen fluoride stays in the molecular form

$[HF] = 0.100$ M $-$ 0.008 M $= 0.092$ M

The pH of 0.10 M HF is considerably higher than that of 0.10 M HCl (pH = 1.00)

pH of 0.10 M HF $= -\log[H^+] = -\log(0.0080) = 2.10$

The two solutions also differ in electrical conductivity (Figure 11.3). Hydrochloric acid (H^+, Cl^- ions) is a *strong electrolyte*, comparable to sodium chloride (Na^+, Cl^- ions). Hydrofluoric acid is a *weak electrolyte*, only slightly dissociated to ions in water.

The behavior of HF in water solution is typical of weak acids. There are a great many such acids, several of them organic (containing carbon). Perhaps the most common organic acid is acetic acid, the acidic component of vinegar. The formula of acetic acid is often written as $HC_2H_3O_2$ to emphasize that the molecule contains one ionizable hydrogen atom. In water solution, acetic acid is partially dissociated to form H^+ ions and acetate ions, $C_2H_3O_2^-$

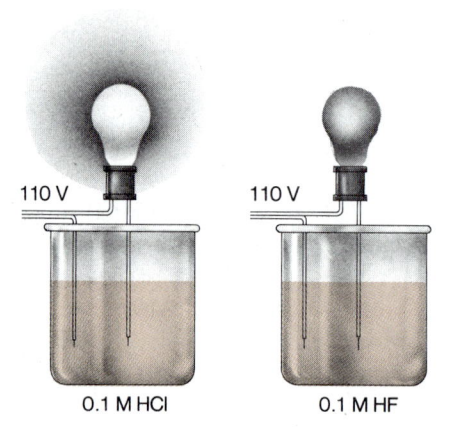

Figure 11.3 A 0.1-M water solution of HCl, a strong acid, is a much better conductor than 0.1 M HF, a weak acid. In one liter of 0.1 M HCl, there is 0.2 mol of ions (0.1 mol H^+, 0.1 mol Cl^-). In one liter of 0.1 M HF, there is only about 0.016 mol of ions (0.008 mol H^+, 0.008 mol F^-).

$$HC_2H_3O_2(aq) \rightleftharpoons H^+(aq) + C_2H_3O_2^-(aq) \qquad (11.14)$$

When we call an acid "weak," we refer only to the fact that its dissociation in water is incomplete. Some weak acids are quite dangerous to work with. Hydrogen fluoride is extremely corrosive; indeed, it is used to etch glass. For that reason, you will not find hydrofluoric acid in the general chemistry laboratory. Another weak acid, hydrogen cyanide, HCN, is a deadly poison. You certainly won't find HCN in the lab. In fact, as Lizzie Borden is reputed to have found, you can't buy it in a drugstore either.

Dissociation Constants of Weak Acids

As we have seen, a weak acid HX in water solution is in equilibrium with its ions, H^+ and X^-.

$$HX(aq) \rightleftharpoons H^+(aq) + X^-(aq)$$

We can write an equilibrium constant for this reaction

$$K_a = \frac{[H^+] \times [X^-]}{[HX]}$$

The equilibrium constant is referred to as the *dissociation constant* of the weak acid.

The magnitude of K_a is a measure of the extent to which a weak acid dissociates in water. The smaller the value of K_a the smaller the extent of dissociation (at a given concentration) and hence the weaker the acid. Looking at Table 11.2, we conclude that $HC_2H_3O_2$ must be a weaker acid than HF

$$K_a \text{ of } HC_2H_3O_2 = \frac{[H^+] \times [C_2H_3O_2^-]}{[HC_2H_3O_2]} = 1.8 \times 10^{-5}$$

TABLE 11.2 Dissociation Constants of Weak Acids

Acid		K_a	
Sulfurous acid	$H_2SO_3(aq) \rightleftharpoons H^+(aq) + HSO_3^-(aq)$	1.7×10^{-2}	
Phosphoric acid	$H_3PO_4(aq) \rightleftharpoons H^+(aq) + H_2PO_4^-(aq)$	7.5×10^{-3}	
Hydrofluoric acid	$HF(aq) \rightleftharpoons H^+(aq) + F^-(aq)$	7.0×10^{-4}	
Nitrous acid	$HNO_2(aq) \rightleftharpoons H^+(aq) + NO_2^-(aq)$	4.5×10^{-4}	
Formic acid	$HCHO_2(aq) \rightleftharpoons H^+(aq) + CHO_2^-(aq)$	1.8×10^{-4}	Decreasing
Acetic acid	$HC_2H_3O_2(aq) \rightleftharpoons H^+(aq) + C_2H_3O_2^-(aq)$	1.8×10^{-5}	acid
Carbonic acid	$H_2CO_3(aq) \rightleftharpoons H^+(aq) + HCO_3^-(aq)$	4.2×10^{-7}	strength
Hydrogen sulfide	$H_2S(aq) \rightleftharpoons H^+(aq) + HS^-(aq)$	1×10^{-7}	
Hypochlorous acid	$HClO(aq) \rightleftharpoons H^+(aq) + ClO^-(aq)$	3.2×10^{-8}	
Boric acid	$H_3BO_3(aq) \rightleftharpoons H^+(aq) + H_2BO_3^-(aq)$	5.8×10^{-10}	
Hydrogen cyanide	$HCN(aq) \rightleftharpoons H^+(aq) + CN^-(aq)$	4.0×10^{-10} ↓	

$$K_a \text{ of HF} = \frac{[H^+] \times [F^-]}{[HF]} = 7.0 \times 10^{-4}$$

The number 1.8×10^{-5} is less than one tenth as large as 7.0×10^{-4}. Acetic acid is indeed the weaker acid of the two. In 0.10 M solution, it is only about 1% ionized, as compared with 8% for HF.

The dissociation constant of a weak acid can be measured in the laboratory. To do this, we start by adding a known amount of the acid to a given volume of water. We then measure the concentration of H^+ ion (or the pH) of the water solution. Using the equation for the dissociation of the weak acid, K_a can then be calculated (Example 11.5).

EXAMPLE 11.5

To determine K_a of a certain weak acid, HX, a student adds 1.00 mol of HX to a liter of water. The following reaction occurs

$$HX(aq) \rightleftharpoons H^+(aq) + X^-(aq)$$

She measures the concentration of H^+ in the solution and finds it to be 0.0030 M. Calculate K_a of the weak acid.

Solution
The expression for K_a is

$$K_a = \frac{[H^+] \times [X^-]}{[HX]}$$

For every mole of H^+ formed, one mole of X^- is formed at the same time. Hence

$$[X^-] = [H^+] = 0.0030 \text{ M}$$

For every mole of H^+ formed, one mole of HX is consumed. We started with 1.00 mol of HX and formed 0.0030 mol of H^+, all in one liter of solution. Hence

$$[HX] = 1.00 \text{ M} - 0.0030 \text{ M} = 1.00 \text{ M (3 significant figures)}$$

Substituting in the expression for K_a

$$K_a = \frac{[H^+] \times [X^-]}{[HX]} = \frac{(3.0 \times 10^{-3})(3.0 \times 10^{-3})}{1.00} = 9.0 \times 10^{-6}$$

Exercise
Suppose the original concentration of HX was 0.50 M and that of H^+ was 0.010 M. Calculate K_a. Answer: 2.0×10^{-4}.

We have just seen that, knowing the original concentration of weak acid and $[H^+]$, it is possible to calculate K_a. This suggests that if we know the original concentration of weak acid and K_a, we should be able to calculate $[H^+]$ in the solution of the weak acid. This can indeed be done as shown in Example 11.6.

EXAMPLE 11.6

A certain weak acid has a dissociation constant of 1.0×10^{-5}. That is

$$HX(aq) \rightleftharpoons H^+(aq) + X^-(aq); K_a = \frac{[H^+] \times [X^-]}{[HX]} = 1.0 \times 10^{-5}$$

Calculate $[H^+]$ in a solution prepared by adding enough HX to water to make $[HX] = 0.10$ M.

Solution
Following the reasoning described in Example 11.5, we can say that

$$[H^+] = [X^-]$$

Substituting in the expression for K_a

$$\frac{[H^+] \times [H^+]}{0.10} = 1.0 \times 10^{-5}$$

Solving

$$[H^+]^2 = 1.0 \times 10^{-6}; [H^+] = 1.0 \times 10^{-3} \text{ M}$$

Exercise
What is the pH of this solution? Answer: 3.00.

The calculation we have just gone through is a very common one. Often, you are asked to calculate the concentration of H^+ or the pH in a solution of a weak acid. Many times, though, the problem is phrased a bit differently. Instead of being told the *equilibrium* concentration of HX as in Example 11.6, you may be given only its *original* concentration. That makes the problem more difficult to set up because you now have to relate [HX] to $[H^+]$. Curiously, though, the answer is usually the same whichever way the problem is worded. Since a weak acid is only slightly dissociated in water, it will ordinarily be true that

orig. conc. HX ≈ [HX]

If you look back at Example 11.5, you will find that this relation was valid there. The same is true of Example 11.6 and indeed for every weak acid problem cited in this text.

11.3 Strong and Weak Bases

A base is a species that reacts to form OH^- ions in water solution. As with acids, we distinguish between strong and weak bases. A strong base is one that dissociates completely to form OH^- ions. In contrast, a weak base is only partially dissociated to give OH^- ions in water.

You will recall that there are only a few strong acids. Similarly, there are few strong bases. The common ones are

—the hydroxides of the Group 1 metals: LiOH, NaOH, KOH, RbOH, CsOH.
—the hydroxides of the Group 2 metals: Mg(OH)$_2$, Ca(OH)$_2$, Sr(OH)$_2$, Ba(OH)$_2$.

These compounds are ionic in the solid state. When they dissolve in water, the ions separate from each other. The solution process can be represented by equations such as

$$NaOH(s) \rightarrow Na^+(aq) + OH^-(aq) \tag{11.15}$$

$$Ba(OH)_2(s) \rightarrow Ba^{2+}(aq) + 2\ OH^-(aq) \tag{11.16}$$

These reactions go to completion. If you add 0.10 mol of NaOH to a liter of water

conc. Na$^+$ = 0.10 M; conc. OH$^-$ = 0.10 M

If you add 0.10 mol of Ba(OH)$_2$ to a liter of water

conc. Ba^{2+} = 0.10 M; conc. OH$^-$ = 0.20 M

EXAMPLE 11.7

Arrange 0.10 M solutions of the following compounds in order of increasing pH

HF, HCl, Ba(OH)$_2$, NaOH

Solution

HCl is a strong acid, completely dissociated in water. A 0.10 M solution of HCl has a pH of 1.00. HF is a weak acid (K_a = 7.0 × 10^{-4}), so we expect a 0.10 M solution of HF to have a pH greater than 1 (but less than 7). The pH of 0.10 M HF turns out to be 2.10. NaOH is a strong base; in 0.10 M NaOH, the concentration of OH$^-$ is 0.10 M and that of H$^+$ is 1.0 × 10^{-14}/(1.0 × 10^{-1}) = 1.0 × 10^{-13} M. Hence the pH of 0.10 M NaOH is 13.00. The pH of 0.10 M Ba(OH)$_2$ is even higher, since the concentration of OH$^-$ is 0.20 M. (You should be able to show that the pH of 0.10 M Ba(OH)$_2$ is 13.30.) The correct order must be

HCl < HF < NaOH < Ba(OH)$_2$

Exercise

Where would 0.10 M HCN (K_a = 4.0 × 10^{-10}) fit in this series? *Answer:* Between HF and NaOH.

Of the common strong bases, the only one you are likely to use in the laboratory is sodium hydroxide, NaOH. It is relatively cheap and very soluble in water. Calcium hydroxide, Ca(OH)$_2$, is even cheaper, but much less soluble. It is often used in those industrial processes where solubility is of minor importance.

There are a great many weak bases. The most important of these, and the only one we will consider here, is ammonia, NH_3. When ammonia is added to water, the following reaction occurs

$$NH_3(aq) + H_2O \rightleftharpoons NH_4^+(aq) + OH^-(aq) \tag{11.17}$$

Notice that the NH_3 molecule has picked up a proton (H^+ ion) from an H_2O molecule. In this way, it is converted to an NH_4^+ ion (Figure 11.4). At the same time, an H_2O molecule is converted to an OH^- ion, which makes the solution basic.

Ammonia is a weak base because Reaction 11.17 does not go to completion. If 0.10 mol of NH_3 is added to a liter of water, only a little more than 1% of it dissociates. The concentration of OH^- ion (and of NH_4^+ ion) is only about 0.0013 M. Contrast this situation with that for the strong base, NaOH

0.10 M NaOH; conc. OH^- = 0.10 M; pH = 13.00

0.10 M NH_3; conc. OH^- = 0.0013 M; pH = 11.11

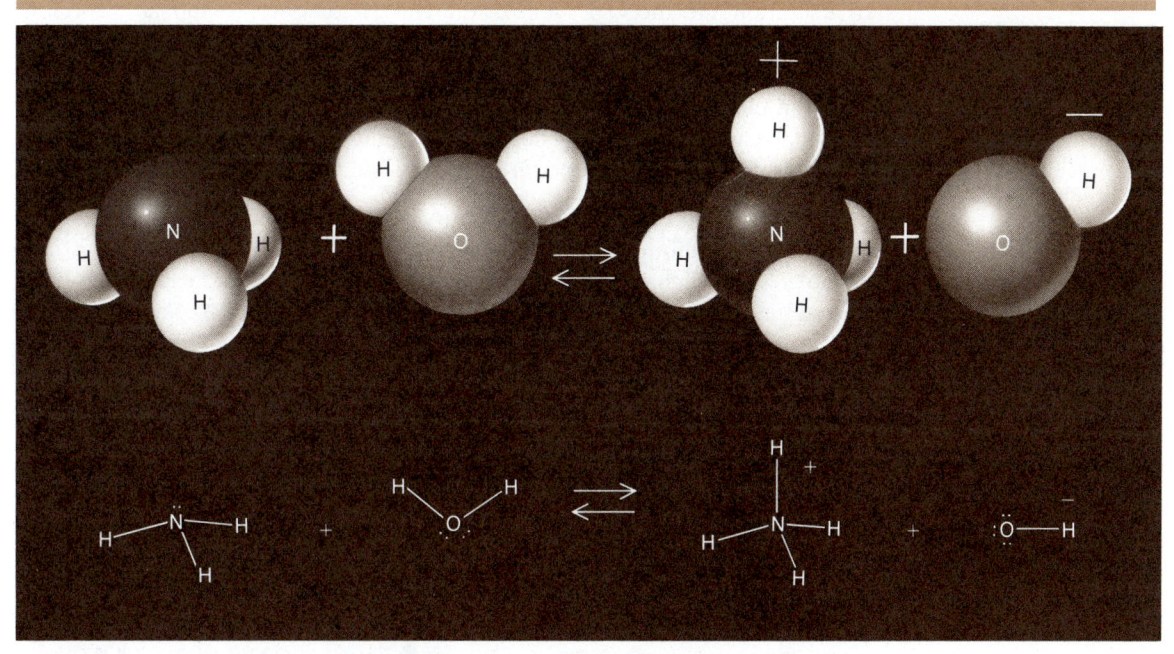

Figure 11.4 If NH_3 molecules are dissolved in water, they tend to remove hydrogen ions from H_2O molecules, forming NH_4^+ ions. This leaves OH^- ions in the solution and makes it basic.

11.4 Acid–Base Reactions

When an acidic water solution is mixed with a solution containing a base, a reaction occurs. The nature of this reaction, and the equation we write for it, depend upon whether the acid and base are strong or weak. In this section, we will look at three different types of acid–base reactions

1. Strong acid–strong base reactions
2. Weak acid–strong base reactions
3. Strong acid–weak base reactions

We will consider, among other things, how these reactions can be used to determine the concentration of an acid or base in solution. This makes use of a procedure called an acid–base titration.

Strong Acid–Strong Base

Consider an experiment (Figure 11.5) in which we add 100 cm^3 of 0.100 M HCl to 100 cm^3 of 0.100 M NaOH. The properties of the solution formed differ greatly from those of the solutions we started with. The H$^+$ ions of the HCl solution and the OH$^-$ ions of the NaOH solution "disappear." The final solution has a pH of 7. The only ions present are Na$^+$ and Cl$^-$. The solution is the same as one we would get by dissolving NaCl in water.

We conclude from this evidence that a reaction takes place when solutions of HCl and NaOH are mixed. To obtain the equation for this reaction, we follow the four-step procedure described in Chapter 9.

1. To decide upon the nature of the reaction, we note that the H$^+$ and OH$^-$ ions have combined with one another. They do this by forming an H$_2$O molecule, which is a product of this acid–base reaction. At the same time, a water solution of sodium chloride, NaCl, forms.
2. The "un-ionized" equation is

   HCl(aq) + NaOH(aq) → H$_2$O + NaCl(aq)

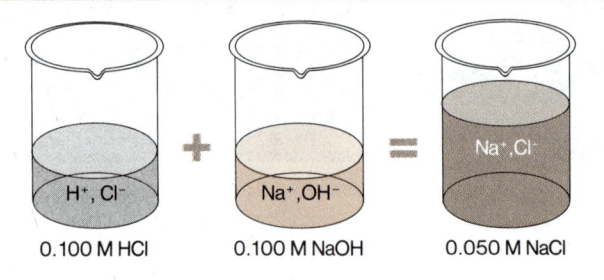

Figure 11.5 When equal volumes of 0.100 M HCl and 0.100 M NaOH are mixed, neutralization occurs. The result is a dilute solution of NaCl.

3. The three compounds HCl, NaOH, and NaCl are all strong electrolytes. They are completely ionized in water (H^+, Cl^- ions; Na^+, OH^- ions; Na^+, Cl^- ions). The "total ionic" equation is

$$H^+(aq) + Cl^-(aq) + Na^+(aq) + OH^-(aq) \rightarrow H_2O + Na^+(aq) + Cl^-(aq)$$

4. Eliminating the "spectator ions," Na^+ and Cl^-, we obtain the **net ionic equation** for the reaction

$$H^+(aq) + OH^-(aq) \rightarrow H_2O \quad (11.18)$$

This reaction is often referred to as **neutralization,** since it results in the formation of a neutral solution, pH 7. It occurs whenever a strong acid is mixed with a strong base. Equation 11.18 applies equally well to the reaction between HCl and NaOH, HNO_3 and $Ba(OH)_2$, or any other strong acid–strong base combination. In each case, H^+ ions of the strong acid react with OH^- ions of the strong base. The other ions (Cl^-, Na^+, NO_3^-, Ba^{2+}, $--$) do not take part in the reaction and so are not included in the net ionic equation.

Weak Acid–Strong Base

When we add 100 cm^3 of 0.100 M HF to 100 cm^3 of 0.100 M NaOH, a reaction occurs that is similar in many ways to that between HCl and NaOH. A water solution of sodium fluoride, NaF, is formed. The "un-ionized" equation is

$$HF(aq) + NaOH(aq) \rightarrow H_2O + NaF(aq)$$

To obtain the "total ionic" equation, we note that NaOH and NaF are completely ionized in solution (Na^+, OH^- ions; Na^+, F^- ions). However, because HF is a weak acid, it is *not* highly ionized in solution. Hydrogen fluoride should be represented as a molecule, HF, since most of it is in that form rather than the ions H^+ and F^-. Hence we write

$$HF(aq) + Na^+(aq) + OH^-(aq) \rightarrow H_2O + Na^+(aq) + F^-(aq)$$

This time, the only spectator ion is Na^+; the F^- ion is formed in the reaction from an HF molecule. The net ionic equation is

$$HF(aq) + OH^-(aq) \rightarrow H_2O + F^-(aq) \quad (11.19)$$

Equations similar to 11.19 can be written for the reaction of any weak acid with any strong base. For the reaction between solutions of acetic acid, $HC_2H_3O_2$, and barium hydroxide, $Ba(OH)_2$, we would write

$$HC_2H_3O_2(aq) + OH^-(aq) \rightarrow H_2O + C_2H_3O_2^-(aq) \quad (11.20)$$

In all equations for weak acid–strong base reactions

1. The reactants are a molecule of the weak acid, HX, and an hydroxide ion, OH^-, both in water solution.

2. The products are an H₂O molecule and the anion of the weak acid, X⁻, in water solution.

The general equation is

$$HX(aq) + OH^-(aq) \rightarrow H_2O + X^-(aq) \quad (11.21)$$

Strong Acid–Weak Base

When 100 cm³ of 0.100 M HCl is added to 100 cm³ of 0.100 M NH₃, an acid–base reaction occurs. The H⁺ ions of the HCl solution "disappear" as do the NH₃ molecules of the ammonia solution. They combine to form an ammonium ion, NH₄⁺. When the reaction is over, we are left with a water solution of ammonium chloride, NH₄Cl.

The "un-ionized" equation for the reaction of hydrochloric acid with a water solution of ammonia is

$$HCl(aq) + NH_3(aq) \rightarrow NH_4Cl(aq)$$

To obtain the "total ionic" equation, we note that HCl is a strong acid, completely dissociated to H⁺ and Cl⁻ ions in solution. Likewise, NH₄Cl is a strong electrolyte, dissociated to NH₄⁺ and Cl⁻ ions in solution. Ammonia is a weak base so it is present mainly as NH₃ molecules. Hence we write

$$H^+(aq) + Cl^-(aq) + NH_3(aq) \rightarrow NH_4^+(aq) + Cl^-(aq)$$

The "spectator ion" is Cl⁻. Eliminating that ion, we arrive at the net ionic equation

$$H^+(aq) + NH_3(aq) \rightarrow NH_4^+(aq) \quad (11.22)$$

We would write this same equation to describe the reaction of *any* strong acid (e.g., HNO₃, HBr, – –) with ammonia.

EXAMPLE 11.8
Write net ionic equations for the reactions of solutions of
a. HI with NH₃   b. HNO₃ with KOH   c. HCN with Ca(OH)₂

Solution
a. Since HI is a strong acid, the equation is the same as 11.22

$$H^+(aq) + NH_3(aq) \rightarrow NH_4^+(aq)$$

b. HNO₃ is a strong acid, KOH a strong base. Hence the reaction is a simple neutralization (Equation 11.18).

$$H^+(aq) + OH^-(aq) \rightarrow H_2O$$

Continued

c. HCN is a weak acid, Ca(OH)$_2$ a strong base. By analogy with Equation 11.21

HCN(aq) + OH$^-$(aq) → H$_2$O + CN$^-$(aq)

Exercise
Write a net ionic equation for the reaction of HF with NH$_3$. Answer: HF(aq) + NH$_3$(aq) → NH$_4^+$(aq) + F$^-$(aq)

Acid–Base Titration

One of the most common experiments in the chemistry laboratory is that of **titration.** Here, we measure the volume of one solution of known concentration required to react exactly with a known volume of a second solution. From these data, and the equation for the reaction involved, we can calculate the concentration of solute in the second solution. In an acid–base titration, one of the solutions is acidic, the other basic.

To illustrate a typical acid–base titration, consider the experiment shown in Figure 11.6. The objective here is to determine the concentration of acetic acid, HC$_2$H$_3$O$_2$, in a sample of vinegar. To do this, we proceed as follows

1. With a pipet, add 2.00 cm^3 of vinegar to the Erlenmeyer flask. For convenience, this is diluted with water.
2. Add one or two drops of phenolphthalein to the flask. This is an *acid–base indicator* used to tell us when the reaction is complete.
3. Add 0.100 M NaOH from a buret until the indicator changes color (from colorless to pink). This indicates that the acid–base reaction between acetic acid and sodium hydroxide is complete. At this point, all the HC$_2$H$_3$O$_2$ molecules originally present have reacted with the OH$^-$ ions in the NaOH that was added.

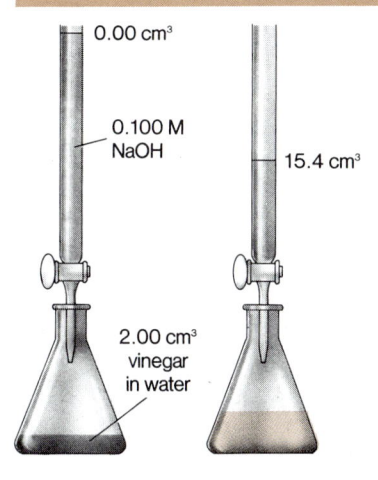

Figure 11.6 Determination of the concentration of acetic acid in a vinegar sample. An acid–base indicator such as phenolphthalein is added so that you can tell when the reaction between HC$_2$H$_3$O$_2$ and OH$^-$ is complete.

4. Calculate the concentration of acetic acid in the vinegar, as shown in Example 11.9.

EXAMPLE 11.9
Suppose you find that 15.4 cm^3 of 0.100 M NaOH is required to react with the 2.00 cm^3 sample of vinegar. What is the concentration of acetic acid in the vinegar?

Solution
A logical way to solve this problem is to first calculate the number of moles of NaOH. Then we can determine the number of moles of HC$_2$H$_3$O$_2$ in the vinegar solution. Finally, we calculate the concentration (molarity) of acetic acid

(1) moles NaOH = $0.100 \dfrac{\text{mol NaOH}}{\text{L}} \times \dfrac{1 \text{ L}}{1000 \text{ cm}^3} \times 15.4 \text{ cm}^3 = 1.54 \times 10^{-3}$ mol

(2) The equation for the acid–base reaction is

$$\text{HC}_2\text{H}_3\text{O}_2(aq) + \text{OH}^-(aq) \rightarrow \text{H}_2\text{O} + \text{C}_2\text{H}_3\text{O}_2^-(aq)$$

One mole of NaOH produces one mole of OH$^-$, which in turn reacts with one mole of HC$_2$H$_3$O$_2$. Hence,

1 mol NaOH ≏ 1 mol HC$_2$H$_3$O$_2$

moles HC$_2$H$_3$O$_2$ = 1.54×10^{-3} mol NaOH $\times \dfrac{1 \text{ mol HC}_2\text{H}_3\text{O}_2}{1 \text{ mol NaOH}} = 1.54 \times 10^{-3}$

(3) M HC$_2$H$_3$O$_2$ = $\dfrac{1.54 \times 10^{-3} \text{ mol HC}_2\text{H}_3\text{O}_2}{2.00 \text{ cm}^3} \times \dfrac{1000 \text{ cm}^3}{1 \text{ L}} =$

$0.770 \dfrac{\text{mol HC}_2\text{H}_3\text{O}_2}{\text{L}}$

Exercise
What volume of 0.100 M NaOH would be required to titrate 5.00 cm^3 of this vinegar? Answer: 38.5 cm^3.

11.5 General Models of Acids and Bases

In this chapter, we have considered an acid to be a substance that produces H$^+$ ions when added to water. Similarly, we have taken a base to be a substance that forms OH$^-$ ions in water solution. These definitions of acids and bases were first proposed by the Swedish chemist Svante Arrhenius in 1884. So long as we are dealing with water solutions, the Arrhenius model of acids and bases is quite adequate.

Svante Arrhenius

Several other, more general models of acids and bases have been proposed over the past century. In this section, we will discuss two such models, both of which were first suggested in 1923. One of these was proposed independently by Brønsted in Denmark and Lowry in England. The other was put forth by the American physical chemist G. N. Lewis.

Brønsted–Lowry Model

According to this model, an acid–base reaction amounts to a proton transfer from one species to another. The species that **donates the proton** (H^+ ion) is referred to as an **acid**. The molecule or ion that **accepts the proton** (H^+ ion) is a **base**.

A simple example of a Brønsted–Lowry acid–base reaction is that between an HF molecule and an OH^- ion

$$HF(aq) + OH^-(aq) \rightarrow H_2O + F^-(aq)$$
$$\text{acid} \quad\quad\, \text{base}$$

Chapter 11

Here, the HF molecule gives up a proton, being converted to a F⁻ ion in the process. Hence, HF is acting as a Brønsted–Lowry acid. The OH⁻ ion accepts a proton to form an H_2O molecule; it is a Brønsted–Lowry base.

You may recall that we referred earlier (Section 11.4) to the HF–OH⁻ reaction. It is an acid–base reaction by the Arrhenius model as well as the Brønsted–Lowry model. In that sense, the proton–transfer model tells us nothing new. Consider, however, the equation we wrote earlier for the reaction of ammonia with water

$$NH_3(aq) + H_2O \rightarrow NH_4^+(aq) + OH^-(aq)$$
$$\quad\text{base} \quad\quad \text{acid}$$

When we discussed this process in Section 11.3, we did not think of it as an acid–base reaction. According to the Brønsted–Lowry model, it is. The NH_3 molecule accepts a proton from an H_2O molecule and hence acts as a base. Since the H_2O molecule donates a proton, it is acting as an acid.

The H_2O molecule can act not only as a Brønsted–Lowry acid but also as a base. In certain reactions, the H_2O molecule accepts a proton to form the H_3O^+ ion. As an example, consider what happens when HCl is added to water. Earlier, we considered the reaction to be a simple dissociation

$$HCl(aq) \rightarrow H^+(aq) + Cl^-(aq)$$

According to the Brønsted–Lowry model, however, the process is a bit more complex. It involves a proton transfer from an HCl molecule to an H_2O molecule (Figure 11.7).

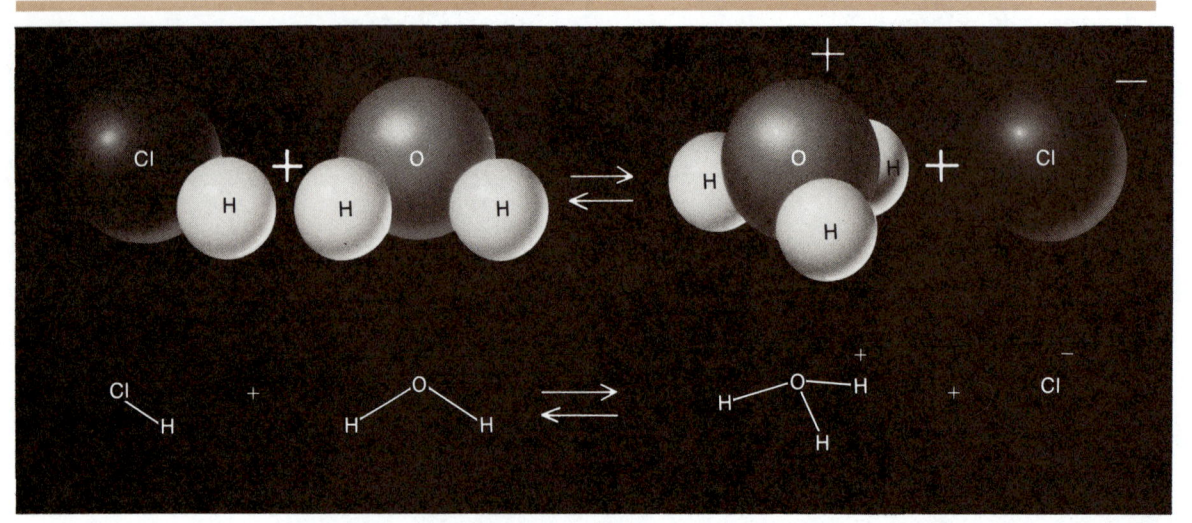

Figure 11.7 When HCl is added to water, there is a proton transfer from an HCl to an H_2O molecule, forming a Cl⁻ and a H_3O^+ ion. In this reaction, HCl acts as a Brønsted–Lowry acid, H_2O as a Brønsted–Lowry base.

$$HCl(aq) + H_2O \rightarrow H_3O^+(aq) + Cl^-(aq) \quad (11.23)$$
acid   base

The products are the hydronium ion, H_3O^+, and the Cl^- ion. The HCl molecule acts as an acid, the H_2O molecule as a base.

EXAMPLE 11.10
Consider the equilibrium system

$$HC_2H_3O_2(aq) + NH_3(aq) \rightleftharpoons NH_4^+(aq) + C_2H_3O_2^-(aq)$$

Identify the two Brønsted–Lowry
a. acids   b. bases

Solution
a. $HC_2H_3O_2$ (donates proton in forward reaction) and NH_4^+ (donates proton in reverse reaction).
b. NH_3 and $C_2H_3O_2^-$

Exercise
The HCO_3^- ion, like H_2O, can be either a Brønsted–Lowry acid or base. To what species is it converted when it acts as an acid? a base? Answer: CO_3^{2-}, H_2CO_3.

Lewis Model

According to this model, an acid–base reaction takes place when one species shares a pair of electrons with another. The species that **accepts the electron pair** is an **acid**. The species that **donates the electron pair** is a **base**.

An example of a Lewis acid–base reaction is that between the H^+ ion (proton) and the NH_3 molecule. The equation for this reaction is

$$H^+(aq) + NH_3(aq) \rightarrow NH_4^+(aq)$$

In terms of Lewis structures

$$H^+ + H\!-\!\underset{\cdot\cdot}{\overset{H}{N}}\!-\!H \rightarrow \left[H\!-\!\underset{H}{\overset{H}{N}}\!-\!H \right]^+$$

acid   base

Notice that the nitrogen atom of the NH_3 molecule is donating a pair of electrons to form a covalent bond with the H^+ ion. Hence, NH_3 is acting as a Lewis base. The H^+ ion accepts the pair of electrons and hence acts as a Lewis acid.

The reaction between H^+ and NH_3 would be considered an acid–base reaction by either the Arrhenius or Brønsted–Lowry (proton transfer) model. However, in many cases, the Lewis model extends the acid–base concept beyond either of these models. Consider, for example, the gas–phase reaction between boron trifluoride and ammonia

$$BF_3(g) + NH_3(g) \rightarrow BF_3NH_3(s) \qquad (11.24)$$

In terms of Lewis structures

$$\begin{array}{ccc} \quad\;\; F \quad\;\; H & & \quad\;\; F \quad\;\; H \\ \quad\;\; | \quad\;\; | & & \quad\;\; | \quad\;\; | \\ F-B \;\; + \;\; :N-H & \rightarrow & F-B-N-H \\ \quad\;\; | \quad\;\; | & & \quad\;\; | \quad\;\; | \\ \quad\;\; F \quad\;\; H & & \quad\;\; F \quad\;\; H \end{array}$$

Here, BF_3 is acting as a Lewis acid by accepting an electron pair to form the B–N covalent bond. The NH_3 molecule, since it contributes an electron pair, acts as a Lewis base. Notice that this would not be classified as an acid–base reaction by the Arrhenius model; it does not take place in water solution. Again, it is not a Brønsted–Lowry acid–base reaction since no proton transfer is involved.

Key Words

acid
acid–base indicator
acidic solution
Arrhenius acid
Arrhenius base
base
basic solution
Brønsted–Lowry acid
Brønsted–Lowry base
dissociation constant
hydronium ion
K_w

Lewis acid
Lewis base
net ionic equation
neutralization
neutral solution
pH
proton
strong acid
strong base
titration
weak acid
weak base

Questions

11.1 List four general properties common to all acidic solutions.

11.2 List four general properties common to all basic solutions.

11.3 What ion is responsible for the properties of acidic water solutions? basic water solutions?

11.4 What is meant by the equilibrium constant K_w? What is it used for?

11.5 Complete the following table (use "greater than" and "less than" signs)

	Acidic Solution	Basic Solution	Neutral Solution
[H$^+$]	___	___	___
[OH$^-$]	___	___	___
pH	___	___	___

11.6 What is the definition of pH? Is pH directly or inversely related to $[H^+]$?

11.7 How does one distinguish between a strong and weak acid? a strong and weak base?

11.8 Give the names and formulas of five strong acids.

11.9 Give the names and formulas of at least five weak acids.

11.10 Give the names and formulas of five strong bases; of one weak base.

11.11 Write the general net ionic equation for the reaction of
   a. a strong acid with a strong base
   b. a weak acid HX with a strong base
   c. a strong acid with ammonia

11.12 What quantities must be measured or known to determine by titration the concentration of an acid or base in solution?

11.13 Define a Brønsted–Lowry acid; a Brønsted–Lowry base.

11.14 Define a Lewis acid; a Lewis base.

Problems

$[H^+]$ vs $[OH^-]$
(Use the relation: $[H^+] \times [OH^-] = 1.0 \times 10^{-14}$.)

11.15 Calculate $[OH^-]$ in a solution in which $[H^+]$ is
   a. 1.0×10^{-3} M   b. 2.0×10^{-6} M
   c. 5.4×10^{-9} M

11.16 Calculate $[H^+]$ in a solution in which $[OH^-]$ is
   a. 1.0×10^{-5} M   b. 2.5×10^{-7} M
   c. 3.4×10^{-10} M

11.17 Classify each of the solutions in Problems 11.15 and 11.16 as acidic or basic.

11.18 Calculate $[OH^-]$ and $[H^+]$ in
   a. 0.12 M NaOH   b. 0.023 M HCl

11.19 A solution is prepared by dissolving 2.8 g of KOH (molar mass = 56 g/mol) to form 100.0 cm³ of solution. Determine
   a. $[OH^-]$ b. $[H^+]$

11.20 A solution is prepared by diluting 10.0 cm³ of 2.05 M HCl to one liter with water. In this solution, determine
   a. $[H^+]$ b. $[OH^-]$

$[H^+]$ and pH
(Use a calculator for logarithms and antilogarithms.)

11.21 Calculate the pH of solutions in which the $[H^+]$ is
   a. 1.0×10^{-6} M   b. 1×10^{-9} M
   c. 2.0×10^{-4} M   d. 5.4×10^{-3} M

11.22 Calculate the pH of solutions in which $[H^+]$ is
   a. 1×10^{-13} M   b. 10 M
   c. 3.3×10^{-5} M   d. 8.2×10^{-9} M

11.23 Calculate $[H^+]$ in solutions with the following pH values
   a. 6   b. 12.0   c. 4.3   d. 6.8

11.24 Calculate $[H^+]$ in solutions of pH
   a. 9.0   b. -1 c. 5.2   d. 11.6

11.25 Classify each of the solutions in Problems 11.23 and 11.24 as acidic or basic.

11.26 Solution A has a pH of 3.0, solution B a pH of 6.0. Which solution is more acidic? more basic? For every H^+ ion in solution B, how many H^+ ions are there in Solution A?

11.27 What is the $[H^+]$ and pH of a solution in which $[OH^-] = 2.4 \times 10^{-6}$ M?

11.28 What is the pH of a 0.050 M solution of nitric acid?

Dissociation of Acids

11.29 Write chemical equations for the dissociation of

a. $HNO_3(aq)$ b. $HBr(aq)$
c. $H_2SO_4(aq)$

11.30 Write chemical equations for the dissociation of
a. $HF(aq)$ b. $H_3PO_4(aq)$
c. $HC_2H_2O_2(aq)$

11.31 Write chemical equations for the dissociation of
a. $HCl(aq)$ b. $H_2CO_3(aq)$
c. $HNO_2(aq)$ d. $HClO_4(aq)$

Dissociation Constants of Weak Acids

11.32 In a certain solution, $[H^+] = 1.0 \times 10^{-3}$ M, $[X^-] = 2.0 \times 10^{-4}$ M, and $[HX] = 2.5 \times 10^{-3}$ M. Calculate K_a for HX.

11.33 A student adds 0.100 mol of a weak acid to a liter of water. He measures $[H^+]$ and finds it to be 2.0×10^{-4} M. Calculate K_a of the weak acid.

11.34 To determine the dissociation constant of HCN, a student adds 1.00 mol to a liter of water. She finds $[H^+]$ in this solution to be 2.0×10^{-5} M. Calculate K_a of HCN.

11.35 The pH of a solution prepared by adding 0.10 mol of H_2S to one liter of water is 4.0. Calculate
a. $[H^+]$ b. K_a of H_2S

11.36 In a certain solution it is found that $[H^+] = 2.9 \times 10^{-6}$ M, $[H_2BO_3^-] = 2.0 \times 10^{-5}$ M. Taking the dissociation constant of boric acid to be 5.8×10^{-10}, calculate $[H_3BO_3]$.

11.37 Calculate $[H^+]$ in a solution prepared by adding enough acetic acid to water to make $[HC_2H_3O_2] = 0.10$ M. Take K_a of $HC_2H_3O_2$ to be 1.8×10^{-5}.

11.38 Calculate $[H^+]$ in a solution prepared by adding 0.10 mol of HClO ($K_a = 3.2 \times 10^{-8}$) to one liter of water.

11.39 0.100 mol of carbonic acid ($K_a = 4.2 \times 10^{-7}$) is dissolved in a liter of water. Calculate
a. $[H^+]$ b. pH
c. the percent of the H_2CO_3 originally present that dissociates

Acid–Base Reactions

11.40 Write balanced net ionic equations for the reactions between solutions of
a. NaOH and HCl   b. $Ca(OH)_2$ and HCl
c. $Ba(OH)_2$ and HNO_3

11.41 Write a balanced net ionic equation for the reaction of solutions of
a. NaOH with $HC_2H_3O_2$ b. HNO_2 with $Ca(OH)_2$ c. HF with KOH

11.42 Write a balanced net ionic equation for the reaction between solutions of
a. KOH and $HC_2H_3O_2$
b. RbOH and $HClO_4$
c. HCl and NH_3

11.43 Write a balanced net ionic equation for the reaction of
a. $NH_3(aq) + H_2SO_4(aq)$
b. $HBr(aq) + Ba(OH)_2(aq)$
c. $HCN(aq) + NaOH(aq)$

Acid–Base Titrations

11.44 What volume of 0.100 M NaOH is required to react with 92.0 cm^3 of 0.200 M HCl?

11.45 It is found that 15.6 cm^3 of 0.119 M NaOH is required to titrate 21.7 cm^3 of an HCl solution. What is the concentration of HCl?

11.46 It is found that 15.6 cm^3 of 0.119 M $Ba(OH)_2$ is required to titrate 21.7 cm^3 of an HCl solution. What is the concentration of the HCl?

11.47 What volume of 0.120 M HCl is required to react with 1.00 g of NaOH (molar mass = 40.0 g/mol)?

Brønsted–Lowry Acids and Bases

11.48 Identify the Brønsted–Lowry acids and bases in the reactions
a. $HNO_3(aq) + H_2O \rightarrow H_3O^+(aq) + NO_3^-(aq)$
b. $F^-(aq) + H_2O \rightarrow HF(aq) + OH^-(aq)$

11.49 Identify the Brønsted–Lowry acids and bases in the equilibrium system
$H_2CO_3(aq) + CO_3^{2-}(aq) \rightleftharpoons 2\, HCO_3^-(aq)$

11.50 Give the formula of the species formed when

a. NH₃ acts as a Brønsted–Lowry base
b. HNO₂ acts as a Brønsted–Lowry acid
c. H₂O acts as a Brønsted–Lowry acid
d. H₂O acts as a Brønsted–Lowry base

Multiple Choice

11.51 Which one of the following could not act as a Lewis base?

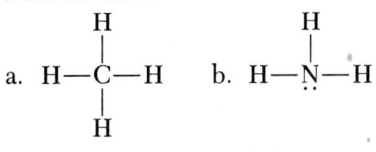

a. H—C(H)(H)—H   b. H—N(H)—H
c. H—Ö—H   d. $(:\ddot{F}:)^-$

11.52 Which one of the following could act as either a Brønsted–Lowry acid or a Brønsted–Lowry base?
a. HCl   b. HSO_4^- c. SO_4^{2-}
d. Cl^-

11.53 What volume of 0.100 M Ca(OH)₂ is required to react with 20.0 cm³ of 0.150 M HCl?
a. 10.0 cm³   b. 15.0 cm³   c. 20.0 cm³
d. 25.0 cm³   e. 30.0 cm³

11.54 The net ionic equation for the reaction between solutions of HBr and NH₃ is
a. HBr(aq) + NH₃(aq) → NH₄⁺(aq) + Br⁻(aq)
b. H⁺(aq) + Br⁻(aq) + NH₃(aq) → NH₄⁺(aq) + Br⁻(aq)
c. H⁺(aq) + NH₃(aq) → NH₄⁺(aq)
d. NH₄⁺(aq) → H⁺(aq) + NH₃(aq)

11.55 The net ionic equation H⁺(aq) + OH⁻(aq) → H₂O would *not* represent the reaction between solutions of

a. HClO₄ and NaOH
b. KOH and H₂SO₄
c. HF and KOH
d. HCl and Ca(OH)₂

11.56 The pH of a 0.10 M solution of an acid with a dissociation constant of 4.0×10^{-7} is about
a. 1.0   b. 3.2   c. 3.7   d. 6.4
e. 7.4

11.57 Which one of the following solutions has the highest pH?
a. 0.10 M NaOH   b. 0.10 M NH₃
c. water   d. 0.10M HF
e. 0.10 M HCl

11.58 Of the solutions in Problem 11.57, which one has the highest H⁺ ion concentration?
a. 0.10 M NaOH   b. 0.10 M NH₃
c. water   d. 0.10 M HF
e. 0.10 M HCl

11.59 What is the pH of a solution prepared by dissolving 3.15 g of HNO₃ (molar mass = 63.0 g/mol) in 2.0×10^2 cm³ of solution?
a. 0.40   b. 0.60   c. 1.30   d. 3.60
e. 7.00

11.60 What is the concentration of OH⁻ in the solution referred to in Problem 11.59?
a. 3.2×10^{-15} M   b. 4.0×10^{-14} M
c. 2.0×10^{-13} M   d. 1.0×10^{-7} M

Oxidation–Reduction Reactions

Learning Objectives
After studying this chapter, you should be able to:

1. Find the oxidation number of an atom in any molecule or polyatomic ion (Example 12.1).
2. Decide, either on the basis of gain or loss of electrons or change in oxidation number, whether an atom has been reduced or oxidized (Example 12.2).
3. Identify the oxidizing agent in a redox reaction; the reducing agent (Example 12.3).
4. Balance a redox equation using the half-equation method (Example 12.4).
5. Using an activity series (Table 12.1), compare the strengths of different oxidizing agents; different reducing agents (Example 12.5).
6. Using an activity series (Table 12.1), predict whether or not a redox reaction will occur spontaneously (Example 12.6).
7. Distinguish between an electrolytic and a voltaic cell; between anode and cathode.
8. Given the nature of the reaction in an electrolytic cell, write balanced half-equations for the cathode and anode reactions (Example 12.7).
9. Given the composition of a voltaic cell, write a balanced equation for the overall reaction; identify the anode and cathode (Example 12.8).

> Therefore, the two processes, that of science and that of art, are not very different. Both science and art form in the course of the centuries a human language by which we can speak about the more remote parts of reality, and the coherent sets of concepts as well as the different styles of art are different words or groups of words in this language.
>
> WERNER HEISENBERG

Photograph by Peter L. Kresan.

CHAPTER 12

So far in this text, we have discussed two different kinds of reactions involving ions in water solution. In Chapter 9, we talked about precipitation reactions; acid–base reactions were considered in Chapter 11. This chapter is devoted to a third type of reaction, one which also deals with ions in water solution. An oxidation–reduction reaction ("redox" for short) involves the transfer of electrons from one species to another.

In discussing redox reactions, we will be interested in obtaining answers to certain basic questions.

1. How do we identify an oxidation–reduction reaction? Given a chemical equation for a reaction in water solution, how do we tell whether it is of the redox type? To accomplish this, we need to establish some basic definitions, particularly that of oxidation number (Section 12.1).
2. How do we balance the equation for an oxidation–reduction reaction? We will consider how to do this in Section 12.2. The basic principle is a simple one; there is no "free lunch" where electrons are concerned. If one species gains a certain number of electrons in a reaction, another species has to lose that same number of electrons.
3. How do we predict whether an oxidation–reduction reaction will occur when the reactants are mixed? We will consider how this is done in Section 12.3. Basically, it involves constructing an "activity series" in which we arrange various species in order of increasing "appetite" for electrons.
4. How do we make practical use of oxidation–reduction reactions in electrical cells? In Section 12.4, we will consider two types of cells. In one, called an electrolytic cell, electrical energy is used to make a redox reaction occur. In the other, a voltaic cell, a redox reaction produces electrical energy.

12.1 Oxidation and Reduction

Consider what happens when a strip of zinc metal is added to a solution containing Cu^{2+} ions (Figure 12.1). As time passes, we observe that

Oxidation—Reduction Reactions

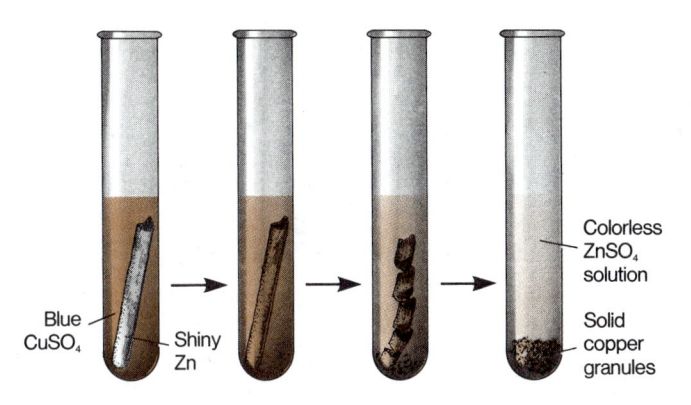

Figure 12.1 The oxidation–reduction reaction between zinc and copper(II) sulfate.

1. The zinc strip, which has a shiny gray appearance to begin with, acquires a reddish-brown coating. This coating can be identified as copper metal. Eventually, the zinc strip completely disintegrates. A spongy deposit of copper collects at the bottom of the container.
2. The blue color of the original solution, which is due to Cu^{2+} ions, fades. Eventually the solution becomes colorless, a property of the Zn^{2+} ion.

From this evidence, we conclude that a reaction has taken place between Zn solid and Cu^{2+} ions in water solution. The products are Zn^{2+} ions in solution and solid Cu. The net ionic equation for the reaction is

$$Zn(s) + Cu^{2+}(aq) \rightarrow Zn^{2+}(aq) + Cu(s) \quad (12.1)$$

This reaction is really the sum of two different "half-reactions." One of these involves the loss of two electrons by a zinc atom

$$Zn(s) \rightarrow Zn^{2+}(aq) + 2\ e^- \quad (12.1a)$$

In the other half-reaction, a Cu^{2+} ion gains two electrons

$$Cu^{2+}(aq) + 2\ e^- \rightarrow Cu(s) \quad (12.1b)$$

The overall reaction represented by Equation 12.1 is simply the sum of these two half-reactions.

A half-reaction such as 12.1a in which a species **loses electrons** is referred to as **oxidation**. We say that a Zn atom has been oxidized to a Zn^{2+} ion. Other examples of oxidation half-reactions are

$$Al(s) \rightarrow Al^{3+}(aq) + 3\ e^-$$

$$Fe^{2+}(aq) \rightarrow Fe^{3+}(aq) + e^-$$

$$2\ I^-(aq) \rightarrow I_2(s) + 2\ e^-$$

A half-reaction such as 12.1b in which a species **gains electrons** is referred to as **reduction**. In Equation 12.1b, we say that a Cu^{2+} ion has been reduced to a Cu atom. Other examples of reduction half-equations are

$$Ag^+(aq) + e^- \rightarrow Ag(s)$$

$$Sn^{4+}(aq) + 2\,e^- \rightarrow Sn^{2+}(aq)$$

$$Cl_2(g) + 2\,e^- \rightarrow 2\,Cl^-(aq)$$

A reaction such as 12.1 in which one species loses electrons and another one gains them is referred to as an **oxidation-reduction** or **redox** reaction. We will refer to many other redox reactions in this chapter. In all cases

1. **Oxidation and reduction occur together.** If one species, such as Zn or Al, loses electrons, another species such as Cu^{2+} or Ag^+ must gain them.
2. **There can be no net loss or gain of electrons.** In Reaction 12.1, the two electrons lost by a Zn atom are gained by a Cu^{2+} ion. In another case, consider what happens when Al reacts with Ag^+ ions. For every Al atom oxidized to Al^{3+}, three Ag^+ ions must be reduced to Ag atoms. Only in this way can we "conserve" electrons.

$$Al(s) \rightarrow Al^{3+}(aq) + 3\,e^-$$
$$\underline{3\,Ag^+(aq) + 3\,e^- \rightarrow 3\,Ag(s)}$$
$$Al(s) + 3\,Ag^+(aq) \rightarrow Al^{3+}(aq) + 3\,Ag(s) \tag{12.2}$$

Oxidation Number

So far, the redox reactions we have considered have been rather simple ones. They involve atoms (Zn, Cu, Al, Ag) or simple monatomic ions (Zn^{2+}, Cu^{2+}, Al^{3+}, Ag^+). Here, it is rather easy to see which species are losing and gaining electrons. However, many redox reactions are more complex. They may involve molecules and polyatomic ions. Consider, for example, the reaction

$$2MnO_4^-(aq) + 5SO_2(g) + 2H_2O \rightarrow 2Mn^{2+}(aq) + 5SO_4^{2-}(aq) + 4H^+(aq) \tag{12.3}$$

In this case, it is by no means obvious which species is losing electrons and which one is gaining electrons. Indeed, just looking at Equation 12.3, it is not at all obvious from what has been said so far that this is even an oxidation–reduction reaction.

To keep track of electrons in complex redox reactions such as 12.3, it is convenient to introduce a new concept known as oxidation number. Each atom in a species can be assigned an oxidation number. This number is a measure of the extent to which the particular atom has gained or lost electrons. Oxidation numbers are assigned according to certain arbitrary rules.

1. **The oxidation number of an atom in any elemental substance is zero.**
   For example, the oxidation number of chlorine in the Cl_2 molecule is 0.
2. **The oxidation number of an atom in a simple, monatomic ion is the charge of that ion.** The oxidation number of zinc in Zn^{2+} is $+2$. That of chlorine in the Cl^- ion is -1.

3. **Certain elements have the same oxidation number in all or nearly all of their compounds.**

   —*the Group 1 metals have an oxidation number of +1 in all their compounds.* Sodium in NaCl has an oxidation number of +1 (it is present as a +1 ion). Again, potassium in KNO$_3$ has an oxidation number of +1.

   —*the Group 2 metals have an oxidation number of +2 in all their compounds.* For example, calcium in CaSO$_4$ has an oxidation number of +2.

   —*hydrogen, in all of its compounds with nonmetals, has an oxidation number of +1.* This is the case, for example, in H$_2$O, NH$_3$, HCl, and H$_2$SO$_4$.

   —*oxygen, with a few exceptions, has an oxidation number of −2 in its compounds.* The oxidation number of oxygen is −2 in such molecules as H$_2$O and H$_2$SO$_4$ and in such polyatomic ions as OH$^-$ and NO$_3^-$.

4. **The sum of the oxidation numbers of all the atoms in a neutral species is zero.** For example

   in H$_2$O: 2 × oxid. no. H + oxid. no. O = 0

   in H$_2$SO$_4$: 2 × oxid. no. H + oxid. no. S + 4 × oxid. no. O = 0

5. **The sum of the oxidation numbers of all the atoms in a polyatomic ion is the charge of the ion.** For example

   in OH$^-$: oxid. no. O + oxid. no. H = −1

   in CrO$_4^{2-}$: oxid. no. Cr + 4 × oxid. no. O = −2

   Using these rules, we can assign oxidation numbers to atoms in many different species. Example 12.1 illustrates how this is done.

EXAMPLE 12.1

What is the oxidation number of
a. N in HNO$_2$ b. Cr in Cr$_2$O$_7^{2-}$

Solution

a. Applying Rule 4: oxid. no. H + oxid. no. N + 2 × oxid. no. O = 0
   According to Rule 3, oxid. no. H = +1, oxid. no. O = −2
   Hence

   +1 + oxid. no N + 2(−2) = 0

   oxid. no. N = 4 − 1 = +3

Continued

b. Applying Rule 5: 2 × oxid. no. Cr + 7 × oxid. no. O = −2
Setting the oxidation number of oxygen to be −2

2 × oxid. no. Cr + 7(−2) = −2

oxid. no. Cr = (−2 + 14)/2 = +6

Exercise
What is the oxidation number of N in nitric acid, HNO_3? In the nitrate ion, NO_3^-?
Answer: +5 in both cases.

The concept of oxidation number leads directly to a working definition of oxidation and reduction:

oxidation = increase in oxidation number

reduction = decrease in oxidation number

These definitions agree with our earlier ones in terms of the loss and gain of electrons. For example, in the half-reaction

$$Zn(s) \rightarrow Zn^{2+}(aq) + 2\ e^-$$

zinc is oxidized, no matter how you look at it. A zinc atom loses two electrons being converted to the Zn^{2+} ion. It also increases in oxidation number from 0 to +2. In the half-reaction

$$Cu^{2+}(aq) + 2\ e^- \rightarrow Cu(s)$$

Cu^{2+} ions are reduced. They gain electrons and the oxidation number of copper drops from +2 to 0.

In many cases where molecules or polyatomic ions are involved, it is by no means easy to keep track of electrons. Here, oxidation and reduction are more readily identified by noting changes in oxidation number (Example 12.2).

EXAMPLE 12.2
Indicate whether the Cr atom in each of the following conversions is being oxidized or reduced.
a. $Cr_2O_3 \rightarrow CrO_4^{2-}$ b. $Cr_2O_7^{2-} \rightarrow Cr^{3+}$

Solution
a. Taking the oxidation number of oxygen to be −2

Cr_2O_3: 2 × oxid. no. Cr + 3(−2) = 0; oxid. no. Cr = +6/2 = +3

CrO_4^{2-}: oxid. no. Cr. + 4(−2) = −2; oxid. no. Cr = −2 + 8 = +6

Continued

Because chromium is increasing in oxidation number (+3 to +6), it is oxidized.

b. As we found in Example 12.1b, the oxidation number of chromium in the $Cr_2O_7^{2-}$ ion is +6. In the Cr^{3+} ion, it is +3. Hence, chromium must be reduced (+6 → +3).

Exercise
Consider the process $CrO_4^{2-} \rightarrow Cr_2O_7^{2-}$. Is this an oxidation or reduction? Answer: Neither.

Oxidizing and Reducing Agents

In discussing redox reactions, we often use the terms "oxidizing agent" and "reducing agent." *The oxidizing agent is the reactant that brings about oxidation. Similarly, the reducing agent is the reactant that causes another species to be reduced.*

To illustrate the use of these terms, consider the reaction

$Zn(s) + Cu^{2+}(aq) \rightarrow Zn^{2+}(aq) + Cu(s)$

In this reaction

Zn is oxidized; Cu^{2+} is reduced

Zn is the reducing agent (it reduces Cu^{2+})

Cu^{2+} is the oxidizing agent (it oxidizes Zn)

EXAMPLE 12.3
In the reaction

$2MnO_4^-(aq) + 5SO_2(g) + 2H_2O \rightarrow 2Mn^{2+}(aq) + 5SO_4^{2-}(aq) + 4H^+(aq)$

identify
a. the species that is reduced   b. the species that is oxidized
c. the oxidizing agent   d. the reducing agent

Solution
a. MnO_4^- ion (oxid. no. Mn = +7 in MnO_4^-, +2 in Mn^{2+})
b. SO_2 molecule (oxid. no. S = +4 in SO_2, +6 in SO_4^{2-})
c. MnO_4^-
d. SO_2

Exercise
In the reaction: $Cl_2(g) + H_2O \rightarrow Cl^-(aq) + ClO^-(aq) + 2H^+(aq)$, what is the oxidizing agent? the reducing agent? Answer: Cl_2 in both cases.

12.2 Balancing Redox Equations

In a balanced oxidation–reduction equation, as in any other balanced equation, there must be

—the same number of atoms of each type on both sides of the equation.
—the same net charge on both sides of the equation.

Sometimes, it is possible to accomplish these objectives very easily. The redox equation

$$Zn(s) + Cu^{2+}(aq) \rightarrow Zn^{2+}(aq) + Cu(s) \tag{12.1}$$

is already balanced. There is one atom of zinc and one atom of copper on both sides. The net charge on both sides is +2.

Most redox equations for reactions in water solution are more complex than Equation 12.1. Recall, for example, Equation 12.3, p. 320. It would be difficult, virtually impossible, to balance this equation by a cut-and-try approach. It is best to follow a systematic approach in balancing redox equations. The one we will use is called the **half-equation** or *ion-electron* method. The first step in this process involves separating the equation into two half-equations, an oxidation and a reduction. These two half-equations are then balanced separately. Finally they are combined in such a way to obtain an overall, balanced equation in which no electrons appear.

We will consider several examples of the half-equation method of balancing redox equations. To start with, we will work with a rather simple reaction so as to illustrate clearly what is involved.

Reaction of Fe with Cl₂

Consider the oxidation–reduction reaction

$$Fe(s) + Cl_2(g) \rightarrow Fe^{3+}(aq) + Cl^-(aq)$$

To balance this equation we follow a three-step path.

1. **Divide the equation into two half-equations, an oxidation and a reduction.** In this case the half-equations are clearly

    $Fe(s) \rightarrow Fe^{3+}(aq)$ oxidation

    $Cl_2(g) \rightarrow Cl^-(aq)$ reduction

    The first half-equation is an oxidation because the oxidation number of iron increases from 0 to +3. The second half-equation is a reduction (oxid. no. chlorine = 0 in Cl_2, −1 in Cl^-).

2. **Balance each half-equation, first with respect to number of atoms and**

then with respect to charge. Use electrons for charge balance. The oxidation half-equation is already balanced with respect to number of atoms (one Fe on both sides). To balance charge, we add three electrons to the right

$$Fe(s) \rightarrow Fe^{3+}(aq) + 3\ e^- \qquad (12.4a)$$

This gives a net charge of zero on both sides. The reduction half-equation is slightly more difficult to balance. We start by writing a coefficient of 2 for Cl^-, so as to get two chlorine atoms on both sides.

$$Cl_2(g) \rightarrow 2\ Cl^-(aq)$$

To balance charge, we add two electrons to the left

$$Cl_2(g) + 2\ e^- \rightarrow 2\ Cl^-(aq) \qquad (12.4b)$$

We now have the same number of atoms of chlorine, 2, and the same charge, -2, on both sides. The reduction half-equation is balanced.

3. **Combine the balanced half-equations so that electrons cancel, leaving none on either side.** This leads to a balanced overall equation in which there is no net loss or gain of electrons.

In Equation 12.4a, we have three electrons on the right; in 12.4b, there are two electrons on the left. To make these cancel, we multiply the first half-equation by two and the second by three

$$2\ Fe(s) \rightarrow 2\ Fe^{3+}(aq) + 6\ e^-$$

$$3\ Cl_2(g) + 6\ e^- \rightarrow 6\ Cl^-(aq)$$

We now add these two half-equations. This gets rid of the electrons, six on both sides. The final balanced equation for the overall reaction is

$$2\ Fe(s) + 3\ Cl_2(g) \rightarrow 2\ Fe^{3+}(aq) + 6\ Cl^-(aq) \qquad (12.4)$$

Reaction Between Cu and NO_3^- (Acidic Solution)

Equation 12.4 was rather easy to balance because only two species were involved. One of these (Fe) was oxidized; the other (Cl_2) was reduced. In many redox reactions in water solution, the situation is a bit more complex. Other elements, in addition to those oxidized or reduced, may be involved. Most commonly, these elements are

—hydrogen, which keeps a $+1$ oxidation number throughout the reaction.
—oxygen, which keeps a -2 oxidation number throughout the reaction.

A reaction of this type occurs when copper metal is added to nitric acid (Figure 12.2). The solution turns blue, indicating that copper atoms have been oxidized to Cu^{2+} ions. At the same time, brown fumes of nitrogen diox-

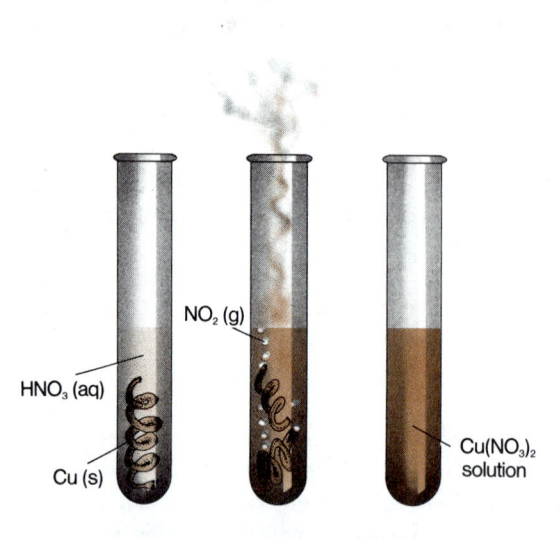

Figure 12.2 When a spiral of copper wire is added to concentrated nitric acid, a redox reaction occurs. The copper is oxidized to Cu^{2+} ions, which give the solution a blue color. The NO_3^- ions of the nitric acid are reduced to NO_2, a brown gas.

ide gas, NO_2, are given off. The NO_3^- ion (oxid. no. N = +5) has been reduced (oxid. no. N = +4 in NO_2). The unbalanced redox equation is

$$Cu(s) + NO_3^-(aq) \rightarrow Cu^{2+}(aq) + NO_2(g) \text{ (acidic solution)}$$

To balance this equation, we follow the same path as before. However, the second step, balancing the individual half-equations, is a bit more complex than it was before.

Step 1 The two half-equations are

$$Cu(s) \rightarrow Cu^{2+}(aq) \text{ oxidation}$$

$$NO_3^-(aq) \rightarrow NO_2(g) \text{ reduction}$$

Step 2 The oxidation half-equation is readily balanced. To do this, we add two electrons to the right, to give a net charge of zero on both sides.

$$Cu(s) \rightarrow Cu^{2+}(aq) + 2\ e^- \qquad (12.5a)$$

The reduction half-equation is more difficult to balance. We start by noting that there is one nitrogen atom on both sides. However, oxygen is not balanced. There are three oxygen atoms on the left and only two on the right. **To balance oxygen, we add H_2O molecules to the appropriate side of the equation.** (This is legitimate because there are plenty of H_2O molecules in any water solution.) In this case, we add one H_2O molecule to the right

$$NO_3^-(aq) \rightarrow NO_2(g) + H_2O$$

Now hydrogen is out of balance; there are 2 H atoms on the right and none on the left. **To balance hydrogen, we add H^+ ions to the appropriate side of**

Oxidation—Reduction Reactions

the equation. (This is legitimate because there are plenty of H^+ ions in any acidic solution.) In this case, we add 2 H^+ ions to the left

$$NO_3^-(aq) + 2\ H^+(aq) \rightarrow NO_2(g) + H_2O$$

Finally, we balance charge by adding one electron to the left

$$NO_3^-(aq) + 2\ H^+(aq) + e^- \rightarrow NO_2(g) + H_2O \quad (12.5b)$$

This is the balanced half-equation for the reduction. There is one N atom, three O atoms, two H atoms, and a net charge of zero on both sides.

Step 3 To eliminate electrons, we multiply Equation 12.5b by two and add to Equation 12.5a

$$Cu(s) \rightarrow Cu^{2+}(aq) + 2\ e^-$$
$$2\ NO_3^-(aq) + 4\ H^+(aq) + 2\ e^- \rightarrow 2\ NO_2(g) + 2\ H_2O$$
$$\overline{Cu(s) + 2\ NO_3^-(aq) + 4\ H^+(aq) \rightarrow Cu^{2+}(aq) + 2\ NO_2(g) + 2\ H_2O} \quad (12.5)$$

Reviewing the process we have just gone through, we should emphasize that, in balancing a half-equation involving oxygen and hydrogen atoms, it is important to follow a certain sequence.

First, balance the number of atoms of the element being oxidized or reduced (adjust coefficients if necessary).
Second, balance oxygen atoms (use H_2O molecules).
Third, balance hydrogen atoms (use H^+ ions).
Fourth, balance charge (use e^-).

EXAMPLE 12.4

Balance the redox equation

$$Fe^{2+}(aq) + Cr_2O_7^{2-}(aq) \rightarrow Fe^{3+}(aq) + Cr^{3+}(aq)$$

Solution

Step 1

$$Fe^{2+}(aq) \rightarrow Fe^{3+}(aq)\ \text{oxidation}$$

$$Cr_2O_7^{2-}(aq) \rightarrow Cr^{3+}(aq)\ \text{reduction}$$

(iron is oxidized from +2 to +3; the oxidation number of chromium decreases from +6 in $Cr_2O_7^{2-}$ to +3 in Cr^{3+}).

Step 2 The oxidation half-equation is readily balanced

$$Fe^{2+}(aq) \rightarrow Fe^{3+}(aq) + e^-$$

To balance the reduction half-equation, we first balance chromium

$$Cr_2O_7^{2-}(aq) \rightarrow 2\ Cr^{3+}(aq)$$

Continued

Second, we balance oxygen by adding H$_2$O molecules to the right

$Cr_2O_7^{2-}(aq) \rightarrow 2\ Cr^{3+}(aq) + 7\ H_2O$

Third, we balance hydrogen by adding H$^+$ ions to the left

$Cr_2O_7^{2-}(aq) + 14\ H^+(aq) \rightarrow 2\ Cr^{3+}(aq) + 7\ H_2O$

Fourth, we balance charges by adding electrons

$Cr_2O_7^{2-}(aq) + 14\ H^+(aq) + 6\ e^- \rightarrow 2\ Cr^{3+}(aq) + 7\ H_2O$

Step 3

$$6\ Fe^{2+}(aq) \rightarrow 6\ Fe^{3+}(aq) + 6\ e^-$$
$$\underline{Cr_2O_7^{2-}(aq) + 14\ H^+(aq) + 6\ e^- \rightarrow 2\ Cr^{3+}(aq) + 7\ H_2O}$$
$$6\ Fe^{2+}(aq) + Cr_2O_7^{2-}(aq) + 14\ H^+(aq) \rightarrow 6\ Fe^{3+}(aq) + 2\ Cr^{3+}(aq) + 7\ H_2O$$

Exercise

How many moles of $Cr_2O_7^{2-}$ are required to produce 1.98 mol of Fe^{3+}? Answer: 0.330 mol.

12.3 Predicting Redox Reactions; Activity Series

We have just seen how to balance equations for redox reactions. We now turn to an even more basic topic, how to predict whether such a reaction will occur when the reactants are mixed. To make such predictions, it is necessary to establish the relative strengths of different oxidizing agents or different reducing agents. Let's start by considering how this might be done.

Strengths of Oxidizing and Reducing Agents; Activity Series

Consider the two molecules Cl$_2$ and Br$_2$. Both of these are capable of acting as oxidizing agents. They can gain electrons to form Cl$^-$ and Br$^-$ ions

$Cl_2(g) + 2\ e^- \rightarrow 2\ Cl^-(aq)$

$Br_2(l) + 2\ e^- \rightarrow 2\ Br^-(aq)$

Suppose we want to determine which of these two molecules is the stronger oxidizing agent. That is, which molecule, Cl$_2$ or Br$_2$, has the greater attraction for electrons?

To answer this question, we might carry out two experiments.

1. In a test tube, shake a water solution containing Cl$_2$(g) with another solution containing Br$^-$ ions, perhaps a water solution of NaBr. When this is done, we see evidence that a reaction is taking place. There is a color

change; an orange color typical of Br_2 molecules in water solution appears. The reaction is

$$Cl_2(g) + 2\ Br^-(aq) \rightarrow 2\ Cl^-(aq) + Br_2(l) \tag{12.6}$$

2. In another test tube, shake a water solution saturated with $Br_2(l)$ with another solution containing Cl^- ions, perhaps a water solution of NaCl. When this is done, there is no evidence of reaction; there is no color change. We conclude that Br_2 molecules do not react with Cl^- ions

$$Br_2(l) + 2\ Cl^-(aq) \rightarrow \text{no reaction}$$

These two experiments indicate that Cl_2 is a stronger oxidizing agent than Br_2. This must be the case, because Cl_2 takes electrons away from Br^- ions while Br_2 does not take electrons away from Cl^- ions.

Table 12.1 extends this comparison to many more species. It is based on experiments of the type just described. A table such as this, which compares the relative strengths of different oxidizing or reducing agents, is often referred to as an *activity series*.

Several features of Table 12.1 are worth emphasis.

TABLE 12.1 Activity Series

	Oxidizing Agent	Reducing Agent	
↑ Increasing Strength Oxidizing Agent ↓	$Li^+(aq) + e^- \rightleftharpoons Li(s)$ $K^+(aq) + e^- \rightleftharpoons K(s)$ $Ca^{2+}(aq) + 2\ e^- \rightleftharpoons Ca(s)$ $Na^+(aq) + e^- \rightleftharpoons Na(s)$ $Mg^{2+}(aq) + 2\ e^- \rightleftharpoons Mg(s)$ $Al^{3+}(aq) + 3\ e^- \rightleftharpoons Al(s)$ $Zn^{2+}(aq) + 2\ e^- \rightleftharpoons Zn(s)$ $Fe^{2+}(aq) + 2\ e^- \rightleftharpoons Fe(s)$ $Co^{2+}(aq) + 2\ e^- \rightleftharpoons Co(s)$ $Ni^{2+}(aq) + 2\ e^- \rightleftharpoons Ni(s)$ $Sn^{2+}(aq) + 2\ e^- \rightleftharpoons Sn(s)$ $2\ H^+(aq) + 2\ e^- \rightleftharpoons H_2(g)$ $Sn^{4+}(aq) + 2\ e^- \rightleftharpoons Sn^{2+}(aq)$ $Cu^{2+}(aq) + 2\ e^- \rightleftharpoons Cu(s)$ $I_2(s) + 2\ e^- \rightleftharpoons 2\ I^-(aq)$ $Fe^{3+}(aq) + e^- \rightleftharpoons Fe^{2+}(aq)$ $NO_3^-(aq) + 2\ H^+(aq) + e^- \rightleftharpoons NO_2(g) + H_2O$ $Ag^+(aq) + e^- \rightleftharpoons Ag(s)$ $Br_2(l) + 2\ e^- \rightleftharpoons 2\ Br^-(aq)$ $O_2(g) + 4\ H^+(aq) + 4\ e^- \rightleftharpoons 2\ H_2O$ $Cr_2O_7^{2-}(aq) + 14\ H^+(aq) + 6\ e^- \rightleftharpoons 2\ Cr^{3+}(aq) + 7\ H_2O$ $Cl_2(g) + 2\ e^- \rightleftharpoons 2\ Cl^-(aq)$ $MnO_4^-(aq) + 8\ H^+(aq) + 5\ e^- \rightleftharpoons Mn^{2+}(aq) + 4\ H_2O$ $F_2(g) + 2\ e^- \rightleftharpoons 2\ F^-(aq)$	↑ Increasing Strength Reducing Agent ↓	

1. All of the species in the left column, at least in principle, can act as oxidizing agents. They do this by acquiring electrons from other species. Note that, in general, there are three different types of oxidizing agents

    —cations, such as Cu^{2+} or Ag^+, which are ordinarily reduced to the corresponding metal (Cu, Ag). In a few cases, the product is a cation with a lower charge (e.g., $Fe^{3+} \rightarrow Fe^{2+}$).
    —nonmetallic elements such as Cl_2 and Br_2, which are ordinarily reduced to the corresponding anion (Cl^-, Br^-).
    —oxyanions such as NO_3^- and MnO_4^- in which the central atom is reduced to a lower oxidation number (NO_2, Mn^{2+}).

2. **Oxidizing strength increases as we move down the left column.** The weakest oxidizing agent listed in Table 12.1 is the Li^+ ion, which has very little tendency to gain an electron to form a Li atom. In contrast, the strongest oxidizing agent is the F_2 molecule, located at the bottom of the left column. Fluorine will take electrons away from just about anything, which explains why it is so dangerous to work with.

3. All of the species in the right column, at least in principle, can act as reducing agents. They do this by supplying electrons to other species. In the case of lithium metal, for example, the half-reaction is

    $Li(s) \rightarrow Li^+(aq) + e^-$

    Of the reducing agents listed in the table, most fall into one of three categories

    —metals such as Li and K, which are oxidized to cations (Li^+, K^+).
    —cations such as Sn^{2+} or Fe^{2+}, which can be oxidized to a higher oxidation number (Sn^{4+}, Fe^{3+}).
    —anions such as I^- and Br^-, which can be oxidized to nonmetallic elements (I_2, Br_2).

4. **Reducing strength increases as we move up the right column.** The strongest reducing agent is the Li atom at the top of the right column. Lithium metal has a strong tendency to lose electrons, which explains why it is so reactive. The weakest reducing agent listed is the fluoride ion, F^-, which has virtually no tendency to give up electrons.

EXAMPLE 12.5
Consider the species

$Br_2(l)$, $Sn^{2+}(aq)$, $Ag^+(aq)$, $Fe(s)$

Using Table 12.1
a. state which are oxidizing agents

Continued

b. state which are reducing agents
c. arrange the oxidizing agents in order of decreasing strength
d. arrange the reducing agents in order of decreasing strength

Solution
a. $Br_2(l)$, $Sn^{2+}(aq)$, $Ag^+(aq)$; all in left column
b. $Sn^{2+}(aq)$ and $Fe(s)$. Note that Sn^{2+} can act as either an oxidizing agent or a reducing agent. It can be reduced to $Sn(s)$ or oxidized to $Sn^{4+}(aq)$.
c. $Br_2(l) > Ag^+(aq) > Sn^{2+}(aq)$
d. $Fe(s) > Sn^{2+}(aq)$

Exercise
Give the formulas of three oxidizing agents that are stronger than Sn^{2+} but weaker than I_2. Answer: H^+, Sn^{4+}, Cu^{2+}.

Predicting Redox Reactions

Using Table 12.1, it is possible to predict whether a reaction will occur when an oxidizing agent is mixed with a reducing agent in water solution. The principle involved is a simple one.

Reaction will take place if the oxidizing agent (left column) is lower in the table than the reducing agent (right column). If this is not the case, reaction will not occur.

According to this principle

— Cl_2 will react with Br^- ions (Cl_2 is below Br^-), but Br_2 will not react with Cl^- ions (Br_2 is above Cl^-). These predictions are indeed correct as we pointed out earlier in this section.

— H^+ ions should react with $Ni(s)$ but not with $Cu(s)$

$$Ni(s) + 2\,H^+(aq) \rightarrow Ni^{2+}(aq) + H_2(g)$$

$$Cu(s) + 2\,H^+(aq) \rightarrow \text{no reaction}$$

Experimentally, we find that *all metals above hydrogen in the activity series react with acid (H^+ ions) to form hydrogen gas*. Metals below hydrogen, such as copper and silver, do not undergo this reaction.

EXAMPLE 12.6
Will the following species react with each other in redox reactions?
a. H^+ and Zn
b. $Cr_2O_7^{2-}$ and Cl^-
c. copper metal and nitric acid

Continued

Solution

a. Yes; H^+ is below Zn. The reaction

$$Zn(s) + 2\,H^+(aq) \rightarrow Zn^{2+}(aq) + H_2(g)$$

is often used to generate hydrogen gas in the laboratory.

b. No; $Cr_2O_7^{2-}$ is above Cl^-

c. Yes; NO_3^- ions from the nitric acid react with Cu(s)

$$Cu(s) + 2\,NO_3^-(aq) + 4\,H^+(aq) \rightarrow Cu^{2+}(aq) + 2\,NO_2(g) + 2\,H_2O$$

(Recall the discussion of this reaction in Section 12.2.)

Exercise

What reaction, if any, occurs when a solution of potassium permanganate is mixed with hydrochloric acid? Answer: $2\,MnO_4^-(aq) + 16\,H^+(aq) + 10\,Cl^-(aq) \rightarrow 2\,Mn^{2+}(aq) + 8\,H_2O + 5\,Cl_2(g)$.

The principle we have just illustrated is sometimes described in slightly different terms. We say that an oxidizing agent (left column, Table 12.1) will react **spontaneously** with a reducing agent that is above it in the right column. By spontaneous we mean that the reaction will take place when the reactants are mixed. The reaction between Cl_2 and Br^- is spontaneous; it occurs when solutions containing these species are mixed

$$Cl_2(g) + 2\,Br^-(aq) \rightarrow 2\,Cl^-(aq) + Br_2(l) \quad \textit{spontaneous}$$

Along these same lines, we would say that the reaction between Br_2 and Cl^- ions is nonspontaneous. It does *not* occur when solutions containing Br_2 and Cl^- are mixed

$$Br_2(l) + 2\,Cl^-(aq) \rightarrow 2\,Br^-(aq) + Cl_2(g) \quad \textit{nonspontaneous}$$

In general, if the oxidizing agent lies above the reducing agent in Table 12.1, the redox reaction between them is nonspontaneous. It does not occur when the reactants are mixed.

12.4 Electrical Cells

There are two types of electrical cells in which redox reactions take place. One of these is an **electrolytic cell**. In such a cell, electrical energy is used to bring about a redox reaction. That reaction is ordinarily one that is nonspontaneous; that is, it does not occur by itself. The other type of electrical cell is referred to as a **voltaic cell**. Such a cell makes use of a spontaneous redox reaction, one which takes place directly when the reactants are mixed. A voltaic cell harnesses such a reaction to produce electrical energy.

Despite their differences, electrolytic and voltaic cells have certain common features. In both cases

1. There are two *electrodes*. Usually, these are wires or strips of metal. At any rate, they are electrical conductors through which electrons can move.
2. At one electrode, an oxidation half-reaction occurs. This electrode is called the anode. At the other electrode, called the cathode, a reduction half-reaction takes place

   Oxidation occurs at the anode

   Reduction occurs at the cathode
3. The two electrodes are separated from one another. Each electrode dips into a liquid, most often a water solution, through which ions can move. Positive ions (cations) migrate through the cell to the cathode. Negative ions (anions) move to the anode.

Electrolytic Cells

Earlier in this text, we mentioned the electrolysis of water (recall Figure 2.1, p. 37). The reaction involved is

$$2 H_2O \rightarrow 2 H_2(g) + O_2(g) \tag{12.7}$$

This reaction is nonspontaneous; liquid water in a saucepan or beaker does not by itself decompose into hydrogen and oxygen. Electrical energy must be supplied to make Reaction 12.7 take place.

In order to carry out an electrolysis, ions must be present to carry the current. To electrolyze water, we add a small amount of an electrolyte such as HCl (H^+, Cl^- ions). In many electrolyses in water solution, ions take part in the electrode reactions. We will consider two such processes, involving water solutions of $CuCl_2$ and NaCl.

Electrolysis of $CuCl_2$ Solution Consider the redox reaction

$$Cu^{2+}(aq) + 2 Cl^-(aq) \rightarrow Cu(s) + Cl_2(g) \tag{12.8}$$

This is a nonspontaneous reaction; if we mix Cu^{2+} and Cl^- ions in water solution, nothing happens. (Note that in Table 12.1, the oxidizing agent, Cu^{2+}, lies above the reducing agent, Cl^-, so we would predict no reaction.) However, Reaction 12.8 can be made to occur by supplying energy from a storage battery (Figure 12.3). The storage battery acts as an "electron pump." It pumps electrons into the cathode, where they reduce Cu^{2+} ions to copper atoms.

Figure 12.3
Electrolysis of $CuCl_2$ solution. In this cell, the electrical energy is furnished by a storage battery. At the cathode, copper metal plates out on the surface of the platinum. At the anode, oxidation of Cl^- ions produces Cl_2 gas.

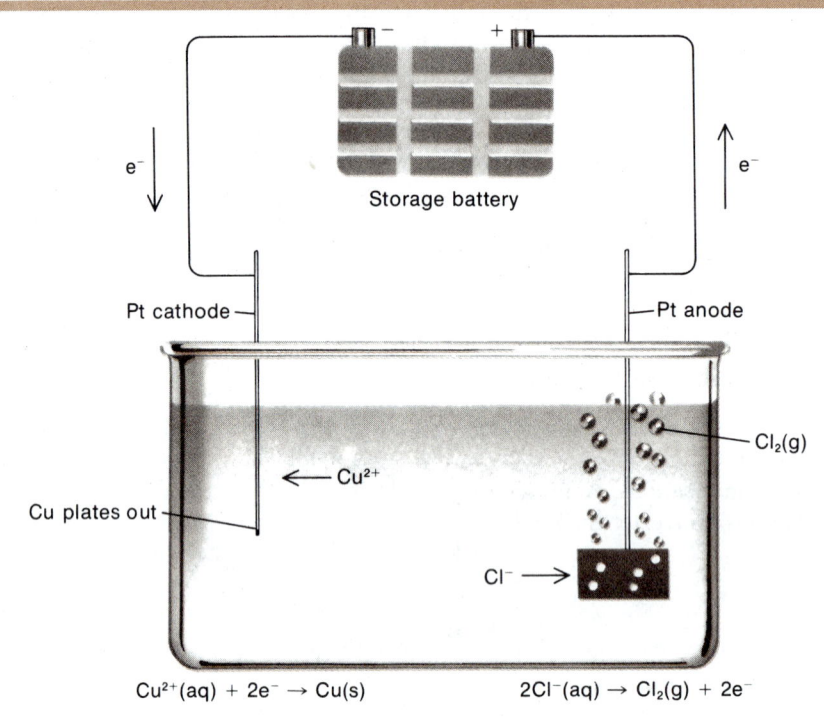

$Cu^{2+}(aq) + 2e^- \rightarrow Cu(s)$ $2Cl^-(aq) \rightarrow Cl_2(g) + 2e^-$

cathode $Cu^{2+}(aq) + 2\ e^- \rightarrow Cu(s)$ reduction   (12.8a)

At the same time, the battery pumps electrons from the anode. These electrons come from Cl^- ions, which are oxidized to Cl_2 molecules.

anode $2\ Cl^-(aq) \rightarrow Cl_2(g) + 2\ e^-$ oxidation   (12.8b)

The overall cell reaction represented by Equation 12.8 is the sum of the two half-reactions occurring at the electrodes (12.8a and 12.8b). As electrolysis proceeds, current is carried through the solution by the migration of ions. The Cu^{2+} ions move to the cathode while Cl^- ions move to the anode.

Electrolysis of NaCl Solution The setup for the electrolysis of a water solution of sodium chloride (Figure 12.4) is very similar to that for $CuCl_2$. The half-reaction at the anode is the same

anode $2\ Cl^-(aq) \rightarrow Cl_2(g) + 2\ e^-$ oxidation   (12.9a)

However, the cathode reaction is quite different. The Na^+ ion lies far above Cu^{2+} in Table 12.1, indicating that it is much more difficult to reduce to the metal. Indeed, the Na^+ ion is more difficult to reduce than an H_2O molecule,

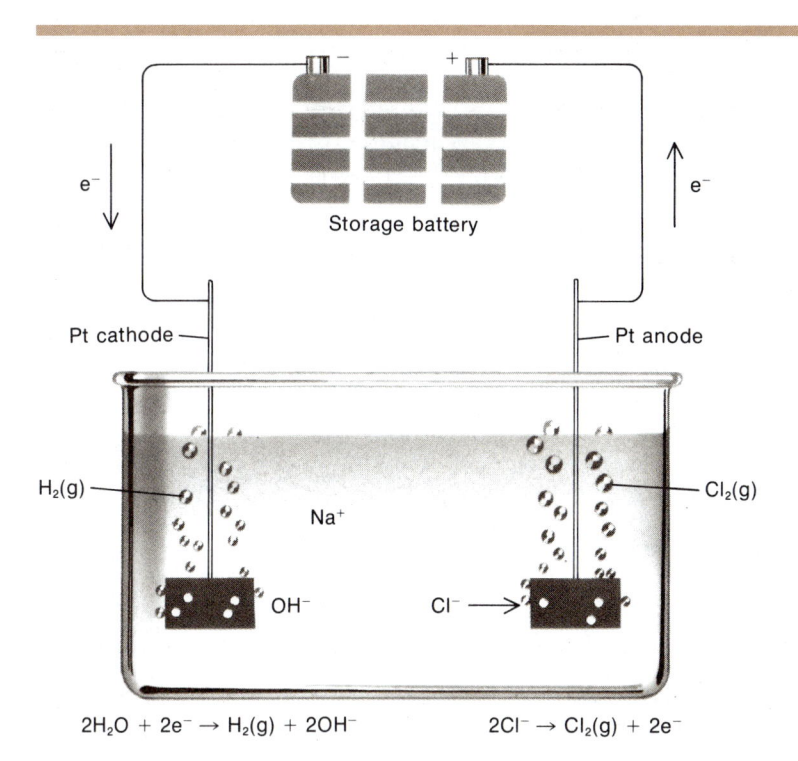

Figure 12.4
Electrolysis of a NaCl solution. At the anode, Cl⁻ ions are oxidized to Cl$_2$(g). At the cathode, H$_2$O molecules are reduced to H$_2$(g) and OH⁻ ions.

$2H_2O + 2e^- \rightarrow H_2(g) + 2OH^-$ $2Cl^- \rightarrow Cl_2(g) + 2e^-$

and there are lots of water molecules around. So, the following half-reaction occurs at the cathode

cathode $2 H_2O + 2 e^- \rightarrow H_2(g) + 2 OH^-(aq)$ reduction   (12.9b)

Water molecules are reduced (oxid. no. H = +1 in H$_2$O, zero in H$_2$). Hydroxide ions are also formed in the reaction; the solution around the cathode becomes strongly basic.

The overall reaction for the electrolysis of a water solution of sodium chloride is obtained by adding Reactions 12.9a and 12.9b

$$2 Cl^-(aq) \rightarrow Cl_2(g) + 2 e^-$$
$$2 H_2O + 2 e^- \rightarrow H_2(g) + 2 OH^-(aq)$$
$$\overline{2 Cl^-(aq) + 2 H_2O \rightarrow Cl_2(g) + H_2(g) + 2 OH^-(aq)}$$ (12.9)

Notice that one effect of this cell reaction is to replace Cl⁻ ions by an equal number of OH⁻ ions. This means that the original solution of NaCl is converted to a solution of NaOH. Commercially, sodium hydroxide is produced this way (1 × 10⁷ metric tons per year in the Untied States) as is chlorine gas.

EXAMPLE 12.7

When a water solution of $CuSO_4$ is electrolyzed, Cu^{2+} ions are reduced at the cathode. At the anode, water molecules are oxidized to oxygen gas. Write equations for
a. the cathode half-reaction
b. the anode half-reaction
c. the overall cell reaction

Solution
a. $Cu^{2+}(aq) + 2\ e^- \rightarrow Cu(s)$
b. The unbalanced half-equation is

$$H_2O \rightarrow O_2(g)$$

Following the procedure described in Section 12.2, we obtain the balanced half-equation

$$2\ H_2O \rightarrow O_2(g) + 4\ H^+(aq) + 4\ e^-$$

c. To obtain the balanced overall equation, we multiply the cathode half-equation by two and add to the anode half-equation

$$2\ Cu^{2+}(aq) + 4\ e^- \rightarrow 2\ Cu(s)$$
$$2\ H_2O \rightarrow O_2(g) + 4\ H^+(aq) + 4\ e^-$$
$$\overline{2\ Cu^{2+}(aq) + 2\ H_2O \rightarrow 2\ Cu(s) + O_2(g) + 4\ H^+(aq)}$$

Exercise

What happens to the pH around the anode during this electrolysis? Answer: Decreases.

Voltaic Cells

As we pointed out earlier, the redox reaction

$$Zn(s) + Cu^{2+}(aq) \rightarrow Zn^{2+}(aq) + Cu(s) \qquad (12.1)$$

is spontaneous. It takes place directly when zinc metal is added to a solution containing Cu^{2+} ions (recall Figure 12.1). When this happens, electrons are transferred directly from Zn atoms to Cu^{2+} ions. A considerable amount of energy is given off as heat.

It is possible to use this redox reaction as a source of electrical energy. To do this, we must make the electron transfer take place over a distance rather than directly. That is, the electrons given up by Zn atoms must travel through an electrical circuit before they reach the Cu^{2+} ions. This way, the electrons can do useful electrical work, such as lighting a bulb. A diagram of a voltaic cell in which this process occurs is shown in Figure 12.5.

Oxidation—Reduction Reactions   337

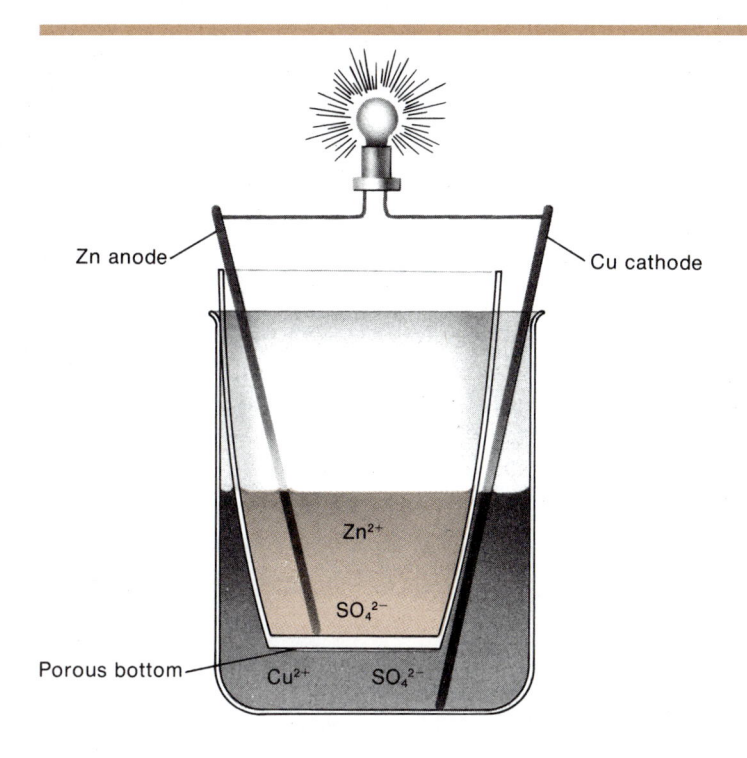

Figure 12.5 A Zn-Cu^{2+} voltaic cell. Electrons flow from the zinc anode through the light bulb to the copper cathode. The zinc electrode is oxidized to Zn^{2+}. The Cu^{2+} ion is reduced to Cu at the copper electrode.

To understand what happens in the Zn-Cu^{2+} voltaic cell, let's trace the path of current through the cell.

1. At the zinc electrode, electrons are given up by zinc atoms

    anode $Zn(s) \rightarrow Zn^{2+}(aq) + 2\ e^-$ oxidation

    The Zn^{2+} ions enter the solution. The electrons pass through the external circuit, including the light bulb.

2. The electrons enter the copper cathode, where they come in contact with Cu^{2+} ions in the solution

    cathode $Cu^{2+}(aq) + 2\ e^- \rightarrow Cu(s)$ reduction

    The overall cell reaction is represented by Equation 12.1, which is the sum of the anode and cathode half-equations.

3. Current is carried through the solution by the movement of ions. Cations (Zn^{2+} and Cu^{2+}) move to the copper cathode. Anions (SO_4^{2-} ions in Figure 12.5) move to the zinc anode.

    There are two points concerning the Zn-Cu^{2+} voltaic cell that deserve emphasis.

    —Cu^{2+} ions must be kept away from the zinc anode. If Cu^{2+} ions came into direct contact with Zn atoms, Reaction 12.1 would occur directly and the cell would "shortcircuit." This explains why the cell shown in Figure 12.5 is separated into two compartments. The $CuSO_4$ solution is in a beaker, separated from the Zn anode in the cup. This way, the reactants are kept apart.

—Ions must be able to move through the cell to complete the electrical circuit. Otherwise, there would be an "open circuit" and no current would flow. This explains why the bottom of the cup is "porous," i.e., permeable to ions.

There are other ways to accomplish these two objectives, i.e., allow ions to move through the cell but prevent Cu^{2+} ions from coming in contact with Zn atoms. One approach uses what is known as a "salt bridge" (Figure 12.6). This is an inverted U-tube containing a solution of an electrolyte such as KNO_3. Ions pass readily through the bridge but Cu^{2+} ions have to wander a long way to reach the zinc electrode.

Another point that should be emphasized is that *the reaction occurring within a voltaic cell must be spontaneous.* Otherwise, the cell will not produce electrical energy. This point is illustrated in Example 12.8.

EXAMPLE 12.8

Suppose that, in the cell shown in Figure 12.5, we replaced the Zn^{2+} ions by Ag^+ ions and the zinc electrode with a silver electrode. The beaker still contains Cu^{2+}

Continued

Figure 12.6 Another type of Zn-Cu^{2+} voltaic cell. The salt bridge allows ions to pass from one solution to the other to complete the circuit. At the same time, it prevents direct contact between Zn atoms and Cu^{2+} ions.

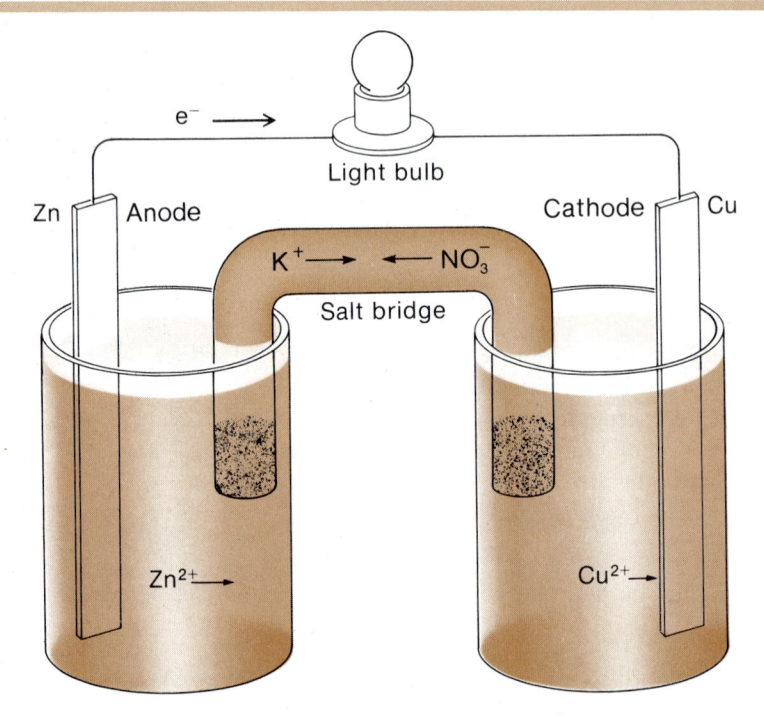

ions and a copper electrode. With this combination, what is
a. the overall cell reaction   b. the anode   c. the cathode

Solution
a. There are two possible redox reactions. We could have a reaction very similar to 12.1, except that Zn is replaced by Ag

$$2\ Ag(s) + Cu^{2+}(aq) \rightarrow 2\ Ag^+(aq) + Cu(s)$$

The other possibility is for the reverse reaction to occur

$$Cu(s) + 2\ Ag^+(aq) \rightarrow Cu^{2+}(aq) + 2\ Ag(s)$$

To decide between these, we must know which reaction is spontaneous. From Table 12.1 we see that Ag^+ ions (oxidizing agent) lie below Cu (reducing agent). Therefore, Ag^+ ions react with Cu metal. By the same argument, Cu^{2+} ions (oxidizing agent) lie above Ag (reducing agent), so Cu^{2+} cannot react spontaneously with Ag(s). Anyway you look at it, the cell reaction must be

$$Cu(s) + 2\ Ag^+(aq) \rightarrow Cu^{2+}(aq) + 2\ Ag(s)$$

b. Looking at the cell reaction, we see that Cu is oxidized to Cu^{2+}; the copper electrode must be the anode.
c. Silver electrode.

Exercise
Suppose we replaced the Cu^{2+} ions by Ag^+ ions and the Cu electrode by Ag, keeping the Zn electrode and Zn^{2+} ions. Under these conditions, what would be the cell reaction? Answer: $Zn(s) + 2\ Ag^+(aq) \rightarrow Zn^{2+}(aq) + 2\ Ag(s)$.

The voltaic cells shown in Figures 12.5 and 12.6 are easy to set up. However, they cannot deliver much power, so they are not a practical source of electrical energy. Commercial voltaic cells, of which there are many, have a quite different design. Perhaps the most familiar of these devices is the lead storage battery shown in Figure 12.7. The electrodes are alternating plates filled with lead metal and lead(IV) oxide, PbO_2. These plates are immersed in a solution of sulfuric acid, H_2SO_4.

When a lead storage battery operates, a spontaneous redox reaction occurs. At the anode, Pb is oxidized to Pb^{2+} ions, which immediately precipitate as insoluble lead sulfate, $PbSO_4$

$$\text{Anode}\quad Pb(s) + SO_4^{2-}(aq) \rightarrow PbSO_4(s) + 2\ e^- \tag{12.10a}$$

At the cathode, PbO_2 is reduced to Pb^{2+} ions, which again precipitate as $PbSO_4$. The balanced half-equation is

$$\text{Cathode}\quad PbO_2(s) + 4H^+(aq) + SO_4^{2-}(aq) + 2e^- \rightarrow PbSO_4(s) + 2H_2O \tag{12.10b}$$

The overall cell reaction is

$$Pb(s) + PbO_2(s) + 4H^+(aq) + 2SO_4^{2-}(aq) \rightarrow 2PbSO_4(s) + 2H_2O \tag{12.10}$$

Figure 12.7 A lead storage battery. A battery like this has a voltage of two volts and can deliver a large amount of electrical energy for a short time. Another advantage of this battery is that it can be recharged. A disadvantage is that it is very heavy.

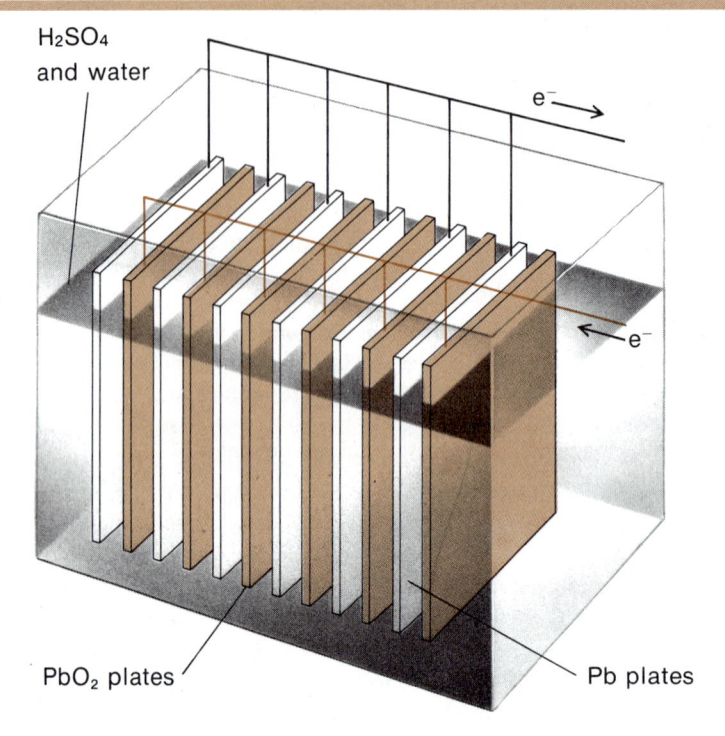

Notice that as the battery operates

— both electrodes become coated with lead sulfate. So long as this material stays on the electrodes, Reaction 12.10 can be reversed. This occurs automatically in an automobile when the car is running. If your battery "runs down," you can get a "quick charge" from a friend, whose charged battery supplies the energy required to reverse Reaction 12.10.

— sulfuric acid is consumed. One mole of H_2SO_4 ($2H^+$, $1\ SO_4^{2-}$) is used up for every mole of $PbSO_4$ formed. As sulfuric acid is used up, its concentration drops and the density of the solution decreases. In a fully charged battery, the density is about 1.30 g/cm^3. If it drops to 1.15 g/cm^3, you're in trouble, at least if you want to start your car on a cold morning.

Key Words

activity series
anode
cathode
electrode

oxidation number
oxidizing agent
redox
reducing agent

Continued

Oxidation—Reduction Reactions

electrolytic cell
half-equation
nonspontaneous reaction
oxidation
reduction
spontaneous reaction
voltaic cell

Questions

12.1 What is meant by oxidation? reduction?

12.2 Why is it that, in the laboratory, oxidation and reduction always occur together?

12.3 What are the common oxidation numbers shown by the following elements in their compounds? sodium _____ magnesium _____ hydrogen _____ oxygen _____

12.4 What is an oxidizing agent? a reducing agent?

12.5 What is meant by a half-equation? How do you balance a half-equation?

12.6 What is an "activity series" and what is it used for?

12.7 What is meant by a "spontaneous" reaction? a "nonspontaneous" reaction?

12.8 Distinguish between an electrolytic and voltaic cell in terms of the types of reactions occurring within them.

12.9 The anode is the electrode at which _____ occurs and toward which _____ move.

12.10 The cathode is the electrode at which _____ occurs and toward which _____ move.

12.11 In the electrolysis of a $CuCl_2$ solution, the products are _____ and _____.

12.12 In the electrolysis of an NaCl solution, the products are _____, _____, and _____.

12.13 What is the purpose of a porous membrane or salt bridge in a voltaic cell?

12.14 In a lead storage battery, the anode contains _____, the cathode contains _____, and the electrolyte is _____.

12.15 Write an equation for the anode half-reaction in the lead storage battery; the cathode half-reaction.

Problems

Oxidation Number

12.16 Give the oxidation number of chlorine in
  a. Cl_2 b. HCl   c. ClO_3^-
  d. ClO_4^-

12.17 Give the oxidation number of chromium in
  a. Cr   b. Cr^{3+} c. CrO_4^{2-}
  d. $Cr_2O_7^{2-}$

12.18 Give the oxidation number of nitrogen in
  a. NH_3 b. N_2 c. N_2H_4 d. NO_3^-

12.19 State the oxidation numbers of all the elements in
  a. $KMnO_4$ b. SO_3 c. Na_2O_2
  d. CH_4

12.20 State the oxidation numbers of all the elements in
  a. Na_2S b. Al_2O_3 c. PO_4^{3-}
  d. H_2SO_4

12.21 Give the formulas of species in which the oxidation number of hydrogen is
  a. 0   b. +1   c. −1

Oxidizing and Reducing Agents

12.22 For the (unbalanced) equations below, identify the species that is oxidized and the

species that is reduced
a. Al(s) + Pb^{2+}(aq) → Al^{3+}(aq) + Pb(s)
b. Zn(s) + NO$_3^-$(aq) + H$^+$(aq) → Zn^{2+}(aq) + NH$_4^+$(aq) + H$_2$O
c. Mn^{2+}(aq) + BiO$_3^-$(aq) + H$^+$(aq) → MnO$_4^-$(aq) + Bi^{3+}(aq) + H$_2$O

12.23 Identify the oxidizing and reducing agents in each reaction in Problem 12.22

12.24 Identify oxidizing and reducing agents in the following (unbalanced) equations
a. Ag(s) + NO$_3^-$(aq) + H$^+$(aq) → Ag$^+$(aq) + NO(g) + H$_2$O
b. Cr$_2$O$_7^{2-}$(aq) + I$^-$(aq) + H$^+$(aq) → Cr^{3+}(aq) + I$_2$(s) + H$_2$O
c. PbO$_2$(s) + H$^+$(aq) + Cl$^-$(aq) → Pb^{2+}(aq) + H$_2$O + Cl$_2$(g)

12.25 Identify the oxidizing and reducing agents in each of the following (unbalanced) equations.
a. MnO$_2$(s) + Cl$^-$(aq) + H$^+$(aq) → Mn^{2+}(aq) + Cl$_2$(g) + H$_2$O
b. Ag$^+$(aq) + Cu(s) → Ag(s) + Cu^{2+}(aq)
c. NO$_3^-$(aq) + Br$^-$(aq) + H$^+$(aq) → NO(g) + Br$_2$(l) + H$_2$O

12.26 Does the ClO$_2^-$ ion act as an oxidizing or reducing agent when it is converted to
a. Cl$_2$ b. Cl$^-$ c. ClO$_3^-$
d. ClO$^-$

12.27 Does the SO$_2$ molecule act as an oxidizing or reducing agent when it is converted to
a. SO$_4^{2-}$ b. H$_2$S   c. S
d. S$_2$O$_3^{2-}$

Balancing Redox Equations

12.28 Balance the redox equations in Problem 12.22.

12.29 Balance the redox equations in Problem 12.24.

12.30 Balance the redox equations in Problem 12.25.

12.31 Balance the following redox equations
a. Cl$_2$(g) + I$^-$(aq) → Cl$^-$(aq) + I$_2$(s)
b. Al(s) + Ni^{2+}(aq) → Al^{3+}(aq) + Ni(s)

c. Sn^{2+}(aq) + Fe^{3+}(aq) → Sn^{4+}(aq) + Fe^{2+}(aq)

12.32 Balance the following redox equations for reactions in acidic solution
a. Cu(s) + NO$_3^-$(aq) → Cu^{2+}(aq) + NO$_2$(aq)
b. ClO$_3^-$(aq) + SO$_2$(g) → ClO$_2$(g) + SO$_4^{2-}$(aq)
c. MnO$_4^-$(aq) + H$_2$S(aq) → Mn^{2+}(aq) + S(s)

12.33 Balance the following redox equations for reactions in acidic solution
a. BrO$_3^-$(aq) + I$^-$(aq) → Br$^-$(aq) + I$_2$(s)
b. Fe^{2+}(aq) + MnO$_4^-$(aq) → Fe^{3+}(aq) + Mn^{2+}(aq)
c. Cr$_2$O$_7^{2-}$(aq) + C$_2$H$_4$(g) → Cr^{3+}(aq) + CO$_2$(g)

12.34 When tin metal is treated with nitric acid, the products include SnO$_2$, which is insoluble, and NO$_2$(g), formed by the reduction of NO$_3^-$ ions. Write a balanced equation for this reaction.

12.35 When a mixture of hydrochloric and nitric acids is heated, two gases, NO$_2$ and Cl$_2$, are formed. Write a balanced
a. half-equation for the oxidation of Cl$^-$ to Cl$_2$
b. half-equation for the reduction of NO$_3^-$ to NO$_2$
c. equation for the overall reaction

Activity Series

(Use Table 12.1 to work out the following problems.)

12.36 Which of the following are reducing agents? oxidizing agents? both oxidizing and reducing agents?
a. Zn^{2+} b. Fe^{2+} c. Sn   d. Br$_2$
e. NO$_3^-$ f. Ag

12.37 Arrange the reducing agents in Problem 12.36 in order of increasing strength; do the same with the oxidizing agents.

12.38 How many of the oxidizing agents listed in

Table 12.1 are stronger than
a. Ca^{2+} b. Sn^{4+} c. O_2 d. F_2

12.39 How many of the reducing agents listed in Table 12.1 are weaker than
a. Zn b. Cl^- c. Ag d. Li

12.40 Will the following react with each other spontaneously?
a. Fe^{2+} and Ni b. Cu^{2+} and Ni
c. Fe^{2+} and Br_2

12.41 Will the following redox reactions take place spontaneously?
a. $O_2(g) + 4 H^+(aq) + 4 I^-(aq) \rightarrow 2 H_2O + 2 I_2(s)$
b. $Cu^{2+}(aq) + 2 Ag(s) \rightarrow Cu(s) + 2 Ag^+(aq)$
c. $Co(s) + Zn^{2+}(aq) \rightarrow Co^{2+}(aq) + Zn(s)$

12.42 Which of the following will react directly with hydrogen gas?
a. Ni^{2+} b. Cu^{2+} c. Ag^+ d. Cl_2

12.43 Which of the following metals will react with Sn^{2+} ions to form tin metal?
a. Zn b. Ag c. Co d. Cu

Electrolytic Cells

12.44 Assuming that solutions of the following ionic compounds behave like $CuCl_2$ on electrolysis, write balanced equations for the anode half-reaction, cathode half-reaction, and the overall reaction.
a. $NiCl_2$ b. $CoBr_2$ c. FeI_2

12.45 Assuming that solutions of the following ionic compounds behave like NaCl on electrolysis, write balanced equations for the anode half-reaction, the cathode half-reaction, and the overall reaction
a. KCl b. NaI c. $CaBr_2$

12.46 Draw a diagram, similar to Figure 12.3, for the electrolysis of a $NiCl_2$ solution. Label anode and cathode. Indicate the direction of flow of electrons outside the cell and of both cations and anions within the cell.

Voltaic Cells

12.47 A voltaic cell consists of a Zn electrode dipping into a solution of Zn^{2+} ions and a Ag electrode dipping into a solution of Ag^+ ions. Using Table 12.1, decide upon the nature of the
a. overall cell reaction
b. anode half-reaction
c. cathode half-reaction

12.48 One half of a voltaic cell contains a Pt electrode surrounded by $Cl_2(g)$ and Cl^- ions. The other half contains a Pt electrode surrounded by a saturated solution of $I_2(s)$ and I^- ions. Use Table 12.1 to decide the
a. overall cell reaction
b. anode half-reaction
c. cathode half-reaction

12.49 The overall reaction in a certain voltaic cell is $Cu(s) + 2 Ag^+(aq) \rightarrow Cu^{2+}(aq) + 2 Ag(s)$. Draw a diagram of the cell, labeling anode and cathode. Show the direction in which electrons flow outside the cell and ions move within the cell.

Multiple Choice

12.50 In the reaction $Cl_2(g) + 2 Br^-(aq) \rightarrow 2 Cl^-(aq) + Br_2(l)$, chlorine
a. acts as an oxidizing agent and is oxidized
b. acts as an oxidizing agent and is reduced
c. acts as a reducing agent and is oxidized
d. acts as a reducing agent and is reduced

Questions 12.51–12.53 require that you balance the following redox equation __ $MnO_4^-(aq)$ + __ $C_2O_4^{2-}(aq)$ + __ $H^+(aq) \rightarrow$ __ $Mn^{2+}(aq)$ + __ $CO_2(g)$ + __ H_2O

12.51 The coefficient of MnO_4^- in the balanced equation is
  a. 1   b. 2   c. 3   d. 4   e. 5
12.52 The coefficient of $C_2O_4^{2-}$ in the balanced equation is
  a. 1   b. 2   c. 3   d. 4   e. 5
12.53 The coefficient of H^+ in the balanced equation is
  a. 2   b. 4   c. 8   d. 16   e. 32
12.54 In all but one of the following compounds, hydrogen has an oxidation number of +1 and oxygen an oxidation number of −2. Identify the compound where this is not the case
  a. H_2O b. HNO_3 c. HNO_2
  d. H_2O_2 e. H_2SO_4
12.55 Nitrogen shows every oxidation number from −3 to +5. Which of the following cannot act as a reducing agent?
  a. NO_2 b. NO   c. N_2O_5 d. N_2O
12.56 In the lead storage battery
  a. PbO_2 is reduced at the cathode
  b. $PbSO_4$ is formed at both electrodes
  c. Pb is oxidized at the anode
  d. all the above are true
  e. none of the above is true
12.57 In a voltaic cell
  a. a nonspontaneous reaction occurs
  b. electrical energy is consumed
  c. oxidation occurs at the anode
  d. all the above are true
  e. none of the above is true
12.58 Zinc is a stronger reducing agent than iron. Hence the reaction $Zn(s) + Fe^{2+}(aq) \rightarrow Zn^{2+}(aq) + Fe(s)$
  a. cannot occur in a voltaic cell
  b. is spontaneous
  c. is nonspontaneous
12.59 In how many of the following species does an element have an oxidation number of +1? CH_4, NaH, MnO_4^-, Ag_2O, H_3O^+
  a. 1   b. 2   c. 3   d. 4   e. 5
12.60 In how many of the following species does an element have an oxidation number of +6? SO_3, H_2SO_4, SO_3^{2-}, CrO_4^{2-}, Cr_2O_3
  a. 1   b. 2   c. 3   d. 4   e. 5

Descriptive Inorganic Chemistry: Metals and Nonmetals

Learning Objectives
After studying this chapter, you should be able to:

1. Describe, with the aid of balanced equations, the reactions of the Group 1 and Group 2 metals with H_2, Cl_2, N_2, S, O_2, and H_2O (Example 13.1).
2. Give the formulas (with charges) and the Lewis structures of the oxide, peroxide, and superoxide ion.
3. Explain, in terms of electron configurations, why transition metals commonly show more than one oxidation number in their compounds and why their compounds are often colored.
4. Given the composition of a complex ion, write its formula, including charge (Example 13.2).
5. Describe, with the aid of balanced equations, the steps in the metallurgy of iron (Example 13.3) and aluminum.
6. Describe the particle structures of the following nonmetals: oxygen, sulfur, nitrogen, phosphorus, and carbon.
7. Explain what is meant by allotropy and give several examples.
8. Describe, with the aid of balanced equations, two methods of preparing hydrogen in the laboratory (Example 13.4) and one method of preparing it commercially.
9. Write balanced equations to describe the reactions of hydrogen with the Group 1 and Group 2 metals, F_2, Cl_2, O_2, and N_2.
10. Discuss the effect of changes in temperature and pressure upon the position of the equilibrium and the rate at which equilibrium is reached in the Haber process.
11. Describe, with the aid of balanced equations, two methods of preparing O_2 in the laboratory; state how O_2 is prepared commercially.
12. Given the formula of a base, give the formula of the corresponding basic anhydride (Example 13.5).
13. Given the formula of an acid, give the formula of the corresponding acid anhydride (Example 13.6).
14. Give the formulas of two different oxides of each of the following nonmetals: sulfur, phosphorus, nitrogen, and carbon.

Alice laughed. "There's no use trying," she said, one can't believe impossible things."
"I daresay, you haven't had much practice," said the Queen. "When I was your age I always did it for half-an-hour a day. Why, sometimes I've believed as many as six impossible things before breakfast."

LEWIS CARROLL

Photograph by Peter L. Kresan.

CHAPTER 13

Throughout Chapters 1–12, we emphasized the basic principles of chemistry. In a first course in chemistry, this emphasis seems appropriate. You must master these principles if you are to benefit from further work in chemistry. However, you should never lose sight of the factual, experimental basis of chemistry. Indeed, the best way to learn chemical principles is to see how they apply in the world around us.

This chapter presents a brief survey of the properties of some of the more common elements. The first part of the chapter deals with metals, which account for approximately 83 of the 108 known elements. We will consider two different types of metals. These are the very reactive metals at the far left of the Periodic Table (Groups 1 and 2, Section 13.1) and the transition metals in the center of the Table (Section 13.2). With that background, we will consider the processes used to extract some of the more common metals from their ores (Section 13.3).

The last three sections of this chapter are devoted to nonmetals. In Section 13.4 we will look at the particle structures of some of the more common nonmetallic elements. Sections 13.5 and 13.6 discuss the chemistry of two important nonmetals, hydroden and oxygen. We will concentrate upon the methods of preparation and the chemical reactions of these elements.

Throughout this chapter, the approach will be descriptive. You will be expected to learn a considerable amount of factual material, much of it summarized in chemical equations. It will help if you can apply chemical principles from earlier chapters to correlate and explain this descriptive material.

13.1 Alkali and Alkaline Earth Metals

As we pointed out in Chapter 11, the strong bases (alkalis) are the hydroxides of the Group 1 and Group 2 metals. This behavior leads to the common names of these groups; alkali metals for Group 1, alkaline earth metals for Group 2.

Descriptive Inorganic Chemistry: Metals and Nonmetals

You may recall from Chapter 12 that the metals at the top of the activity series are those in Groups 1 and 2 (e.g., Li, K, Ca, Na, Mg in Table 12.1, p. 329). As their position in this table implies, these metals are generally very reactive. All the alkali metals react vigorously with water and oxidize rapidly on exposure to air. These metals are stored under kerosene or mineral oil to protect them from air and water. The heavier metals in Group 2 (Ca, Sr, Ba) are about as reactive as the alkali metals. Magnesium is somewhat less reactive. It is not, for example, affected by cold water, although it does react with steam. Of all the metals in these two groups, beryllium is clearly the least reactive. It does not react with water and is only slowly oxidized in air.

Table 13.1 lists some of the physical properties of the alkali and alkaline earth metals. Table 13.2 summarizes the reactions of these metals with the more common nonmetals and with water. For the most part the entries in these tables are self-explanatory. However, there are a few points that perhaps deserve special emphasis.

1. Compounds of several of these metals impart distinctive colors to a Bunsen flame. The bright yellow color that you see when a piece of glass tubing is heated comes from sodium compounds in the glass. The outer 3s electron is excited to a higher sublevel when the temperature increases. When this electron drops back, it gives off light at 589 nm in the yellow region. The flame colors of other alkali and alkaline earth metals are less intense than

TABLE 13.1 Properties of the Alkali and Alkaline Earth Metals

	Li	Na	K	Rb	Cs
Atomic Number	3	11	19	37	55
Atomic Mass	6.94	22.99	39.10	85.47	132.91
Outer Electron Configuration	$2s^1$	$3s^1$	$4s^1$	$5s^1$	$6s^1$
Melting Point (°C)	186	98	64	39	29
Boiling Point (°C)	1326	889	774	688	690
Density (g/cm^3)	0.534	0.971	0.862	1.53	1.87
Ionization Energy (kJ/mol)	527	502	427	410	377
Natural Abundance (%)	0.0065	2.6	2.4	0.031	0.0007
Oxidation Numbers	+1	+1	+1	+1	+1
Flame Color	red	yellow	violet	purple	blue

	Be	Mg	Ca	Sr	Ba
Atomic Number	4	12	20	38	56
Atomic Mass	9.01	24.30	40.08	87.62	137.33
Outer Electron Configuration	$2s^2$	$3s^2$	$4s^2$	$5s^2$	$6s^2$
Melting Point (°C)	1283	650	845	770	725
Boiling Point (°C)	2970	1120	1420	1380	1640
Density (g/cm^3)	1.85	1.74	1.55	2.60	3.51
Ionization Energy (kJ/mol)	904	745	594	556	510
Natural Abundance (%)	0.0006	1.9	3.4	0.030	0.025
Oxidation Numbers	+2	+2	+2	+2	+2
Flame Color	—	—	red	crimson	green

TABLE 13.2 Reactions of Alkali Metals and Alkaline Earth Metals

Combining Substance	Reaction	Metal
	Group 1	
Hydrogen	$2\ M(s) + H_2(g) \rightarrow 2\ MH(s)$	All
Halogens	$2\ M(s) + X_2 \rightarrow 2\ MX(s)$	All
Nitrogen	$6\ M(s) + N_2(g) \rightarrow 2\ M_3N(s)$	Li
Sulfur	$2\ M(s) + S(s) \rightarrow M_2S(s)$	All
Oxygen	$4\ M(s) + O_2(g) \rightarrow 2\ M_2O(s)$	Li
	$2\ M(s) + O_2(g) \rightarrow M_2O_2(s)$	Na
	$M(s) + O_2(g) \rightarrow MO_2(s)$	K, Rb, Cs
Water	$2\ M(s) + 2\ H_2O(l) \rightarrow 2\ M^+(aq) + 2\ OH^-(aq) + H_2(g)$	All
	Group 2	
Hydrogen	$M(s) + H_2(g) \rightarrow MH_2(s)$	Ca, Sr, Ba
Halogens	$M(s) + X_2 \rightarrow MX_2(s)$	All
Nitrogen	$3\ M(s) + N_2(g) \rightarrow M_3N_2(s)$	Mg, Ca, Sr, Ba
Sulfur	$M(s) + S(s) \rightarrow MS(s)$	Mg, Ca, Sr, Ba
Oxygen	$2\ M(s) + O_2(g) \rightarrow 2\ MO(s)$	Be, Mg, Ca, Sr, Ba
	$M(s) + O_2(g) \rightarrow MO_2(s)$	Ba
Water	$M(s) + 2\ H_2O(l) \rightarrow M^{2+}(aq) + 2\ OH^-(aq) + H_2(g)$	Ca, Sr, Ba
	$M(s) + H_2O(g) \rightarrow MO(s) + H_2(g)$	Mg

that of sodium, but can often be used to identify the element. Compounds of Na, K, Sr, and Ba are used in fireworks to give brilliant colors (yellow, violet, crimson, green).

2. Beryllium, the least metallic element in these two groups, is the only one that commonly forms molecular compounds. You may recall, for example, that $BeF_2(g)$ is molecular, with a non-octet structure (Chapter 8). The compounds of all the other metals in Groups 1 and 2 are ionic. In these compounds, the alkali metals are present as +1 ions (Li^+, Na^+, K^+, Rb^+, Cs^+), the alkaline earths as +2 ions (Mg^{2+}, Ca^{2+}, Sr^{2+}, Ba^{2+}). This means that

—the hydrogen compounds of these metals are ionic and contain the *hydride* ion, H^-. This anion has the electron configuration of the noble gas helium

(H:)$^-$

In the alkali hydrides, general formula MH, one hydride ion (H^-) is balanced by one alkali metal cation (M^+). In the alkaline earth hydrides, MH_2, two H^- ions are required to balance one alkaline earth cation, M^{2+}.

—the nitrogen compounds of these metals are ionic and contain the *nitride* ion, N^{3-}. This anion has the Lewis structure

Descriptive Inorganic Chemistry: Metals and Nonmetals

$(:\ddot{\text{N}}:)^{3-}$

The alkali nitrides have the general formula M_3N; three M^+ ions are required to balance one N^{3-} ion. The alkaline earth nitrides have the general formula M_3N_2; three M^{2+} ions balance two N^{3-} ions.

3. These metals form a variety of different compounds with oxygen (Table 13.2). As we would expect, the metals in all of these compounds are present as +1 ions (Group 1) or +2 ions (Group 2). However, oxygen can be in the form of any one of three different anions

—the simple *oxide* ion, O^{2-}, with the Lewis structure

$(:\ddot{\text{O}}:)^{2-}$

This is the anion found in Li_2O (two Li^+ ions per O^{2-} ion) and CaO (one Ca^{2+} ion per O^{2-} ion).

—the *peroxide* ion, O_2^{2-}. This is a polyatomic ion that has the Lewis structure

$(:\ddot{\text{O}}-\ddot{\text{O}}:)^{2-}$

It is found in sodium peroxide, Na_2O_2 (two Na^+ ions per O_2^{2-} ion).

—the *superoxide* ion O_2^-. This is a polyatomic ion that has the non-octet structure

$(:\ddot{\text{O}}-\ddot{\text{O}}:)^-$

It is found in potassium superoxide KO_2 (one K^+ ion per O_2^- ion).

EXAMPLE 13.1

Consider the alkaline earth metal barium. Using Table 13.2 if necessary
a. Write balanced equations for the reactions of barium with chlorine, sulfur, nitrogen, and oxygen (two different reactions with oxygen).
b. Identify the cation and anion in each of the compounds formed in (a).

Solution
a. $Ba(s) + Cl_2(g) \rightarrow BaCl_2(s)$
   $Ba(s) + S(s) \rightarrow BaS(s)$
   $3\ Ba(s) + N_2(g) \rightarrow Ba_3N_2(s)$
   $2\ Ba(s) + O_2(g) \rightarrow 2\ BaO(s)$
   $Ba(s) + O_2(g) \rightarrow BaO_2(s)$
b. In each case, the cation is Ba^{2+}. The anions are, in order, Cl^-, S^{2-}, N^{3-}, O^{2-}, and O_2^{2-} (peroxide).

Exercise
What are the formulas of the oxide, peroxide, and superoxide of cesium? Answer: Cs_2O, Cs_2O_2, CsO_2.

All the reactions listed in Table 13.2 are strongly exothermic. This means that the products are ordinarily formed at high temperatures, often creating a safety hazard. Consider, for example, the reactions of sodium and calcium with liquid water and of magnesium with steam

$$2Na(s) + 2H_2O(l) \rightarrow 2Na^+(aq) + 2\ OH^-(aq) + H_2(g); \Delta H = -368 \text{ kJ} \quad (13.1)$$

$$Ca(s) + 2H_2O(l) \rightarrow Ca^{2+}(aq) + 2\ OH^-(aq) + H_2(g); \Delta H = -431 \text{ kJ} \quad (13.2)$$

$$Mg(s) + H_2O(g) \rightarrow MgO(s) + H_2(g); \Delta H = -360 \text{ kJ} \quad (13.3)$$

In each case, enough heat is given off to cause the hydrogen to ignite. Firefighters trying to put out a magnesium fire have discovered this, often with tragic results.

13.2 Transition Metals

The properties of the metals in the first transition series ($_{21}$Sc $\rightarrow$ $_{30}$Zn) are listed in Table 13.3. Comparing Tables 13.1 and 13.3, you will note that the transition metals, in comparison to those in Groups 1 and 2 are

—*higher melting.* Of the ten metals in Groups 1 and 2, only one (Be) melts

TABLE 13.3 Properties of Transition Metals in the Fourth Period

	Sc	Ti	V	Cr	Mn
Atomic Number	21	22	23	24	25
Atomic Mass	44.96	47.90	50.94	52.00	54.94
Outer Electron Configuration	$4s^2 3d^1$	$4s^2 3d^2$	$4s^2 3d^3$	$4s^1 3d^5$	$4s^2 3d^5$
Melting Point (°C)	1541	1660	1890	1857	1244
Boiling Point (°C)	2831	3287	3380	2672	1962
Density (g/cm^3)	3.0	4.51	6.11	7.19	7.43
Ionization Energy (kJ/mol)	632	661	653	653	715
Abundance (%)	0.0005	0.58	0.015	0.020	0.10
Oxidation Numbers	+3	+2, +3, +4	+2, +3, +4, +5	+2, +3, +6	+2, +3, +4, +6, +7

	Fe	Co	Ni	Cu	Zn
Atomic Number	26	27	28	29	30
Atomic Mass	55.85	58.93	58.69	63.55	65.38
Outer Electron Configuration	$4s^2 3d^6$	$4s^2 3d^7$	$4s^2 3d^8$	$4s^1 3d^{10}$	$4s^2 3d^{10}$
Melting Point (°C)	1535	1495	1453	1083	420
Boiling Point (°C)	2750	2870	2732	2567	907
Density (g/cm^3)	7.87	8.92	9.91	8.94	7.13
Ionization Energy (kJ/mol)	761	757	736	745	904
Abundance (%)	4.7	0.0023	0.0080	0.0070	0.013
Oxidation Numbers	+2, +3	+2, +3	+2	+1, +2	+2

above 1000°C. Of the ten metals in the first transition series, only one (Zn) melts below 1000°C.

—*more dense.* All these transition metals except scandium have a density greater than 4.0 g/cm^3. None of the metals in Groups 1 and 2 has a density as high as 4.0 g/cm^3.

Three characteristics above all others distinguish the transition metals from those in Groups 1 and 2. We will now consider these properties and the rationale behind them.

Multiple Oxidation Numbers

In contrast to the alkali and alkaline earth metals, transition metals commonly form more than one cation. Thus, although calcium in its compounds is always present as the Ca^{2+} ion, iron can form either Fe^{2+} or Fe^{3+}. To explain this difference in behavior, consider the electron configurations

$$Ca\ 1s^2 2s^2 2p^6 3s^2 3p^6 4s^2 \rightarrow Ca^{2+}\ 1s^2 2s^2 2p^6 3s^2 3p^6$$

$$Fe\ 1s^2 2s^2 2p^6 3s^2 3p^6 4s^2 3d^6 \rightarrow Fe^{2+}\ 1s^2 2s^2 2p^6 3s^2 3p^6 3d^6$$

Both Ca^{2+} and Fe^{2+} are formed by the loss of two 4s electrons by the corresponding atoms. However, from that point on, the outlook is quite different. To form a Ca^{3+} ion, an electron would have to be removed from a 3p sublevel. This is much closer to the nucleus than the 4s sublevel. The energy requirement for removal of this third electron is too great and Ca^{3+} does not form. In contrast, to form Fe^{3+}, we need only remove a 3d electron. The 3d and 4s sublevels are nearly equal in energy. Hence, it is relatively easy to convert Fe^{2+} to Fe^{3+}. In general, the availability of several electrons of comparable energy (4s, 3d) allows transition metals to form more than one cation.

When a transition metal shows an oxidation number above +3, it is not present as a cation. Instead, a transition metal atom is covalently bonded to a nonmetal atom, most commonly oxygen. This point is illustrated in Table 13.4, which shows typical species with various oxidation numbers for chromium and manganese. The oxyanions CrO_4^{2-} (chromate), $Cr_2O_7^{2-}$ (dichromate), and MnO_4^- (permanganate) are often encountered in the general chemistry laboratory.

TABLE 13.4 Oxidation Numbers of Chromium and Manganese

Cr		Mn	
+6	CrO_4^{2-}, $Cr_2O_7^{2-}$	+7	MnO_4^-
+3	Cr^{3+}	+6	MnO_4^{2-}
+2	Cr^{2+}	+4	MnO_2
		+3	Mn^{3+}
		+2	Mn^{2+}

Colored Compounds

Compounds of the Group 1 and Group 2 metals are usually white solids. They dissolve in water to form colorless solutions. In contrast, transition metal compounds and their water solutions are often brightly colored. Examples include

K_2CrO_4 yellow

$K_2Cr_2O_7$ red

$KMnO_4$ purple

$Fe(OH)_3$ red

$CoCl_2 \cdot 6H_2O$ pink

$NiCl_2 \cdot 6H_2O$ green

$CuSO_4 \cdot 5H_2O$ blue

The colors of transition metal compounds are readily explained in terms of electronic structure. In most transition metal ions, there are unfilled or partially filled d orbitals. These differ by only small amounts of energy from orbitals that are occupied by electrons. The energy difference is often comparable to the energy of visible light. By absorbing light of a particular color, an electron can move from a lower to a higher orbital. This absorption of light accounts for the color we see when we look at a transition metal compound or its solution (Table 13.5).

To illustrate how Table 13.5 is interpreted, consider potassium permanganate, $KMnO_4$. A water solution of this compound absorbs strongly in the wavelength region 490–580 nm (Figure 13.1). We see the light that is *not* absorbed, i.e., the light that passes through the solution. This is rich in the blue (<490 nm) and red (>580 nm) regions. Hence, a water solution of $KMnO_4$ appears purple, a mixture of blue and red. This same color appears in glass bottles exposed to sunlight for many, many years. Slow oxidation of manganese compounds forms MnO_4^- ions, which impart a purple tint to the glass.

TABLE 13.5 Color of Substances that Absorb Light in the Visible Region

Wavelength Region	Color Absorbed	Color Transmitted
<400 nm	Ultraviolet	Colorless
400–450	Violet	Red, orange, yellow
450–490	Blue	
490–550	Green	Purple
550–580	Yellow	
580–650	Orange	Blue, green
650–700	Red	
>700 nm	Infrared	Colorless

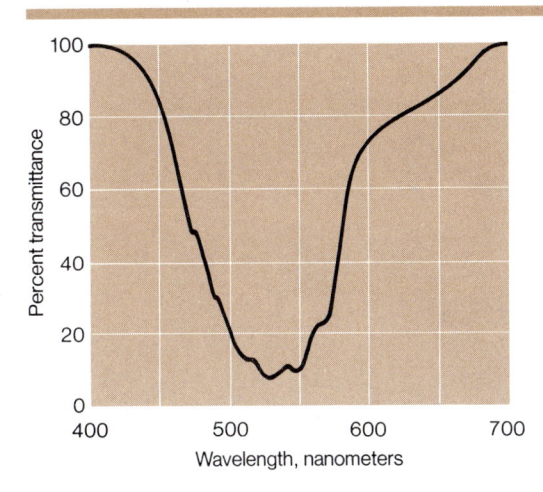

Figure 13.1 Graph of percent of light transmitted by a solution of KMnO₄ versus wavelength. Notice that the transmittance is high in the blue region (< 490 nm) and the red region (> 580 nm). This explains why the solution appears purple.

Complex Ions

Transition metal cations in water are not present as simple monatomic ions such as Cu^{2+}, Zn^{2+}, or Cr^{3+}. Instead, they exist as *complex ions*, in which a central metal cation is bonded to molecules and/or anions. In water solutions of $Cu(NO_3)_2$, $Zn(NO_3)_2$, and $Cr(NO_3)_3$, the positive ions are

$Cu(H_2O)_4^{2+}$, $Zn(H_2O)_4^{2+}$, $Cr(H_2O)_6^{3+}$

In each of these species, a central metal ion (Cu^{2+}, Zn^{2+}, Cr^{3+}) is bonded to water molecules. The geometries of these complex ions are shown in Figure 13.2, p. 356. Note that

— in the $Cu(H_2O)_4^{2+}$ ion, the four H_2O molecules are located at the corners of a square with the Cu^{2+} ion at the center.
— in the $Zn(H_2O)_4^{2+}$ ion, the H_2O molecules are located at the corners of a tetrahedron (Chapter 8), with the Zn^{2+} ion at the center.
— in the $Cr(H_2O)_6^{3+}$ ion, the H_2O molecules are located at the corners of an **octahedron**, with the Cr^{3+} ion at the center. An octahedron has eight sides, each of which is an equilateral triangle. More important here, it has six corners, each the same distance from the center.

A wide variety of molecules and anions can bond to a central metal ion to form a complex ion. The molecule or anion bonded to the metal is called a **ligand**. Some common ligands are

H_2O, NH_3, OH^-, Cl^-, CN^-

Examples of complex ions containing these ligands are

Figure 13.2
Geometries of the three complex ions $Cu(H_2O)_4^{2+}$, $Zn(H_2O)_4^{2+}$, and $Cr(H_2O)_6^{3+}$. The four water molecules are located at the corners of a square in $Cu(H_2O)_4^{2+}$ and a regular tetrahedron in $Zn(H_2O)_4^{2+}$. In the $Cr(H_2O)_6^{3+}$ complex, the six water molecules are located at the corners of a regular octahedron. In each case, the cation (Cu^{2+}, Zn^{2+}, Cr^{3+}) is at the center of the figure.

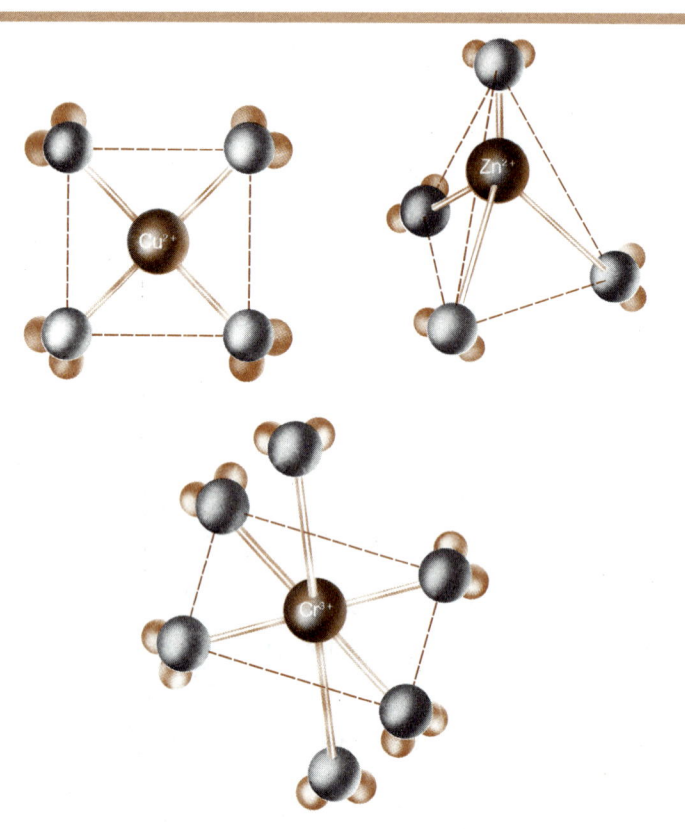

$Zn(H_2O)_4^{2+}$	$Zn(NH_3)_4^{2+}$	$Zn(OH)_4^{2-}$
$Cu(H_2O)_4^{2+}$	$Cu(NH_3)_4^{2+}$	$CuCl_4^{2-}$
$Cr(H_2O)_6^{3+}$	$Cr(NH_3)_6^{3+}$	$Cr(CN)_6^{3-}$

Complex ions containing ligands other than water can often be formed by adding the appropriate anion or molecule to a water solution of a transition metal compound. Consider, for example, what happens when a water solution of copper(II) sulfate is treated with an excess of hydrochloric acid, HCl. There is a pronounced color change, from blue to green. The reaction involved is

$$Cu(H_2O)_4^{2+}(aq) + 4\,Cl^-(aq) \rightarrow CuCl_4^{2-}(aq) + 4\,H_2O \qquad (13.4)$$
$$\text{blue} \hspace{6.5cm} \text{green}$$

The water molecules bonded to the Cu^{2+} ion are replaced by Cl^- ions of the HCl.

The number of bonds formed by the central metal ion in a complex is referred to as the **coordination number**. This number is most frequently 6 or 4. A coordination number of 2 is less common; it is shown only by Cu^+ in the first transition series (e.g., $Cu(NH_3)_2^+$). Odd coordination numbers (1, 3, 5, – –) are very rare.

EXAMPLE 13.2
Give the formulas, including charges, of all the complexes formed by Zn^{2+} (coordination number = 4) with H_2O molecules and/or OH^- ions.

Solution
With four H_2O molecules, the species in $Zn(H_2O)_4^{2+}$. Each time we substitute an OH^- ion for an H_2O molecule, the charge decreases by one unit. The possible species are

$Zn(H_2O)_4^{2+}$, $Zn(H_2O)_3(OH)^+$, $Zn(H_2O)_2(OH)_2$, $Zn(H_2O)(OH)_3^-$, $Zn(OH)_4^{2-}$

These species are formed, in succession, when a strongly acidic solution of a compound such as $Zn(NO_3)_2$ is made basic.

Exercise
Write a balanced net ionic equation for the reaction by which $Zn(H_2O)_4^{2+}$ is converted to $Zn(OH)_4^{2-}$ on addition of NaOH. Answer: $Zn(H_2O)_4^{2+}(aq) + 4\ OH^-(aq) \rightarrow Zn(OH)_4^{2-}(aq) + 4\ H_2O$.

Notice from Example 13.2 that the charge of a complex ion is the sum of the charges of the central metal ion and the anionic ligands. (A molecule, being neutral makes no contribution to the charge.) Other examples of this principle, beyond those complex ions referred to in Example 13.2, include

$CuCl_4^{2-}$ (Cu^{2+} ion, 4 Cl^- ions)   charge = +2 + 4(−1) = −2

$Cr(CN)_6^{3-}$ (Cr^{3+} ion, 6 CN^- ions)   charge = +3 + 6(−1) = −3

$Cr(NH_3)_6^{3+}$ (Cr^{3+} ion, 6 NH_3 molecules)   charge = +3

13.3 Extraction of Metals from their Ores: Metallurgy

Metals occur in nature in many different chemical forms (Figure 13.3). Most often they are found in one of the following forms.

1. *Uncombined elements.* A few of the very inactive transition metals occur as the element rather than a compound. Gold is the best-known example. Ocassionally it is found as almost pure nuggets. More often, gold occurs in veins of quartz or other rocky material.
2. *Oxides.* In these ores, the metal is most often present as a cation (Al^{3+},

Fe^{3+}) balanced by oxide ions, O^{2-}. The best known oxide ores are hematite (Fe_2O_3) and bauxite (Al_2O_3). Several of the more reactive transition metals occur as oxide ores. These include titanium (TiO_2) and manganese (MnO_2).

3. *Sulfides.* These are perhaps more common than any other type of ore. Most of the transition metals toward the right of the Periodic Table occur as sulfides. Among these are nickel (NiS), copper (Cu_2S), zinc (ZnS), and mercury (HgS). The metalloids arsenic and antimony are extracted from their sulfides (As_2S_3, Sb_2S_3). So too are the post-transition metals bismuth and lead (Bi_2S_3, PbS).

4. *Chlorides.* Sodium is obtained from rock salt, impure NaCl. Huge deposits of this mineral are found in underground beds along the Gulf Coast in Texas and Louisiana. Most of the other alkali metals are also found primarily as chlorides (KCl, RbCl, CsCl).

5. *Carbonates.* The alkaline earth metals occur commonly as carbonates. Among these are limestone, $CaCO_3$, and dolomite, a mixture of $CaCO_3$ and $MgCO_3$.

The study of the processes used to extract metals from their ores is referred to as *metallurgy*. We will not attempt to survey the metallurgy of the many different types of ores shown in Figure 13.3. Instead we will concentrate upon two important metals, iron and aluminum.

Fe from Fe_2O_3

The principal high-grade ore of iron, hematite, is mostly iron(III) oxide, Fe_2O_3, mixed with silicon dioxide, SiO_2. The extraction of iron is carried out

Figure 13.3 Sources of the metals. Most metals occur as either oxides or sulfides, but some occur uncombined or as other types of compounds.

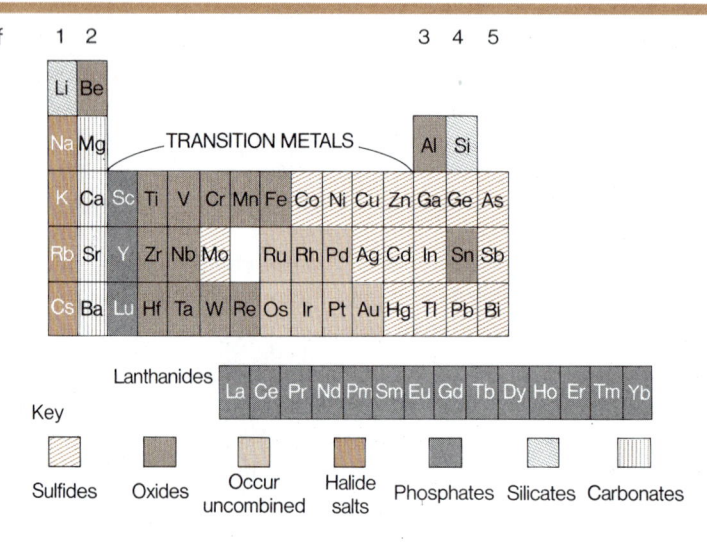

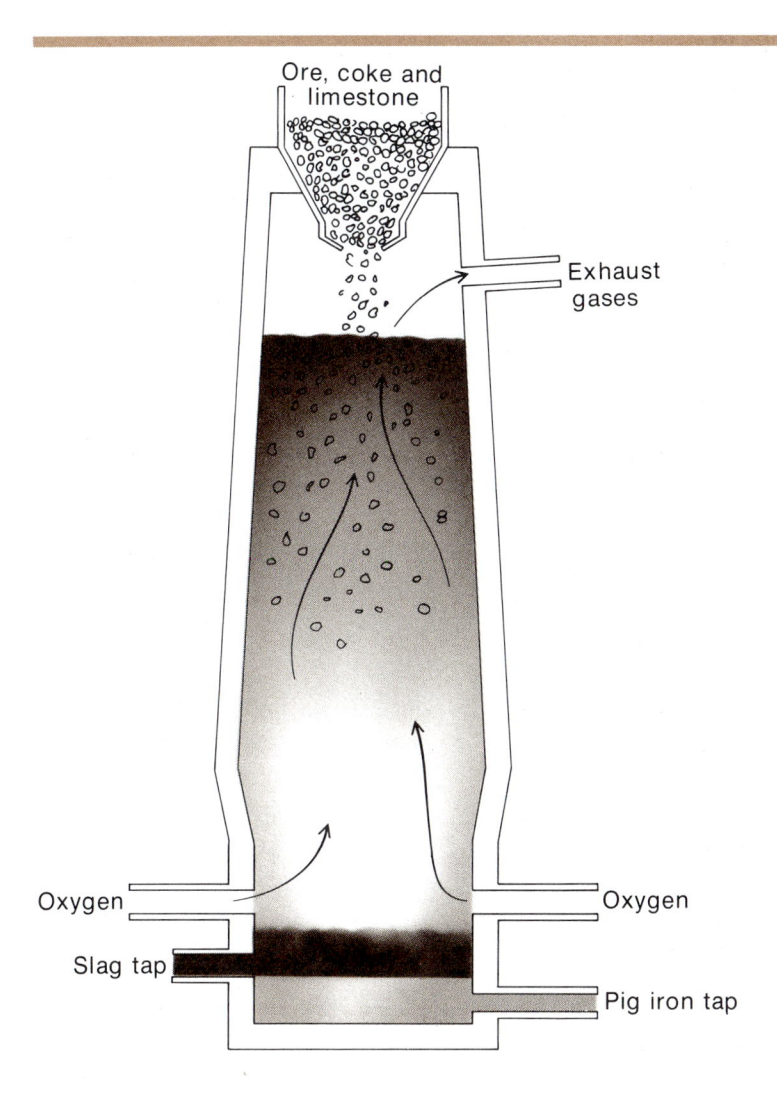

Figure 13.4 Blast furnace. Oxygen reacting with coke furnishes the high temperatures needed for reduction of the iron oxide. The ore appears to be reduced by CO, formed by reaction of CO_2 with coke in the upper parts of the furnace. Using pure oxygen instead of air greatly increases the efficiency of the process.

in a "blast" furnace (Figure 13.4) roughly 30 m high and 10 m in diameter. A mixture of three different materials is introduced at the top of the furnace. These are iron ore (Fe_2O_3, SiO_2), limestone ($CaCO_3$), and "coke," which is nearly pure carbon. To get things started, air or pure O_2 at 500°C is blown into the furnace. A complex series of reactions takes place. The three most important reactions are

1. *Conversion of carbon to carbon monoxide*

   $$2\ C(s) + O_2(g) \rightarrow 2\ CO(g) \qquad (13.5)$$

   This reaction is strongly exothermic ($\Delta H = -221$ kJ). It produces enough heat to raise the temperature in the furnace to 500–1000°C.

2. *Reduction of iron(III) oxide to iron*

$$Fe_2O_3(s) + 3\ CO(g) \rightarrow 2\ Fe(l) + 3\ CO_2(g) \tag{13.6}$$

The reducing agent is carbon monoxide (oxid. no. C = +2 in CO, +4 in CO_2). Molten iron collects at the bottom of the furnace.

3. *Formation of a glassy "slag," which contains calcium silicate*

$$CaCO_3(s) + SiO_2 \rightarrow CaSiO_3(l) + CO_2(g) \tag{13.7}$$

The slag floats on the surface of the iron. It is withdrawn from time to time through a tap located above that used for the molten iron.

A typical blast furnace produces about 1000 metric tons of iron each day. This iron, commonly referred to as "pig iron," is highly impure. The major impurity is carbon (about 4%), which comes from the coke used to reduce the iron ore. Smaller amounts of several other elements are present, including silicon, phosphorus, and manganese.

The pig iron that comes out of the blast furnace is converted to *steel*, which contains from 0 to 2% carbon. To make steel, the impure iron is treated with oxygen gas at a high temperature. At least half of the carbon burns off as carbon dioxide

$$C(s) + O_2(g) \rightarrow CO_2(g) \tag{13.8}$$

Other impurities are converted to oxides (SiO_2, P_4O_{10}, MnO_2) which are later separated as a slag.

EXAMPLE 13.3
Write balanced equations for
a. The reaction of phosphorus, P_4, with oxygen in steel making
b. The reaction of silicon with oxygen
c. The formation of a slag when limestone is added to react with the product in (b).

Solution
a. $P_4(s) + 5\ O_2(g) \rightarrow P_4O_{10}(s)$
b. $Si(s) + O_2(g) \rightarrow SiO_2(s)$
c. $CaCO_3(s) + SiO_2(s) \rightarrow CaSiO_3(l) + CO_2(g)$

Exercise
Another iron ore is magnetite, Fe_3O_4. Write a balanced equation for the reduction of this ore to iron by carbon monoxide. Answer: $Fe_3O_4(s) + 4\ CO(g) \rightarrow 3\ Fe(l) + 4\ CO_2(g)$.

The properties of steel depend upon the percent of carbon present. "Mild" steel, containing less than 0.2% carbon, is very ductile and malleable.

It is used to make wire, pipe, and sheet for automobile bodies. Medium steel (0.2–0.6% C) is less ductile but stronger. It is used in rails, machine parts, girders, and bridge supports. High-carbon steel (0.6–1.5% C) is hard and brittle. Springs, razor blades, and surgical instruments are made from high-carbon steel.

Another way to adjust the properties of steel is to add small amounts of other metals. Stainless steel, containing about 15% chromium and 8% nickel, is very resistant to corrosion. Steel armor plate contains 10% or more of manganese, which makes it hard and tough. High-speed cutting tools are made with steel containing small amounts of tungsten or molybdenum.

Al from Al_2O_3

Aluminum occurs as bauxite ore, which contains about 50% of aluminum oxide, Al_2O_3. The major impurities are iron(III) oxide, Fe_2O_3, and silicon dioxide, SiO_2. To purify the ore, it is treated with a concentrated solution of sodium hydroxide, NaOH. The OH^- ions convert Al^{3+} ions in Al_2O_3 to a complex ion, $Al(OH)_4^-$. The overall reaction is

$$Al_2O_3(s) + 3\ H_2O + 2\ OH^-(aq) \rightarrow 2\ Al(OH)_4^-(aq) \tag{13.9}$$

As a result of this reaction, aluminum is brought into solution. In contrast, the impurities, Fe_2O_3 and SiO_2, do not go into solution. They are filtered off, leaving a solution containing only Na^+ ions and $Al(OH)_4^-$ ions. Dilution of this solution with water followed by strong heating in effect reverses Reaction 13.9. The product is pure white aluminum oxide, Al_2O_3.

There is no chemical reducing agent that is capable of reducing Al_2O_3 to aluminum metal. Instead, the Al^{3+} ions are reduced to Al atoms by electrolysis. Electrical energy is used to bring about the strongly endothermic reaction

$$2\ Al_2O_3(l) \rightarrow 4\ Al(l) + 3\ O_2(g)\ ;\ \Delta H = +3340\ kJ \tag{13.10}$$

Molten aluminum is produced by reduction of Al^{3+} ions at the iron cathode. Oxygen gas is formed by oxidation of O^{2-} ions at the anode (Figure 13.5, p. 362). It attacks the carbon anode, slowly oxidizing it to CO and CO_2.

The process used to make aluminum was worked out in 1886 by Charles Hall, a graduate student at Oberlin College. The problem he faced was to find a practical way to electrolyze Al_2O_3, which has an extremely high melting point, about 2000°C. To do this, he applied a principle described in Chapter 9. By adding a mineral called cryolite, Na_3AlF_6, Hall obtained a mixture with a much lower melting point, about 1000°C. Curiously, the same process was discovered, also in 1886, by Heroult in France. Both men were 22 years old at the time; they were unaware of each other's work.

Figure 13.5
Electrolytic preparation of aluminum. Aluminum, being more dense than cryolite, collects at the bottom of the cell and is protected from oxidation by the air.

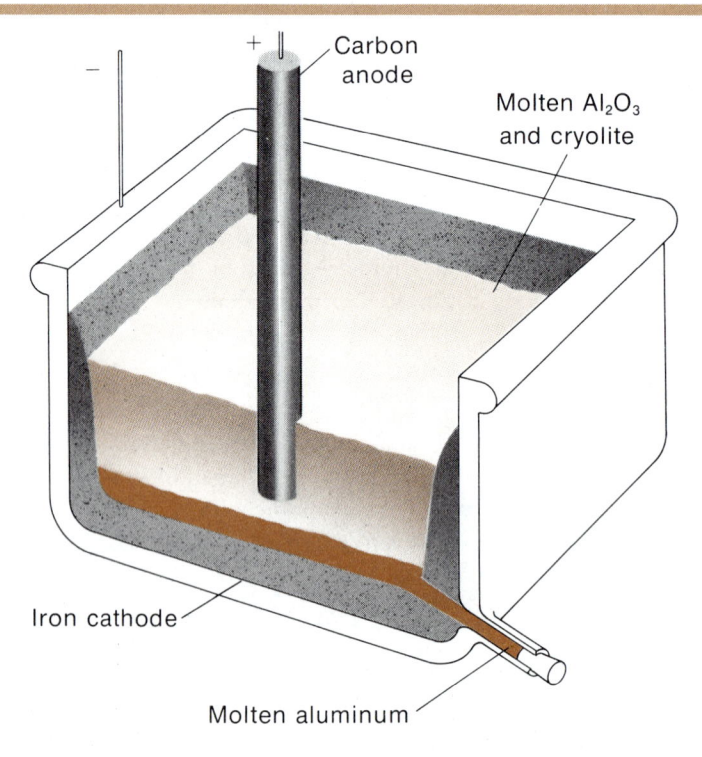

13.4 Nonmetallic Elements; Molecular Structure

There are only a few elements that are classified as nonmetals. Those that fall clearly in this category include

—the Group 8 elements (noble gases)
—the Group 7 elements (halogens)
—oxygen and sulfur in Group 6
—nitrogen and phosphorus in Group 5
—carbon in Group 4

In this section, we will look at the particle structures of these elements.

As pointed out in Chapter 8, the noble gases are the only elements that consist of isolated atoms. The Lewis structure of neon is typical

$:\ddot{Ne}:$

The halogens all exist as diatomic molecules (F_2, Cl_2, Br_2, I_2). In each case, the octet rule is followed. For F_2 we have

$:\ddot{F}—\ddot{F}:$

Descriptive Inorganic Chemistry: Metals and Nonmetals

The molecular structures of the other nonmetals cannot be described as simply as those of the noble gases and halogens. Let's consider, in turn, the structures of the nonmetals in Groups 6, 5, and 4.

Oxygen and Sulfur in Group 6

Under ordinary conditions, elemental oxygen exists as the diatomic molecule O_2. Surprisingly, it is not possible to write a simple Lewis structure that satisfactorily explains the properties of O_2. There appears to be a double bond in the molecule, which suggests the structure

$$:\ddot{O}=\ddot{O}:$$

However, experiment shows that there are two unpaired electrons in the O_2 molecule, which is inconsistent with this structure.

Gaseous oxygen can exist in another form called ozone, molecular formula O_3. The Lewis structure of O_3 is similar to that of SO_2.

Ozone can be formed by the reaction of O_2 molecules with oxygen atoms

$$O_2(g) + O(g) \rightarrow O_3(g) \tag{13.11}$$

This reaction, occurring in the upper atmosphere, is responsible for the "ozone layer" that exists at an altitude of about 30 km. The ozone absorbs a considerable amount of harmful ultraviolet radiation from the sun, preventing this radiation from reaching the earth's surface. Concern for maintaining the ozone layer has led to the elimination of two aerosol propellants, $CFCl_3$ and CF_2Cl_2. Both of these organic compounds react with ozone and might reduce the concentration of O_3 in the upper atomsphere.

The two species O_2 and O_3 are referred to as **allotropes** of the element oxygen. The term allotropy is used to describe the situation where an element exists in two different forms in the same physical state.

Many elements in addition to oxygen show allotropy. One of these is sulfur, directly below oxygen in the Periodic Table. The two allotropic forms of solid sulfur (Figure 13.6) differ only in crystal structure; both contain S_8 molecules. If liquid sulfur is cooled slowly to 119°C, thin needles of monoclinic sulfur separate from the melt. This allotrope is stable between 119 and 96°C. Below 96°C, rhombic sulfur is stable. However, at room temperature, the change from monoclinic to rhombic sulfur takes place very slowly. Crystals of the monoclinic form, obtained by rapid cooling from 119 to 25°C, can be kept around for several days.

Another, unstable form of solid sulfur is often referred to as "plastic" sulfur. This is made by heating sulfur past its melting point to approximately

Rhombic sulfur

Monoclinic sulfur

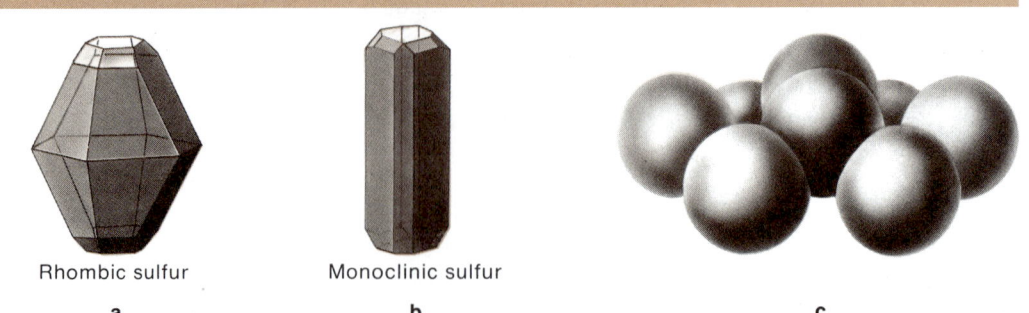

a b c

Figure 13.6 The two allotropes of solid sulfur, rhombic (a) and monoclinic (b), differ only in the way in which S_8 molecules are packed in the crystal. These molecules consist of eight-membered, puckered rings (c).

160°C. At this point, the liquid sulfur becomes viscous and takes on a deep red-brown color. If the liquid is cooled quickly by pouring into water, it forms a sticky brown solid that looks like molasses and acts somewhat like rubber (Figure 13.7). After standing at room temperature for a while, plastic sulfur converts to yellow crystals of rhombic sulfur.

Figure 13.7 Plastic sulfur. The plastic sulfur was obtained by heating sulfur until it became so viscous it was reluctant to flow. It was then poured into the water in the large tank, where it cooled to a rubbery material. In the photo on the right the material shows its elastic properties. Within about 24 hours this material reverted to hard crystalline sulfur.

Nitrogen and Phosphorus in Group 5

The element nitrogen exists as diatomic molecules, N_2. The Lewis structure of this molecule is

:N≡N:

The nitrogen-to-nitrogen triple bond is one of the strongest known covalent bonds. To dissociate one mole of N_2 into atoms, 941 kJ of energy must be absorbed. This high energy requirement is largely responsible for the lack of reactivity of elemental nitrogen.

Solid phosphorus has several allotropes. The two most common are

1. *White phosphorus,* which consists of P_4 molecules. This allotrope is a soft, waxy substance with a low melting point (44°C) and boiling point (280°C). The reactivity of P_4 is so great that it is stored under water to protect it from oxygen. A piece of white phosphorus exposed to air in a dark room glows because of the light given off in the reaction

    $$P_4(s) + 5\ O_2(g) \rightarrow P_4O_{10}(s) \qquad (13.12)$$

    (The word phosphorescence, used nowadays in a wider context, was coined to describe this property of phosphorus.) White phosphorus is extremely toxic. As little as 0.1 g taken into the body can be fatal. Direct contact of P_4 with the skin produces painful burns.

2. *Red phosphorus,* which has properties quite different from those of white phosphorus. It is much higher melting (mp = 590°C at 43 atm) and is less toxic than the white form. It is also less reactive. Red phosphorus must be heated to 250°C to react with oxygen. The detailed structure of this allotrope has not been established, but it is known to be *macromolecular* (Chapter 8). That is, atoms of phosphorus are linked to one another by a network of covalent bonds that extends across an entire crystal. There are no small discrete molecules in red phosphorus.

Carbon in Group 4

The element carbon has two crystalline allotropes, diamond and graphite. Both of these are macromolecular. However, as you can see from Figure 13.8, p. 366, the bonding patterns are quite different.

1. In graphite, the carbon atoms are covalently bonded in two dimensions. Each atom is bonded to three other atoms (to one by a double bond, to the others by single bonds)

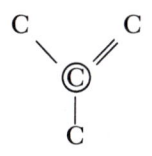

Chapter 13

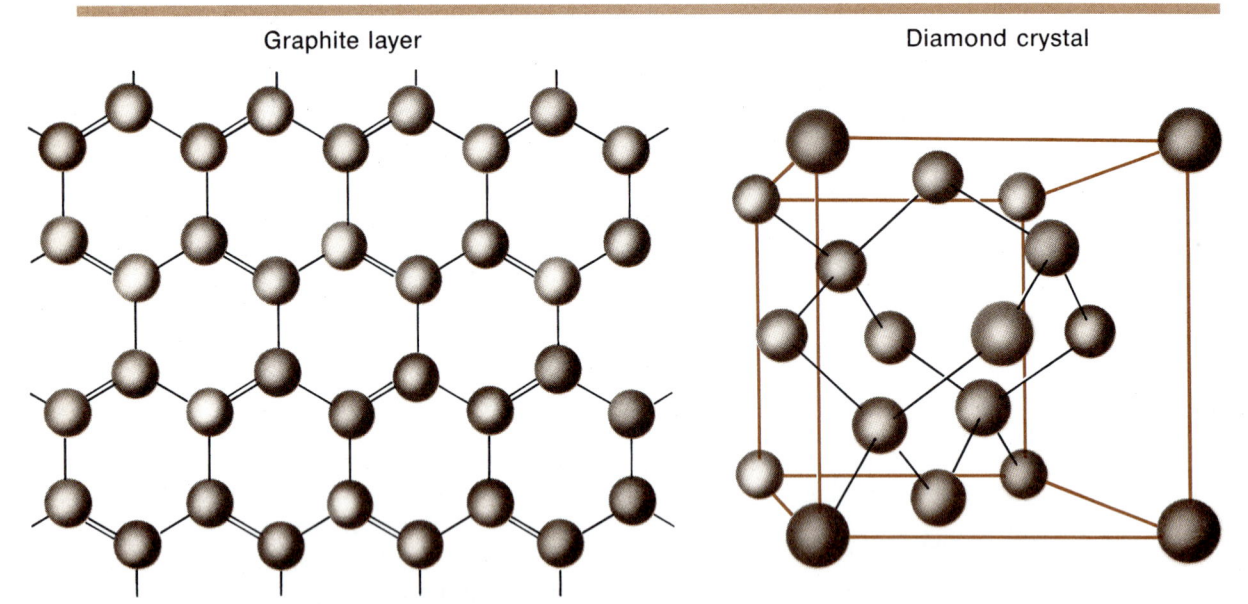

Graphite layer

Diamond crystal

Figure 13.8 In diamond, each carbon atom is at the center of a tetrahedron, on each corner of which is another carbon atom. In graphite, the carbon atoms are linked together in planes of hexagons. Within a layer, a carbon atom is bonded to three other carbon atoms; one third of the bonds are double bonds. The layers are held to one another by weak dispersion forces.

There are no chemical bonds between carbon atoms in adjacent two-dimensional layers.
2. In diamond, the carbon atoms are bonded in a three-dimensional pattern. Each carbon atom is at the center of a tetrahedron, singly bonded to four other carbon atoms.

The properties of diamond and graphite are readily explained in terms of their structures. Since both are macromolecular, they are very high melting (>3500°C). In both allotropes, covalent bonds must be broken for melting to occur. However, it is easy to separate layers from one another in graphite, since there are no bonds between layers. This accounts for the use of graphite in "lead" pencils. As you write, layers of carbon atoms rub off on the paper.

Descriptive Inorganic Chemistry: Metals and Nonmetals

In contrast, diamond, with a three-dimensional structure, is the hardest substance known. High-speed cutting tools often have diamond tips.

13.5 Hydrogen

Elemental hydrogen is never found in nature. The most common compound of hydrogen is water, H_2O. All organic compounds contain hydrogen as well as carbon atoms. All in all, about one of every six atoms in the world around us are hydrogen atoms. Since the hydrogen atom is very light, the abundance of the element on a mass basis is much lower, of the order of 1%.

The major use of H_2 gas is the manufacture of ammonia, NH_3, by a process discussed later in this section. Smaller amounts of hydrogen are used to convert liquid fats (oils) to solids. In the presence of a nickel catalyst, H_2 molecules add to double bonds in the "unsaturated" fat molecules, converting them to single bonds ("saturated fats"). The general process can be represented by the equation

$$\text{C=C} + \text{H—H} \rightarrow \text{—C—C—} \quad \text{with H, H substituents} \tag{13.13}$$

Preparation

As pointed out in Chapter 2, hydrogen gas can be prepared in the laboratory by the electrolysis of water

$$2\ H_2O(l) \rightarrow 2\ H_2(g) + O_2(g)\ ;\ \Delta H = +572\ \text{kJ} \tag{13.14}$$

The energy required to bring about this endothermic reaction is supplied by a battery or other source of direct electric current. Hydrogen gas is produced by reduction at the cathode of the electrolytic cell (ox. no. H = +1 in H_2O, 0 in H_2). Half as much oxygen, by volume, is produced at the anode.

Small amounts of hydrogen are often made in the laboratory by reacting a metal with acid (H^+ ions). In principle, any metal above hydrogen in the activity series (Table 12.1, p. 329) can be used. The metal most often used is zinc

$$Zn(s) + 2\ H^+(aq) \rightarrow Zn^{2+}(aq) + H_2(g) \tag{13.15}$$

The metal atoms are oxidized to cations while H^+ ions are reduced to H_2. The acid used is most often hydrochloric acid (H^+, Cl^- ions) or sulfuric acid (H^+, HSO_4^- ions).

> **EXAMPLE 13.4**
> Another metal that reacts with acid to form hydrogen gas is aluminum. Write a balanced net ionic equation for the reaction of aluminum with hydrochloric acid.
>
> **Solution**
> As in Equation 13.15, H^+ ions are reduced to H_2 gas. Al atoms are oxidized to Al^{3+} ions. The balanced net ionic equation is
>
> $$2\ Al(s) + 6\ H^+(aq) \rightarrow 2\ Al^{3+}(aq) + 3\ H_2(g)$$
>
> Notice that the Cl^- ions of the hydrochloric acid do not appear in the equation. They are "spectator ions" which take no part in the reaction.
>
> **Exercise**
> Write a balanced net ionic equation for the reaction of aluminum with sulfuric acid.
> Answer: $2\ Al(s) + 6\ H^+(aq) \rightarrow 2\ Al^{3+}(aq) + 3\ H_2(g)$

Neither Reaction 13.14 nor 13.15 is suitable for the commercial (industrial) preparation of H_2. The electrolysis of water uses too much energy and the zinc used in Reaction 13.15 is too expensive. Most of the hydrogen produced in the United States today is made by reacting steam at 600–1000°C with natural gas, which is mostly methane, CH_4

$$CH_4(g) + H_2O(g) \rightarrow CO(g) + 3\ H_2(g) \tag{13.16}$$

A nickel catalyst is used to speed up the reaction. The mixture of CO and H_2 formed is separated by first converting the CO to CO_2 and then passing through water. Since CO_2 is much more soluble in water than H_2, the gas leaving the water is nearly pure hydrogen.

Reactions of Hydrogen

As pointed out earlier, hydrogen reacts with the alkali and alkaline earth metals to form ionic compounds containing the hydride ion H^-. The reactions with sodium and calcium are typical and occur at temperatures of about 200°C.

$$2\ Na(s) + H_2(g) \rightarrow 2\ NaH(s) \tag{13.17}$$

$$Ca(s) + H_2(g) \rightarrow CaH_2(s) \tag{13.18}$$

When hydrogen reacts with nonmetals, the compounds formed are molecular. In all these compounds, hydrogen has an oxidation number of +1. The reactions with fluorine, chlorine, and oxygen occur readily

$$H_2(g) + F_2(g) \rightarrow 2\ HF(g) \tag{13.19}$$

$$H_2(g) + Cl_2(g) \rightarrow 2\ HCl(g) \tag{13.20}$$

$$2\ H_2(g) + O_2(g) \rightarrow 2\ H_2O(l) \tag{13.21}$$

The reaction with oxygen may occur unexpectedly when hydrogen gas is made in the laboratory. Air containing as little as 4% hydrogen forms an explosive mixture readily ignited by a spark or flame. If you prepare hydrogen in the laboratory, make sure there are no burners operating nearby.

From a commercial standpoint, the most important reaction of hydrogen is that with nitrogen to form ammonia, NH_3. The reaction reaches a position of equilibrium

$$N_2(g) + 3\ H_2(g) \rightleftharpoons 2\ NH_3(g)\ ;\ \Delta H = -92\ kJ \tag{13.22}$$

From the principles discussed in Chapter 10, we can predict that

1. An increase in pressure will have a favorable effect on the position of the equilibrium, shifting it to the right to form more NH_3. This is true because, in the forward reaction, 4 mol of gas are converted to 2 mol. An increase in pressure should also speed up the reaction, because it increases the number of molecules per unit volume (i.e., the concentrations of N_2 and H_2).
2. An increase in temperature will have an unfavorable effect on the position of the equilibrium. Because the forward reaction is exothermic, the equilibrium will shift to the left at high temperatures, reducing the yield of NH_3. On the other hand, an increase in temperature will speed up the reaction.
3. An appropriate catalyst should speed up the reaction without affecting the position of the equilibrium.

Combining these factors, it seems that, to get a good yield of ammonia in a short time, we should use

—a high pressure
—a moderate temperature
—a catalyst

Fritz Haber, a German chemist, came to these same conclusions 75 years ago. He worked out the conditions for the preparation of ammonia by Reaction 13.22. These have been modified relatively little over the years. Today, in the Haber process, a pressure of about 400 atm is used. The temperature is close to 400°C; the catalyst is a mixture of Fe, Al_2O_3, and K_2O.

Haber was interested in NH_3 as a starting material for explosives. The Haber process first went into production in 1913, one year before the outbreak of World War I. Today, large amounts of ammonia are still converted to nitric acid, HNO_3, which is used to make such explosives as trinitrotoluene (TNT) and nitroglycerine. However, most of the ammonia made by the Haber process is used for fertilizers. Liquid NH_3 can be used directly. More often, ammonia is converted to a solid nitrogen compound. One useful fertilizer (and explosive) is ammonium nitrate, NH_4NO_3. This is made by an acid–base reaction between ammonia and nitric acid

$$NH_3(g) + HNO_3(l) \rightarrow NH_4NO_3(s)$$
   base   acid   salt

13.6 Oxygen

Oxygen is our most abundant element; about one out of every two atoms in the world is an oxygen atom. Elemental O_2 accounts for approximately 21 mole percent of air. Most of the oxygen at or below the surface of the earth is in the form of H_2O, SiO_2, or complex aluminosilicate rocks such as feldspar or granite.

Elemental oxygen is essential to life. Exothermic reactions of O_2 with foods such as glucose supply us with energy

$$C_6H_{12}O_6(aq) + 6\ O_2(g) \rightarrow 6\ CO_2(g) + 6\ H_2O \qquad (13.23)$$

Pure oxygen is kept available in hospitals for people who have difficulty breathing. The main use of the element is in making steel; about 1×10^7 metric tons of O_2 are used annually in the United States for this purpose.

Preparation

Oxygen gas can be made in the laboratory by the electrolysis of water (Reaction 13.14). A variety of chemical reactions can also be used to produce small amounts of O_2. Perhaps the simplest and safest of these is the decomposition of hydrogen peroxide, H_2O_2. When a little MnO_2 is added to a 3% water solution of H_2O_2 at 25°C, the following reaction occurs

$$2\ H_2O_2(aq) \rightarrow 2\ H_2O + O_2(g) \qquad (13.24)$$

The maganese(IV) oxide acts as a catalyst.

Industrially, oxygen is prepared by the *fractional distillation* of liquid air (Figure 13.9). The first step in this process is the removal of carbon dioxide and water vapor from air. The air is then compressed, cooled to room temperature, and allowed to expand suddenly. When this happens its temperature drops. By repeating this sequence of compression, cooling, and expansion several times, a temperature of $-200°C$ can be reached. At that point, the air liquifies.

When liquid air is allowed to warm up slowly, the gas that comes off first is mostly N_2 (bp = $-196°C$). The liquid that remains behind is mostly O_2 (bp = $-183°C$) with some argon (bp = $-186°C$). By repeated fractionation, very pure oxygen can be obtained.

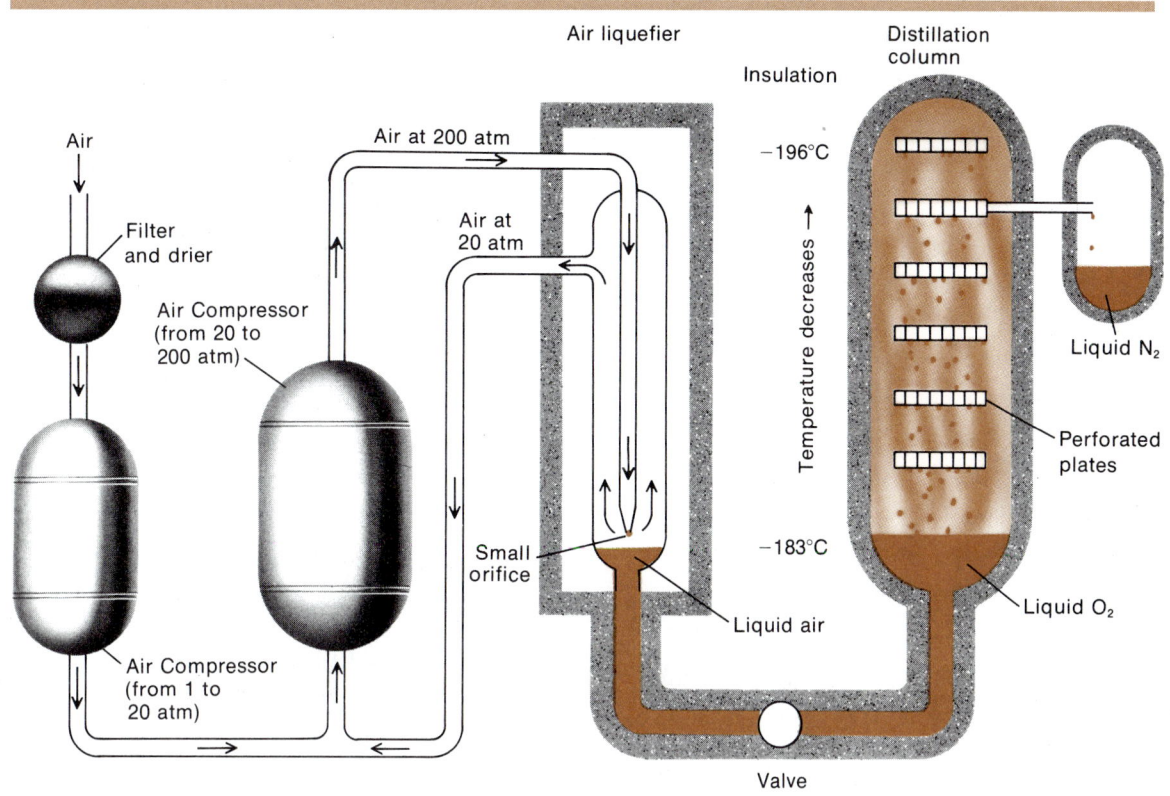

Figure 13.9 Liquid air is made in enormous quantities by the process outlined in the drawing. The method is based on the fact that air tends to cool when expanded rapidly. Air thus cooled is passed by incoming gas, cooling it further, until finally liquid air forms. The liquid air is then distilled on a fractionating column, where N_2, which boils at a lower temperature than O_2, can be separated from the liquid. Both N_2 and O_2 have large industrial applications.

Reaction of Oxygen with Metals

When oxygen reacts with a metal, the product is ordinarily a simple metal oxide containing the oxide ion, O^{2-}, which has the Lewis structure

$(:\ddot{O}:)^{2-}$

Lithium, calcium, and aluminum react in this way

$$4\,Li(s) + O_2(g) \rightarrow 2\,Li_2O(s); \quad Li^+, O^{2-} \text{ ions} \tag{13.25}$$

$$2\,Ca(s) + O_2(g) \rightarrow 2\,CaO(s); \quad Ca^{2+}, O^{2-} \text{ ions} \tag{13.26}$$

$$4\,Al(s) + 3\,O_2(g) \rightarrow 2\,Al_2O_3(s); \quad Al^{3+}, O^{2-} \text{ ions} \tag{13.27}$$

Oxides of the Group 1 metals and the heavier Group 2 metals (Ca, Sr, Ba) react with water to form metal hydroxides

$$Li_2O(s) + H_2O(l) \rightarrow 2\ LiOH(s) \tag{13.28}$$

$$CaO(s) + H_2O(l) \rightarrow Ca(OH)_2(s) \tag{13.29}$$

Compounds such as Li_2O and CaO which react with water in this way are referred to as **basic anhydrides**; upon reacting with water they form strong bases. Oxides of other metals such as MgO and Al_2O_3 do not react in this way, in part because they are very insoluble in water.

EXAMPLE 13.5

Consider the two strong bases NaOH and $Ba(OH)_2$.
a. What are the formulas of the corresponding basic anhydrides?
b. Write balanced equations for the reactions of these basic anhydrides with water.

Solution

a. Basic anhydrides are metal oxides. They contain metal cations electrically balanced by oxide ions, O^{2-}. The oxides of sodium and barium have the formulas Na_2O (two Na^+ ions per O^{2-} ion) and BaO (one Ba^{2+} ion per O^{2-} ion).
b. $Na_2O(s) + H_2O(l) \rightarrow 2\ NaOH(s)$; compare Equation 13.28
   $BaO(s) + H_2O(l) \rightarrow Ba(OH)_2(s)$; compare Equation 13.29

Exercise

Scandium oxide, Sc_2O_3, slowly reacts with water to form scandium hydroxide, $Sc(OH)_3$. Write a balanced equation for this reaction. Answer: $Sc_2O_3(s) + 3\ H_2O(l) \rightarrow 2\ Sc(OH)_3(s)$.

Reaction of Oxygen with Nonmetals

Some of the more common nonmetal oxides are listed, along with their physical properties, in Table 13.6. As you can see from the table, many nonmetals form more than one oxide. Indeed, nitrogen forms seven (N_2O_5, N_2O_4, NO_2, N_2O_3, N_2O_2, NO, N_2O) of which only two are listed.

Several nonmetal oxides act as **acid anhydrides**; that is, they react with water to form acids. When this happens, the oxidation number of the central nonmetal atom remains the same. Thus, CO_2 (oxid. no. C = +4) is the acid anhydride of carbonic acid, H_2CO_3 (oxid. no. C = +4)

$$CO_2(g) + H_2O(l) \rightarrow H_2CO_3(aq) \tag{13.30}$$

Again, P_4O_{10} (oxid. no. P = +5) is the acid anhydride of phosphoric acid, H_3PO_4 (oxid. no. P = +5)

$$P_4O_{10}(s) + 6\ H_2O(l) \rightarrow 4\ H_3PO_4(l) \tag{13.31}$$

Table 13.6 Properties of Some Common Nonmetal Oxides

Formula	Common Name	Melting Point (°C)	Boiling Point (°C)	Physical State 25°C, 1 atm
CO	Carbon monoxide	−207	−192	gas
CO₂	Carbon dioxide	*	*	gas
NO	Nitric oxide	−161	−151	gas
NO₂	Nitrogen dioxide	−9	21	gas
P₄O₆	Phosphorus trioxide	22	173	liquid
P₄O₁₀	Phosphorus pentoxide	*	*	solid
SO₂	Sulfur dioxide	−76	−10	gas
SO₃	Sulfur trioxide	17	45	liquid

*Liquid CO₂ is not stable at 1 atm; the solid (Dry Ice) sublimes (passes directly to vapor) at −79°C. Similarly, P₄O₁₀ sublimes at 250°C.

EXAMPLE 13.6

Consider the two strong acids H_2SO_4 and HNO_3.
a. What is the oxidation number of sulfur in H_2SO_4? nitrogen in HNO_3?
b. What are the formulas of the acid anhydrides of H_2SO_4 and HNO_3?
c. Write balanced equations for the reactions of these gaseous acid anhydrides with water.

Solution

a. Recall the discussion in Section 12.1, Chapter 12. Oxid. no. S in H_2SO_4 = +6. Oxid. no. N in HNO_3 = +5.
b. The oxide of sulfur in which S has an oxidation number of +6 is SO_3. The oxide of nitrogen in which N has an oxidation number of +5 is N_2O_5.
c. $SO_3(g) + H_2O(l) \rightarrow H_2SO_4(l)$
   $N_2O_5(g) + H_2O(l) \rightarrow 2\ HNO_3(l)$

Exercise

What is the formula of the acid anhydride of $HClO_4$? Answer: Cl_2O_7.

Several gaseous nonmetal oxides are serious air pollutants. One of these is sulfur dioxide, SO_2. It can be produced by burning sulfur in air

$$S(s) + O_2(g) \rightarrow SO_2(g) \tag{13.32}$$

More commonly, sulfur dioxide is formed as a byproduct in the combustion of fuels (coal, heating oil) containing sulfur compounds as impurities. Sulfur dioxide has a choking odor and can cause severe breathing problems for people who suffer from asthma, bronchitis, or emphysema. Moreover, SO_2 in the air is slowly oxidized to sulfur trioxide, SO_3

$$2\ SO_2(g) + O_2(g) \rightarrow 2\ SO_3(g) \tag{13.33}$$

Sulfur trioxide, as we saw in Example 13.6, is the anhydride of sulfuric acid, H_2SO_4. Hence, it reacts readily with water vapor in the air

$$SO_3(g) + H_2O(g) \rightarrow H_2SO_4(l) \tag{13.34}$$

Sulfuric acid formed in this way is largely responsible for the "acid rain" (pH < 5) that falls in parts of the northeastern United States and Canada. As the pH of lakes and streams drops, fish and other forms of marine life die.

Another serious air pollutant is carbon monoxide, CO. This can be formed by the combustion of carbon at high temperatures in a limited amount of O_2 (recall Equation 13.5). More commonly, CO is formed when hydrocarbon fuels such as octane, C_8H_{18}, burn in a limited amount of air

$$C_8H_{18}(l) + 17/2\ O_2(g) \rightarrow 8\ CO(g) + 9\ H_2O(l) \tag{13.35}$$

This reaction and others like it occur when gasoline, a mixture of hydrocarbons, burns in an automobile engine. Small amounts of carbon monoxide are also present in cigarette smoke. As little as 0.02% CO can cause unconsciousness; 0.1% CO in air can be fatal.

In principle, carbon monoxide should readily be oxidized to carbon dioxide

$$2\ CO(g) + O_2(g) \rightarrow 2\ CO_2(g);\ K_c = 10^{91}\ \text{at}\ 25°C \tag{13.36}$$

However, this reaction takes place quite slowly under ordinary conditions. The catalytic converter in automobiles contains a mixture of Pt and Pd, which catalyzes Reaction 13.36. This greatly reduces the concentration of carbon monoxide coming out of the exhaust. Even so, you should never leave a car engine running in a closed space such as a garage.

The oxides of nitrogen are major culprits in smog formation. Nitric oxide, NO, is formed by direct reaction of the elements at high temperatures

$$N_2(g) + O_2(g) \rightarrow 2\ NO(g);\ \Delta H = +181\ \text{kJ} \tag{13.37}$$

The equilibrium constant for this reaction is very small at room temperature, about 1×10^{-30}. However, since Reaction 13.37 is endothermic, K_c increases with temperature, becoming about 0.1 at 2000°C. Significant amounts of NO are formed in an automobile engine. When the nitric oxide passes into the atmosphere, it is oxidized to nitrogen dioxide, NO_2

$$2\ NO(g) + O_2(g) \rightarrow 2\ NO_2(g) \tag{13.38}$$

Nitrogen dioxide has a brownish color and an acrid, unpleasant odor.

Key Words

acid anhydride
acid rain

alkali metal
alkaline earth metal

Continued

allotropy
basic anhydride
complex ion
coordination number
electrolysis
fractional distillation
Haber process
halogen

ligand
macromolecule
metal
metallurgy
octahedron
slag
tetrahedron
transition metal

Questions

13.1 Of the metals in Group 2, which one is the least reactive? Which one forms molecular compounds?

13.2 What is the flame color of Na? K? Ca? Ba?

13.3 Give the Lewis structure of the hydride ion; the nitride ion.

13.4 Give the formula and Lewis structure of the oxide ion; the peroxide ion; the superoxide ion.

13.5 As compared to the Group 1 and 2 metals, the transition metals have _____ melting points and _____ densities.

13.6 Explain, in terms of electron configurations, why the transition metals commonly show more than one oxidation number.

13.7 Give the formulas of species in which
   a. Cr shows an oxidation number of +6
   b. Mn shows an oxidation number of +7; of +4

13.8 Cite several examples of colored transition metal compounds.

13.9 Explain, in terms of electron configurations, why compounds containing transition metal ions are commonly colored.

13.10 Explain what is meant by a complex ion.

13.11 What is a ligand? What is meant by coordination number?

13.12 Describe the geometry of complex ions in which the coordination number is six.

13.13 List four different anions commonly found in ores of metals.

13.14 What materials are put into a blast furnace? What substances are taken out of it?

13.15 How do the properties of steel depend upon its carbon content?

13.16 What is the purpose of adding cryolite in the electrolysis of Al_2O_3?

13.17 Write the Lewis structure of ozone.

13.18 Explain what is meant by allotropy and cite four different examples.

13.19 What is plastic sulfur? rhombic sulfur? monoclinic sulfur?

13.20 How do white and red phosphorus differ in molecular structure? in physical and chemical properties?

13.21 How do diamond and graphite differ in structure? properties?

13.22 Cite two different uses of elementary hydrogen.

13.23 Discuss the effect of temperature and pressure upon the position of the equilibrium and the rate of the Haber process for making ammonia.

13.24 Describe two different uses of ammonia.

13.25 Explain what is meant by the "fractional distillation" of liquid air.

13.26 Give the formulas of five different nonmetal oxides that are air pollutants.

Problems

Reactions of Group 1, Group 2 Metals

13.27 Write balanced equations for the reaction of potassium with
   a. H_2 b. Cl_2 c. S   d. H_2O

13.28 Write balanced equations for the reaction of strontium with
   a. H_2 b. Br_2 c. N_2 d. S

13.29 Write balanced equations for the reactions of the following metals with oxygen (see Table 13.2).
   a. Li   b. Na   c. Rb   d. Mg   e. Ca

13.30 Give the formula of
   a. lithium nitride
   b. potassium peroxide
   c. potassium oxide
   d. potassium superoxide

13.31 Give the formula of
   a. barium oxide   b. barium peroxide
   c. strontium nitride   d. calcium hydride

Complex Ions

13.32 Give the formulas, including charges, of all the complexes formed by Zn^{2+} (coordination number = 4) with NH_3 molecules and/or Cl^- ions.

13.33 Give the formulas and charges of complex ions in which Co^{3+} is bonded to
   a. six NH_3 molecules
   b. three NH_3 molecules, three H_2O molecules
   c. four NH_3 molecules, two Cl^- ions
   d. two NH_3 molecules, four Cl^- ions

13.34 Give the charge of the central metal cation in each of the following complexes (the ligands are NH_3 molecules and Cl^- ions).
   a. $Cr(NH_3)_3Cl_3$ b. $Cu(NH_3)_3Cl^+$
   c. $Pd(NH_3)_4^{2+}$ d. $Pt(NH_3)Cl_3^+$
   e. $Cu(NH_3)Cl$

13.35 State the coordination number in each of the complexes listed in Problems 13.33 and 13.34.

Metallurgy of Iron and Aluminum

13.36 Write balanced equations for the
   a. formation of CO in the blast furnace
   b. reduction of Fe_2O_3 to iron in the blast furnace

13.37 Write a balanced equation for
   a. slag formation in the blast furnace
   b. removal of carbon in making steel

13.38 Write a balanced equation for
   a. dissolving Al_2O_3 in concentrated NaOH
   b. electrolysis of Al_2O_3

13.39 Write balanced equations for the reduction of the following oxides to the metals using carbon monoxide
   a. SnO_2 b. ZnO   c. Cr_2O_3
   d. Co_3O_4

Preparation and Reactions of Hydrogen and Oxygen

13.40 Write balanced equations for
   a. two different laboratory preparations of hydrogen gas
   b. the commercial preparation of hydrogen

13.41 Write balanced equations for the reaction of hydrogen with
   a. sodium   b. fluorine   c. nitrogen
   d. barium

13.42 Write balanced equations for two different laboratory preparations of oxygen gas.

13.43 Write balanced equations for the reaction of oxygen with
   a. aluminum   b. sulfur   c. nitrogen

13.44 Of the compounds referred to in Problems 13.41 and 13.43, which are ionic? molecular?

13.45 Give the formulas of
 a. four different oxides of nitrogen
 b. two different oxides of sulfur
 c. two different oxides of phosphorous

Anhydrides

13.46 State the formulas of the basic anhydrides of
 a. $Ca(OH)_2$ b. $LiOH$ c. $Cr(OH)_3$

13.47 Write balanced equations for the reaction with water of each of the anhydrides referred to in Problem 13.46.

13.48 Consider the acid $H_2CO_3(aq)$
 a. What is the oxidation number of carbon in H_2CO_3?
 b. What is the acid anhydride of H_2CO_3?
 c. Write a balanced equation for the reaction of this acid anhydride with water.

13.49 Give the formula of the acid anhydride of
 a. H_2SO_3 b. H_2SO_4 c. H_3PO_4
 d. HNO_3

13.50 Of the compounds listed in Problems 13.46 and 13.49, which are ionic? molecular?

Multiple Choice

13.51 The products of the reaction of magnesium with steam are
 a. MgH_2 and O_2
 b. MgO and H_2
 C. $Mg(OH)_2$ and H_2
 d. $Mg(OH)_2$ and O_2

13.52 When sodium reacts with oxygen at room temperature the principal product is
 a. Na_2O b. Na_2O_2 c. NaO_2
 d. $NaOH$

13.53 Which one of the following anions does *not* have a noble gas structure?
 a. H^- b. N^{3-} c. O^{2-} d. O_2^{2-}
 e. O_2^-

13.54 The MnO_4^- ion has a purple color. This is explained by the fact that it absorbs radiation in the
 a. ultraviolet   b. green and yellow
 c. blue and red   d. infrared

13.55 A complex ion formed from Fe^{3+} with three chloride ions, a carbonate ion, and a water molecule, has a charge of
 a. +3   b. +2   c. +1   d. −1
 e. −2

13.56 Most transition metal compounds found in nature are either
 a. chlorides or bromides
 b. chlorides or carbonates
 c. oxides or hydroxides
 d. oxides or sulfides
 e. sulfides or selenides

13.57 In the electrolysis of aluminum oxide
 a. aluminum is formed at the cathode
 b. oxygen is formed at the anode
 c. 4 mol of Al forms for every 3 mol of O_2
 d. all the above are true
 e. none of the above is true

13.58 Stainless steel, in addition to iron, contains
 a. Mn and P   b. Al and Ni   c. Cr and Ni   d. Zn and Na   c. Mg and Al

13.59 Which one of the following oxides is *not* an air pollutant?
 a. SO_2 b. SO_3 c. CO d. CO_2
 e. NO

13.60 The stable allotropic form of sulfur at room temperature is
 a. plastic sulfur   b. rhombic sulfur
 c. monoclinic sulfur   d. SO_2 e. SO_3

Introduction to Organic Chemistry

Learning Objectives
After studying this chapter, you should be able to:

1. Draw the structural formulas of all the isomeric alkanes corresponding to a given molecular formula (Example 14.1).
2. State the IUPAC name of an alkane, given its structural formula (Example 14.2); give the structural formula corresponding to a IUPAC name (Example 14.3).
3. Draw a portion of the structural formula of a polymer derived from a given monomer (Example 14.4).
4. Determine, from the molecular formula, whether a hydrocarbon is an alkane, alkene, or alkyne (Example 14.5).
5. Describe how simple hydrocarbons including CH_4, C_2H_6, C_2H_4, and C_2H_2 are obtained commercially.
6. Describe the structure and bonding in benzene.
7. Use the structural formula of an organic oxygen compound to determine the functional group present and the class of compound (Example 14.6).
8. Draw the structural formulas of all the isomers of an organic oxygen compound, knowing the molecular formula and the class of compound (Example 14.7).
9. Describe how the following compounds are prepared: methyl alcohol, ethyl alcohol, and acetic acid.
10. Show by means of a balanced equation how an ester can be prepared from a carboxylic acid and an alcohol (Example 14.8).
11. Explain what a fat is; give the structure of a typical fat.

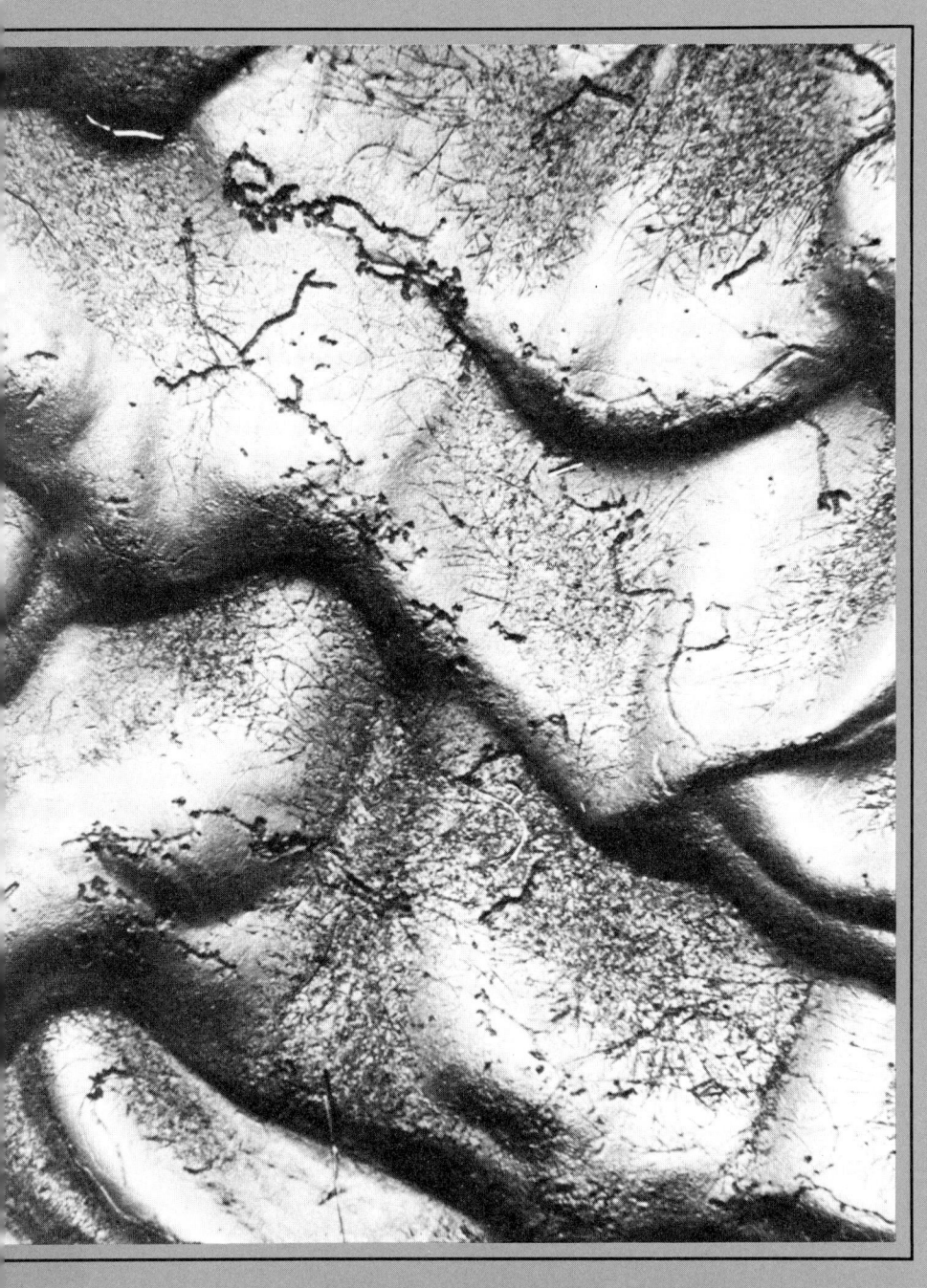

> In science there are unending horizons to be explored and no limit to the discoveries to be made. It is difficult to explain to one who has never experienced it, the incomparable thrill, excitement, and satisfaction of original discovery. In science this satisfaction is frequent.
>
> GEO. W. BEADLE

Photoitgraph by Peter L. Kresan.

CHAPTER 14

"Organic chemistry nowadays drives one mad. To me it appears like a primeval forest full of the most remarkable things, a dreadful endless jungle which one dares not enter, for there seems no way out." When the German chemist Friedrich Wöhler wrote this in 1835, it was an apt description of this branch of chemistry. Today, organic chemistry is no longer a "dreadful jungle." Instead it is a highly organized discipline with rules and principles that can be grasped by a beginning student. One objective of this chapter will be to outline some of the basic concepts that underlie organic chemistry.

Organic chemistry deals with the compounds of carbon, of which there are literally millions. More than 99% of all known compounds contain carbon atoms. There is a simple explanation for this remarkable fact. Carbon atoms bond to one another to a far greater extent than do atoms of any other element. Carbon atoms may link together to form chains or rings

$$-\overset{|}{\underset{|}{C}}-\overset{|}{\underset{|}{C}}- \quad \underset{/}{\overset{\backslash}{C}}=\overset{/}{\underset{\backslash}{C}} \quad -C\equiv C- \quad -\overset{|}{\underset{|}{C}}-\overset{|}{\underset{|}{C}}-\overset{|}{\underset{|}{C}}-\overset{|}{\underset{|}{C}}-$$

(ring of 6 C atoms)

The bonds may be single (one electron pair), double (two electron pairs), or triple (three electron pairs.)

In this chapter, we will consider a variety of different organic compounds. They will have quite different structures and properties. However, all these substances have certain features in common. In particular

1. *Organic compounds are molecular rather than ionic.* Most of the compounds we will discuss consist of small, discrete molecules. Many of them are gases or liquids at room temperature. A few, generally of high molar mass, are solids.

2. *Each carbon atom forms a total of four covalent bonds.* This is illustrated by the structures written above. A particular carbon atom may form four single bonds, two single bonds and a double bond, or one single bond and a triple bond. One way or another, though, the bonds add up to four.
3. *Carbon atoms may be bonded to each other or to other nonmetal atoms, most commonly hydrogen, oxygen, or a halogen (Group 7 element).* In stable organic compounds

—a hydrogen atom forms one covalent bond, —H.
—an oxygen atom forms two covalent bonds, —O— or =O.
—a halogen atom (F, Cl, Br, I) forms one covalent bond, —X.

We start this chapter by discussing hydrocarbons (Sections 14.1–14.3). These are compounds that contain only the two elements carbon and hydrogen. Then we go on (Section 14.4) to consider some of the simpler compounds that contain oxygen in addition to carbon and hydrogen.

14.1 Saturated Hydrocarbons (Alkanes)

In a saturated hydrocarbon or **alkane,** all the carbon—carbon bonds are single bonds. There is a series of such compounds differing from one another in the number of carbon atoms per molecule. The alkanes can be represented by

1. *Structural formulas,* which show all the bonds in the molecule. The structural formulas of the first three alkanes are

```
      H            H  H           H  H  H
      |            |  |           |  |  |
   H—C—H        H—C—C—H        H—C—C—C—H
      |            |  |           |  |  |
      H            H  H           H  H  H

   methane        ethane           propane
```

These formulas can be somewhat misleading so far as geometry is concerned. Recall from Chapter 8 that the methane molecule is *tetrahedral.* The carbon atom is at the center of a regular tetrahedron with a hydrogen atom at each corner. This tetrahedral geometry is maintained in all the alkanes. All the carbon atoms in these molecules form four bonds directed toward the corners of a regular tetrahedron. All the bond angles are the tetrahedral angle, 109.5°.

The geometries of methane, ethane, and propane are shown in Figure 14.1. Notice that although we often refer to propane (and higher members of the series) as "straight-chain" hydrocarbons, the chain of carbon

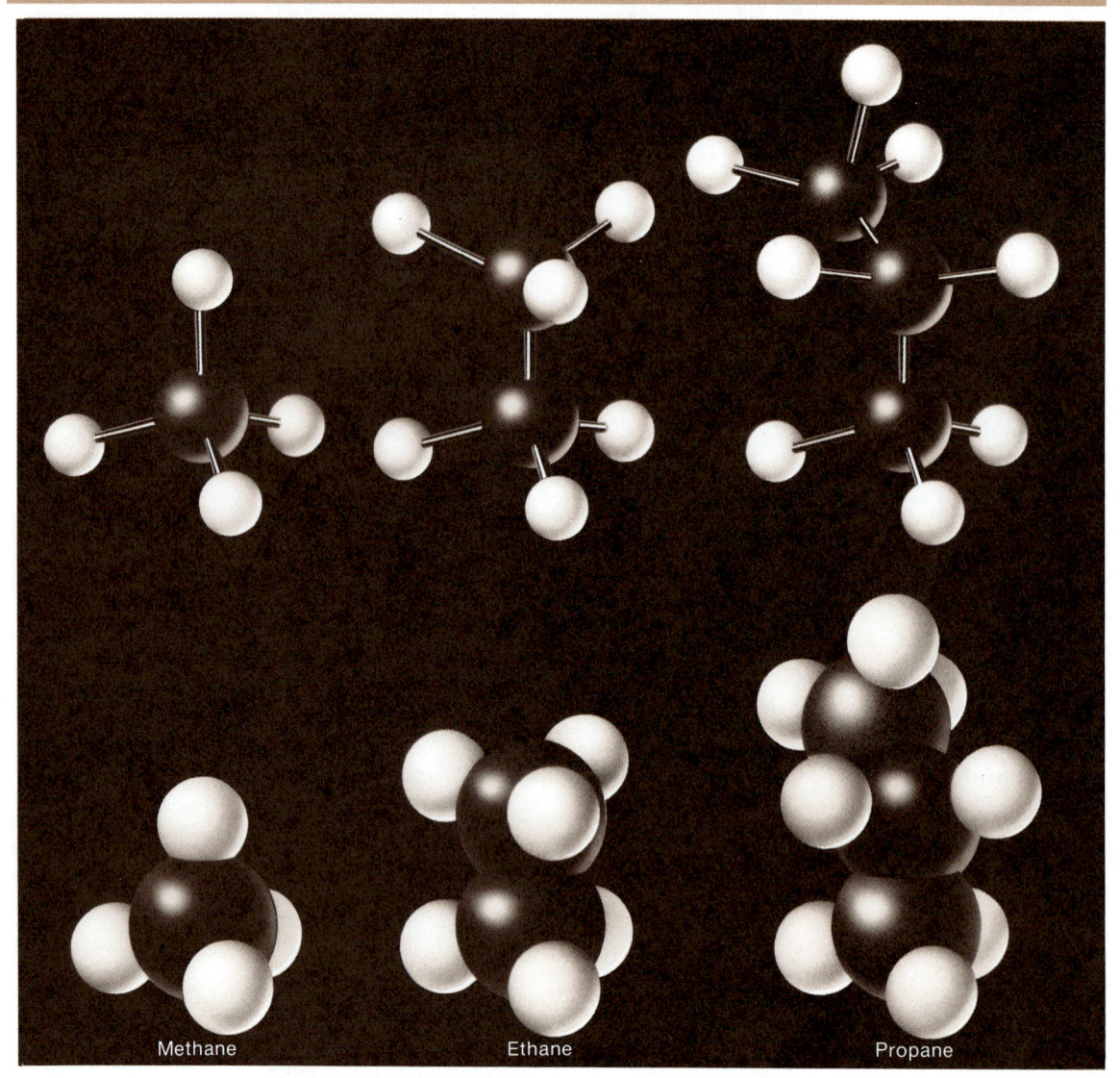

Figure 14.1 Ball-and-stick and space-filling models of CH_4, C_2H_6, and C_3H_8. Note the zig-zag chain in the propane molecule.

atoms really has a zig-zag shape. The carbon atoms are arranged in the pattern

$$C-C-C$$
$$109.5°C$$

2. *Condensed structural formulas.* To save space, we often abbreviate a structural formula, showing only the number of hydrogen atoms bonded to each carbon. The condensed structural formula of propane can be written as

$$CH_3-CH_2-CH_3 \quad \text{or} \quad CH_3CH_2CH_3$$

In these formulas, it is understood that the hydrogen atoms are bonded to the carbon atom that precedes them in the formula.

Condensed structural formulas of several "straight-chain" alkanes are given in Table 14.1 along with some of their physical properties. Notice that boiling point and melting point increase with molar mass. As the number of carbon atoms increases, we go from hydrocarbons that are gases at 25°C and 1 atm to liquids and finally to solids.

3. *Molecular formulas.* The molecular formulas of methane, ethane, and propane are

$$CH_4 \quad C_2H_6 \quad C_3H_8$$

methane   ethane   propane

It is possible to write a general molecular formula that applies to all alkanes. To see what that formula is, look at the entries in Table 14.1. Notice that in all these alkanes, each carbon atom is bonded to at least two hydrogens (CH_2 groups). In addition, there are two "extra" hydrogen atoms at the ends of the molecule (i.e., two CH_3 groups). Hence the general molecular formula is

$$C_nH_{2n+2}$$

where n is the number of carbon atoms in the molecule. For propane

Table 14.1 Properties of Straight-Chain Alkanes

	Condensed Structural Formula	MP(°C)	BP(°C)	State (25°C, 1 atm)	No. of Isomers
Methane	CH_4	−183	−162	Gas	—
Ethane	CH_3CH_3	−172	−89	Gas	—
Propane	$CH_3CH_2CH_3$	−187	−42	Gas	—
Butane	$CH_3(CH_2)_2CH_3$	−135	0	Gas	2
Pentane	$CH_3(CH_2)_3CH_3$	−130	36	Liquid	3
Hexane	$CH_3(CH_2)_4CH_3$	−94	69	Liquid	5
Heptane	$CH_3(CH_2)_5CH_3$	−90	98	Liquid	9
Octane	$CH_3(CH_2)_6CH_3$	−57	126	Liquid	18
Nonane	$CH_3(CH_2)_7CH_3$	−54	151	Liquid	35
Decane	$CH_3(CH_2)_8CH_3$	−30	174	Liquid	75
Octadecane	$CH_3(CH_2)_{16}CH_3$	28	308	Solid	60,523
Eicosane	$CH_3(CH_2)_{18}CH_3$	36		Solid	366,319

Chapter 14

n = 3; 2n + 2 = 6 + 2 = 8; molecular formula = C_3H_8

The next member of the series, butane, has the molecular formula C_4H_{10}.

Isomerism in Alkanes

There are two different alkanes with the molecular formula C_4H_{10}. One of these, called butane, is listed in Table 14.1. Its normal boiling point is 0°C. The other, called 2-methylpropane, boils at −10°C. In general, these two compounds have distinctly different physical and chemical properties. They also have different structural formulas. Writing these in condensed form

$$CH_3-CH_2-CH_2-CH_3 \qquad CH_3-\underset{\underset{CH_3}{|}}{\overset{\overset{H}{|}}{C}}-CH_3$$

butane 2-methylpropane

Notice that in butane there is a continuous chain of four carbon atoms. In 2-methylpropane, the longest chain has only three carbon atoms. The molecular shapes of the two isomers are quite different (Figure 14.2). Relatively, 2-methylpropane is a short, compact molecule; butane is a long, skinny molecule.

Compounds with the same molecular formula but different physical and chemical properties are called **isomers**. If they differ in structure, as is the

Figure 14.2 Models showing the structures of butane and 2-methylpropane. These are the two isomers of C_4H_{10}.

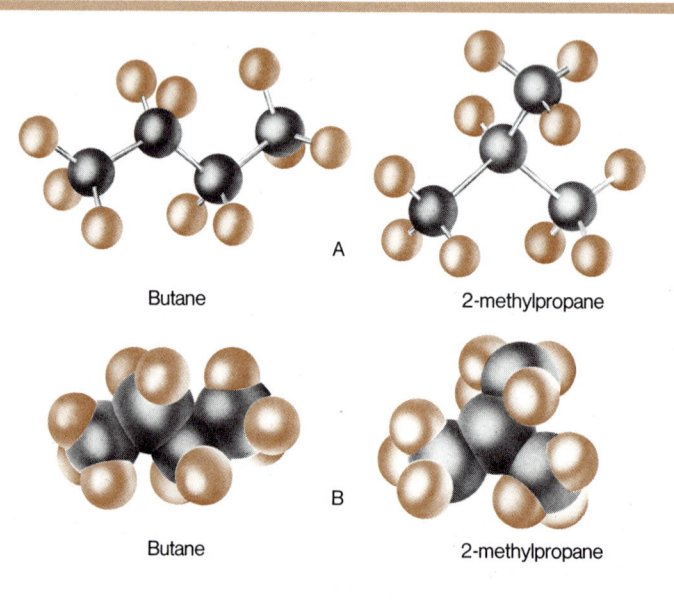

A
Butane   2-methylpropane

B
Butane   2-methylpropane

Introduction to Organic Chemistry

case here, they are called **structural isomers.** We would say that butane and 2-methylpropane are structural isomers. Isomerism of this type is very common in organic chemistry. It helps to explain why there are so many different organic compounds. As the number of carbon atoms in the molecule increases, the number of structural isomers goes up rapidly (Table 14.1).

EXAMPLE 14.1
Draw condensed structural formulas for the three isomers of C_5H_{12}.

Solution
It is best to proceed systematically. Start with the isomer in which there is a five-carbon chain

$$CH_3—CH_2—CH_2—CH_2—CH_3 \quad (1)$$

Next, consider a four-carbon chain. Here, there is one more isomer

$$\begin{array}{c} H \\ | \\ CH_3—C—CH_2—CH_3 \\ | \\ CH_3 \end{array} \quad (2)$$

In the third isomer, the longest chain contains only three carbon atoms

$$\begin{array}{c} CH_3 \\ | \\ CH_3—C—CH_3 \\ | \\ CH_3 \end{array} \quad (3)$$

The most common mistake, in problems of this type, is to draw too many isomers. You might be tempted to draw (A) and (B) as isomers of C_5H_{12}

$$\begin{array}{cc} CH_3—CH_2—CH_2—CH_2 & CH_3—CH_2—\overset{\overset{\displaystyle H}{|}}{C}—CH_3 \\ | & | \\ CH_3 & CH_3 \\ (A) & (B) \end{array}$$

However, if you think about it for a moment, you should realize that

— (A) is identical to (1). Both contain a five-carbon chain. It doesn't matter whether we show the terminal —CH$_3$ group at the "east" or "south" end of the chain. In three dimensions, the two positions are equivalent.
— (B) is identical to (2). Both contain a four-carbon chain. If you rotate (B) 180° through the plane of the paper, it becomes (2).

Exercise
How many isomers are there of C_3H_7Cl, a compound in which one of the hydrogen atoms of propane has been replaced by a chlorine atom? Answer: two.

Naming Alkanes

To distinguish one organic compound from another, we need a systematic way of naming them. The International Union of Pure and Applied Chemistry (IUPAC) has established a system of nomenclature that can be applied to all organic compounds. We will illustrate how this system works by applying it to alkanes.

For "straight-chain" alkanes such as

$$CH_3—CH_2—CH_2—CH_3 \qquad CH_3—CH_2—CH_2—CH_2—CH_3$$

butane   pentane

the IUPAC name consists of a single word. These names, up to ten carbon atoms, are listed in Table 14.1. Notice that beyond four carbons the name is formed by adding *-ane* to the Greek prefix that designates the number of carbon atoms. Thus we have *hex*ane (six carbon atoms), *hept*ane (seven carbon atoms) and so on.

With alkanes containing a branched chain, such as

$$\begin{array}{c} H \\ | \\ CH_3—C—CH_3 \\ | \\ CH_3 \end{array}$$

2-methylpropane

the name is a bit more complex. It consists of three parts. These are, in reverse order

1. A **"family name"** or *suffix*, which gives the number of carbon atoms in the longest chain. For a three-carbon chain, the family name is propane. This is the name of the straight-chain alkane with three carbon atoms. A four-carbon chain is a butane, a five-carbon chain a pentane, and so on.
2. A **"surname"** or *prefix*, which denotes the alkyl group that forms the branch on the chain. An alkyl group is the residue left when a hydrogen atom is removed from a hydrocarbon molecule. The names of a few alkyl groups are given in Table 14.2.
3. A **number**, which shows the carbon atom in the chain at which branching occurs. In 2-methylpropane, the methyl branch is located at the second carbon atom from the end of the chain.

$$\begin{array}{c} C_1—C_2—C_3 \\ | \end{array}$$

Following this system, we can obtain the names of the three isomers of C_5H_{12}.

Table 14.2 Names of Some Alkyl Groups

CH₃—	methyl
CH₃—CH₂— or C₂H₅—	ethyl
CH₃—CH₂—CH₂— or C₃H₇—	propyl

CH₃—CH₂—CH₂—CH₂—CH₃   CH₃—CH(CH₃)—CH₂—CH₃

pentane   2-methylbutane

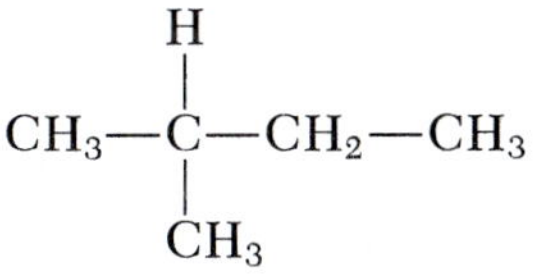

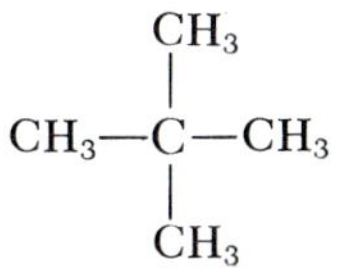

CH₃—C(CH₃)(CH₃)—CH₃

2,2-dimethylpropane

Notice that

— if there is the same alkyl group at two branches, the prefix "di" is used (2,2-*di*methylpropane). If there were three methyl branches, we would write *tri*methyl, and so on.
— the number in the name is made as small as possible. Thus we write 2-methylbutane, numbering the chain from the left rather than the right

C₁—C₂—C₃—C₄

EXAMPLE 14.2
Assign IUPAC names to the following

a. 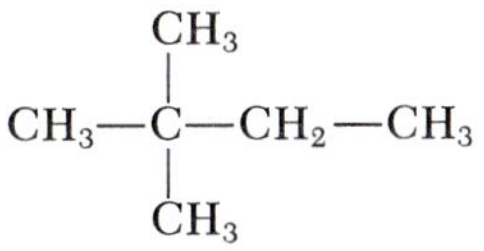 CH₃—C(CH₃)(CH₃)—CH₂—CH₃   b. 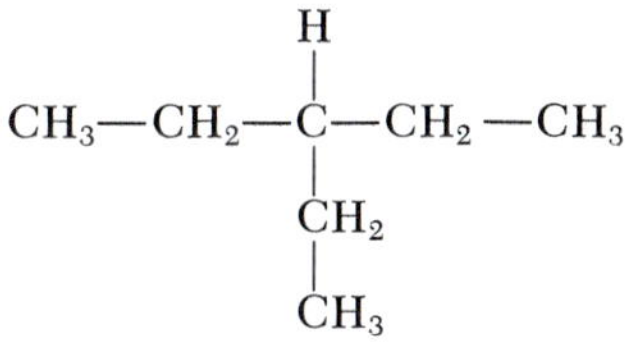 CH₃—CH₂—CH(CH₂CH₃)—CH₂—CH₃

Solution
a. The longest chain contains four carbon atoms (butane). There are two CH₃ (methyl) groups branching at the second carbon from the end of the chain (2). The correct name is

2,2-dimethylbutane

Continued

b. The longest chain, however you count it, contains five carbon atoms. There is a CH₃—CH₂ branch at the number three carbon, whichever end of the chain you start from. The IUPAC name is

3-ethylpentane

Exercise

How would you name the compound:

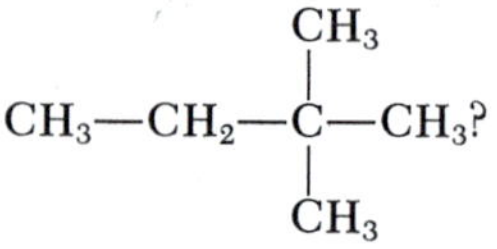

$$CH_3-CH_2-\underset{\underset{CH_3}{|}}{\overset{\overset{CH_3}{|}}{C}}-CH_3?$$

Answer: 2,2-dimethylbutane.

There is one bonus in using IUPAC names. Naming compounds properly can help you identify isomers and avoid writing extra ones. The principle here is a simple one. *Each isomer must have a different name.* If you obtain the same name for two structures, they are identical. Consider, for example, two of the structures referred to in Example 14.2

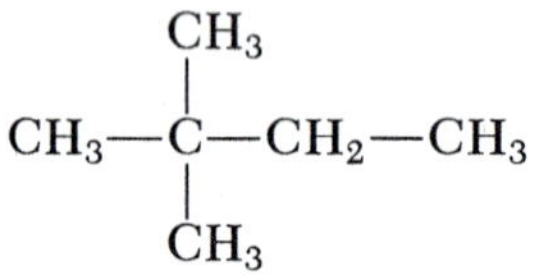

$$CH_3-\underset{\underset{CH_3}{|}}{\overset{\overset{CH_3}{|}}{C}}-CH_2-CH_3 \quad \text{and} \quad CH_3-CH_3-\underset{\underset{CH_3}{|}}{\overset{\overset{CH_3}{|}}{C}}-CH_3$$

2,2-dimethylbutane   2,2-dimethylbutane

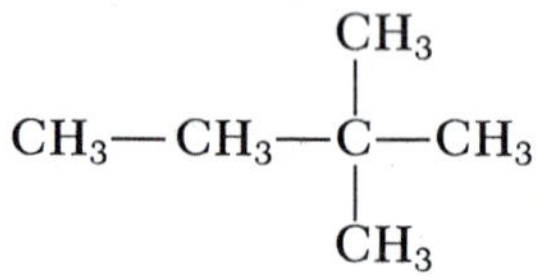

Because the names are the same, the structures must be identical. They represent the same compound. On the other hand

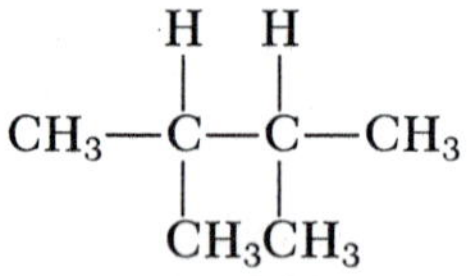

$$CH_3-\underset{\underset{CH_3}{|}}{\overset{\overset{H}{|}}{C}}-\underset{\underset{CH_3}{|}}{\overset{\overset{H}{|}}{C}}-CH_3$$

2,3-dimethylbutane

is a quite different compound, as indicated by the fact that it has a different name. It, along with one of the two structures drawn above, is an isomer of C₆H₁₄.

EXAMPLE 14.3

Write the condensed structural formula of the alkane whose IUPAC name is 3-ethylpentane.

Solution

We start by drawing a five-carbon chain

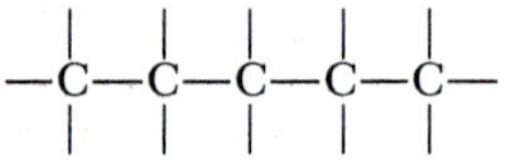

$$-\overset{|}{\underset{|}{C}}-\overset{|}{\underset{|}{C}}-\overset{|}{\underset{|}{C}}-\overset{|}{\underset{|}{C}}-\overset{|}{\underset{|}{C}}-$$

Continued

We then attach an ethyl group at the number three carbon, the one in the center of the chain.

```
   |   |   |   |   |
 —C—C—C—C—C—
   |   |   |   |   |
          —C—
             |
          —C—
             |
```

The condensed structural formula is

$$CH_3—CH_2—\underset{\underset{CH_3}{\overset{|}{CH_2}}}{\overset{\overset{H}{|}}{C}}—CH_2—CH_3$$

Exercise
A student incorrectly names an isomer of the above molecule as 2-ethylpentane. What is the correct name of the isomer? Answer: 3-methylhexane.

Sources of Alkanes

There are two major natural sources of alkanes:

1. *Natural gas.* This gaseous fuel, found above petroleum deposits, consists largely of methane (82 mol %) and ethane (10 mol %). It contains smaller amounts of propane (4 mol %) and butane (2 mol %). If natural gas is compressed to about 10 atm at 25°C, the propane and butane separate as a liquid mixture. This is sold as "bottled gas."
2. *Petroleum.* Crude oil is a complex mixture consisting mostly of alkanes with five or more carbon atoms per molecule. Other hydrocarbons are present in smaller amounts as are organic compounds containing sulfur, oxygen, or nitrogen. It is possible, but very expensive, to separate pure alkanes from petroleum by fractional distillation. More commonly, petroleum is separated into fractions of the type indicated in Table 14.3.

Table 14.3 Distillation Fractions from a Typical Petroleum

Fraction	Percent of Total	Composition (C Atoms Per Molecule)	Boiling Range (°C)	Uses
Gas	2	C_1–C_5	<20	gaseous fuel
Petroleum ether	2	C_5–C_7	30–100	solvents
Gasoline	32	C_6–C_{12}	40–200	motor fuel
Kerosene	18	C_{12}–C_{15}	175–275	diesel, jet fuel
Fuel oil	20	C_{15}–C_{20}	250–400	heating
Lubricating oil, residue	26	>C_{20}	>300	lubricants, paraffin wax, asphalt

14.2 Unsaturated Hydrocarbons: Alkenes and Alkynes

In an unsaturated hydrocarbon, there is at least one multiple carbon—carbon bond in the molecule. Here, we will consider two different kinds of unsaturated hydrocarbons

1. Alkenes, where there is one double bond (C=C) in the molecule.
2. Alkynes, where there is one triple bond (C≡C) in the molecule.

Alkenes

When we introduce a double bond into a hydrocarbon molecule, two hydrogen atoms are eliminated

$$\begin{array}{c} H\ H \\ |\ \ | \\ -C-C- \\ |\ \ | \\ H\ H \end{array} \rightarrow \begin{array}{c} H\ H \\ |\ \ | \\ -C=C- \\ \end{array} + 2\ H$$

This means that a molecule of an alkene must contain two fewer hydrogen atoms than the corresponding alkane molecule. Recall that the general molecular formula of an alkane is C_nH_{2n+2}. Subtracting two hydrogen atoms, we arrive at the general formula of an alkene

C_nH_{2n}

Successive alkenes have the molecular formulas C_2H_4 (n = 2), C_3H_6 (n = 3), and so on.

The first member of the alkene series is ethylene, C_2H_4 (IUPAC name ethene). This compound is a gas at 25°C and 1 atm (bp = −104°C). Its structural formula is

$$\begin{array}{c} H \qquad\quad H \\ \diagdown \quad\ \diagup \\ C=C \\ \diagup \quad\ \diagdown \\ H \qquad\quad H \end{array}$$

The molecule is written this way to indicate its geometry. Each of the two carbon atoms is at the center of an equilateral triangle. All the angles between the bonds are 120°. The molecule is *planar*. That is, all six atoms are in a single plane. Following this pattern, we write the structural formula of propylene, C₃H₆ (IUPAC name propene) as

$$\begin{array}{c}\text{H}\\ \diagdown\\ \text{C}=\text{C}\\ \diagup\\ \text{H}\end{array}\begin{array}{c}\text{CH}_3\\ \diagup\\ \\ \diagdown\\ \text{H}\end{array}$$

Here, a CH₃ group has been substituted for one of the hydrogen atoms of ethylene. Propylene, like ethylene, is a gas at 25°C and 1 atm; its boiling point is −48°C. The three-dimensional structures of ethylene and propylene are shown in Figure 14.3.

None of the alkenes occurs naturally. They are made from the alkanes in petroleum by a process called *cracking*. When a saturated hydrocarbon is heated to 500–700°C in the presence of a solid catalyst, it decomposes. The products include alkanes and alkenes of lower molar mass. A typical reaction is

$$C_{10}H_{22}(l) \rightarrow C_5H_{12}(l) + C_2H_4(g) + C_3H_6(g) \tag{14.1}$$

decane   pentane   ethylene   propylene

Reactions such as these serve as the industrial source of ethylene and propylene along with other alkenes.

Alkenes are much more reactive than alkanes. From a commercial standpoint, the most important reaction of alkenes is *polymerization*. In this reaction, small molecules such as C₂H₄ add to one another to form much larger molecules. As an example, consider the polymerization of ethylene.

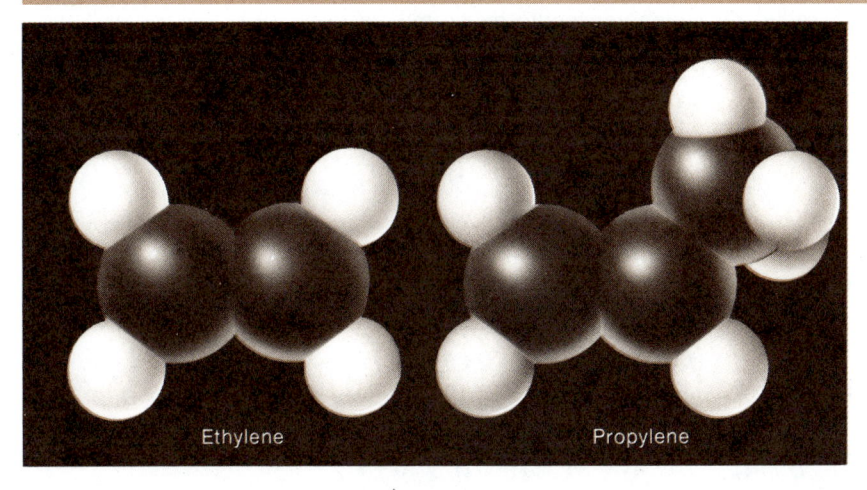

Figure 14.3 Space-filling models for ethylene and propylene. The ethylene molecule is planar, because the six nuclei lie in a plane. The carbon–carbon double bond contains four electrons. Propylene contains three carbon atoms, two of which are attached by a double bond.

The first step is

$$\begin{array}{c}H\\ \\ H\end{array}\!\!\!\!\diagdown\!\!\!C\!\!=\!\!C\!\!\diagup\!\!\!\!\begin{array}{c}H\\ \\ H\end{array} + \begin{array}{c}H\\ \\ H\end{array}\!\!\!\!\diagdown\!\!\!C\!\!=\!\!C\!\!\diagup\!\!\!\!\begin{array}{c}H\\ \\ H\end{array} \longrightarrow \begin{array}{cccc}H&H&H&H\\ |&|&|&|\\ -C-C-C-C-\\ |&|&|&|\\ H&H&H&H\end{array}$$

This process can continue, with ethylene molecules adding to both ends of the carbon chain

$$\begin{array}{c}H\\ \\ H\end{array}\!\!\!\!\diagdown\!\!\!C\!\!=\!\!C\!\!\diagup\!\!\!\!\begin{array}{c}H\\ \\ H\end{array} + \begin{array}{cccc}H&H&H&H\\ |&|&|&|\\ -C-C-C-C-\\ |&|&|&|\\ H&H&H&H\end{array} + \begin{array}{c}H\\ \\ H\end{array}\!\!\!\!\diagdown\!\!\!C\!\!=\!\!C\!\!\diagup\!\!\!\!\begin{array}{c}H\\ \\ H\end{array} \longrightarrow \begin{array}{cccccccc}H&H&H&H&H&H&H&H\\ |&|&|&|&|&|&|&|\\ -C-C-C-C-C-C-C-C-\\ |&|&|&|&|&|&|&|\\ H&H&H&H&H&H&H&H\end{array}$$

Eventually we arrive at a huge molecule containing a thousand or more carbon atoms. Such molecules are found in polyethylene, which is a familiar polymer used in squeeze bottles, film, plastic toys, and many other products. You should keep in mind that

1. Polyethylene has the same chemical composition as ethylene. In both species, there are two hydrogen atoms per carbon atom. The simplest formula in both cases is CH_2.
2. Polyethylene has a much higher molar mass than ethylene. This explains why polyethylene is a solid even though ethylene is a gas.
3. Polyethylene molecules are coiled (Figure 14.4), bending and turning in many directions. This explains why polyethylene film is so flexible and elastic.
4. There are no double bonds in polyethylene. As a result, this polymer is much less reactive than ethylene.

    In principle at least, any alkene can be polymerized. The starting material, called the monomer, could be propylene

$$\begin{array}{c}H\\ \\ H\end{array}\!\!\!\!\diagdown\!\!\!C\!\!=\!\!C\!\!\diagup\!\!\!\!\begin{array}{c}CH_3\\ \\ H\end{array}$$

rather than ethylene. Propylene molecules add to one another to give an *addition polymer* called polypropylene. The structure of this polymer is very similar to that of polyethylene, except that one fourth of the hydrogen atoms are replaced by CH_3 groups

$$\begin{array}{cccccccc}H&CH_3&H&CH_3&H&CH_3&H&CH_3\\ |&|&|&|&|&|&|&|\\ -C-C-C-C-C-C-C-C-\\ |&|&|&|&|&|&|&|\\ H&H&H&H&H&H&H&H\end{array}$$

Introduction to Organic Chemistry

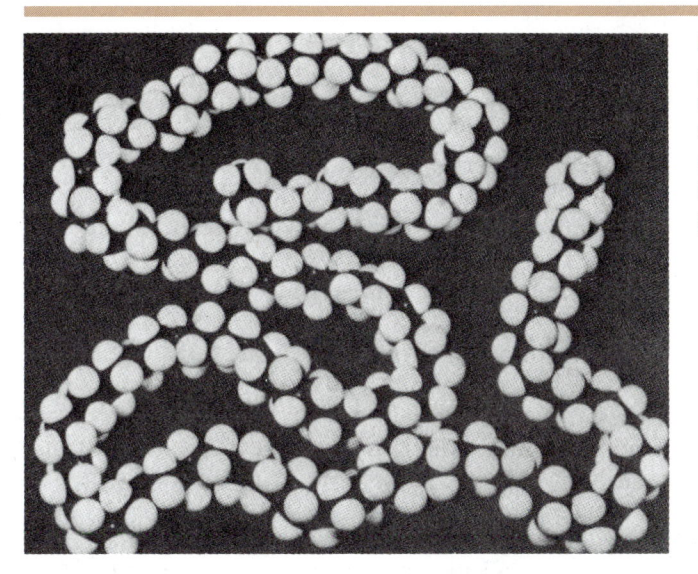

Figure 14.4 In polyethylene, the carbon chain is coiled as shown here. This gives a flexible, somewhat elastic polymer.

Many of the most important addition polymers are made from derivatives of alkenes in which one or more hydrogen atoms are replaced by halogen atoms. Two such polymers, polyvinylchloride (PVC) and Teflon, are considered in Example 14.4.

EXAMPLE 14.4

Show the structure of the polymer formed from vinyl chloride

$$\begin{array}{c} H \\ \diagdown \\ \\ \diagup \\ H \end{array} C = C \begin{array}{c} Cl \\ \diagup \\ \\ \diagdown \\ H \end{array}$$

Solution

The structure is very similar to that of polypropylene, with a Cl atom substituted for a CH$_3$ group.

$$\begin{array}{c} HClHClHClHCl \\ |||||||| \\ -C-C-C-C-C-C-C-C- \\ |||||||| \\ HHHHHHHH \end{array}$$

Exercise

What is the monomer from which Teflon

Continued

$$\begin{array}{c} \text{F} \;\; \text{F} \;\; \text{F} \;\; \text{F} \;\; \text{F} \;\; \text{F} \;\; \text{F} \;\; \text{F} \\ | \;\; | \;\; | \;\; | \;\; | \;\; | \;\; | \;\; | \\ -\text{C}-\text{C}-\text{C}-\text{C}-\text{C}-\text{C}-\text{C}-\text{C}- \\ | \;\; | \;\; | \;\; | \;\; | \;\; | \;\; | \;\; | \\ \text{F} \;\; \text{F} \;\; \text{F} \;\; \text{F} \;\; \text{F} \;\; \text{F} \;\; \text{F} \;\; \text{F} \end{array}$$

is made? Answer: $F_2C=CF_2$.

Alkynes

When an alkyne is formed from an alkene, two hydrogen atoms are lost

$$\begin{array}{c} \text{H} \;\; \text{H} \\ | \;\; | \\ -\text{C}=\text{C}- \end{array} \rightarrow -\text{C}\equiv\text{C}- \; + \; 2\,\text{H}$$

Since the general molecular formula of an alkene is C_nH_{2n}, it follows that the formula of an alkyne must be

C_nH_{2n-2}

EXAMPLE 14.5
Classify each of the following as an alkane, alkene, or alkyne
a. $C_{12}H_{24}$ b. C_6H_{14} c. C_9H_{16}

Solution
a. alkene (n = 12; 24 = 2n)
b. alkane (n = 6; 14 = 2n + 2)
c. alkyne (n = 9; 16 = 2n − 2)

Exercise
What are the formulas of the alkane, alkene, and alkyne with ten carbon atoms per molecule? Answer: $C_{10}H_{22}$, $C_{10}H_{20}$, $C_{10}H_{18}$.

Of the alkynes, by far the most important is acetylene, C_2H_2 (IUPAC name ethyne). The acetylene molecule is linear; all four atoms are in a straight line

$H-C\equiv C-H$

You may be most familiar with acetylene as a gaseous fuel used in welding. When mixed with pure oxygen in a torch, acetylene burns at temperatures above 2000°C. The heat comes from the reaction

$$C_2H_2(g) + \frac{5}{2}O_2(g) \rightarrow 2\,CO_2(g) + H_2O(l); \; \Delta H = -1300 \text{ kJ} \qquad (14.2)$$

Acetylene is made commercially from the controlled oxidation of methane at 1500°C

Introduction to Organic Chemistry

$$6\ CH_4(g) + O_2(g) \rightarrow 2\ C_2H_2(g) + 2\ CO(g) + 10\ H_2(g) \tag{14.3}$$

An older synthesis is still used on a small scale to make acetylene in the laboratory. This uses the reaction between water and calcium carbide, CaC_2

$$CaC_2(s) + H_2O(l) \rightarrow C_2H_2(g) + CaO(s) \tag{14.4}$$

The water is allowed to drip slowly on solid lumps of calcium carbide. The gas given off is nearly pure acetylene, but has a garlic-like odor due to traces of phosphine, PH_3.

14.3 Aromatic Hydrocarbons: Benzene

The simplest aromatic hydrocarbon is benzene, molecular formula C_6H_6. Benzene is a pleasant smelling ("aromatic") liquid that boils at 80°C and freezes at 5°C. The fact that there is only one hydrogen atom per carbon atom in benzene would suggest a considerable amount of unsaturation. Yet, benzene does not behave chemically like an alkene or alkyne. Moreover, all six carbon atoms in the benzene molecule appear to be equivalent. The same is true of the six hydrogen atoms. This behavior is difficult to explain in terms of a chain-type structure.

Reflecting on these facts, the German chemist Friedrich Kekulé in 1865 proposed a ring structure for benzene

In this structure, all the carbon atoms are equivalent; the same is true of the hydrogens. Moreover, as this structure predicts, benzene is known to be planar and the carbon atoms are located at the corners of a regular hexagon. This feature is emphasized in the abbreviated Kekulé structure

In this structure, it is understood that there is a carbon atom at each corner of the hexagon. The hydrogen atoms are not shown. There is a hydrogen atom bonded to each ring carbon.

There is one difficulty with the Kekulé structure of benzene. It shows two different kinds of carbon–carbon bonds, single and double. From experiment, it appears that all the bonds are of the same type. In particular, all the carbon–carbon distances are the same, 0.139 nm. One way to explain this situation is to assume that

— each carbon atom forms three single bonds, one to a hydrogen atom and two to adjacent carbon atoms.
— the remaining six electrons (the ones used to form double bonds in the Kekulé structure) are spread over the entire molecule instead of being tied down to particular carbon atoms.

The structure written to correspond to this model of benzene is

The circle represents the six electrons shared by the molecule as a whole. Here, as in the Kekulé structure, it is understood that there is a carbon atom at each corner of the hexagon. Again, the six hydrogen atoms, each bonded to a carbon, are not shown.

All aromatic hydrocarbons are derivatives of benzene. They may contain a single benzene ring with one or more alkyl groups substituted for hydrogen atoms. Examples of aromatic hydrocarbons of this type include

common name	toluene	ortho xylene	meta xylene	para xylene
systematic name	methyl benzene	1,2-dimethyl benzene	1,3-dimethyl benzene	1,4-dimethyl benzene
boiling point	111°C	144°C	139°C	139°C

In another type of aromatic hydrocarbon, two or more benzene rings are fused together. Naphthalene, a solid melting at 80°C, contains two such rings. Anthracene (mp = 218°C) contains three benzene rings.

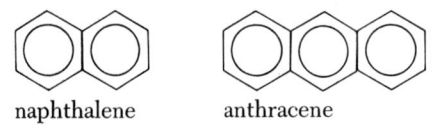

naphthalene   anthracene

Aromatic hydrocarbons are commonly obtained from either of two sources.

1. Coal tar, a product obtained by heating coal to 1000°C in the absence of air. This is the major source of naphthalene, anthracene, and other more complex aromatic hydrocarbons.
2. Gasoline. The amount of aromatic hydrocarbons in straight-run gasoline is quite small. However, benzene, toluene, and the xylenes, can be made from the alkanes in gasoline by a catalytic process.

14.4 Oxygen Compounds

Many of the most familiar organic compounds contain the three elements carbon, hydrogen, and oxygen. Some of these, like ethyl alcohol and acetic acid (the active component of vinegar), have rather simple structures

$$CH_3CH_2OH \qquad CH_3-\underset{\underset{O}{\|}}{C}-OH$$

ethyl alcohol   acetic acid

Others, like aspirin and Vitamin C, are more complex

acetylsalicylic acid (aspirin)   ascorbic acid (Vitamin C)

In this section, we will look at the structures and properties of some of the simpler organic oxygen compounds. It will be convenient to organize our discussion around the type of functional group found in the molecule.

Functional Groups

A **functional group** is an atom or group of atoms that confers certain properties on an organic molecule. Compounds containing a given functional group comprise a particular class of organic compounds. For example, all the compounds referred to as alcohols contain the —OH group. The first alcohol is methyl alcohol (IUPAC name methanol), which has the structural formula

$$\begin{array}{c} \text{H} \\ | \\ \text{H—C—OH} \\ | \\ \text{H} \end{array} \quad \text{or} \quad \text{CH}_3\text{OH}$$

The next member is ethyl alcohol (IUPAC name ethanol)

$$\begin{array}{cc} \text{H} & \text{H} \\ | & | \\ \text{H—C—C—OH} \\ | & | \\ \text{H} & \text{H} \end{array} \quad \text{or} \quad \text{CH}_3\text{CH}_2\text{OH} \quad \text{or} \quad \text{C}_2\text{H}_5\text{OH}$$

The geometries of these molecules are shown in Figure 14.5.

Table 14.4 lists the functional groups present in several different classes of organic oxygen compounds. Note that

— each class contains a large number of compounds, only two of which are listed in Table 14.2. For example, additional aldehydes would include

$$\text{CH}_3\text{CH}_2\text{—C—H} \quad \text{CH}_3\text{CH}_2\text{CH}_2\text{—C—H} \quad \text{CH}_3\text{CH}_2\text{CH}_2\text{CH}_2\text{—C—H}$$
$$\quad\quad\quad \| \quad\quad\quad\quad\quad\quad\quad \| \quad\quad\quad\quad\quad\quad\quad\quad\quad \|$$
$$\quad\quad\quad \text{O} \quad\quad\quad\quad\quad\quad\quad \text{O} \quad\quad\quad\quad\quad\quad\quad\quad\quad \text{O}$$

—the functional group is ordinarily bonded to an alkyl group. For example, in methyl alcohol, the —OH group is bonded to a methyl group, CH$_3$.

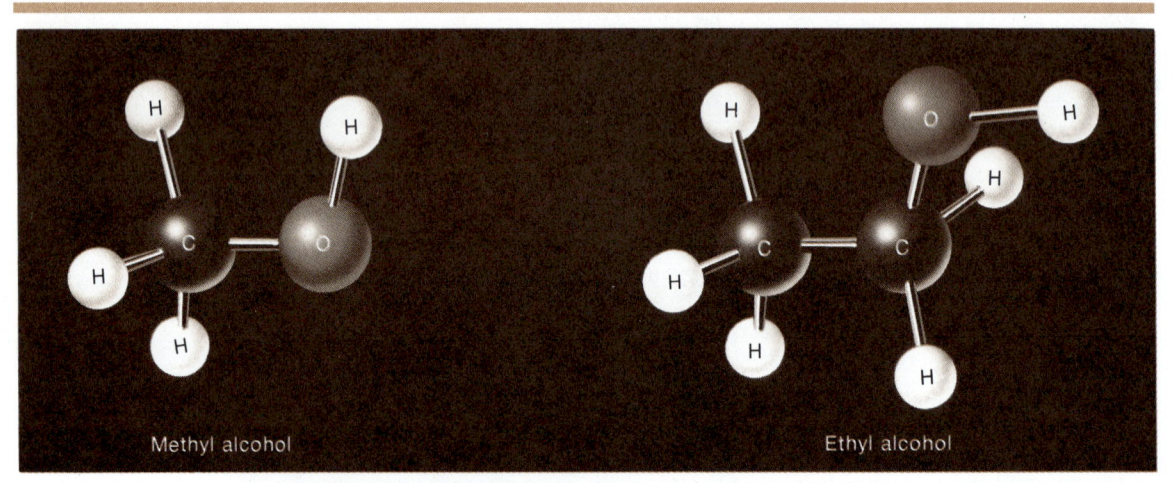

Figure 14.5 Ball-and-stick models of CH$_3$OH and C$_2$H$_5$OH. In both of these molecules, an —OH group is bonded to an alkyl group. The two compounds have very similar chemical properties.

In acetone, the $\diagdown$C=O$\diagup$ group is bonded to two methyl groups.

—in a few cases, the functional group may be bonded to a hydrogen atom. This is the case with formaldehyde and formic acid, the first aldehyde and carboxylic acid, respectively.

TABLE 14.4 Classes of Organic Oxygen Compounds

Class	Functional Group	First Member	Other Example
Alcohol	—OH	CH$_3$OH methyl alcohol	CH$_3$CH$_2$OH ethyl alcohol
Aldehyde	—C(=O)—H	H—C(=O)—H formaldehyde	CH$_3$—C(=O)—H acetaldehyde
Carboxylic acid	—C(=O)—OH	H—C(=O)—OH formic acid	CH$_3$—C(=O)—OH acetic acid
Ester	—C(=O)—O—	H—C(=O)—OCH$_3$ methyl formate	CH$_3$—C(=O)—OCH$_2$CH$_3$ ethyl acetate
Ether	—O—	CH$_3$—O—CH$_3$ dimethyl ether	CH$_3$CH$_2$—O—CH$_2$CH$_3$ diethyl ether
Ketone	—C(=O)—	CH$_3$—C(=O)—CH$_3$ acetone	CH$_3$—C(=O)—CH$_2$CH$_3$ methyl ethyl ketone

---

EXAMPLE 14.6

Identify the class of compound represented by each of the following molecules

a. CH$_3$CH$_2$—C(=O)—CH$_3$ b. CH$_3$CH$_2$—O—CH$_3$

Solution

a. *Ketone*, since it contains the functional group $\diagdown$C=O$\diagup$. Note that in a ketone two alkyl groups are bonded to the $\diagdown$C=O$\diagup$ group. The general structure of a ketone is

Continued

$$R-\underset{\underset{O}{\|}}{C}-R'$$

where R and R' are alkyl groups that may be the same or different.

b. *Ether*, since it contains the functional group —O—. In an ether, as in a ketone, two alkyl groups are bonded to the functional group. The general formula of an ether is

$$R-O-R'$$

where R and R' are alkyl groups that may be the same or different.

Exercise
Give the structural formula of a carboxylic acid containing three carbon atoms per molecule. Answer: $CH_3CH_2-\underset{\underset{O}{\|}}{C}-OH$.

In the remainder of this section, we will concentrate upon three classes of organic oxygen compounds. They are

—alcohols
—carboxylic acids
—esters

Alcohols

The first member of this class of compounds, methyl alcohol, CH_3OH, is often referred to as "wood alcohol." At one time it was made by heating hardwoods such as birch or maple to about 300°C in the absence of air. Today, methyl alcohol is made from carbon monoxide and hydrogen

$$CO(g) + 2\ H_2(g) \rightarrow CH_3OH(l) \quad (14.5)$$

This reaction is carried out at high temperatures and pressures (300–400°C, 200–300 atm) with a solid catalyst. Methyl alcohol is a colorless, volatile liquid (bp = 65°C). It is extremely toxic; small amounts taken internally can cause blindness or death.

Ethyl alcohol, C_2H_5OH (bp = 78°C), can be made from ethylene

$$C_2H_4(g) + H_2O \rightarrow C_2H_5OH(aq) \quad (14.6)$$

The reaction is carried out in dilute sulfuric acid; H^+ ions act as a catalyst. The ethyl alcohol used in beverages is made by fermentation of grains or other vegetable matter. The early stages of fermentation produce a simple sugar called glucose, molecular formula $C_6H_{12}O_6$. This in turn breaks down to ethyl alcohol and carbon dioxide.

$$C_6H_{12}O_6(aq) \rightarrow 2\ C_2H_5OH(aq) + 2\ CO_2(g) \quad (14.7)$$

Introduction to Organic Chemistry

Each step in the fermentation process requires a specific organic catalyst known as an enzyme. The process stops when the concentration of alcohol in the water solution reaches 12%. At higher alcohol contents, the enzyme molecules are destroyed. Beverages containing more than 12% alcohol are made by distillation (Table 14.5). The proof of an alcoholic beverage is twice the volume percent of ethyl alcohol.

There are two isomeric alcohols containing three carbon atoms. In one of these, the —OH group is bonded to the carbon atom at the end of the chain. In the other, the —OH group is joined to the carbon atom in the center of the chain.

$$CH_3-CH_2-CH_2-OH \qquad CH_3-\underset{OH}{\overset{H}{C}}-CH_3$$

common name	propyl alcohol	isopropyl alcohol
IUPAC name	1-propanol	2-propanol
boiling point	97°C	82°C

EXAMPLE 14.7

Draw structural formulas for all the isomeric alcohols containing four carbon atoms, C_4H_9OH.

Solution

You will recall that there are two alkanes with four carbon atoms

$$CH_3-CH_2-CH_2-CH_3 \quad \text{and} \quad CH_3-\underset{CH_3}{\overset{H}{C}}-CH_3$$

The alcohols are formed by replacing an H atom of one of the hydrocarbons by an

Continued

Table 14.5 Common Alcoholic Beverages

Beverage	Major Source of Carbohydrate	Volume % Alcohol	Proof
Beer	Barley, rice, wheat	3–9	6–18
Wine	Grapes, other fruit	10–12	20–24
Brandy	Wine	40–45	80–90
Whiskey	Barley, rye, corn	40–55	80–110
Rum	Molasses	45	90
Gin	Barley, other grains	35–45	70–90
Vodka	Grains, potatoes, sugar beets	35–45	70–90

—OH group. In the straight-chain hydrocarbon, there are two different kinds of hydrogen atoms

—those on carbon atoms at the end of the chain (CH_3 groups)
—those on carbon atoms in the middle of the chain (CH_2 groups)

So, there are two isomeric alcohols derived from this hydrocarbon

$$CH_3-CH_2-CH_2-CH_2-OH \qquad CH_3-\underset{\underset{OH}{|}}{\overset{\overset{H}{|}}{C}}-CH_2-CH_3$$

isomer 1   isomer 2

A similar situation applies with the branched-chain hydrocarbon. The hydrogen atom bonded to the central carbon differs from those in the CH_3 groups. This leads to two more isomeric alcohols

$$CH_3-\underset{\underset{CH_3}{|}}{\overset{\overset{OH}{|}}{C}}-CH_3 \qquad CH_3-\underset{\underset{CH_3}{|}}{\overset{\overset{H}{|}}{C}}-CH_2-OH$$

isomer 3   isomer 4

There are four and only four alcohols with the formula C_4H_9OH.

Exercise
How many isomers are there of C_4H_9Cl? Answer: four (as with C_4H_9OH).

Certain alcohols contain more than one —OH group per molecule. Two of the best known compounds of this type are ethylene glycol, used in antifreeze, and glycerol, a starting material for making drugs, plastics, and explosives

$$H-\underset{\underset{OH}{|}}{\overset{\overset{H}{|}}{C}}-\underset{\underset{OH}{|}}{\overset{\overset{H}{|}}{C}}-H \qquad H-\underset{\underset{OH}{|}}{\overset{\overset{H}{|}}{C}}-\underset{\underset{OH}{|}}{\overset{\overset{H}{|}}{C}}-\underset{\underset{OH}{|}}{\overset{\overset{H}{|}}{C}}-H$$

ethylene glycol   glycerol

Carboxylic Acids

The functional group common to all carboxylic acids is called the carboxyl group. It combines the carbonyl group, $\ce{C=O}$, found in aldehydes and ketones, with the —OH group of alcohols. The general formula of a carboxylic acid is

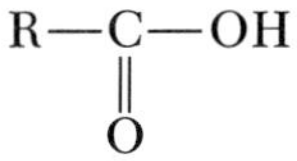

$$R-\underset{\underset{O}{\|}}{C}-OH$$

In the first member of this class, R is a hydrogen atom

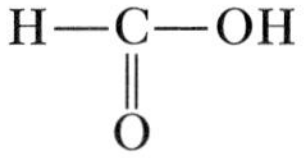

$$H-\underset{\underset{O}{\|}}{C}-OH$$

formic acid (bp = 101°C)

In all the other carboxylic acids, R is an alkyl group

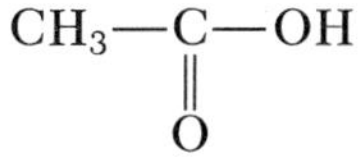

$$CH_3-\underset{\underset{O}{\|}}{C}-OH$$

acetic acid (bp = 118°C)

The carboxylic acids act as weak acids in water. The general equation for their dissociation can be written

$$RCOOH(aq) \rightleftharpoons H^+(aq) + RCOO^-(aq) \qquad (14.8)$$

Formic acid ($K_a = 1.8 \times 10^{-4}$) is about 4% ionized in 0.1 molar solution. Acetic acid is somewhat weaker ($K_a = 1.8 \times 10^{-5}$). In 0.1 molar solution, it is only a little more than 1% dissociated to H^+ ions and acetate ions.

When wine or cider is allowed to stand in contact with air, the ethyl alcohol present is slowly oxidized to acetic acid

$$CH_3CH_2OH(aq) + O_2(g) \rightarrow CH_3COOH(aq) + H_2O \qquad (14.9)$$

In this way, vinegar containing about 4% acetic acid is produced. Pure acetic acid is ordinarily made by the oxidation of acetaldehyde

$$CH_3CHO(l) + \frac{1}{2} O_2(g) \rightarrow CH_3COOH(l) \qquad (14.10)$$

Several carboxylic acids occur naturally in fruits and vegetables. Typically, they contain more than one —COOH group and/or an —OH group in the molecule. Examples include

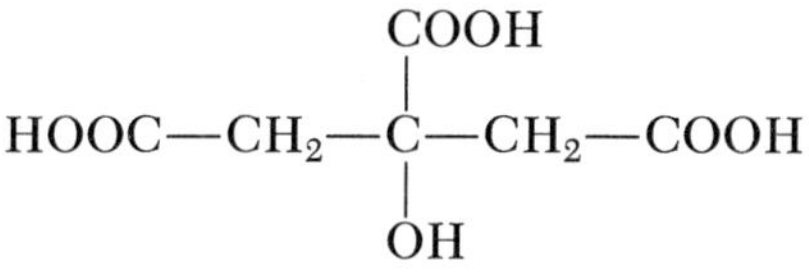

$$HOOC-CH_2-\underset{\underset{OH}{|}}{\overset{\overset{COOH}{|}}{C}}-CH_2-COOH \qquad HOOC-CH_2-\underset{\underset{OH}{|}}{\overset{\overset{H}{|}}{C}}-COOH$$

citric acid
(lemons, oranges)

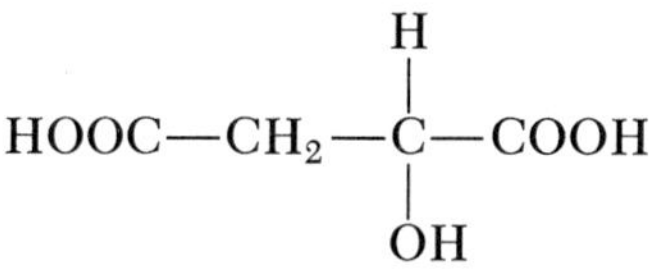

malic acid
(apples)

$$\text{HOOC}-\underset{\underset{\text{OH}}{|}}{\overset{\overset{\text{H}}{|}}{\text{C}}}-\underset{\underset{\text{OH}}{|}}{\overset{\overset{\text{H}}{|}}{\text{C}}}-\text{COOH} \qquad \text{HOOC}-\text{COOH}$$

tartaric acid   oxalic acid
(grapes)   (rhubarb, spinach)

Esters

An ester is produced when a carboxylic acid reacts with an alcohol. The reaction between acetic acid and methyl alcohol is typical. It forms the ester called methyl acetate; water is a byproduct.

$$\text{CH}_3-\underset{\underset{\text{O}}{\|}}{\text{C}}-\boxed{\text{OH} + \text{H}}\text{O}-\text{CH}_3 \rightarrow \text{CH}_3-\underset{\underset{\text{O}}{\|}}{\text{C}}-\text{O}-\text{CH}_3 + \text{H}_2\text{O}$$

acetic acid   methyl alcohol   methyl acetate

$$\text{CH}_3\text{COOH}(aq) + \text{CH}_3\text{OH}(aq) \rightarrow \text{CH}_3\text{COOCH}_3(aq) + \text{H}_2\text{O} \qquad (14.11)$$

The preparation is carried out in water solution. Hydrochloric acid or sulfuric acid is added to furnish H^+ ions, which act as a catalyst. The product, methyl acetate, is a volatile liquid (bp = 57°C) with a pleasant, fruity odor. Esters similar to methyl acetate are commonly used in fruit flavorings and perfumes.

EXAMPLE 14.8

Write an equation, similar to 14.11, for the preparation of ethyl formate, which has the structure

$$\text{H}-\underset{\underset{\text{O}}{\|}}{\text{C}}-\text{O}-\text{CH}_2\text{CH}_3$$

Solution
The reactants are formic acid and ethyl alcohol

$$\text{H}-\underset{\underset{\text{O}}{\|}}{\text{C}}-\text{OH} + \text{HO}-\text{CH}_2\text{CH}_3 \rightarrow \text{H}-\underset{\underset{\text{O}}{\|}}{\text{C}}-\text{O}-\text{CH}_2\text{CH}_3 + \text{H}_2\text{O}$$

formic acid   ethyl alcohol   ethyl formate

$$\text{HCOOH}(aq) + \text{CH}_3\text{CH}_2\text{OH}(aq) \rightarrow \text{HCOOCH}_2\text{CH}_3(aq) + \text{H}_2\text{O}$$

Continued

Introduction to Organic Chemistry

Exercise
Draw the structural formula of the ester formed by the reaction of acetic acid with propyl alcohol. Answer: $CH_3-\underset{\underset{O}{\|}}{C}-O-CH_2CH_2CH_3$.

A *fat* is an important type of naturally occurring ester. In principle at least, fats can be made from the reaction of long-chain carboxylic acids with glycerol. The general equation for this process can be written

$$\begin{array}{l} R-CO\underline{|OH} \quad \overline{H|}O-CH_2 \\ R'-CO\underline{|OH} + \overline{H|}O-CH \\ R''-CO\underline{|OH} \quad \overline{H|}O-CH_2 \end{array} \rightarrow \begin{array}{l} R-COO-CH_2 \\ R'-COO-CH \\ R''-COO-CH_2 \end{array} + 3\,H_2O$$

$$\text{glycerol} \qquad \qquad \text{a fat}$$

One molecule of glycerol reacts with three carboxylic acid molecules. The products include a molecule of a fat and three water molecules.

The groups R, R', and R'' are long-chain alkyl groups. They may be the same or different. A typical fat molecule might have the structure

$$\begin{array}{l} CH_3(CH_2)_{14}-COO-CH_2 \\ CH_3(CH_2)_7CH=CH(CH_2)_7-COO-CH \\ CH_3(CH_2)_{16}-COO-CH_2 \end{array}$$

The three carboxylic acids from which this fat is derived are

—palmitic acid, $CH_3(CH_2)_{14}COOH$.

—oleic acid, $CH_3(CH_2)_7CH=CH(CH_2)_7COOH$.

—stearic acid, $CH_3(CH_2)_{16}COOH$.

These acids are typical of those found in fats. Some "fatty acids" are *saturated*, like palmitic and stearic acid; the hydrocarbon chain contains no multiple bonds. Others, like oleic acid are *unsaturated*; there is one or more carbon–carbon multiple bond in the molecule.

So-called "saturated fats" contain relatively few carbon–carbon multiple bonds. They are solids at room temperature and are commonly found in animal products such as lard and butter. "Unsaturated fats" contain a higher proportion of multiple bonds. They are liquids at room temperature and are found in such vegetable products as corn oil and cottonseed oil. Unsaturated fats can be converted to saturated fats by a process called hydrogenation. This involves adding H_2 molecules across a carbon–carbon multiple bond. The reaction is carried out under pressure in the presence of a nickel catalyst

$$\overset{\diagdown}{\underset{\diagup}{C}}=\overset{\diagup}{\underset{\diagdown}{C}} + H-H \rightarrow -\overset{\overset{H}{|}}{\underset{|}{C}}-\overset{\overset{H}{|}}{\underset{|}{C}}-$$

Recently, there has been a trend toward the use of unsaturated as opposed to saturated fats. It appears that saturated fats may raise the level of cholesterol in the blood, perhaps contributing to the risk of heart attacks and other circulatory problems.

Key Words

alcohol
aldehyde
alkane
alkene
alkyl group
alkyne
aromatic hydrocarbon
carboxylic acid
ester
ether
fat

functional group
hydrocarbon
isomer
ketone
monomer
polymer
saturated hydrocarbon
structural formula
structural isomer
unsaturated hydrocarbon

Questions

14.1 In organic compounds, how many bonds are formed by a carbon atom? hydrogen atom? chlorine atom? oxygen atom?

14.2 In alkanes, all the bond angles are _____°.

14.3 The melting points and boiling points of hydrocarbons and other organic compounds ordinarily _____ as molar mass increases.

14.4 The general molecular formula of an alkane is _____.

14.5 The first alkane to show structural isomerism is _____.

14.6 Name the straight chain alkane containing the following number of carbon atoms per molecule
a. 1   b. 3   c. 5   d. 7   e. 9

14.7 Name the straight chain alkane containing the following number of carbon atoms per molecule
a. 2   b. 4   c. 6   d. 8   e. 10

14.8 Draw the structure of the methyl group; the ethyl group; the propyl group

14.9 Describe the natural sources of alkanes.

14.10 The general molecular formula of an alkene is _____; that of an alkyne is _____.

14.11 Describe the geometry of the ethylene molecule; the acetylene molecule.

14.12 Describe the process by which alkanes are obtained industrially.

14.13 What is meant by a polymer? a monomer?

14.14 Write balanced equations for
  a. the commercial preparation of acetylene
  b. the preparation of acetylene from calcium carbide
14.15 Describe the geometry of the benzene molecule.
14.16 In the structural formula of benzene, what does the circle represent?
14.17 What are the common sources of aromatic hydrocarbons?
14.18 What is meant by a functional group?
14.19 Show the functional group present in a(an)
  a. alcohol   b. carboxylic acid
  c. ester
14.20 Show the functional group present in a(an)
  a. aldehyde   b. ether   c. ketone
14.21 Write balanced equations for the preparation of
  a. CH_3OH b. C_2H_5OH (two different methods)
14.22 Draw the structural formula of acetic acid and describe two different methods of preparing it.
14.23 Describe the molecule structure of a fat.
14.24 What is meant by a "saturated" fat? an "unsaturated" fat?

Problems

General Formulas
14.25 Write the molecular formula of a hydrocarbon with seven carbon atoms that is an
  a. alkane   b. alkene   c. alkyne
14.26 Write the molecular formula of a hydrocarbon with 14 hydrogen atoms that is an
  a. alkane   b. alkene   c. alkyne
14.27 Which of the following represent alkanes? alkenes? alkynes?
  a. C_5H_{12} b. C_5H_{10} c. C_5H_8
  d. C_6H_{10} e. $C_{12}H_{26}$
14.28 Which of the following represent alkanes? alkenes? alkynes?
  a. C_9H_{18} b. C_8H_{18} c. C_9H_{16}
  d. C_2H_2 e. C_3H_6
14.29 Give the molecular formula of
  a. an alkane with six carbon atoms
  b. an alkene with six hydrogen atoms
  c. an alkyne with four carbon atoms
  d. an alkane with ten hydrogen atoms

Isomerism
14.30 Draw the condensed structural formulas of all the isomers of hexane, C_6H_{14}.
14.31 Which of the following structural formulas represent the same molecule?

  a. $CH_3-\underset{\underset{CH_3}{|}}{\overset{\overset{H}{|}}{C}}-CH_2-CH_3$

  b. $CH_3-CH_2-\underset{\underset{CH_3}{|}}{\overset{\overset{H}{|}}{C}}-CH_3$

  c. $CH_3-\underset{\underset{CH_3}{|}}{\overset{\overset{CH_3}{|}}{C}}-CH_3$

  d. $CH_3-\underset{\underset{CH_3}{|}}{\overset{\overset{H}{|}}{C}}-\underset{\underset{CH_3}{|}}{\overset{\overset{H}{|}}{C}}-H$

14.32 Draw the structural isomers of C_4H_9Cl, a compound in which a chlorine atom is substituted for one of the hydrogen atoms of C_4H_{10}.
14.33 Draw the structural formulas of all the isomers of the alkyne C_4H_6.
14.34 Draw the structural formulas of all the isomers of $C_3H_5Cl_3$, a compound in which

three of the hydrogen atoms of propane have been replaced by chlorine atoms.

14.35 What is the first carboxylic acid to show structural isomerism?

14.36 Show the structural formulas of all the ethers containing four carbon atoms per molecule.

14.37 Show the structural formulas of all the isomers of $C_6H_3Cl_3$, in which three of the hydrogen atoms of benzene have been replaced by chlorine atoms.

Nomenclature

14.38 Give the IUPAC names of all the isomers of hexane shown in Problem 14.30.

14.39 Name the compounds whose structural formulas are shown in Problem 14.31.

14.40 Give the structural formulas corresponding to the following IUPAC names
   a. 2-methylhexane
   b. 2,2-dimethylpentane
   c. 2-methyl, 3-ethylhexane
   d. octane
   e. 4-ethylheptane

14.41 Alkyl chlorides are named like alkanes except that the prefix "chloro" is used. Name the following alkyl chlorides

a. $CH_3-CH_2-\underset{\underset{Cl}{|}}{\overset{\overset{H}{|}}{C}}-CH_3$

b. $CH_3-\underset{\underset{Cl}{|}}{\overset{\overset{H}{|}}{C}}-\underset{\underset{Cl}{|}}{\overset{\overset{H}{|}}{C}}-CH_3$

c. $CH_3-CH_2-CH_2-CCl_3$

d. $H-\underset{\underset{Cl}{|}}{\overset{\overset{H}{|}}{C}}-CH_2-\underset{\underset{Cl}{|}}{\overset{\overset{H}{|}}{C}}-H$

Polymerization

14.42 Identify the monomer from which the following polymer is made

$-\underset{\underset{H}{|}}{\overset{\overset{H}{|}}{C}}-\underset{\underset{CN}{|}}{\overset{\overset{H}{|}}{C}}-\underset{\underset{H}{|}}{\overset{\overset{H}{|}}{C}}-\underset{\underset{CN}{|}}{\overset{\overset{H}{|}}{C}}-\underset{\underset{H}{|}}{\overset{\overset{H}{|}}{C}}-\underset{\underset{CN}{|}}{\overset{\overset{H}{|}}{C}}-$

14.43 Sketch a portion of the addition polymer formed by
   a. propylene, C_3H_6 b. C_2F_4
   c. styrene, $H-\underset{\underset{H}{|}}{\overset{\overset{H}{|}}{C}}=\underset{\underset{H}{|}}{\overset{}{C}}-\bigcirc$

14.44 What is the simplest formula of polyethylene? What are the mass percents of carbon and hydrogen?

Functional Groups

14.45 Identify the class to which each of the following compounds belongs

a. $CH_3-CH_2-\overset{\overset{O}{\|}}{C}-H$

b. $HO-\overset{\overset{O}{\|}}{C}-H$

c. $CH_3-\overset{\overset{O}{\|}}{C}-O-CH_3$

d. $CH_3-O-\bigcirc$

e. $HO-CH_2-CH_3$

14.46 Draw the structural formula of a molecule containing three carbon atoms which represents a(an)
   a. alcohol   b. ether   c. aldehyde
   d. ketone   e. acid   f. ester

14.47 What is the smallest number of carbon atoms that can be present in a molecule of a(an)
   a. alcohol   b. ether   c. aldehyde
   d. ketone   e. acid   f. ester

Preparation of Esters

14.48 Give the structure of the ester formed by reacting
  a. formic acid with propyl alcohol
  b. acetic acid with isopropyl alcohol

14.49 Give the structural formulas of all the esters that can be formed by reacting methyl or ethyl alcohol with formic or acetic acid.

14.50 Identify the alcohol and acid used to form

a. $\text{H}-\underset{\underset{\text{O}}{\|}}{\text{C}}-\text{O}-\text{CH}_2-\text{CH}_3$

b. $\text{CH}_3-\underset{\underset{\text{O}}{\|}}{\text{C}}-\text{O}-\text{CH}_2-\text{CH}_2-\text{CH}_3$

c. $\text{CH}_3-\underset{\underset{\text{O}}{\|}}{\text{C}}-\text{O}-\underset{\underset{\text{CH}_3}{|}}{\overset{\overset{\text{H}}{|}}{\text{C}}}-\text{CH}_3$

Multiple Choice

14.51 In benzene and in ethylene, all the bond angles are
  a. 90°   b. 109.5°   c. 120°   d. 180°
  e. 360°

14.52 A certain aromatic hydrocarbon has the structural formula

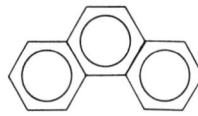

Its molecular formula is
  a. $C_{10}H_8$ b. $C_{12}H_{10}$ c. $C_{14}H_{10}$
  d. $C_{14}H_{12}$ e. $C_{18}H_{18}$

14.53 In which one of the following classes are there always two oxygen atoms per molecule?
  a. alcohol   b. aldehyde   c. carboxylic acid   d. ether   e. ketone

14.54 How many ketones are there with four carbon atoms per molecule?
  a. 1   b. 2   c. 3   d. 4   e. 5

14.55 The ester made from formic acid and propyl alcohol is isomeric with the ester made from
  a. formic acid and ethyl alcohol
  b. acetic acid and ethyl alcohol
  c. acetic acid and isopropyl alcohol
  d. acetic acid and propyl alcohol

14.56 Of the six classes of organic oxygen compounds discussed in this chapter, how many contain the C=O group?
  a. 2   b. 3   c. 4   d. 5   e. 6

14.57 How many isomeric heptanes (C_7H_{16}) have a six-membered carbon chain as the longest carbon chain in the molecule?
  a. 1   b. 2   c. 3   d. 4   e. 5

14.58 Which one of the following is not a correct name for an isomeric hexane, C_6H_{14}?
  a. hexane
  b. 2-methylpentane
  c. 3-methylpentane
  d. 4-methylpentane
  e. 2,2-dimethylbutane

14.59 Which one of the following is not a correct name for an isomeric octane, C_8H_{18}?
  a. 2-methylheptane
  b. 3-methylheptane
  c. 4-methylheptane
  d. 2-ethylhexane

14.60 Successive members of the alkane series differ in molar mass by how many units (grams per mole)?
  a. 10   b. 12   c. 14   d. 16
  e. 24

Nuclear Chemistry

Learning Objectives
After studying this chapter, you should be able to:

1. Describe the properties of the three types of radiation given off in natural radioactivity.
2. Given the identity of all but one of the nuclei taking part in a nuclear reaction, find the mass number and atomic number of that isotope; write an equation for the reaction (Examples 15.1–15.3).
3. Explain what is meant by a rad; a rem.
4. Relate the fraction of a radioactive nucleus remaining to the number of half-lives elapsed (Examples 15.4, 15.5).
5. Using Figure 15.3, relate fraction or amount left to time in the decay of a radioactive nucleus (Example 15.6).
6. Explain how nuclear reactions are used to find the age of a rock; the age of organic material.
7. Explain what is meant by nuclear fission and write a balanced nuclear equation for a typical fission reaction.
8. Explain what is meant by nuclear fusion and write a balanced nuclear equation for a typical fusion reaction.
9. Relate Δm in grams for a process to ΔE in kilojoules (Examples 15.7, 15.8).

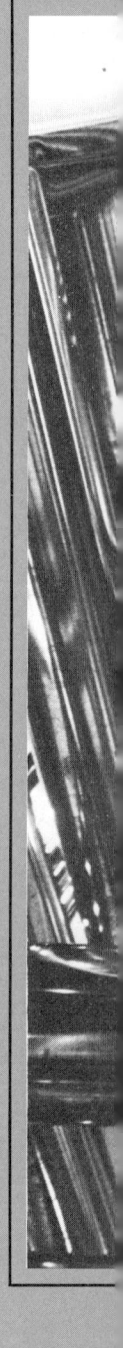

> Knowledge is one. Its division into subjects is a concession to human weakness.
>
> HALFORD JOHN MACKINDER

Photograph by William T. Moore.

CHAPTER 15

Up to this point, we have dealt with "ordinary" chemical reactions. In such reactions, atoms or molecules change only their outer electronic structures. This chapter considers nuclear reactions. Such reactions result in changes in the nucleus, involving protons and neutrons.

Nuclear reactions differ from ordinary chemical reactions in several ways.

1. In ordinary reactions, isotopes of an element behave in the same way. Thus the two isotopes $^{12}_{6}C$ (6 protons + 6 neutrons) and $^{14}_{6}C$ (6 protons + 8 neutrons) both react with oxygen in the same way to produce carbon dioxide. Moreover, the chemical behavior of CO_2 is the same regardless of the relative amounts of $^{12}_{6}C$ and $^{14}_{6}C$ present. In contrast, the nuclear reactions of these two isotopes are quite different. The carbon-12 nucleus is extremely stable while the carbon-14 nucleus decomposes spontaneously.
2. The energy change in nuclear reactions exceeds by several powers of ten that for ordinary chemical reactions. Consider, for example, the nuclear decomposition of carbon-14 (to form nitrogen-14 and an electron). For every gram of carbon-14 that reacts, about one million kilojoules of energy is released. In contrast, when one gram of carbon dioxide is formed by the combustion of carbon, only about nine kilojoules of energy is evolved.
3. In nuclear reactions, the number of protons in the nucleus usually changes. When this happens, the identity of the element changes. For example, the decomposition of the carbon-14 nucleus (six protons) leads to the formation of nitrogen-14 (seven protons). In this way, nuclear reactions achieve the dream of the alchemists, the transmutation of one element to another. In ordinary chemical reactions, elements never change their identity. There, only electron rearrangements are involved; the nuclei remain unchanged.

In this chapter, we will look at three different types of nuclear reactions. We start (Sections 15.1, 15.2) with a discussion of radioactivity. This was the first type of nuclear reaction identified; it was discovered late in the 19th century. Another type of nuclear reaction, fission (Section 15.3) was discovered in 1938. About ten years later, nuclear fusion (Section 15.4) was first carried out (in a test of the "hydrogen" bomb). The source of energy in all these reactions is a decrease in mass. The products of nuclear reactions weigh significantly less than the reactants (Section 15.5).

15.1 Radioactivity

Certain unstable nuclei decompose by themselves, giving off energy in the process. A few such nuclei are found in nature. Their decomposition *("decay")* is referred to as *natural radioactivity.* Many more unstable nuclei have been prepared in the laboratory. The process by which such nuclei decompose is called *induced radioactivity.*

Natural Radioactivity

This process was discovered by the French scientist Henri Becquerel in 1896. He found that uranium salts give off high-energy radiation capable of blackening a photographic plate. Becquerel showed that the rate of emission of radiation was directly proportional to the percent of uranium present. There was one exception to this rule. An ore of uranium called pitchblende gave off radiation four times as rapidly as it should have on the basis of its uranium content. Marie and Pierre Curie, colleagues of Becquerel at the Sorbonne, concluded that pitchblende must contain a new element more intensely radioactive than uranium. In 1898, they isolated a fraction of a gram of that element from a ton of pitchblende. It was named polonium, after Marie Curie's native country. A short time later the Curies isolated another new element, also intensely radioactive, which they called radium.

The radiation given off in natural radioactivity is separated by an electrical or magnetic field into three distinct parts (Figure 15.1, p. 414).

1. **Alpha radiation** consists of a stream of positively charged particles. Each **alpha particle** has a charge of +2 and a mass number of four. This means that an alpha particle is a helium nucleus and has the nuclear symbol $^4_2$He.

    When an alpha particle is emitted from a nucleus, the atomic number decreases by two units, the mass number by four units. Consider, for example, the radioactive decay of the most common isotope of uranium, $^{238}_{92}$U. When a uranium-238 nucleus gives off an alpha particle it is converted to a nucleus of atomic number 90 and mass number 234. This is an isotope of the element thorium (at. no. Th = 90). The nuclear equation for the process can be written

    $$^{238}_{92}\text{U} \rightarrow {}^{4}_{2}\text{He} + {}^{234}_{90}\text{Th} \qquad (15.1)$$

Figure 15.1 A beam of radiation from a radioactive source can be resolved into three components by an electric field. These components, called α, β, and γ rays, consist of helium nuclei, electrons, and short wavelength x-rays respectively.

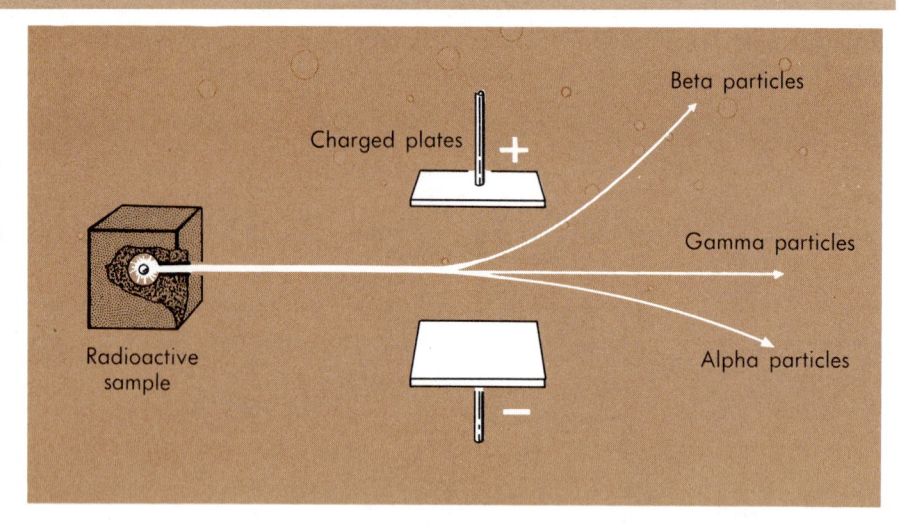

Here, as in all nuclear equations, there is a balance of

atomic number (2 + 90 = 92)

mass number (4 + 234 = 238)

on the two sides of the equation.

EXAMPLE 15.1

Another radioactive isotope that decays by giving off an alpha particle is radium-226, which has the nuclear symbol $^{226}_{88}Ra$.
a. What is the symbol of the nucleus formed by alpha decay?
b. Write a balanced equation for the nuclear reaction.

Solution
a. The atomic number decreases by two units, the mass number by four. The nucleus formed has an atomic number of 88 − 2 = 86. Its mass number is 226 − 4 = 222. From the Periodic Table, we see that the element of atomic number 86 is radon, Rn. The nuclear symbol of this isotope of radon is $^{222}_{86}Rn$.
b. $^{226}_{88}Ra \rightarrow\ ^{222}_{86}Rn +\ ^{4}_{2}He$

Exercise
When an alpha particle is emitted, what happens to the number of protons in the nucleus? the number of neutrons? Answer: Both decrease by two.

Alpha particles given off in radioactive decay have high energies. As they pass through matter, they transfer this energy to molecules with which they collide. This may cause a molecule to dissociate into atoms or ionize, forming a cation and an electron. Either way, the molecule is destroyed. This accounts for the radiation damage caused by alpha particles (and other types of radiation as well).

Alpha particles have a short range. Most of their energy is given up in collisions that take place close to their source. A sheet of paper can protect you from alpha particles passing through the air. However, if a radioactive source of alpha particles enters your body, the situation is entirely different. Alpha particles burn and destroy tissue with which they come in contact.

2. **Beta radiation** consists of a stream of negatively charged particles, originally called beta particles, now known to be **electrons.** An electron has a charge of −1. It has a very small mass, nearly zero on the atomic mass scale. Its nuclear symbol is

$$_{-1}^{0}e$$

Beta radiation is produced by a reaction within the nucleus. A neutron (at. no. = 0, mass no. = 1) decomposes to emit an electron and leave behind a proton (at. no. = 1, mass no. = 1). Hence, when an electron is given off the atomic number increases by one unit while the mass number remains unchanged. Consider, for example, what happens when thorium-234 decays by electron emission

$$^{234}_{90}\text{Th} \rightarrow {}^{234}_{91}\text{Pa} + {}^{0}_{-1}e \qquad (15.2)$$

The mass number of the product nucleus is the same as that of the reactant, 234; the atomic number is one unit greater.

EXAMPLE 15.2
Carbon-14 decays by electron emission. Write a balanced nuclear equation for this reaction.

Solution
The mass number of the product nucleus is the same as that of carbon-14; the atomic number is one unit greater. The product must then be an isotope of nitrogen (at. no. = 7). The equation is

$$^{14}_{6}\text{C} \rightarrow {}^{0}_{-1}e + {}^{14}_{7}\text{N}$$

Exercise
The nucleus $^{234}_{91}$Pa decomposes by electron emission. The product of this decomposition is also radioactive, decaying by alpha emission. What is the nuclear symbol of the species formed by alpha emission? Answer: $^{230}_{90}$Th.

The electrons produced by radioactive decay have much greater penetrating power than do alpha particles. They readily pass through the walls of glass test tubes or beakers. Indeed, they can move as far as one centimeter through human tissue. On the other hand, they do less damage to tissue than do alpha particles of the same energy.

3. **Gamma radiation** is a form of high-energy radiation of very short wavelength (0.0005 to 0.1 nm). Gamma rays are emitted in most nuclear reactions. Because gamma emission involves no change in either atomic or mass number, it is ordinarily ignored in writing nuclear equations.

Of the three types of radiation, gamma radiation is by far the most penetrating. In human tissue, it can cause damage over a range 10 to 20 times as great as electrons. Heavy lead shielding is used to protect against gamma rays.

Induced Radioactivity; Bombardment Reactions

Only about 70 naturally radioactive nuclei have been identified. In contrast, more than 1500 radioactive nuclei have been made in the laboratory. They are prepared by bombardment reactions in which a high-energy particle collides with a stable nucleus. The product formed as a result of that collision is an unstable, radioactive nucleus.

A typical reaction sequence is one which starts with the stable isotope of aluminum, aluminum-27. When this nucleus is struck by a neutron, an unstable nucleus, aluminum-28 is formed. This decays by electron emission to form silicon-28. The two steps in the process can be represented by the nuclear equations

$$^{27}_{13}Al + ^{1}_{0}n \rightarrow ^{28}_{13}Al \tag{15.3}$$

$$^{28}_{13}Al \rightarrow ^{28}_{14}Si + ^{0}_{-1}e \tag{15.4}$$

The radioactivity associated with Reaction 15.4 is referred to as induced radioactivity. It is "induced" by the bombardment reaction, Reaction 15.3.

Radioactive nuclei formed by bombardment reactions have found applications in many different areas. Cobalt-60 is used in cancer treatment. Carbon-14 serves as a "tracer" to follow the path of organic and biochemical reactions. One of the most interesting uses of bombardment reactions is in preparing the so-called transuranium elements. During the past 40 years, a total of 16 elements of atomic number greater than 92 have been made in the laboratory. Much of this work has been done by a group at the University of California in Berkeley. Other groups in Russia and, most recently, in West Germany, have made important contributions as well.

There are two general approaches followed in preparing elements of very high atomic number.

1. *Neutron bombardment.* The reactions here are very similar to 15.3 and 15.4. For example, neptunium (at. no. = 93) is made in two steps, starting with the neutron bombardment of uranium (at. no. = 92). The nuclear equations are

$$^{238}_{92}U + ^{1}_{0}n \rightarrow ^{239}_{92}U \tag{15.5}$$

$$^{239}_{92}U \rightarrow ^{239}_{93}Np + ^{0}_{-1}e \tag{15.6}$$

This type of reaction is effective in making isotopes of elements of atomic number 93–95.

2. *Positive ion bombardment.* A positive ion such as a proton or alpha particle can be effective in bringing about nuclear reactions if accelerated to high speeds. The high speeds are reached in a device called a cyclotron (Figure 15.2). An alpha particle emerging from a cyclotron can bring about a reaction such as

$$^{239}_{94}Pu + ^{4}_{2}He \rightarrow ^{242}_{96}Cm + ^{1}_{0}n \tag{15.7}$$

Figure 15.2 The cyclotron contains two oppositely charged "dees" placed between the poles of an electromagnet. Positive ions, formed in the center corridor, enter the dee at the left, which originally carries a negative charge. They pass through the dee, following a curved path (color). At the instant they emerge into the center corridor, the polarity of the dees is reversed. This causes the positive ions to enter the dee at the right at an increased speed. This process is repeated over and over again. The positive ions move faster and faster in curved paths of larger and larger radius. Eventually they are deflected from the outside wall of one of the dees to focus upon their target.

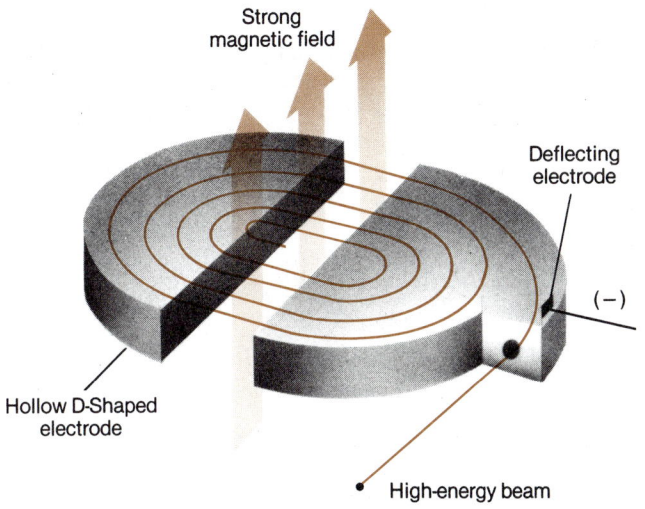

EXAMPLE 15.3
When a $^{238}_{92}U$ nucleus is bombarded by a $^{12}_{6}C$ nucleus, the products are a new, heavy nucleus and four neutrons.
a. What is the atomic number and mass number of the new nucleus?
b. Write a balanced nuclear equation for the reaction.

Solution
a. The reactants are $^{238}_{92}U$ and $^{12}_{6}C$, giving a total atomic number of 98 and a total mass number of 250. The four neutrons ($^{1}_{0}n$) have a total atomic number of 0 and a total mass number of 4. Hence

   atomic number of nucleus = 98 − 0 = 98

   mass number of nucleus = 250 − 4 = 246

b. The element californium, symbol Cf, has an atomic number of 98. The balanced equation is

$$^{238}_{92}U + {}^{12}_{6}C \rightarrow {}^{246}_{98}Cf + 4\,{}^{1}_{0}n$$

Exercise
Suppose you wanted to make an isotope of element 104 with a mass number of 257. Assume the reaction is similar to the one just considered, using $^{12}_{6}C$ as a bombarding particle and forming four neutrons as a byproduct. What heavy nucleus should you start with? Answer: $^{249}_{98}Cf$.

Effects of Radiation

Radiation is harmful because it can destroy the organic molecules that make up our body cells. The extent of damage depends upon two factors.

1. *The amount of radiation absorbed.* This is usually expressed in **rads**. A rad corresponds to the absorption of 10^{-2} joules of energy per kilogram of tissue. This means that a person weighing 60 kg exposed to radiation with an energy of one joule would absorb

$$\frac{1.0 \text{ J}}{60 \text{ kg}} = 1.7 \times 10^{-2} \text{ J/kg} = 1.7 \text{ rad}$$

2. *The type of radiation absorbed.* Alpha particles do ten times as much damage to body cells as do electrons of equal energy.

Taking these two factors into account, we commonly express the biological effect of radiation in terms of the number of **rems** absorbed

number of rems = f (number of rads) (15.8)

Here, f is a factor that varies from one for electrons and gamma radiation to ten for alpha particles and fast neutrons.

Effects of radiation begin to show up when a person is exposed to a single dose of radiation of 25 rems. Doses of 100–200 rems can be fatal. Small doses of radiation repeated over long periods of time can also have serious effects. Many of the early workers in radioactivity, including Marie Curie, developed cancer. It is not unusual for malignancies to appear 20, 30, or even 40 years after initial exposure. Studies have shown an unusually large number of cases of leukemia among survivors of the atom bomb blasts at Hiroshima and Nagasaki.

On the average, a person in the United States is exposed to approximately 0.2 rem of radiation per year. The sources of this radiation are shown in Table 15.1. Several features of the data in this table are of interest.

1. About two thirds of radiation exposure comes from natural sources. Here, the level depends upon the area in which you live. Cosmic radiation is much more intense at high elevations. A resident of Denver is exposed to about 100 millirems/year from this source. This is twice the national average.
2. The largest exposure to manmade sources of radiation comes from x-rays. The numbers quoted in Table 15.1 assume only occasional x-ray examinations. A single dental x-ray subjects you to 20 millirems, a chest x-ray to as much as 200 millirems. Clearly, excessive use of x-rays should be avoided.

TABLE 15.1 Typical Radiation Exposures in the United States (1 Millirem = 10^{-3} Rem)

Natural Sources	Millirems per Year
A. External to the body	
1. From cosmic radiation	50
2. From the earth	47
3. From building materials	3
B. Inside the body	
1. Inhalation of air	5
2. In human tissues (mostly $^{40}_{19}K$)	21
Total from natural sources	126
Manmade Sources	
A. Medical procedures	
1. Diagnostic x-rays	50
2. Radiotherapy x-rays, radioisotopes	10
3. Internal diagnosis, therapy	1
B. Nuclear power industry	0.2
C. Luminous watch dials, TV tubes, industrial wastes	2
D. Radioactive fallout	4
Total from manmade sources	67
Total	193

3. On the average, exposure to radiation from nuclear power plants is very low, about 0.2 millirem/year. This number is somewhat higher in the vicinity of the plant. Still, a person sitting on a fence at the boundary of a plant all year long would receive only 5 millirems of radiation.

15.2 Rate of Radioactive Decay

One of the most important characteristics of radioactive decay is the rate at which it occurs. This is most often described by quoting the **half-life** of the process. Half-life is exactly what it sounds like. It is the time required for one half of a sample to decompose. Half-life is inversely related to rate. A process that takes place slowly has a long half-life. By the same token, if the rate is very fast, the time required for half of a sample to decompose is very small.

Half-Lives of Radioactive Nuclei

The rate law for radioactive decay can be stated very simply: **The half-life of a radioactive nucleus is a constant, independent of concentration, amount, or any other factor.**

To appreciate what this statement means, let's consider a specific nucleus. The radioactive nucleus iodine-131 decays by electron emission

$$^{131}_{53}I \rightarrow {}^{131}_{54}Xe + {}^{0}_{-1}e \tag{15.9}$$

The half-life for this process is eight days. If we start with a 1.00-g sample, after eight days we will still have

$$\frac{1}{2}(1.00 \text{ g}) = 0.500 \text{ g}$$

of iodine-131. The other 0.500 g has been converted to xenon-131. Suppose now that we wait another eight days and measure how much iodine-131 remains at that point. Because half-life is independent of amount, we should have

$$\frac{1}{2}(0.500 \text{ g}) = 0.250 \text{ g}$$

of iodine-131 left after two half-lives, i.e., 16 days. Over the next eight days, half of the 0.250 g of iodine-131 will decay. So, after 24 days (three half-lives), the amount of this radioactive isotope left is

$$\frac{1}{2}(0.250 \text{ g}) = 0.125 \text{ g}$$

This process continues through successive half-lives. After every eight-day period, there is just half as much iodine-131 left as there was at the beginning of the period.

This discussion of the decay of iodine-131 is summarized in Table 15.2. Looking at the last two columns of the table, you can see that there is a simple relation between the fraction left and the number of half-lives. This relation is

$$\text{fraction left} = \left(\frac{1}{2}\right)^n \tag{15.10}$$

where n is the number of half-lives that have passed. For the data in Table 15.2, where we started with 1.00 g of iodine-131, the amount left in grams is numerically equal to the fraction left. More generally

$$\text{amount left} = (\text{fraction left}) \times (\text{original amount}) \tag{15.11}$$

The use of Equations 15.10 and 15.11 is illustrated in Examples 15.4 and 15.5.

EXAMPLE 15.4

The isotope $^{90}_{38}\text{Sr}$ is radioactive. It decays with a half-life of 29 years. Suppose we start with a sample weighing 0.0200 g. After 58 years,
a. What fraction of the sample is left?
b. How many grams of strontium-90 remain?

Solution
a. 58 years is two half-lives (58/29 = 2). So the fraction left will be

$$\left(\frac{1}{2}\right)^2 = \frac{1}{4}$$

b. $\frac{1}{4}(0.0200 \text{ g}) = 0.00500 \text{ g}$

Exercise
What fraction will be left after 116 years? Answer: 1/16.

TABLE 15.2 Rate of Decay of $^{131}_{53}\text{I}$ (Half-life = 8 days)

Time (Days)	Amount Left (g)	Fraction Left	Number of Half-Lives
0	1.00		
8	0.500	½	1
16	0.250	(½)² = ¼	2
24	0.125	(½)³ = ⅛	3
32	0.0625	(½)⁴ = ¹⁄₁₆	4
8n		(½)ⁿ	n

EXAMPLE 15.5

The common isotope of radium, $^{226}_{88}$Ra, has a half-life of 1.6×10^3 years. Suppose we start with a sample weighing 0.200 g. How long will it take for the amount to drop to 0.0250 g?

Solution

We first find the fraction left

$$\text{fraction left} = \frac{\text{amount left}}{\text{original amount}} = \frac{0.0250 \text{ g}}{0.200 \text{ g}} = \frac{1}{8}$$

Now we apply Equation 15.10 to find the number of half-lives that have elapsed

$$\left(\frac{1}{2}\right)^n = \frac{1}{8}$$

Clearly, n = 3 since ½ × ½ × ½ = ⅛. We conclude that three half-lives have passed. Because the half-life is 1600 years,

time required = 3(1600 yr) = 4800 yr

Exercise

How long will it take for the amount to drop to 0.0125 g? Answer: 6400 years.

As we have just seen, Equation 15.10 is readily applied when n is a whole number. It is relatively easy to relate fraction left to time when we are dealing with an integral number of half-lives, i.e., 1, 2, 3, - - -. Things are a bit more complex if the time involved does not represent a whole number of half-lives. To illustrate the problem, suppose that in Example 15.4 you had been asked to obtain the amount left after 20 years. This is not an integral number of half-lives (half-life = 29 years). How would you carry out that kind of calculation? To do it, you need a general relation between fraction left and time.

There are several ways to obtain a general relation between fraction of a radioactive isotope left and time. A simple approach is to use Equation 15.10 to construct a graph relating these two quantities. Figure 15.3 shows such a plot. On the vertical axis (y axis), we show the fraction left. This varies from 1.00 at the top (all of the reactant left) to 0.00 at the bottom (all of the reactant gone). On the horizontal axis (x axis) we show the number of half-lives, n. The points shown as filled circles correspond to whole-number values of n. Thus we have

Point	y value	x value
1	fraction left = ½	one half-life (n = 1)
2	fraction left = ¼	two half-lives (n = 2)
3	fraction left = ⅛	three half-lives (n = 3)
4	fraction left = 1/16	four half-lives (n = 4)

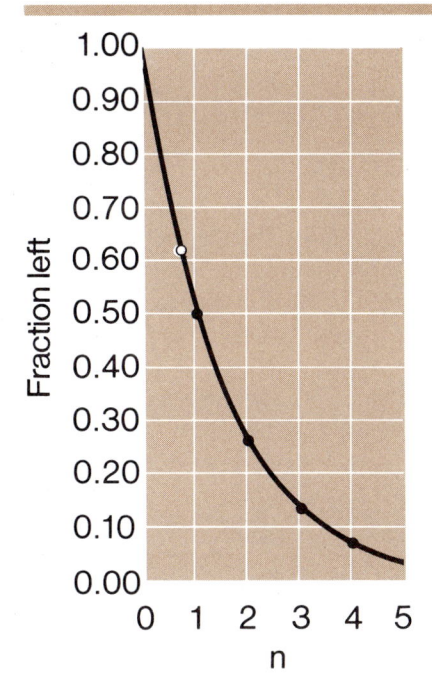

Figure 15.3 Graph of fraction of a radioactive species left versus number of half-lives, n.

EXAMPLE 15.6
Referring back to Example 15.5, what fraction of a radium sample will be left after 2.0×10^3 years? (Use Figure 15.3.)

Solution
Find the number of half-lives and then the fraction left.
1. n = 2000 yr/1600 yr = 1.25
   In other words, 1¼ half-lives have passed.
2. To find the fraction left from the graph:
   a. Draw a vertical line up from the x axis at n = 1.25. Mark the intersection of this line with the curve.
   b. From this intersection, draw a horizontal line across to the y axis. You should find that this line touches the y axis at 0.42. The fraction left must then be 0.42.

   It is a good idea to check to make sure that your answer is reasonable. Since n = 1.25, between one and two half-lives have passed. The fraction left should be between 0.50 (one half-life) and 0.25 (two half-lives). The calculated value, 0.42, falls between these two limits. It is closer to 0.50 than to 0.25, which makes sense; 1.25 is closer to 1 than to 2.

Continued

> **Exercise**
> Suppose the fraction left were 0.60. Using Figure 15.3, estimate the number of half-lives in that case. Answer: 0.74.

Age of Rocks

Certain radioactive isotopes act as "natural clocks" to date events that took place long ago. The principle is a simple one. By measuring the fraction of the isotope that remains today, we can calculate how much time has passed since the event occurred. This method has been applied to determine the age of uranium-bearing rocks. In this case, "age" means the time that has passed since the rock solidified.

To illustrate how this method works, let's consider the overall process for the radioactive decay of uranium. The nuclear equation is

$$^{238}_{92}U \rightarrow {}^{206}_{82}Pb + 8\,{}^{4}_{2}He + 6\,{}^{0}_{-1}e \tag{15.12}$$

half-life = 4.5×10^9 yr

Notice that

— the half-life is 4.5×10^9 years. In other words, it takes 4.5 billion years for half of the uranium originally in the rock to decay.
— the ultimate product of the decay is lead-206. Every atom of lead-206 that we find in a rock today is assumed to have come from an atom of uranium-238.

Suppose now that a certain rock is found to contain equal numbers of atoms of U-238 and Pb-206. To have some numbers to work with, suppose there are in the sample

1.0×10^{23} atoms U-238 and 1.0×10^{23} atoms of Pb-206

Because every Pb-206 atoms came from a U-238 atom, it follows that 1.0×10^{23} U-238 atoms must have decayed to produce the Pb-206. In other words, we originally had

1.0×10^{23} U-238 atoms + 1.0×10^{23} U-238 atoms = 2×10^{23} U-238 atoms

    (still present)   (decayed to lead)   (total originally)

Clearly, one half-life has passed. There are half as many uranium atoms present now as there were originally. This means that the age of the rock is about 4.5×10^9 years.

Ages of rocks determined by this and other methods involving radioactive nuclei range from $(3 \text{ to } 4.5) \times 10^9$ years. The larger number is often taken as an approximate value for the age of the earth. Analyses of rock samples from the moon indicate ages in this same range. This suggests that the earth and the moon were formed at about the same time.

Age of Organic Material

It is possible to estimate the age of a carbon-containing object by measuring its carbon-14 content. This isotope of carbon is formed in the atmosphere when neutrons from cosmic radiation collide with nitrogen-14 nuclei

$$^{14}_{7}N + ^{1}_{0}n \rightarrow ^{14}_{6}C + ^{1}_{1}H \tag{15.13}$$

The carbon-14 formed by this reaction is converted into carbon dioxide. In atmospheric CO_2, there is a constant ratio of C-14 to C-12, about $1:10^{12}$. Living plants, which take in carbon dioxide from the air, show this same ratio of C-14/C-12. The same is true for animals, including human beings, since they eat these plants.

When a plant or animal dies, the intake of carbon-14 stops. From this point on, the radioactive decay of carbon-14 takes over

$$^{14}_{6}C \rightarrow ^{14}_{7}N + ^{0}_{-1}e; \text{ half-life} = 5720 \text{ years} \tag{15.14}$$

The ratio of C-14/C-12 drops off steadily as time passes. By measuring this ratio in a plant or animal relic and comparing it to what it must have been originally, we can estimate how much time has passed since the plant or animal died.

This approach might be used to date an artifact such as a piece of charcoal from an ancient cave. Suppose we find that the C-14/C-12 ratio in the charcoal is only one fourth of that in plants living today. The rest of it has decayed via Reaction 15.14. Applying Equation 15.10,

$$\text{fraction left} = \frac{1}{4} = \left(\frac{1}{2}\right)^n$$

Clearly the number of half-lives, n, is 2. The age of the charcoal is

$2(5720 \text{ yr}) \approx 11,000 \text{ yr}$

We conclude that the tree from which the charcoal was made must have died about 1.1×10^4 years ago.

15.3 Nuclear Fission

Fifty years ago, the only nuclear reaction known was radioactivity. Then, in 1938, two German scientists, Otto Hahn and Fritz Strassman, discovered an entirely new type of nuclear reaction. They found that when uranium (at. no. = 92) was exposed to neutrons, a reaction took place to form barium (at. no. = 56). A colleague of Hahn and Strassman, Lise Meitner, interpreted this to mean that uranium atoms had undergone nuclear fission. That is, a uranium atom had split into fragments of smaller mass.

The discovery of nuclear fission came on the eve of World War II, which broke out in 1939. The fission process quickly became the object of intense research in the United States, Canada, Great Britain, and Germany. Within a few months, several crucial facts were established.

1. Nuclear fission produces enormous amounts of energy. For every gram of uranium that undergoes fission, about 8×10^7 kJ of energy is evolved. This is considerably greater than the energy given off in radioactive decay, which is of the order of 2×10^6 kJ/g. It is much, much greater than the energy evolved when an ordinary explosive like TNT is detonated (about 3 kJ/g). Putting it another way, the fission of one gram of uranium yields as much energy as the explosion of 3×10^7 g (30 metric tons) of TNT.

2. Only one isotope of uranium, $^{235}_{92}U$, splits apart when struck by neutrons. This isotope makes up only 0.7% of natural uranium. The more abundant isotope, $^{238}_{92}U$, does not itself undergo fission.

3. The fission of uranium-235 gives many different products. One atom may split into isotopes of barium (at. no. = 56) and krypton (at. no. = 36). At the same time, another atom may break down to give isotopes of cesium (at. no. = 55) and rubidium (at. no. = 37). Still another atom of uranium-235 may yield isotopes of lanthanum (at. no. = 57) and bromine (at. no. = 35)

$$^{235}_{92}U + ^{1}_{0}n \rightarrow ^{146}_{56}Ba + ^{87}_{36}Kr + 3\,^{1}_{0}n \tag{15.15}$$

$$^{235}_{92}U + ^{1}_{0}n \rightarrow ^{144}_{55}Cs + ^{90}_{37}Rb + 2\,^{1}_{0}n \tag{15.16}$$

$$^{235}_{92}U + ^{1}_{0}n \rightarrow ^{146}_{57}La + ^{87}_{35}Br + 3\,^{1}_{0}n \tag{15.17}$$

In other words, fission is not a single, unique reaction. Instead, it consists of a family of reactions of which 15.15–15.17 are typical. In general, fission leads to the formation of two lighter nuclei from one heavy nucleus. One of the products has an atomic number in the vicinity of 56, the other an atomic number close to 36.

4. Fission produces excess neutrons. Notice from Equations 15.15–15.17 that 2–3 neutrons are produced for every one that reacts. This means that once a few U-235 nuclei split, enough neutrons are produced to split many more nuclei. In this way, we get a "chain reaction," which continues as long as there is any U-235 left. Unless some of the neutrons are removed, the rate of reaction increases rapidly, resulting in an explosion. This is what happens in an atomic bomb such as the one exploded over Hiroshima in 1945.

Fission Reactors

At the present time, there are about 75 nuclear power plants operating in the United States. They supply between 10 and 15% of our electrical energy. This energy comes from the fission of uranium-235. Energy is given off as heat which is used to convert water to steam. The steam in turn drives an electrical generator. The reactor is designed so that the rate of fission is constant. Energy is given off at a constant rate, eliminating the possibility of the kind of explosion that takes place in an atomic bomb.

Figure 15.4 shows the type of nuclear reactor that is in common use in the United States. The nuclear fuel consists of pellets of uranium oxide enriched in U-235. These pellets are contained within stainless steel tubes, surrounded by control rods. The rods are made of cadmium or boron, both of which absorb neutrons readily. The position of the control rods can be adjusted. In an emergency, they can be lowered to absorb more neutrons and stop the fission process.

The heat produced by fission is transferred to water that is pressurized to keep it from boiling. This water, which becomes radioactive, circulates

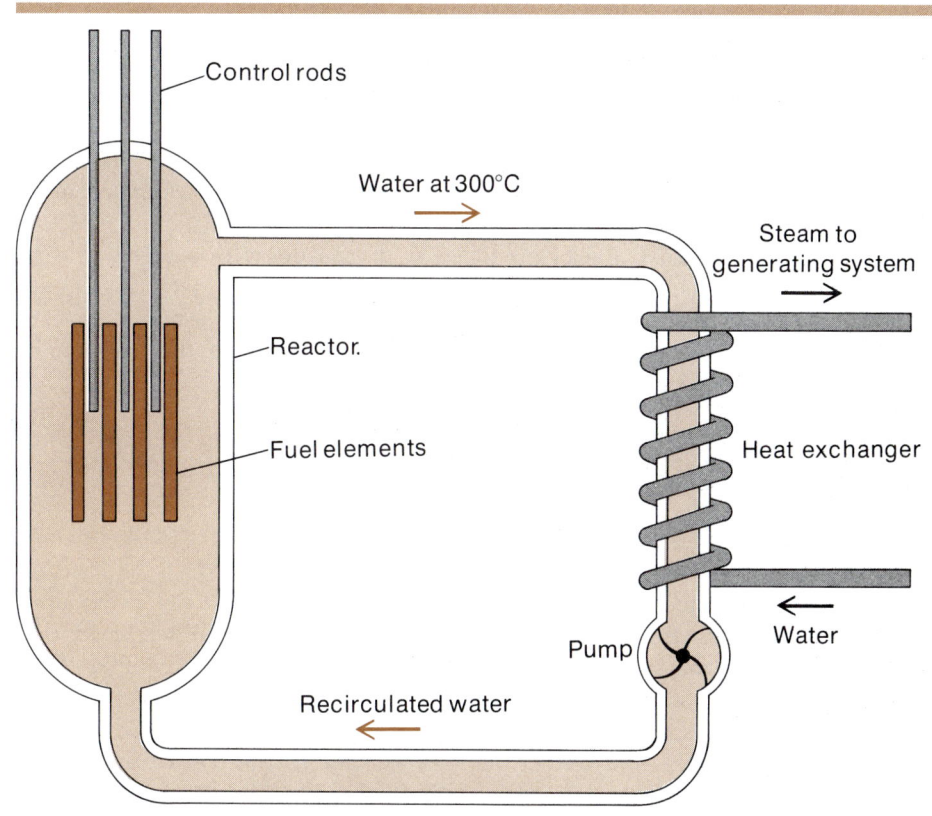

Figure 15.4 Nuclear reactor of the pressurized water type. The control rods are made of a material such as cadmium, which absorbs neutrons very effectively. The rate of fission is carefully monitored and controlled.

through a closed loop at a temperature of about 300°C. Heat is given off to another stream of water (right of Figure 15.4), which is not radioactive. It is this water that is converted to steam for generating electrical energy.

Nuclear reactors have two major advantages over conventional power plants using coal or oil.

1. The level of air pollution is reduced virtually to zero. In contrast, a coal-burning plant produces large amounts of sulfur dioxide. As pointed out in Chapter 13, SO_2 is toxic, particularly to people who have chronic lung disorders. It has been estimated that several thousand people in the United States die prematurely each year from the air pollution produced by conventional power plants.
2. Fission reactors allow us to reduce the consumption of petroleum. Since a large fraction of our petroleum is imported, conservation makes sense both economically and politically.

Balanced against these advantages of fission reactors are several drawbacks. All of these are related to the fact that the products of fission are unstable and highly radioactive. Looking back at Equations 15.15–15.17, you will note that the product nuclei contain many more neutrons than do the stable isotopes of the same elements. Consider, for example, Equation 15.16. Cesium-144 has a mass number 11 units higher than the atomic mass of the element (at. mass Cs = 133). Rubidium-90, the other nucleus formed in this reaction, also contains excess neutrons (at. mass Rb = 85.5).

The radioactive nuclei produced by fission decay in a series of steps. Consider, for example, what happens to rubidium-90. It undergoes three successive electron emissions. In each step, a neutron in the nucleus is converted to a proton. The atomic number increases by one unit while the mass number remains unchanged.

$$^{90}_{37}Rb \rightarrow\ ^{90}_{38}Sr +\ ^{0}_{-1}e; \text{ half-life} = 2.8 \text{ min} \tag{15.18}$$

$$^{90}_{38}Sr \rightarrow\ ^{90}_{39}Y +\ ^{0}_{-1}e; \text{ half-life} = 29 \text{ yr} \tag{15.19}$$

$$^{90}_{39}Y \rightarrow\ ^{90}_{40}Zr +\ ^{0}_{-1}e; \text{ half-life} = 64 \text{ h} \tag{15.20}$$

Strontium-90 is particularly hazardous. In the form of $SrCO_3$, it is readily incorporated into the bones, replacing calcium carbonate. Since it has a half-life of 29 years, strontium stays around for a long time. A child exposed to Sr-90 could still be suffering from its effects 60 years later. About ¼ of this dangerous isotope would remain (recall Example 15.4).

Because fission products are radioactive, a fire or explosion at a nuclear power plant could be extremely serious. Despite all the precautions taken to prevent accidents, there is no guarantee that they will not happen. You will probably recall the "near-miss" at the Three Mile Island reactor in Pennsylvania a few years ago. There, a set of valves in the steam generating system accidentally closed. Within seconds, the temperature and pressure within the reactor increased to the danger point. The problem was compounded when the operators cut off the supply of emergency cooling water. For some time the fuel elements were uncovered and there was a danger that they might "melt down." Had that happened, the release of radiation could have been many times greater than the small amount that actually escaped.

It is essential that we develop methods for the safe, long-term storage of the radioactive wastes that fission reactors produce. So far these wastes have been stored as water solutions in underground storage tanks. These tanks are heavily shielded and cooled to absorb the heat given off by radioactive decay. At present, about 5×10^8 L of liquid waste is in storage in the United States. The plan is to evaporate this material to obtain solids, which can be stored more safely. One proposal would put the solids in abandoned salt mines deep beneath the surface of the earth.

Breeder Reactors

Fission reactors now in use in the United States rely upon the rare isotope of uranium, U-235. Supplies of this isotope are limited. For nuclear fission to make a long-term contribution to our energy needs, it will be necessary to use the more abundant isotope, U-238. There is about 150 times as much of this isotope as of U-235.

Uranium-238 does not undergo fission directly. However, neutron bombardment converts some of the U-238 to plutonium

$$^{238}_{92}U + ^{1}_{0}n \rightarrow ^{239}_{94}Pu + 2\,^{0}_{-1}e \qquad (15.21)$$

This isotope of plutonium can undergo fission when struck by neutrons. A typical reaction is

$$^{239}_{94}Pu + ^{1}_{0}n \rightarrow ^{147}_{56}Ba + ^{90}_{38}Sr + 3\,^{1}_{0}n \qquad (15.22)$$

Reaction 15.22, like Reactions 15.15–15.17, evolves a large amount of energy. It also produces an excess of neutrons. These can react with U-238 to give more Pu-239. In that sense, this process "breeds" fissionable material. Reactors using this process are called breeder reactors.

The feasibility of this two-step process was shown many years ago. The bomb exploded over Nagasaki in 1945 contained Pu-239 made from U-238. The first nuclear reactor to produce electrical energy (1951) was of the breeder type. However, there are no breeder reactors in use in the United States today. A host of technical problems and safety hazards have slowed their development. Economic concerns have also played a role. Construction costs for all types of nuclear reactors have increased greatly in recent years, considerably outpacing inflation.

A major concern with breeder reactors is the behavior of plutonium-239. For one thing, it is highly radioactive. As little as 1×10^{-5} g can cause cancer if it enters the body. Fears have also been expressed that plutonium from breeder reactors might be diverted to produce atomic bombs.

15.4 Nuclear Fusion

The process by which small nuclei combine with one another to form a larger nucleus is referred to as fusion. A typical fusion reaction is

$$^2_1H + {}^2_1H \rightarrow {}^4_2He \tag{15.23}$$

This process, like the fission of very heavy nuclei, gives off energy. Indeed, the energy change in fusion is about five to ten times as great as that for fission. About 6×10^8 kJ of energy is evolved per gram of deuterium (^2_1H) involved in fusion.

Fusion has many advantages as an energy source. In contrast to fission, the products of fusion are stable, nonradioactive nuclei. This means that the radiation hazard is greatly reduced. Of equal importance, deuterium is much more abundant than uranium. One of every 6000 hydrogen atoms is a deuterium atom, and there are a lot of hydrogen atoms around, mostly in water molecules. The oceans contain about 2×10^{23} g of deuterium. This is enough to meet our energy needs for the next thousand years.

Unfortunately, there is a catch. Fusion reactions, unlike fission, have a very high activation energy. Two deuterium nuclei, each with a +1 charge, repel each other strongly at close distances. To overcome this repulsion, the nuclei must have very high kinetic energies, of the order of 10^7 kJ/mol. This is more than a million times the average kinetic energy of deuterium atoms at room temperature.

In the hydrogen bomb, the activation energy for fusion came from fission of a small atomic bomb. To make a fusion reactor work, it will be necessary to trigger the fusion process in some other way. It turns out that it is not all that difficult to supply the required activation energy to deuterium nuclei.

Figure 15.5 One way to carry out a controlled fusion reaction is to confine very light nuclei in a strong magnetic field. The problem is to extend the time during which the reaction occurs to one second or more.

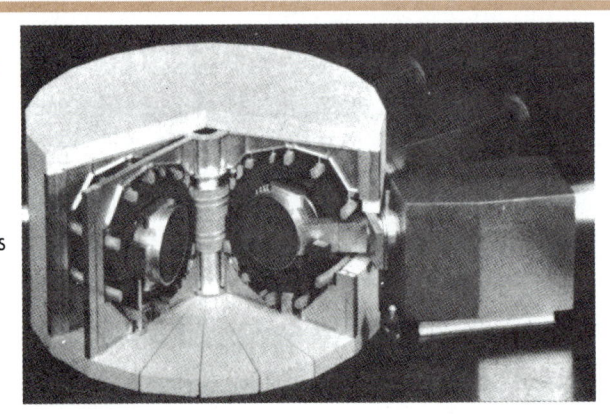

The trick is to maintain these high energies long enough for fusion to take place. In any ordinary container, the nuclei would quickly lose energy by colliding with the container walls. This stops the fusion process.

A possible design for a fusion reactor is shown in Figure 15.5. Here, the positively charged nuclei are confined in a strong magnetic field. This way, they cannot lose energy by colliding with the container walls. So far, with this technique, it has been possible to maintain fusion for only a fraction of a second. If fusion is to be a practical source of energy, this period will have to be extended to at least one second. It appears unlikely that fusion reactors will be available commercially before the year 2000 at the earliest.

15.5 Mass–Energy Relations

We have emphasized throughout this chapter that large amounts of energy are evolved in nuclear reactions. The energy change, ΔE, can be related to the mass change, Δm, by the Einstein equation

$$\Delta E = c^2 \, \Delta m \qquad (15.24)$$

Here, c is the velocity of light. In nuclear reactions, we find that ΔE and Δm are both negative quantities. This means that energy is evolved and that there is a decrease in mass. In radioactivity, fission, or fusion, the products weigh less than the reactants.

If we substitute in Equation 15.24 the value of c, 3.00×10^8 m/s, we can relate ΔE in *joules* to Δm in *kilograms*

$$\Delta E \text{ (in joules)} = (3.00 \times 10^8)^2 \times \Delta m \text{ (in kilograms)}$$
$$= 9.00 \times 10^{16} \times \Delta m \text{ (in kilograms)}$$

Usually we want ΔE in *kilojoules* corresponding to a mass change, Δm, in *grams*. Using the relations

$$1 \text{ kJ} = 10^3 \text{ J}; \; 1 \text{ g} = 10^{-3} \text{ kg}$$

we obtain

$$\Delta E \text{ (in kilojoules)} = 9.00 \times 10^{10} \times \Delta m \text{ (in grams)} \qquad (15.25)$$

The use of Equation 15.25 is illustrated in Examples 15.7 and 15.8.

EXAMPLE 15.7

Radium undergoes radioactive decay, emitting an alpha particle

$$^{226}_{88}\text{Ra} \rightarrow ^{222}_{86}\text{Rn} + ^{4}_{2}\text{He}$$

Continued

For every gram of radium that reacts, the total mass of products is 0.999 977 g. Calculate

a. Δm b. ΔE, using Equation 15.25

Solution
a. Δm = mass products − mass reactants
= 0.999 977 g − 1.000 000 g = −0.000 023 g = −2.3 × 10^{-5} g

b. ΔE (in kilojoules) = 9.00 × 10^{10} × (−2.3 × 10^{-5}) = −2.1 × 10^6 kJ

We conclude that 2.1 × 10^6 kJ of energy is evolved when one gram of radium undergoes radioactive decay.

Exercise
What is ΔE when one mole of Ra (226 g) decays? Answer: −4.7 × 10^8 kJ.

EXAMPLE 15.8
In the combustion of carbon

$$C(s) + O_2(g) \rightarrow CO_2(g)$$

9.0 kJ of energy is evolved for every gram of CO_2 that is formed. Calculate Δm for this reaction using Equation 15.25.

Solution
Solving Equation 15.25 for Δm

$$\Delta m \text{ (in grams)} = \frac{\Delta E \text{ (in kilojoules)}}{9.00 \times 10^{10}}$$

Substituting ΔE = −9.0 kJ,

$$\Delta m \text{ (in grams)} = \frac{-9.0}{9.00 \times 10^{10}} = -1.0 \times 10^{-10} \text{ g}$$

In other words, there is a decrease in mass in this reaction of 1.0 × 10^{-10} g. This means that the mass of the reactants (carbon and oxygen) exceeds that of the product (carbon dioxide) by 1.0 × 10^{-10} g. Taking the mass of CO_2 to be one gram, the mass of reactants is

1.000 000 000 1 g

As you might guess, this mass is so close to one gram that the difference cannot be detected, even with the most sensitive balance.

Exercise
How many grams of CO_2 would have to be formed to make Δm = −1.0 × 10^{-6} g? Answer: 10^4 g.

In terms of the magnitude of the mass change, Δm, we can distinguish between four different types of reactions.

TABLE 15.3 Mass-Energy Conversions

Process	Δm/gram reactant	ΔE/gram reactant
Combustion of carbon	-1.0×10^{-10} g	-9.0 kJ
Radioactive decay of Ra	-2.3×10^{-5} g	-2.1×10^6 kJ
Fission of U-235	-9.1×10^{-4} g	-8.2×10^7 kJ
Fusion of H-2	-6.4×10^{-3} g	-5.7×10^8 kJ

1. *Ordinary chemical reactions,* such as the combustion of carbon (Example 15.8). Here, Δm is too small to be detected. The fractional change in mass, Δm/m, where m is the mass of reactants, is of the order of -1×10^{-10}.
2. *Radioactive decay,* such as that of radium (Example 15.7). Here, Δm is still small, but large enough to be detected. The fractional change in mass is of the order of -2×10^{-5}.
3. *Fission of U-235.* In fission, Δm is much larger than in radioactive decay. The quantity Δm/m is of the order of -1×10^{-3}. In other words, about 0.1% of the original mass is "lost" (i.e., converted to energy) when uranium undergoes fission.
4. *Fusion* of very light isotopes such as $^{2}_{1}H$. Here, Δm has its largest value. The fractional change in mass is about -6×10^{-3}. About 0.6% of the total mass of reactants is converted to energy when deuterium undergoes fission (Table 15.3).

Key Words

alpha particle
beta radiation
cyclotron
Einstein equation
electron
fission
fusion
gamma radiation

half-life
isotope
neutron
nucleus
proton
rad
radioactivity
rem

Questions

15.1 List three general features in which nuclear reactions differ from ordinary chemical reactions.

15.2 What does alpha radiation consist of? beta radiation? gamma radiation?

15.3 When an alpha particle is emitted, atomic number _____ by _____ units; mass number _____ by _____ units.

15.4 What happens to atomic number and to mass number when an electron is emitted?

15.5 What effect does the emission of gamma radiation have on atomic number? mass number?

15.6 Compare alpha, beta, and gamma radiation as to range (penetrating power).

15.7 How does induced radioactivity differ from natural radioactivity?

15.8 Describe two different ways in which elements of very high atomic number can be produced.

15.9 What is meant by a rad? a rem?

15.10 For a given amount of energy absorbed, which is the more harmful to body tissue, alpha or beta radiation?

15.11 The average exposure to radiation of a person in the United States is about _____ rem.

15.12 The largest exposure to "manmade" radiation comes from _____ .

15.13 Explain what is meant by half-life.

15.14 Explain how it is possible to determine the age of rocks by a radioactive method.

15.15 Explain how it is possible to determine the age of organic material by a radioactive method.

15.16 What is meant by nuclear fission?

15.17 List two isotopes that undergo fission directly.

15.18 Explain how a nuclear reactor produces electrical energy.

15.19 What are some of the advantages of nuclear reactors over conventional power plants? the disadvantages?

15.20 What is meant by a breeder reactor?

15.21 What is nuclear fusion?

15.22 Compare the degree of radioactivity of the products of fission vs fusion reactions.

15.23 How does the energy from nuclear fusion compare to that from fission?

15.24 Why are there no fusion reactors operating at the present time?

15.25 When one gram of mass is converted to energy, _____ kJ of energy is produced.

15.26 Compare the mass changes in ordinary chemical reactions, radioactivity, fusion, and fission.

Problems

Nuclear Equations

(Use the Periodic Table where necessary.)

15.27 Write a balanced nuclear equation for
  a. emission of an alpha particle by uranium-235
  b. emission of an electron by cesium-137

15.28 Write a balanced nuclear equation for
  a. emission of an alpha particle by thorium-232
  b. emission of an electron by strontium-90

15.29 Complete the following nuclear equations
  a. $^{226}_{88}Ra \rightarrow \,^{4}_{2}He\, +$ _____
  b. $^{3}_{1}H \rightarrow \,^{0}_{-1}e\, +$ _____

15.30 Complete the following nuclear equations
  a. $^{196}_{83}Bi \rightarrow \,^{4}_{2}He\, +$ _____
  b. $^{28}_{13}Al \rightarrow \,^{0}_{-1}e\, +$ _____

15.31 Complete the following nuclear equations
  a. $^{32}_{16}S\, +$ _____ $\rightarrow \,^{32}_{15}P\, +\, ^{1}_{1}H$
  b. $^{9}_{4}Be\, +\, ^{4}_{2}He \rightarrow \,^{1}_{0}n\, +$ _____

15.32 Complete the following nuclear equations
  a. $^{24}_{12}Mg\, +$ _____ $\rightarrow \,^{27}_{14}Si\, +\, ^{1}_{0}n$
  b. $^{10}_{5}B\, +\, ^{1}_{1}H \rightarrow \,^{4}_{2}He\, +$ _____

15.33 Complete the following nuclear equations
  a. $^{239}_{94}Pu\, +\, ^{1}_{0}n \rightarrow \,^{90}_{37}Rb\, +$ _____ $+\, 3^{1}_{0}n$
  b. $^{3}_{1}H\, +\, ^{6}_{3}Li \rightarrow$ _____

15.34 A certain radioactive series starts with U-238 and ends with Pb-206. Each step in the series involves the emission of either an alpha particle or an electron. In the series, how many alpha particles are emitted? how many electrons?

15.35 A certain radioactive series starts with Np-237 and ends with Bi-209. How many alpha particles are emitted? how many electrons?

Rate of Radioactive Decay

15.36 Tritium, ^3_1H, has a half-life of 12 yr. Starting with 1.60 g of tritium, how much will be left after
a. 12 yr   b. 24 yr   c. 48 yr

15.37 The half-life of tritium, ^3_1H, is 12 yr. How long will it take for a sample weighing 2.80 g to decay to
a. 1.40 g   b. 0.35 g

15.38 The half-life of Be-7 is 53 days. What fraction of a sample of Be-7 will be left after
a. 53 days   b. 424 days

15.39 Taking the half-life of Be-7 to be 53 days, determine the time required for
a. ¾ of a sample to decay
b. ¹⁵⁄₁₆ of a sample to decay

15.40 The half-life of carbon-14 is 5720 yr. Using Figure 15.3, estimate the fraction of carbon-14 left after
a. 2000 yr   b. 8000 yr

15.41 Taking the half-life of C-14 to be 5720 yr, use Figure 15.3 to estimate the time required for a C-14 sample to decay to
a. 0.75 of its original value
b. 0.40 of its original value

15.42 Suppose the C-14/C-12 ratio in an ancient artifact is 0.65 of its value in living plants. Taking the half-life of C-14 to be 5720 yr and using Figure 15.3, estimate the age of the artifact.

15.43 Assume that the Shroud of Turin is genuine and was made in 30 A.D. How would the C-14/C-12 ratio in the shroud compare to that in living plants? Take the half-life of C-14 to be 5720 yr and use Figure 15.3.

Mass–Energy Relations

15.44 Calculate ΔE for processes in which Δm is
a. -1.4×10^{-8} g   b. -2.5×10^{-6} g

15.45 Calculate Δm for processes in which ΔE is
a. -3.4×10^6 kJ   b. -8.9×10^8 kJ

15.46 In the radioactive decay of Th-230, one gram of starting material is converted to 0.999 978 g of product. Calculate
a. Δm b. ΔE

15.47 In the radioactive decay of K-40, one gram of reactant is converted to 0.999 965 g of product. Calculate
a. Δm b. ΔE

15.48 In the combustion of hydrogen, 16.0 kJ of energy is evolved for every gram of water formed. Calculate Δm for this reaction.

15.49 In the decomposition of TNT, about 2.8 kJ of energy is evolved for every gram that explodes. Calculate Δm when one gram of TNT explodes.

15.50 When one gram of deuterium undergoes fusion, Δm is -6.4×10^{-3} g. For the fusion of one mole (2.02 g) of deuterium, calculate
a. Δm b. ΔE

Multiple Choice

15.51 The half-life of a certain radioactive species is 48 days. The fraction decayed after 96 days is
a. ⅛   b. ¼   c. ½   d. ¾   e. ⅞

15.52 It is found that 66 days is required for ⅞ of a certain radioactive species to decay. The half-life is
a. 22 d   b. 44 d   c. 66 d   d. 88 d
e. 132 d

15.53 A certain rock is found to contain 1.10 mol of U-238 and 1.05 mol of Pb-206. The half-life of U-235 is 4.5×10^9 yr. The age of the rock is
   a. less than 4.5×10^9 yr
   b. 4.5×10^9 yr
   c. greater than 4.5×10^9 yr

15.54 In the fission of $^{235}_{92}U$, the mass numbers of the two heavy nuclei produced add to
   a. less than 235   b. 235   c. more than 235

15.55 Among the advantages of nuclear fusion over fission are
   a. the products are not radioactive
   b. the evolution of energy is greater
   c. the starting materials are more abundant
   d. all of the above
   e. none of the above

15.56 When one gram of uranium-235 undergoes fission, the amount of energy evolved is
   a. 9×10^{10} kJ
   b. less than 9×10^{10} kJ
   c. more than 9×10^{10} kJ

15.57 Emission of how many of the following particles would result in a decrease in both atomic number and mass number?
   alpha particle, electron, proton, neutron
   a. 0   b. 1   c. 2   d. 3   e. 4

15.58 Which one of the following particles would not be accelerated in a cyclotron?
   a. proton   b. deuteron
   c. alpha particle   d. neutron

15.59 Exposure to radiation would increase least if you were to
   a. move from New York to Denver
   b. move closer to a nuclear power plant
   c. have a dental x-ray taken
   d. receive Co-60 treatment

15.60 Carbon-14, in its chemical behavior, would most closely resemble
   a. N-14   b. C-12   c. O-16
   d. F-19

Glossary

A

acid A substance that dissolves in water to produce a solution in which [H^+] is greater than 10^{-7} M. Examples of acids include HCl, HF, and HNO_3.

acid anhydride A nonmetal oxide that reacts with water to form an acid.

acid–base indicator A substance that changes color when pH changes; used to detect the point at which an acid–base reaction is complete.

acid dissociation constant See *dissociation constant*.

acid rain Rainfall in which the pH is less than 5, caused largely by dissolved sulfuric and nitric acids.

acidic solution Water solution in which [H^+] is greater than 10^{-7} M and the pH is less than 7.

actinides Elements of atomic number 89 (Ac) through 102 (No).

activation energy, E_a The minimum energy that molecules must have if collision is to result in reaction.

activity series A table, such as Table 12.1, in which oxidizing agents and reducing agents are listed in order of increasing (or decreasing) strength.

actual yield The amount of product obtained in a reaction.

alcohol An organic oxygen compound containing a —OH functional group. Examples include CH_3OH and CH_3CH_2OH.

aldehyde An organic oxygen compound containing an

$$-\overset{H}{\underset{}{C}}=O$$

functional group. An example is

$$CH_3-\overset{H}{\underset{}{C}}=O.$$

alkali metal A metal in Group 1 of the Periodic Table (Li, Na, K, Rb, Cs, Fr).

alkaline earth metal A metal in Group 2 of the Periodic Table (Be, Mg, Ca, Sr, Ba, Ra).

alkane A hydrocarbon in which all the carbon–carbon bonds are single, of general formula C_nH_{2n+2}. Examples include CH_4, C_2H_6, and C_4H_{10}.

alkene A hydrocarbon in which there is one carbon–carbon double bond in the molecule, of general formula C_nH_{2n}. Examples include C_2H_4 and C_4H_8.

alkyl group A hydrocarbon group such as $CH_3—$, $CH_3CH_2—$, or $CH_3CH_2CH_2—$.

alkyne A hydrocarbon in which there is one carbon–carbon triple bond in the molecule, of general formula C_nH_{2n-2}. Examples include C_2H_2 and C_4H_6.

allotropy Existence of an element in two or more forms with different properties in the same physical state. Example: $O_2(g)$ and $O_3(g)$ are allotropes of oxygen.

alpha particle A helium nucleus, $^4_2He^{2+}$.

anhydride A substance derived from another by removal of water. SO_3 is the anhydride of H_2SO_4; CaO is the anhydride of $Ca(OH)_2$.

anion A negatively charged ion such as Cl^- or NO_3^-.

anode An electrode at which oxidation occurs and toward which anions move. In the electrolysis of $CuCl_2(aq)$, Cl^- ions move to the anode where they are oxidized to Cl_2.

aromatic hydrocarbon A hydrocarbon containing one or more benzene rings.

Arrhenius acid A species which, upon addition to water, increases [H^+].

Arrhenius base A species which, upon addition to water, increases [OH^-].

atmosphere A unit of pressure equal to the pressure exerted by a mercury column 760 mm high; 1 atm = 760 mm Hg = 101.3 kPa.

atom The smallest particle of an element.

atomic mass A number that gives the relative mass of an atom of an element. Atomic masses are expressed relative to that of a C-12 atom, taken to have an atomic mass of exactly 12. A hydrogen atom (at. mass = 1.008) is 1.008/12 as heavy as a C-12 atom.

atomic number The number of protons in the nucleus of an atom. Every hydrogen atom has one proton in the nucleus; the atomic number of hydrogen is 1.

atomic spectrum A collection of lines at different wavelengths representing the light given off when electrons of excited atoms drop to lower energy levels.

Avogadro's Law A principle stating that equal volumes of different gases at the same temperature and pressure contain equal numbers of molecules.

Avogadro's number 6.022×10^{23}, the number of units in a mole.

B

balanced equation An equation representing a chemical reaction in which there is the same number of atoms on both sides. Usually, the equation with the smallest whole-number coefficients is referred to as *the* balanced equation.

barometer A device used to measure the pressure of the atmosphere.

base A substance that dissolves in water to produce a solution in which [OH$^-$] is greater than 10^{-7} M. Examples of bases include NaOH, Ca(OH)$_2$, and NH$_3$.

basic anhydride A metal oxide that reacts with water to form a base.

basic solution A water solution in which [OH$^-$] is greater than 10^{-7} M and the pH is greater than 7.

bent molecule A molecule in which there are three atoms that are not in a straight line. The H$_2$O molecule is bent.

beta radiation A stream of electrons emitted by radioactive nuclei.

Bohr model Model of the hydrogen atom derived by Niels Bohr.

boiling point The temperature at which vapor bubbles form in a liquid. At the boiling point, the vapor pressure becomes equal to the pressure over the liquid. The *normal* boiling point is that observed when the pressure is one atmosphere.

boiling point elevation Increase in the boiling point of a liquid caused by adding a nonvolatile solute. For water solutions of nonelectrolytes, the boiling point elevation is 0.52°C × m, where m is the molality.

bond A linkage between two atoms.

Boyle's Law A relation stating that for a sample of gas at constant temperature, volume is inversely proportional to pressure, i.e., V = k/P.

Brønsted–Lowry acid A species that donates a proton to another species. In the reaction HCl(aq) + H$_2$O → H$_3$O$^+$(aq) + Cl$^-$(aq), HCl is acting as a Brønsted–Lowry acid.

Brønsted–Lowry base A species that accepts a proton from another species. In the reaction just cited, H$_2$O acts as a Brønsted–Lowry base.

C

calorie A unit of energy equal to 4.184 joules.

carboxylic acid An organic oxygen compound containing the functional group
$$-\overset{\overset{\displaystyle O}{\|}}{C}-OH.$$
An example is acetic acid $CH_3-\overset{\overset{\displaystyle O}{\|}}{C}-OH$.

catalyst A substance that changes the rate of a reaction without being consumed by the reaction.

cathode An electrode at which reduction occurs and toward which cations move. In the electrolysis of CuCl$_2$, Cu^{2+} ions move to the cathode where they are reduced to Cu(s).

cation A positively charged ion such as Na$^+$, Fe^{3+}, or NH$_4^+$.

Celsius scale A temperature scale based on there being 100° between the freezing point of water (0°C) and the normal boiling point of water (100°C).

centi- Metric prefix meaning one hundredth. Example: 1 cm = 0.01 m = 10^{-2} m.

Charles' Law A relation stating that the volume of a gas sample at constant pressure is directly proportional to its absolute temperature.

chemical change A change in which a new substance is produced; commonly referred to as a "reaction."

chemical property Property of a substance observed when it undergoes a chemical change.

coefficient A number written before a formula in a chemical equation. The coefficients in an equation give the relative numbers of moles of reactants and products.

colligative property A physical property of a solution that depends primarily on the concentration of solute particles rather than the type of solute particles. Freezing point lowering and boiling point elevation are colligative properties.

complex ion An ion containing a central metal cation bonded to two or more anions or molecules called ligands. Examples include $Ag(NH_3)_2^+$, $Zn(H_2O)_4^{2+}$, and $Cr(H_2O)_6^{3+}$.

compound A substance that contains more than one kind of atom.

concentrated solution A solution that contains a relatively large amount of solute.

conversion factor A ratio, numerically equal to one, by which an initial quantity is multiplied to give a final quantity. Example: the conversion factor 10^3 g/1 kg, when multiplied by a mass in kilograms, converts that mass to grams.

coordination number The number of bonds formed by the central metal ion in a complex ion.

covalent bond A chemical link between two atoms, consisting of a pair of electrons shared by the two atoms. In the H_2 molecule, there is a covalent bond between the two hydrogen atoms.

cubic centimeter A volume unit equal to the volume of a cube 1 cm on an edge; 1 cm^3 = 10^{-3} L.

cyclotron A device used to accelerate charged particles to very high velocities (see Figure 15.2).

D

Dalton's Law A relation stating that the total pressure of a gas mixture is the sum of the partial pressures of the components of the mixture.

deci- Metric prefix meaning one tenth. Example: 1 dm = 0.1 m = 10^{-1} m.

density A property of a substance equal to its mass per unit volume. At room temperature, water has a density of about 1.00 g/cm^3, mercury a density of 13.6 g/cm^3.

deuterium A heavy isotope of hydrogen, ^2_1H.

dilute solution A solution that contains a relatively small amount of solute.

dissociation constant, K_a The equilibrium constant for the dissociation of a weak acid. For the weak acid HA,

$$K_a = \frac{[H^+] \times [A^-]}{[HA]}$$

distillation A procedure in which a liquid is heated to form a vapor that is then condensed and collected.

double bond Two shared electron pairs that form two covalent bonds between two atoms.

E

Einstein equation The relation $\Delta E = \Delta mc^2$ relating mass and energy changes.

electrode A general name for anode or cathode.

electrolysis A process in which a direct electric current is used to make a redox reaction take place at two electrodes of a cell.

electrolyte A substance that exists as ions in water solution. NaCl, KNO_3, and HCl are electrolytes; their water solutions conduct an electric current.

electrolytic cell The cell in which electrolysis occurs; electrical energy is used to bring about a redox reaction.

electron A component of an atom carrying a -1 charge; located outside the nucleus.

electron cloud A region of negative charge in an atom, associated with an orbital.

electron configuration A statement of the number of electrons in each sublevel of an atom. The electron configuration of neon is $1s^22s^22p^6$, which means that the neon atom contains two 1s electrons, two 2s electrons, and six 2p electrons.

electronegativity A property of an atom that is a measure of its attraction for electrons in a bond. Fluorine has a higher electronegativity than hydrogen (4.0 versus 2.1), so the electrons in the H–F bond are closer to F than to H.

element A substance containing only one kind of atom. Unlike a compound, an element cannot be broken down into two or more simpler substances.

endothermic A process that absorbs heat; ΔH for an endothermic process is positive.

enthalpy change, ΔH Difference in enthalpy between reactants and products. ΔH for a process is equal to the amount of heat absorbed (ΔH positive) or evolved (ΔH negative) in a process.

equilateral triangle molecule A molecule, such as BF_3, in which one atom is at the center of an equilateral triangle and three other atoms are at the corners of the triangle.

equilibrium A state of dynamic balance in which forward and reverse processes are occurring at the same rate, so the system does not change with time. The two processes may be physical changes (vaporization of a liquid versus condensation) or reactions ($Cl_2 \rightarrow 2Cl$ versus $2Cl \rightarrow Cl_2$).

equilibrium constant, K_c A number that expresses the relationship between the equilibrium concentrations of products and reactants. Example: for the system

$$Cl_2(g) \rightleftharpoons 2\,Cl(g) \text{ at } 25°C, K_c = [Cl]^2/[Cl_2]$$
$$= 1 \times 10^{-38}.$$

ester An organic oxygen compound containing the
$$-\underset{\underset{O}{\|}}{C}-O-$$
group. An ester can be produced by the reaction of an alcohol with a carboxylic acid.

ether An organic oxygen compound containing the —O— group. The simplest ether has the structural formula CH_3-O-CH_3.

excited state A state in which an electron has a higher energy than it does in the ground state.

exothermic A process that evolves heat; ΔH for an exothermic process is negative.

exponential notation A way of expressing numbers in the form $C \times 10^n$, where C is a number between one and ten and n is an integer which may be positive or negative (e.g., 2, 1, −1).

F

Fahrenheit scale A temperature scale in which there are 180° between the freezing point of water (32°F) and the normal boiling point of water (212°F).

fat A particular type of ester. A fat molecule can be decomposed into one molecule of glycerol and three molecules of long-chain carboxylic acids.

fission A nuclear reaction in which a heavy nucleus, ordinarily $^{235}_{92}U$ or $^{239}_{94}Pu$, is split by neutrons into two lighter nuclei, with the evolution of very large amounts of energy.

formula Expression showing the composition of a substance; a molecular formula uses subscripts to show the number of atoms of each type in the molecule.

formula mass Sum of the atomic masses of the atoms in a formula.

fractional distillation Procedure used to separate two volatile liquids that involves selective distillation and condensation.

freezing point The temperature at which a liquid changes to solid upon cooling.

freezing point lowering Decrease in the freezing point of a liquid caused by adding a solute. For water solutions of nonelectrolytes, the freezing-point lowering is 1.86°C × m, where m is the molality.

functional group A small group of atoms in an organic molecule which gives the molecule its distinctive chemical behavior. Example: the functional group —OH is common to all alcohols.

fusion A nuclear reaction in which two small nuclei combine to form a heavier nucleus, with the evolution of tremendous amounts of energy. Example: $^2_1H + ^2_1H \rightarrow ^4_2He$.

G

gamma radiation High-energy radiation emitted by radioactive nuclei.

gas law See *Ideal Gas Law*.

gram A unit of mass equal to the mass of one cubic centimeter of water at 4°C.

ground state The lowest energy state for an electron in an atom.

group In the Periodic Table, a vertical column of elements such as Li, Na, K, Rb, Cs, Fr (Group 1).

H

Haber process Industrial process used to make ammonia, NH_3, from the elements.

half-equation Equation written to describe a half-reaction of oxidation or reduction. The half-equation for the reduction of Cu^{2+} is $Cu^{2+}(aq) + 2e^- \rightarrow Cu(s)$.

half-life The time required for one half of a reactant to be converted to product.

halogen An element in Group 7 of the Periodic Table (F, Cl, Br, I, At).

heat of combustion ΔH when one mole of a substance burns in oxygen.

heat of formation ΔH when one mole of a compound is formed from the elements.

heat of fusion ΔH when unit amount, one gram or one mole, of a solid melts.

heat of vaporization ΔH when unit amount, one gram or one mole, of a liquid is converted to vapor.

heterogeneous Having a non-uniform composition.

homogeneous Having a uniform composition; a solution is a homogeneous mixture.

hydrate A solid containing H_2O molecules in the crystal lattice. Examples include $BaCl_2 \cdot 2H_2O$ (two moles of water per mole $BaCl_2$) and $CuSO_4 \cdot 5H_2O$ (five moles of water per mole of $CuSO_4$).

hydrocarbon A compound containing only hydrogen and carbon atoms.

hydrogen bond An attractive force between molecules in which a hydrogen atom is bonded to F, O, or N. The hydrogen bond exists between the hydrogen atom of one such molecule and the F, O, or N atom of another molecule.

hydronium ion The H_3O^+ ion.

I

Ideal Gas Law Relation between the pressure (P), volume (V), number of moles (n), and Kelvin temperature (T) of a gas: $PV = nRT$, where R is a constant with the value 0.0821 L·atm/(mol·K).

indicator, acid–base See *acid–base indicator*.

ion A charged species such as Cl^-, Na^+, Cr^{3+} or NO_3^-.

ionic bond The strong electrical force that holds oppositely charged ions such as Na^+ and Cl^- together in an ionic solid.

ionization energy The energy that must be absorbed to remove the outermost electron from a gaseous atom, usually expressed in kilojoules per mole.

isomer A species that has the same kind and number of atoms as another species, but with different properties. Example: CH_3-O-CH_3 (dimethyl ether) is isomeric with CH_3CH_2OH (ethyl alcohol).

isotope An atom with the same number of protons as another atom but a different number of neutrons; ^1_1H and ^2_1H are isotopes of hydrogen.

J

joule A unit of energy in the metric system; 4.18 J is required to raise the temperature of one gram of water one degree Celsius.

K

K_a See *dissociation constant*.

K_c See *equilibrium constant*.

K_w Equilibrium constant for the dissociation of water: $K_w = [H^+] \times [OH^-] = 1.0 \times 10^{-14}$ at 25°C.

Kelvin scale Absolute temperature scale in which 0 is taken to be the lowest possible temperature; the size of the Kelvin degree is the same as that of the Celsius degree.

ketone An organic oxygen compound containing the $-\overset{\overset{\displaystyle O}{\|}}{C}-$ functional group. The simplest ketone is acetone, $CH_3-\overset{\overset{\displaystyle O}{\|}}{C}-CH_3$.

kilo Metric prefix meaning one thousand; 1 km = 1000 m = 10^3 m.

kilojoule Unit of energy equal to 1000 joules.

kilopascal A pressure unit: 101.3 kPa = 1 atm. On a typical day, atmospheric pressure is approximately 100 kPa.

kinetic molecular theory Model of molecular motion used to explain many of the properties of gases.

L

lanthanides Elements of atomic number 57 (La) through 70 (Yb).

Law of Combining Volumes Relation that states that the volumes of gases taking part in a reaction are in a ratio of small whole numbers, given by the coefficients of the balanced equation (all gases at same T and P).

Law of Conservation of Energy Statement that energy is neither created nor destroyed in any process.

Law of Conservation of Mass Statement that in any process the mass of products is equal to that of the starting materials.

Law of Constant Composition Statement that a given compound always contains the same elements with the same mass percents. Example: All samples of water contain 11.19% hydrogen and 88.81% oxygen.

Le Chatelier's Principle Statement that when a system at equilibrium is disturbed, it will respond in such a way as to partially counteract the change.

Lewis acid A species that can accept a pair of electrons. In the reaction $BF_3(g) + NH_3(g) \rightarrow BF_3NH_3$, BF_3 acts as a Lewis acid because it accepts a pair of electrons from the nitrogen atom of NH_3.

Lewis base A species that donates a pair of electrons in an acid–base reaction. In the reaction just cited, NH_3 acts as a Lewis base.

Lewis structure A diagram of an atom or molecule that shows how the valence electrons are distributed, either as bonds or unshared electron pairs.

ligand A molecule or anion bonded to a central metal cation in a complex.

limiting reactant The reactant whose amount sets the limit for the amount of product that can be formed. Theoretical yield is based on the amount of limiting reactant.

linear molecule A molecule in which all the atoms are in a straight line.

liter A unit of volume; 1 L = 1000 cm^3.

M

macromolecule A huge molecule in which a large number of atoms are bonded to one another by a network of covalent bonds.

main-group element An element in one of the numbered groups of the Periodic Table (Groups 1–8).

mass A property that reflects the amount of matter in a sample.

mass number An integer equal to the sum of the protons and neutrons in the nucleus of an atom. The mass number of $^{35}_{17}Cl$ (17 protons, 18 neutrons) is 35.

mass percent The mass of a component in a 100-g sample.

matter Anything that has mass and occupies space.

mega- Metric prefix meaning one million; 1 Mg = 10^6 g.

melting point The temperature at which a solid liquifies upon heating.

metal An element that is a good electrical conductor in the solid state. Metals readily lose electrons to form cations.

metallic character A measure of the extent to which an element shows the properties of metals.

metalloid An element whose properties are intermediate between those of metals and nonmetals.

metallurgy The science and processes of extracting metals from their ores.

meter Unit of length in the metric system.

micro- Metric prefix meaning one millionth; 1 μm = 10^{-6} m.

milli- Metric prefix meaning one thousandth; 1 mm = 10^{-3} m.

millimeter of mercury Unit of pressure, equal to the pressure exerted by a column of mercury one millimeter high.

molality A concentration unit defined as the number of moles of solute per kilogram of solvent. A solution made by dissolving 0.10 mol of NaCl in 500 g of water would have a molality of 0.20.

molar mass Mass of one mole of a substance, expressed in grams per mole. The substance O_2 has a molar mass of 32.00 g/mol.

molar volume Volume of one mole of a substance. The molar volume of a gas at standard temperature and pressure is 22.4 L/mol.

molarity A concentration unit defined as the number of moles of solute per liter of solution. A solution made by dissolving 0.10 mol of NaCl to form 500 cm^3 of solution would have a molarity of 0.20.

mole Avogadro's number of items, 6.022 × 10^{23}. One mole of the element carbon contains 6.022 × 10^{23} carbon atoms and weighs 12.01 g. One mole of H_2O contains 6.022 × 10^{23} H_2O molecules and weighs 18.02 g.

molecular formula An expression that gives the number and kind of each atom present in a molecule. The molecular formula of ethane is C_2H_6; there are two carbon and six hydrogen atoms in a molecule of ethane.

molecule A small group of atoms held together by covalent bonds. A molecule of hydrogen (H$_2$) contains two hydrogen atoms; a molecule of water (H$_2$O) contains two hydrogen atoms and one oxygen atom.

monomer The small molecule from which a polymer is formed.

N

nano- Metric prefix meaning one billionth; 1 nm = 10^{-9} m.

net ionic equation A chemical equation for a reaction in which only those species which take part in the reaction are included. The net ionic equation for the reaction between solutions of HCl and NaOH is H$^+$(aq) + OH$^-$(aq) → H$_2$O.

neutral solution A water solution in which [H$^+$] = [OH$^-$]; pH = 7.

neutralization The reaction between a strong acid and a strong base, represented by the equation H$^+$(aq) + OH$^-$(aq) → H$_2$O.

neutron A particle in an atomic nucleus which has zero charge and a mass of one unit.

noble gas Element in Group 8 at the far right of the Periodic Table.

nonelectrolyte A substance that exists as molecules rather than ions in water solution. A solution of a nonelectrolyte does not conduct an electric current.

nonmetal An element that does not show metallic properties. Nonmetals are located toward the upper right of the Periodic Table.

nonpolar Not having + and − poles. Nonpolar molecules include H$_2$ and CH$_4$, in both of which the electron distribution is symmetrical.

nonspontaneous reaction A reaction that cannot occur by itself without work being supplied from an outside source. The decomposition of water to the elements is a nonspontaneous reaction.

normal boiling point The temperature at which a liquid boils when the pressure above it is one atmosphere.

nuclear symbol Symbol used to show the mass number and atomic number of a nucleus. Examples include $^1_1$H and $^2_1$H.

nucleus The small, dense, positively charged region at the center of an atom.

O

octahedron A solid with six vertices and eight sides, all of which are equilateral triangles (see Figure 13.2).

octet rule The general principle that atoms in a molecule tend to be surrounded by eight electrons.

orbital A region of space in which an electron is likely to be found. Each orbital (s, p, d, or f) has a capacity of two electrons.

orbital diagram A sketch that shows the number of electrons in each orbital of an atom and the spin of each electron.

organic A compound that contains carbon, hydrogen, and possibly other atoms.

oxidation Loss of electrons or increase in oxidation number. The conversion of Zn to Zn^{2+} is an oxidation.

oxidation number A number assigned to an atom in a molecule or ion which reflects the number of electrons controlled by that atom. Oxidation numbers are ordinarily assigned by applying rather arbitrary rules (e.g., oxid. no. H = +1, oxid. no. O = −2).

oxidizing agent A species that accepts electrons from another species. In the reaction Zn(s) + Cu^{2+}(aq) → Zn^{2+}(aq) + Cu(s), the Cu^{2+} ion is an oxidizing agent.

oxyacid An acid that contains oxygen such as HNO$_3$ or H$_2$SO$_4$.

oxyanion A negative ion that contains oxygen, derived from an oxyacid. Examples include NO$_3^-$, derived from HNO$_3$, and SO$_4^{2-}$, derived from H$_2$SO$_4$.

P

partial pressure The pressure a component of a mixture would exert if it occupied the entire volume by itself at the same temperature.

percent composition Percents by mass of the elements in a compound.

percent yield Quantity equal to 100 times the ratio of the actual yield to the theoretical yield.

period A horizontal row in the Periodic Table.

Periodic Table An arrangement of elements, in order of increasing atomic number, in horizontal periods of such length that elements with similar chemical properties fall directly beneath one another in vertical groups.

pH A quantity equal to $-\log[H^+]$.

physical change A change that does not involve the formation of a new substance.

physical property A property of a substance observed when it undergoes a physical change.

polar Having + and − poles. The HF molecule is polar, because the electrons in the H–F bond are not symmetrically distributed. There is a positive pole at the H atom and a negative pole at the F atom.

polyatomic ion An ion that contains more than one atom. Examples include OH^- and NO_3^-.

polymer A macromolecule made of many small molecules (monomers) bonded to one another. The polymer derived from ethylene is called polyethylene.

posttransition metal Metal in Groups 3, 4, and 5, following the transition metals in Periods 4, 5, and 6. Ga, Sn, and Bi are posttransition metals.

precipitate A solid that settles out when two water solutions are mixed.

principal energy level Main energy level in which electrons are located. Each principal level has a total capacity of $2n^2$ electrons, where n is the quantum number designating the level.

product A species formed as the result of a chemical reaction. Formulas of products appear on the right side of the equation written to represent a reaction.

proton A particle in an atomic nucleus which has a charge of +1 and a mass of one unit.

pyramidal molecule A molecule such as NH_3 in which one atom is located at the apex of a pyramid and three other atoms form a triangular base.

Q

quantum mechanics Approach used to determine the energies and approximate locations of electrons in atoms.

quantum number A number used to designate the energy state of an electron in an atom.

R

rad A unit of absorbed radiation equal to 10^{-2} joules per kilogram of tissue.

radioactivity The process by which an element undergoes a nuclear reaction to form a more stable nucleus.

rate of reaction A ratio that tells how the concentration of a species in a reaction changes with time.

reactant The starting material in a reaction. Formulas of reactants are written on the left side of the equation written to represent a reaction.

redox Abbreviation for oxidation–reduction.

reducing agent A species that furnishes electrons to another species. In the reaction, $Zn(s) + Cu^{2+}(aq) \rightarrow Zn^{2+}(aq) + Cu(s)$, Zn acts as a reducing agent.

reduction Gain of electrons or decrease in oxidation number. The conversion of Cu^{2+} to Cu is a reduction.

rem Unit of absorbed radiation. The number of rems absorbed is found by multiplying the number of rads by a factor varying from one to ten.

S

saturated hydrocarbon A hydrocarbon in which all the carbon–carbon bonds are single bonds; an alkane.

saturated solution A solution that contains the maximum amount of solute that the solvent can hold at equilibrium.

significant figure Meaningful digit in an experimental measurement. A mass of 1.243 g, measured to the nearest milligram, contains four significant figures.

simplest formula Formula used to show the relative numbers of different kinds of atoms in a compound. The simplest formula of water is H_2O; that of hydrogen peroxide is HO.

single bond A pair of electrons shared between two atoms.

slag Molten waste product formed in metallurgy. In making iron, SiO_2 impurities are removed as a slag of $CaSiO_3$.

solubility rules Rules used to classify ionic compounds as to their water solubility (see Table 9.2).

solute Solution component present in smaller amount than a solvent. When solids or gases are dissolved in a liquid, the solid or gas is always called the solute.

solution Homogeneous mixture of two or more substances.

solvent A substance, most often a liquid, in which another substance, called a solute, is dissolved.

specific gravity The ratio of the density of a substance to that of a reference substance, ordinarily water. The specific gravity of a substance is numerically equal to its density in grams per cubic centimeter.

spectator ion An ion present in solution which does not take part in a chemical reaction and so is not included in the net ionic equation written to represent the reaction.

spin A property of an electron loosely related to its spin around an axis. An electron can have either of two different spins.

spontaneous reaction A reaction that takes place by itself without the input of work from the outside. The reaction of hydrogen with oxygen to form water is spontaneous.

stoichiometry Having to do with the relation between masses (moles, grams) of substances taking part in a reaction.

STP Standard temperature (0°C) and pressure (1 atm).

strong acid An acid that is completely dissociated to form H^+ ions in water solution.

strong base A base that is completely dissociated to form OH^- ions in water solution.

structural formula A formula that shows the pattern in which atoms are bonded in a molecule.

structural isomer A substance that has the same molecular formula as another substance but a different structural formula.

sublevel A subdivision of a principal energy level. Sublevels are designated by the letters s, p, d, and f.

supersaturated solution A solution that contains more solute than the saturated solution.

symbol One or two letters used to represent an element. The element helium has the symbol He.

T

tetrahedral molecule A molecule in which the four bonds around a central atom are directed toward the corners of a regular tetrahedron.

tetrahedron A figure with four sides, all of which are equilateral triangles, and four vertices.

theoretical yield The amount of product that would be formed in a reaction if the limiting reactant were completely consumed.

thermochemical equation A chemical equation in which the value of ΔH is specified.

titration A process in which the volume of one solution required to react with a known volume of another solution is measured. Knowing the concentration of one of the solutions, it is possible to calculate the concentration of the other solution.

transition element A metal in the central groups of the fourth, fifth, sixth, and seventh periods of the Periodic Table. In the first transition series, the metals start with Sc (at. no. = 21) and go through Zn (at. no. = 30).

triple bond Three electron pairs shared between two atoms.

U

unsaturated hydrocarbon A hydrocarbon in which there are carbon–carbon double or triple bonds.

unsaturated solution A solution that contains less solute than the saturated solution.

unshared pair A pair of electrons not involved in bond formation.

V

valence electron An s or p electron in the outermost principal energy level.

vapor A gas formed by vaporization of a liquid.

vapor pressure The pressure exerted by a vapor in equilibrium with the corresponding liquid. The vapor pressure of water is 24 mm Hg at 25°C and increases with temperature.

voltaic cell A device in which a spontaneous redox reaction produces electrical energy.

volatile Easily vaporized.

W

weak acid An acid that is only partially dissociated to H^+ ions in water solution.

weak base A base that is only partially dissociated to OH^- ions in water solution.

APPENDIX 1

Review of Mathematics

The mathematics required for this course is relatively simple. Frequently you will carry out operations involving exponential numbers. These are discussed in Chapter 1, pp. 1–7. Beyond that, you will often deal with the following topics discussed in this appendix.

1. *Percent,* referred to in Chapters 2, 4, 5, and 9.
2. *Proportionality,* both direct and inverse, used in Chapter 6 and, to a lesser extent, in Chapter 11.
3. *Algebraic Equations,* used in Chapters 6, 9, 10, and 11.
4. *Graphs,* considered in Chapter 6 and, especially, in Chapter 15. (All of these topics and several others are discussed in greater detail in *Mathematical Preparation for General Chemistry,* by Masterton and Slowinski, Saunders College Publishing, 2nd edition, 1982.)

Before considering these topics, let's look at what has become an essential tool for the general chemistry course: a hand-held calculator.

Calculators

Many different calculator models are available, varying in price and versatility. For general chemistry, the most useful type is one referred to as a "scientific" or "slide rule" calculator. It sells for $20–30 and can be used for many purposes beyond simple arithmetic operations. You can use a scientific calculator to

—enter and perform operations on numbers expressed in exponential notation.
—find a common (base 10) logarithm or antilogarithm (number corresponding to a given logarithm).
—raise a number to any power, n, or extract the nth root of a number.

Scientific calculators vary somewhat from one model to another. The first thing you should do after buying a calculator is to read the instruction manual that comes with it. At the least you should be sure that you can

1. *Detect when the batteries are getting weak.* All calculators have some

Appendix 1

sort of "early warning" system that indicates this. Find out what the system is and pay attention to it. The last thing you need is a calculator that dies midway through an examination.

2. *Enter successive numbers.* Most calculators do this in a standard way, referred to as "algebraic notation." Throughout this discussion, we will assume this is the system used with your calculator. If you happen to have a Hewlett-Packard model (very few students do nowadays), the sequence of operations is somewhat different.

3. *Remove entries.* There are at least two different ways to "clear" or "erase" on a calculator. One is used when you press the wrong key in the middle of a calculation and want to make a single correction. Another, more powerful clearing operation is used to erase an entire calculation and start over again.

Arithmetic Operations

Among the simplest calculator operations are those of *addition, subtraction, multiplication,* and *division*. Make sure you know how to carry out these operations by working through Example 1 on your calculator.

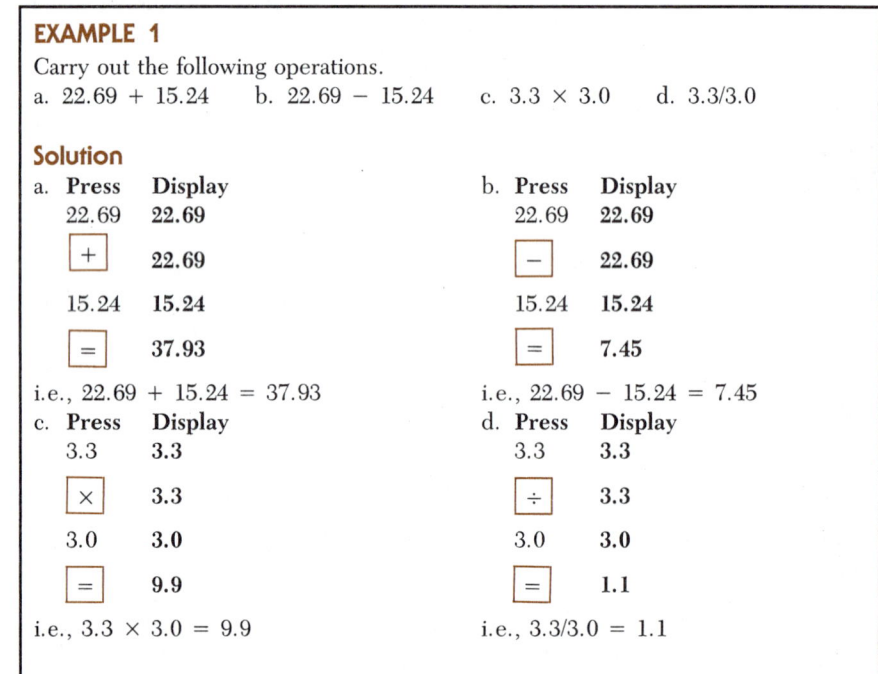

EXAMPLE 1
Carry out the following operations.
a. $22.69 + 15.24$ b. $22.69 - 15.24$ c. 3.3×3.0 d. $3.3/3.0$

Solution

a. | Press | Display |
   |-------|---------|
   | 22.69 | 22.69 |
   | $+$ | 22.69 |
   | 15.24 | 15.24 |
   | $=$ | 37.93 |

i.e., $22.69 + 15.24 = 37.93$

b. | Press | Display |
   |-------|---------|
   | 22.69 | 22.69 |
   | $-$ | 22.69 |
   | 15.24 | 15.24 |
   | $=$ | 7.45 |

i.e., $22.69 - 15.24 = 7.45$

c. | Press | Display |
   |-------|---------|
   | 3.3 | 3.3 |
   | $\times$ | 3.3 |
   | 3.0 | 3.0 |
   | $=$ | 9.9 |

i.e., $3.3 \times 3.0 = 9.9$

d. | Press | Display |
   |-------|---------|
   | 3.3 | 3.3 |
   | $\div$ | 3.3 |
   | 3.0 | 3.0 |
   | $=$ | 1.1 |

i.e., $3.3/3.0 = 1.1$

Exercise
Carry out the following operations, expressing your answers to four significant figures: 22.69×15.24; $22.69/15.24$. Answer: 345.8, 1.489.

Appendix 1 **449**

Successive Multiplications and Divisions

Many calculations in chemistry require more than one step. Frequently you will need to carry out two or more operations in succession, most often multiplication and division. This can be done on your calculator in a single, continuous operation. Example 2 illustrates the method used.

EXAMPLE 2

Carry out the following operations, expressing all answers to three significant figures.

a. $1.21 \times 2.62 \times 1.95$ b. $\dfrac{1.21 \times 2.62}{2.18}$ c. $\dfrac{1.21 \times 2.62}{2.18 \times 1.42}$

Solution

a. | Press | Display |
|---|---|
| 1.21 | 1.21 |
| $\times$ | 1.21 |
| 2.62 | 2.62 |
| $\times$ | 3.17. . |
| 1.95 | 1.95 |
| = | 6.18. . . |

i.e., $1.21 \times 2.62 \times 1.95 = 6.18$

b. | Press | Display |
|---|---|
| 1.21 | 1.21 |
| $\times$ | 1.21 |
| 2.62 | 2.62 |
| $\div$ | 3.17. . |
| 2.18 | 2.18 |
| = | 1.45. . . . . |

i.e., $\dfrac{1.21 \times 2.62}{2.18} = 1.45$

c. | Press | Display |
|---|---|
| 1.21 | 1.21 |
| $\times$ | 1.21 |
| 2.62 | 2.62 |
| $\div$ | 3.17. . |
| 2.18 | 2.18 |
| $\div$ | 1.45. . . . . |
| 1.42 | 1.42 |
| = | 1.02. . . . . |

i.e., $\dfrac{1.21 \times 2.62}{2.18 \times 1.42} = 1.02$

Exercise

Evaluate: $\dfrac{1.90 \times 6.14 \times 3.92}{8.06 \times 1.47}$. Answer: 3.86.

Appendix 1

Raising to a Power or Extracting a Root

All scientific calculators have special keys for

—squaring a number; use the $\boxed{x^2}$ key.

—taking the reciprocal of a number; use the $\boxed{1/x}$ key.

—taking the square root of a number; use the $\boxed{\sqrt{x}}$ key.

EXAMPLE 3
Evaluate
a. $(4.0)^2$ b. $(4.0)^{-1}$ c. $(4.0)^{1/2}$

Solution
a. Press 4.0, then $\boxed{x^2}$; the answer appears as 16.
b. Here you use the fact that $x^{-1} = 1/x =$ reciprocal of x.
   Press 4.0, then $\boxed{1/x}$; the answer is 0.25.
c. To evaluate $(4.0)^{1/2}$, note that $x^{1/2} = \sqrt{x}$.
   Press 4.0, then $\boxed{\sqrt{x}}$; the answer is 2.0.

Exercise
Evaluate, to three significant figures, $(7.25)^{-1}$ and $(7.25)^{1/2}$ Answer: 0.138, 2.69.

In this course, you will seldom have occasion to find powers and roots other than those just described. In case you do, there is a general approach that you can use. It involves using the $\boxed{y^x}$ key found on all scientific calculators.

EXAMPLE 4
Find the value of
a. $(5.00)^3$ b. $(5.00)^{-3}$ c. $(5.00)^{1/3}$

Solution

a. | Press | Display |
   |---|---|
   | 5.00 | 5.00 |
   | $\boxed{y^x}$ | 5.00 |
   | 3 | 3 |
   | $\boxed{=}$ | 125 |

b. | Press | Display |
   |---|---|
   | 5.00 | 5.00 |
   | $\boxed{y^x}$ | 5.00 |
   | 3 | 3 |
   | $\boxed{+/-}$ | −3 |

Continued

i.e., $(5.00)^3 = 125$

| | = | 0.00800 |

i.e., $(5.00)^{-3} = 0.00800$

Note that to raise to a negative power such as -3 you press the $\boxed{+/-}$ key to change the sign of the exponent.

c. Here, a simple approach starts by expressing ⅓ as a decimal.

$$\frac{1}{3} = 0.33333333$$

Press	Display
5.00	5.00
$\boxed{y^x}$	5.00
0.33333333	0.33333333
$\boxed{=}$	1.71. . . . .

i.e., $(5.00)^{1/3} = 1.71\ldots$

Exercise

Evaluate, to three significant figures, $(7.25)^{10}$, $(7.25)^{-10}$, $(7.25)^{1/5}$. Answer: 4.01×10^8, 2.49×10^{-9}, 1.49.

Exponential Notation

In working the Exercise following Example 4, the first two answers appeared on your calculator as

4.01 . . . 08 and 2.49. . −09

These are interpreted as the exponential numbers 4.01×10^8 and 2.49×10^{-9}. A scientific calculator automatically shifts to exponential notation when the answer is too large or too small to fit in the display as an ordinary number.

It is possible to enter numbers directly in exponential notation on a scientific calculator. To do this you use a key which may be labeled $\boxed{EE \downarrow}$ (most common) or $\boxed{EEX}$ or $\boxed{EXP}$. To enter the numbers 6.2×10^4 and 6.2×10^{-4}, you proceed as follows

Press	Display		Press	Display
6.2	6.2		6.2	6.2
$\boxed{EE \downarrow}$	6.2 00		$\boxed{EE \downarrow}$	6.2 00
4	6.2 04		4	6.2 04
			$\boxed{+/-}$	6.2 −04

Once you have entered a number in exponential notation, you can carry out any operation with it in the usual way. To multiply or divide 6.2×10^{-4} by 3.1×10^6, you would proceed as follows

Appendix 1

Multiplication			Division	
Press	Display		Press	Display
6.2	6.2		6.2	6.2
EE ↓	6.2 00		EE ↓	6.2 00
4	6.2 04		4	6.2 04
+/−	6.2 −04		+/−	6.2 −04
×	6.2 −04		÷	6.2 −04
3.1	3.1		3.1	3.1
EE ↓	3.1 00		EE ↓	3.1 00
6	3.1 06		6	3.1 06
=	1.9 03		=	2.0 −10

i.e., $(6.2 \times 10^{-4}) \times (3.1 \times 10^6) = 1.9 \times 10^3$

i.e., $\dfrac{(6.2 \times 10^{-4})}{(3.1 \times 10^6)} = 2.0 \times 10^{-10}$

These procedures are tedious; nine steps are involved in each case. Multiplication and division of exponential numbers can be shortened considerably by applying the rules cited in Chapter 1 (to multiply, *add* exponents; to divide, *subtract* exponents).

Exponential notation on your calculator is perhaps most useful in two situations

—adding or subtracting exponential numbers when the exponents differ (Example 5a).

—taking the square root of an exponential number when the exponent is not evenly divisible by two (Example 5b).

EXAMPLE 5
Using exponential notation on your calculator, evaluate
a. $2.6 \times 10^4 + 3 \times 10^3$ b. $(3.6 \times 10^{-5})^{1/2}$

Solution

a. | Press | Display |
    |-------|---------|
    | 2.6 | 2.6 |
    | EE ↓ | 2.6 00 |
    | 4 | 2.6 04 |
    | + | 2.6 04 |
    | 3 | 3 |
    | EE ↓ | 3   00 |

b. | Press | Display |
    |-------|---------|
    | 3.6 | 3.6 |
    | EE ↓ | 3.6 00 |
    | 5 | 3.6 05 |
    | +/− | 3.6 −05 |
    | √x | 6.0 −03 |

i.e., $(3.6 \times 10^{-5})^{1/2} = 6.0 \times 10^{-3}$

Continued

3	3	03
$\boxed{=}$	2.9	04

i.e., $2.6 \times 10^4 + 3 \times 10^3 = 2.9 \times 10^4$

Exercise
Evaluate $(4.5 \times 10^{-3} - 5 \times 10^{-4})$ and $(2.8 \times 10^7)^{1/2}$. Answer: 4.0×10^{-3}, 5.3×10^3.

Logarithms and Antilogarithms

You can find the logarithm of a number by pressing the $\boxed{\text{LOG}}$ key on your calculator. In this way, you should find that

$\log 2.00 = 0.301..$ $\log 0.0100 = -2.000$
$\log 5.00 = 0.699..$ $\log 0.129 = -0.889.....$
$\log 13.00 = 1.114...$ $\log 0.900 = -0.046.....$

It is also possible to take the logarithm of an exponential number of the form

$C \times 10^n$

on your calculator. Here you have two choices

1. Enter the number in exponential notation and press $\boxed{\text{LOG}}$.
2. Find log C on your calculator and apply the rule:
   $\log (C \times 10^n) = \log C + n$.

EXAMPLE 6
Find the logarithm of 6.4×10^{-3} by the two methods just described.

Solution

Method 1
Press	Display
6.4	6.4
$\boxed{\text{EE}\downarrow}$	6.4 00
3	6.4 03
$\boxed{+/-}$	6.4 -03
$\boxed{\text{LOG}}$	$-2.19..$

i.e., $\log (6.4 \times 10^{-3}) = -2.19$

Method 2
Press	Display
6.4	6.4
$\boxed{\text{LOG}}$	$0.81......$

$\log (6.4 \times 10^{-3}) = \log 6.4 - 3 = 0.81 - 3 = -2.19$

Exercise
Find the logarithm of 5.2×10^{-9}. Answer: -8.28.

Appendix 1

You can also use a scientific calculator to find an antilogarithm, i.e., the number corresponding to a given logarithm. The way this is done depends on the type of calculator you have. With some models, you press the $\boxed{10^x}$ key; with others you press $\boxed{\text{INV}}$ and then $\boxed{\text{LOG}}$. Thus, to find the antilogarithm of 0.301 you either

Press	Display		Press	Display
0.301	0.301		0.301	0.301
		or		
$\boxed{10^x}$	2.00.		$\boxed{\text{INV}}$	0.301
			$\boxed{\text{LOG}}$	2.00. . . . .

Either way, you conclude that the number 2.00 has a logarithm of 0.301.

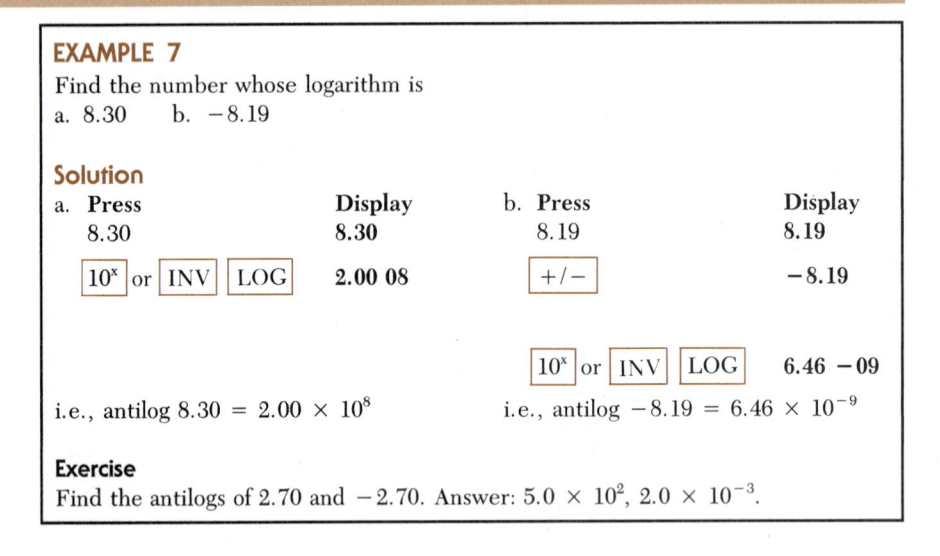

EXAMPLE 7
Find the number whose logarithm is
a. 8.30   b. -8.19

Solution

a. Press   Display   b. Press   Display
    8.30   8.30   8.19   8.19

    $\boxed{10^x}$ or $\boxed{\text{INV}}$ $\boxed{\text{LOG}}$ 2.00 08 $\boxed{+/-}$ -8.19

                                                                            $\boxed{10^x}$ or $\boxed{\text{INV}}$ $\boxed{\text{LOG}}$ 6.46 -09

i.e., antilog $8.30 = 2.00 \times 10^8$ i.e., antilog $-8.19 = 6.46 \times 10^{-9}$

Exercise
Find the antilogs of 2.70 and -2.70. Answer: 5.0×10^2, 2.0×10^{-3}.

As these examples show, a calculator is a powerful tool for carrying out operations in mathematics. However, you must be aware of its limitations. In particular, many of the digits shown in the display are not meaningful. Ordinarily, you have to round off your answer by following the rules of significant figures discussed in Chapter 1.

Percent

Percent means "parts per hundred." A candidate who receives 51% of the vote in an election gets 51/100 of the total vote. In general, the percent of a component A in a sample can be determined by comparing the amount of A to the total amount of the sample. The general relation is

$$\% \text{ of A} = \frac{\text{amount of A}}{\text{total amount}} \times 100 \qquad (1)$$

The percents of all the components (A, B, C, – – –) in a sample when added, give 100

$$\% \text{ of A} + \% \text{ of B} + \% \text{ of C} + -- = 100 \qquad (2)$$

This equation tells us that if we add up the percent of the votes obtained by all the candidates in an election, the sum must be 100.

EXAMPLE 8

Analysis of a sample of salt solution weighing 12.64 g shows that it contains 11.45 g of water. What are the mass percents of water and salt in the solution?

Solution

To find the mass percent of water we apply Equation 1

$$\% \text{ water} = \frac{\text{amount water}}{\text{total amount sample}} \times 100 = \frac{11.45 \text{ g}}{12.64 \text{ g}} \times 100 = 90.59$$

To find the mass percent of salt, we apply Equation 2

$\%$ water + $\%$ salt = 100; $\%$ salt = 100.00 – $\%$ water = 100.00 – 90.59 = 9.41

We conclude that the sample contains 90.59% water and 9.41% salt.

Exercise

A sample of 2.193 g of water upon electrolysis gives 1.948 g of oxygen; the only other element present in water is hydrogen. What are the mass percents of oxygen and hydrogen in water? Answer: 88.83% O, 11.17% H.

Two other useful relations can be obtained from Equation 1. We can solve for "amount of A" to obtain

$$\text{amount of A} = \text{total amount} \times \frac{\% \text{ of A}}{100} \qquad (3)$$

Alternatively, we can solve for "total amount"

$$\text{total amount} = \text{amount of A} \times \frac{100}{\% \text{ of A}} \qquad (4)$$

The use of Equations 3 and 4 is shown in Example 9.

EXAMPLE 9

The compound sodium chloride contains 39.34% by mass of sodium.

Continued

> a. How much sodium can be obtained by the electrolysis of 16.25 g of molten sodium chloride?
> b. How much sodium chloride is required to produce 6.19 g of sodium?
>
> **Solution**
> a. Applying Equation 3
>
> $$\text{amount of Na} = \text{total amount NaCl} \times \frac{\%\text{ of Na}}{100} = 16.25 \text{ g} \times \frac{39.34}{100} = 6.393 \text{ g}$$
>
> b. Applying Equation 4
>
> $$\text{total amount of NaCl} = \text{amount Na} \times \frac{100}{\%\text{ of Na}} = 6.19 \text{ g} \times \frac{100}{39.34} = 15.7 \text{ g}$$
>
> **Exercise**
> Using the answer to the Exercise following Example 8, calculate the mass of hydrogen formed from 1.000 g of water; the mass of water required to form 1.000 g of hydrogen. Answer: 0.1117 g H; 8.953 g H$_2$O.

As we have seen, the relative amount of a component in a sample can be expressed in terms of percent, defined by Equation 1. Alternatively, we may cite the *fraction* of the component in the sample

$$\text{fraction A} = \frac{\text{amount of A}}{\text{total amount}} \tag{5}$$

Comparing Equations 1 and 5, we see that

$$\text{fraction A} = \frac{\%\text{ of A}}{100} \tag{6}$$

In other words

—since the percent of sodium in sodium chloride is 39.34, the fraction of sodium is 0.3934.
—if a salt solution contains 90.59% water, the fraction of water is 0.9059.

Just as the percents of all the components in a sample must add to 100, so the fractions must add to one.

$$\text{fraction A} + \text{fraction B} + \text{fraction C} + \text{---} = 1 \tag{7}$$

Proportionality

Direct Proportionality

When we say that a quantity, y, is directly related to another quantity, x, we mean that the two quantities change in the same direction. If x increases, y increases; if x decreases, y decreases. By saying that y is directly proportional to x, we go a step further. Not only does y change in the same direction as x; it changes in the same ratio. If y is directly proportional to x, then

y is doubled (e.g., from 2 to 4) by doubling x (e.g., from 1 to 2).

y is cut in half (e.g., from 2 to 1) by cutting x in half (e.g., from 1 to $\frac{1}{2}$).

Mathematically, the statement that y is directly proportional to x means that y is equal to a constant times x. That is

$$y = kx \tag{8}$$

where k is a constant that is independent of the values of y or x. Alternatively, we can say that the ratio y/x is the same for all values of y and x. That is

$$\frac{y}{x} = k \tag{9}$$

If we know the value of the constant k, we can calculate y corresponding to any value of x. Suppose, for example, that k is 1. Equation 8 then becomes

y = x

and we have

| y | 0 | 1 | 2 | 3 | 4 | 5 |
| x | 0 | 1 | 2 | 3 | 4 | 5 |

In another case, k might be 2. If so

| y | 0 | 2 | 4 | 6 | 8 | 10 |
| x | 0 | 1 | 2 | 3 | 4 | 5 |

If y is directly proportional to x, then a graph of y versus x is a straight line passing through the origin (Figure A.1). The value of the constant k determines the slope of the straight line. In Figure A.1, the two straight lines correspond to

lower line: y = x; k = 1

upper line: y = 2x; k = 2

The slope is greater for the upper line.

Equations 8 and 9 can be very useful even if we do not know the value of the constant k. We can obtain a "two-point" equation relating "final" and "initial" values of y and x. To do this, we write Equation 8 twice, first for initial conditions (y_1, x_1), then for final conditions (y_2, x_2).

$y_1 = kx_1$

$y_2 = kx_2$

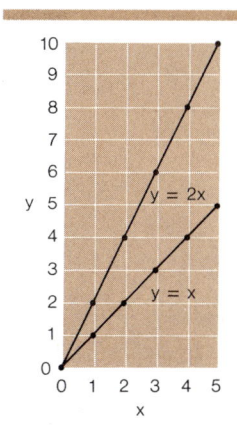

Figure A.1 Graph of direct proportionality.

If we divide the second equation by the first, k cancels and we obtain

$$\frac{y_2}{y_1} = \frac{x_2}{x_1} \tag{10}$$

This is the two-point equation for a direct proportionality.

EXAMPLE 10
A certain quantity y is directly proportional to x.
a. If x doubles (i.e., $x_2 = 2x_1$), what happens to y?
b. Suppose y = 3.16 when x = 1.00. Calculate y when x = 1.50.

Solution
a. From Equation 10 we see that:

$$\frac{y_2}{y_1} = \frac{x_2}{x_1} = \frac{2x_1}{x_1} = 2$$

In other words, y is twice as large as it was originally. This is characteristic of a direct proportionality; doubling x doubles y as well.

b. The initial values of y and x are 3.16 and 1.00, i.e., $y_1 = 3.16$, $x_1 = 1.00$. The final value of x is 1.50; $x_2 = 1.50$. We need to find the final value of y, y_2. Substituting in Equation 10

$$\frac{y_2}{3.16} = \frac{1.50}{1.00}; \quad y_2 = 3.16 \times \frac{1.50}{1.00} = 4.74$$

Exercise
What is the value of the constant k for the relation in (b)? Answer: 3.16.

Inverse Proportionality

When we say that y is inversely related to x, we mean that the two quantities change in opposite directions. If x increases, y decreases; if x decreases, y increases. If y is inversely proportional to x, we go one step further. This means that the ratio by which y changes is the inverse of that by which x changes. For example, if y is inversely proportional to x, then

y is doubled (e.g., from 2 to 4) when x is cut in half (e.g., from 1 to $\frac{1}{2}$).

y is cut in half (e.g., from 2 to 1) when x is doubled (e.g., from 1 to 2).

Mathematically, the statement that y is inversely proportional to x means that y is equal to a constant divided by x. That is

$$y = \frac{k}{x} \tag{11}$$

where k is a constant independent of the values of y or x. Alternatively, we could say that the product yx is a constant

$$yx = k \tag{12}$$

If we know the value of k, we can calculate y corresponding to any given x. Suppose, for example, that k is 2. Equation 11 then becomes

$$y = \frac{2}{x}$$

and we have

y	4	2	1	$\frac{2}{3}$	$\frac{1}{2}$
x	$\frac{1}{2}$	1	2	3	4

Figure A.2 shows a plot of this equation. The graph is referred to as a *hyperbola*; it is typical of an inverse proportionality.

To obtain the two-point equation for an inverse proportionality, we work with Equation 11. Writing it twice, once for initial conditions (y_1, x_1) and then for final conditions (y_2, x_2)

$$y_1 = \frac{k}{x_1}$$

$$y_2 = \frac{k}{x_2}$$

Dividing the second equation by the first

$$\frac{y_2}{y_1} = \frac{k}{x_2} \div \frac{k}{x_1} = \frac{k}{x_2} \times \frac{x_1}{k} = \frac{x_1}{x_2} \tag{13}$$

The equation $y_2/y_1 = x_1/x_2$ is the two-point equation for an inverse proportionality (Example 11).

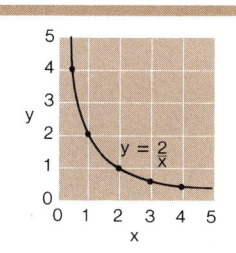

Figure A.2 Graph of inverse proportionality.

EXAMPLE 11

The Ideal Gas Law (Chapter 6) is $PV = nRT$, where P is the pressure, V is the volume, n is the number of moles, T is the temperature, and R is a constant independent of P, V, n or T.
a. If the volume is doubled (i.e., $V_2 = 2V_1$) at constant n and T, what happens to the pressure?
b. If $P = 1.00$ atm when $V = 2.00$ L, what is P when V becomes 6.00 L at constant n and T?

Continued

Solution
a. At constant n and T, the entire right side of the equation is constant. That is

$PV = k$

We conclude that P is inversely proportional to V. Hence Equation 13 applies

$$\frac{P_2}{P_1} = \frac{V_1}{V_2} = \frac{V_1}{2V_1} = \frac{1}{2}$$

The pressure is one half its original value, as we would expect for an inverse proportionality.

b. $P_1 = 1.00$ atm, $V_1 = 2.00$ L, $V_2 = 6.00$ L, $P_2 = ?$ Substituting in Equation 13

$$\frac{P_2}{1.00 \text{ atm}} = \frac{2.00 \text{ L}}{6.00 \text{ L}}; \quad P_2 = 1.00 \text{ atm} \times \frac{2.00 \text{ L}}{6.00 \text{ L}} = 0.333 \text{ atm}$$

Exercise
Calculate R if $V = 24.6$ L when $P = 1.00$ atm, $n = 1.00$ mol, $T = 300$ K. Answer: 0.0820 L·atm/mol·K.

Algebraic Equations

The equations you will need to solve in this course are quite simple. Typically, they involve a single unknown raised to the first power. These equations are solved by using a basic principle of algebra:

An equation remains valid if the same operation is performed on both sides. In particular, we can

1. Add the same quantity to both sides.
2. Subtract the same quantity from both sides.
3. Multiply both sides by the same quantity.
4. Divide both sides by the same quantity.

EXAMPLE 12
Solve the following equations for the quantity indicated.
a. $d = m/V$; (solve for m)   b. °F = 1.8(°C) + 32; (solve for °C)

Solution
a. This is the equation relating density, d, to mass, m, and volume, V. To solve for m, we multiply both sides by V

$$dV = m \times \frac{V}{V} = m$$

Continued

(mass = density × volume)

b. This is the equation relating temperature in degrees Fahrenheit (°F) to temperature in degrees Celsius (°C). To solve °C, we follow a two-step path

  (1) Subtract 32 from both sides of the equation

  $$°F - 32 = 1.8(°C) + 32 - 32$$

  $$°F - 32 = 1.8(°C)$$

  (2) Divide both sides by 1.8

  $$\frac{°F - 32}{1.8} = \frac{1.8(°C)}{1.8} = °C$$

Exercise
Solve the equation PV = nRT for V; for n. Answer: V = nRT/P; n = PV/RT.

Notice the general approach followed in Example 12. *We get the quantity we want to solve for, the "unknown" by itself on one side of the equation.* To do this, we apply the rules cited at the beginning of this section, carrying out the same operation on both sides of the equation. This approach can be used to solve any algebraic equation of this type even if the unknown is in the denominator. In that case, it is usually simplest to start by clearing fractions (Example 13).

EXAMPLE 13

Solve the following equations for the unknown indicated.

a. $d = \frac{m}{V}$; (solve for V)   b. $\frac{V_2}{T_2} = \frac{V_1}{T_1}$; (solve for T_2)

Solution

a. First multiply both sides of the equation by V

  $$dV = m$$

  Now divide both sides by d

  $$V = \frac{m}{d}$$

b. First clear of fractions by multiplying both sides of the equation by the product $T_2 \times T_1$

  $$T_2 \times T_1 \times \frac{V_2}{T_2} = T_2 \times T_1 \times \frac{V_1}{T_1}$$

  simplifying

Continued

Appendix 1

$$T_1V_2 = T_2V_1$$

Now divide both sides of the equation by V_1

$$T_2 = \frac{T_1V_2}{V_1}$$

Exercise

Solve the equation $\dfrac{P_2V_2}{T_2} = \dfrac{P_1V_1}{T_1}$ for T_2. Answer: $T_2 = \dfrac{T_1P_2V_2}{P_1V_1}$.

Graphs

Several of the figures in this text are graphs, drawn to show the relation between two variables, y and x. A graph is a collection of points connected by a smooth curve. Points on the graph are located with respect to two different axes, which are perpendicular to each other. Values of y increase as we move up along the vertical axis. Values of x increase as we move from left to right along the horizontal axis.

To see how a graph is constructed, consider the following data relating the vapor pressure of water (y) to temperature (x).

vapor pressure H$_2$O (mm Hg)	5	9	18	32	55	92
temperature (°C)	0	10	20	30	40	50

A graph of these data is shown in Figure A.3. The six points shown as filled circles are those listed above. The points are located by the following general procedure.

1. To locate the first point (y = 5, x = 0), we first look for "5" on the vertical axis. It falls halfway between the major divisions marked "0" and "10." An x value of 0 is easy to find; it lies on the vertical axis. So we decide that the first point lies on the y axis, halfway from 0 to 10.
2. To locate the third point (y = 18, x = 20), we start by finding 18 on the vertical axis. Notice that each small division on the y axis represents two units (i.e., 2 mm Hg). Hence, 18 must be located one small division below the major division marked 20. You may find it helpful to enter a light hash mark at this point on the y axis. Now move to the right from this mark, parallel to the x axis (dotted line in Figure A.3), until you come to the major division x = 20. Point 3 is located at this intersection, y = 18, x = 20.

The other points are located similarly. A smooth curve is then drawn through all six points (Figure A.3).

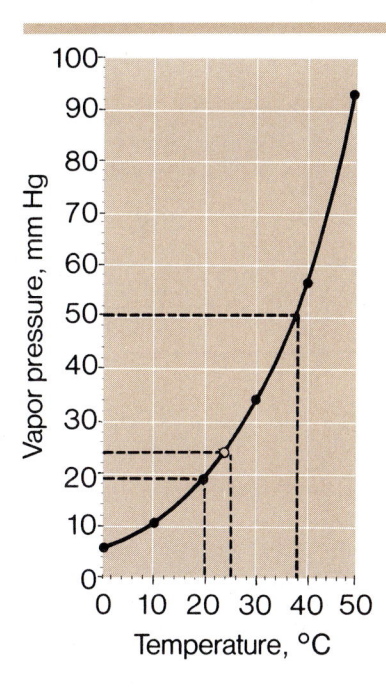

Figure A.3 Vapor pressure of water from 0–50°C.

Once you understand how a graph is constructed, it is easy to interpret it. The approach followed is described in Example 14.

EXAMPLE 14

Using Figure A.3, estimate
a. the vapor pressure of water at 25°C
b. the temperature at which the vapor pressure of water is 50 mm Hg

Solution
a. Three steps are involved.
   (1) draw a vertical line up from the x axis at 25°C, halfway between the major divisions corresponding to 20 and 30°C. This is shown as a dotted line in Figure A.3.
   (2) Mark the intersection of this vertical line with the curve. This is shown as an open circle in Figure A.3.
   (3) Draw a horizontal line across from the intersection (open circle) to the y axis. This line meets the y axis at about 24 mm Hg (two small divisions above y = 20). We conclude that

   vapor pressure of water at 25°C = 24 mm Hg

Continued

b. Again we follow a three-step path.
   (1) Draw a horizontal line across from the y axis at 50 mm Hg.
   (2) Mark the intersection of this horizontal line with the curve.
   (3) Draw a vertical line from this intersection down to the x axis. This line meets the x axis at about 38°C (one small division before x = 40). This means that

   38°C = temperature at which vapor pressure water = 50 mm Hg

Exercise
Using Figure A.3, estimate the vapor pressure of water at 34°C; at 46°C.
Answer: 40 mm Hg, 76 mm Hg.

APPENDIX 2

Answers to Problems and Multiple Choice

CHAPTER 1

1.14 a. 281   b. 0.000416   c. 38,000,000   d. 0.571
1.15 a. 3180   b. 0.000029   c. 6.12   d. 0.0018
1.16 a. 1.12×10^{-5} b. 3.152×10^3 c. 1.902×10^{-3} d. 2.865×10^2
1.17 a. 6.7131×10^4 b. 8.8×10^{-2} c. 3.15×10^{-2} d. 4.9×10^1
1.18 a. 9.3×10^3 b. 4.0×10^{20} c. 1.6×10^{12} d. 5.8×10^4
1.19 a. 6.4×10^{-5} b. 2.5×10^{-11} c. 2.4×10^5 d. 6.4×10^{-11}
1.20 a. 0.153 m   b. 24 m   c. 1.53×10^9 nm
1.21 a. 8.1×10^5 nm   b. 2.3×10^{-2} m   c. 1.5×10^{-4} cm
   d. 3.5×10^{-5} m
1.22 a. 1.52×10^5 cm   b. 1.52×10^2 cm   c. 1.52×10^{-1} cm
   d. 1.52×10^{-7} cm
1.23 a. 2.78×10^{-2} L   b. 82 mL   c. 6.19×10^{-3} L
1.24 a. 0.352 L   b. 0.0261 L   c. 2.19×10^1 L
1.25 a. 5×10^2 g   b. 0.514 g   c. 4.5×10^{-6} kg
1.26 a. 0.0170 g   b. 1.4 g   c. 6.2×10^{-5} g
1.27 a. $-15°C$ b. $-57°C$ c. 338°F   d. 261 K
1.28 a. $-27°C$ b. $-28°C$ c. 37°C   d. $-174°C$
1.29 $-105.5°F$, 197 K
1.30 80°C, 176°F
1.31 $-459°F$
1.32 $-40°C$, 233 K
1.33 0.877 g/cm^3
1.34 0.795 g/cm^3
1.35 2.3 g/cm^3
1.36 a. 12.6 g   b. 20.3 cm^3
1.37 2.0×10^1 g
1.38 0.0735 cm^3
1.39 a. 19.6 cm^3 b. 1.16 g/cm^3
1.40 a. 48 cm^3 b. 1.8 g/cm^3
1.41 a. 2   b. 3   c. 3
1.42 a. 2   b. 2   c. 4   d. 4
1.43 a. 2.2 cm   b. 0.83 cm^2 c. 2.95 g/cm^3
1.44 a. 4.0 g   b. 4.05 g   c. 4.05 g
1.45 a. 0.56 cm^3 b. 0.556 cm^3 c. 0.556 cm^3
1.46 65 g
1.47 152 g

Appendix 2

1.48 d	1.52 b	1.55 b	1.58 b
1.49 c	1.53 d	1.56 c	1.59 a
1.50 c	1.54 c	1.57 b	1.60 e
1.51 c			

CHAPTER 2

2.21 a. C b. HM c. S d. E e. E
2.22 a. E b. M c. M d. C e. E
2.23 1.023 g, 1.578 g
2.24 272.9 g
2.25 2.38 g
2.26 a. 47.55% b. 52.45%
2.27 a. P b. C c. C d. P e. P
2.28 Test to see if properties of copper and sulfur remain
2.29 a. C b P c. P d. C
2.30 2.664 g
2.31 0.1119 g
2.32 a. 2.59×10^3 J b. 2.1×10^6 cal c. 1.77×10^2 kJ d. 2.4×10^2 J
2.33 a. 1.45×10^{-1} kJ b. 2.467×10^9 kJ c. 4.594 kJ
2.34 79.6 cal/g
2.35 32.8 kJ
2.36 a. 2.00×10^3 J b. 4.22×10^3 J
2.37 2.810×10^2 J
2.38 a. 86.2 g b. 3.38 g
2.39 151 J/g
2.40 1.24×10^5 cal

2.41 c	2.46 b	2.51 b	2.56 e
2.42 a	2.47 b	2.52 c	2.57 d
2.43 b	2.48 b	2.53 c	2.58 c
2.44 c	2.49 d	2.54 a	2.59 c
2.45 b	2.50 a	2.55 c	2.60 d

CHAPTER 3

3.15 a. $^{55}_{26}$Fe b. $^{60}_{27}$Co c. $^{30}_{15}$P d. $^{242}_{94}$Pu
3.16 a. $^{14}_{6}$C b. $^{19}_{9}$F c. $^{90}_{38}$Sr d. $^{137}_{56}$Ba
3.17 a. $^{15}_{7}$N b. $^{40}_{19}$K c. $^{108}_{47}$Ag d. $^{206}_{82}$Pb
3.18 a. 26, 29 b. 27, 33 c. 15, 15 d. 94, 148
3.19 7, 14, 7; $^{207}_{82}$Pb, 125; 13, 27, 14; $^{45}_{21}$Sc, 21
3.20 $^{88}_{38}$Sr, 50; 19, 39, 20; $^{75}_{33}$As, 33; 8, 16, 8
3.21 a. 12, 10 b. 35, 36 c. 21, 18 d. 29, 28
3.22 a. O^{2-} b. F^- c. Na^+ d. Mg^{2+}
3.23 $^{14}_{7}N^{3-}$, −3; $^{23}_{11}Na^+$, +1; 35, 36, 45, −1; 26, 23, 30, +3
3.24 29, 28, 35, +1; $^{84}_{36}$Kr, 36; $^{31}_{15}P^{3-}$, −3
3.25 a. 1.0165 b. 0.94763 c. 0.9516
3.26 a. 6.887 b. 1.734 c. 0.7702

3.27 4.867
3.28 12.152
3.29 a. 1.674×10^{-24} g   b. 6.647×10^{-24} g   c. 1.153×10^{-23} g
3.30 a. 3.35×10^{-21} g   b. 1.357×10^{-4} g   c. 3.995 g
3.31 a. 2.62×10^{22} b. 1.08×10^{22} c. 3.09×10^{21}
3.32 a. 4.30×10^{25} b. 3.58×10^{16} c. 3.763×10^{22}
3.33 5.4×10^{-22} g; 4.33×10^{22}; 8.2×10^{-4} g
3.34 9.55×10^{-18} g; 3.33×10^{-4} g; 2.91×10^{21}
3.35 a. 40.08 g/mol   b. 56.08 g/mol   c. 110.99 g/mol   d. 100.09 g/mol
3.36 a. 161.97 g/mol   b. 180.16 g/mol   c. 465.96 g/mol
3.37 a. 1.673   b. 1.3664   c. 1.1548
3.38 73.8 g/mol
3.39 a. 0.504 g   b. 8.00 g   c. 4.50 g   d. 8.50 g
3.40 a. 0.496 mol   b. 0.0312 mol   c. 0.0555 mol   d. 0.0294 mol
3.41 a. 0.248 mol   b. 6.295 mol   c. 10.1 mol
3.42 a. 12 g   b. 59.5 g   c. 2.5 g
3.43 a. 3.40×10^{2} g   b. 1.69 mol   c. 1.02×10^{24}
3.44 1.50 mol
3.45 c < a < b
3.46 c   3.50 b   3.54 c   3.58 b
3.47 d   3.51 c   3.55 e   3.59 c
3.48 c   3.52 e   3.56 d   3.60 c
3.49 c   3.53 b   3.57 d

CHAPTER 4

4.19 a. LiI   b. MgS   c. Al_2O_3
4.20 a. $CaBr_2$ b. Na_2O c. BaS
4.21 a. ZnI_2 b. MnS   c. Ag_2O
4.22 a. $FeCl_2$, $FeCl_3$ b. CuCl, $CuCl_2$ c. $CoCl_2$, $CoCl_3$
4.23 a. NH_4Cl b. K_2CO_3 c. Na_3PO_4
4.24 a. $(NH_4)_2SO_4$ b. $Mg(ClO_3)_2$ c. Li_2SO_4
4.25 a. NaI   b. SrI_2 c. ZnI_2 d. AgI   e. CdI_2
4.26 a. $Ca(NO_3)_2$, $CaSO_4$ b. $Al(NO_3)_3$, $Al_2(SO_4)_3$ c. $Ni(NO_3)_2$, $NiSO_4$
     d. $AgNO_3$, Ag_2SO_4
4.27 a. sodium bromide   b. potassium sulfide   c. calcium chloride
4.28 a. copper(II) sulfate   b. copper(I) sulfate   c. iron(II) iodide
     d. iron(III) bromide
4.29 a. aluminum sulfate   b. calcium nitrate   c. potassium carbonate
     d. magnesium chromate
4.30 a. $CrCl_2$ b. CrI_3 c. $Co(NO_3)_3$ d. $Co(NO_2)_2$
4.31 a. sodium hydrogen carbonate   b. magnesium hydrogen carbonate
     c. potassium hydrogen sulfite   d. calcium hydrogen sulfate
4.32 a. NaH_2PO_4 b. $Cu_3(PO_4)_2$ c. $Ca(HCO_3)_2$ d. $Co_2(SO_4)_3$
4.33 a. hydrogen chloride   b. nitrogen trichloride
     c. tetraphosphorus decoxide
4.34 a. dinitrogen oxide   b. dinitrogen trioxide   c. dinitrogen tetroxide
     d. dinitrogen pentoxide

Appendix 2

4.35 a. XeF$_4$ b. UF$_6$ c. NO d. TeO$_2$
4.36 a. SeO$_3$ b. I$_2$O$_5$ c. BF$_3$ d. SF$_4$
4.37 a. 69.68% K, 28.52% O, 1.796% H
  b. 21.20% N, 6.103% H, 24.26% S, 48.43% O
  c. 52.14% C, 13.13% H, 34.73% O
4.38 36.08
4.39 a. 51.43% Cu, 9.720% C, 38.85% O; 2.57 g Cu
  b. 33.88% Cu, 14.94% N, 51.18% O; 1.69 g Cu
  c. 29.74% Cu, 15.01% S, 52.42% O, 2.831% H; 1.49 g Cu
4.40 62.97% H$_2$O, 37.03% Na$_2$CO$_3$, 1.259 g H$_2$O
4.41 a. 0.2919 g Fe b. 0.4647 g Fe c. 0.3130 g Co
4.42 a. CH$_4$ b. Mg$_2$P$_2$O$_7$
4.43 a. KNO$_2$ b. C$_5$H$_7$N
4.44 a. 42.1% Na, 18.9% P, 39.0% O b. Na$_3$PO$_4$
4.45 a. 78.77 % Sn, 21.23% O b. SnO$_2$
4.46 4   4.50 C$_6$H$_8$O$_6$ 4.54 b   4.58 d
4.47 6   4.51 a   4.55 d   4.59 e
4.48 C$_5$H$_5$ 4.52 c   4.56 c   4.60 d
4.49 CH; C$_6$H$_6$ 4.53 a   4.57 e

CHAPTER 5

5.11 a. 2 Na(s) + O$_2$(g) → Na$_2$O$_2$(s)
  b. 2 Na(s) + 2 H$_2$O(l) → 2 NaOH(s) + H$_2$(g)
  c. SnO$_2$(s) + 2 H$_2$(g) → Sn(s) + 2 H$_2$O(l)
5.12 a. 3 MnO$_2$(s) → Mn$_3$O$_4$(s) + O$_2$(g)
  b. SiO$_2$(s) + 4 HF(g) → SiF$_4$(g) + 2 H$_2$O(l)
  c. 2 NaClO$_3$(s) → 2 NaCl(s) + 3 O$_2$(g)
5.13 a. Fe(s) + S(s) → FeS(s)
  b. Xe(g) + 2 F$_2$(g) → XeF$_4$(s)
  c. 3 H$_2$(g) + N$_2$(g) → 2 NH$_3$(g)
  d. 2 Cu$_2$O(s) → 4 Cu(s) + O$_2$(g)
5.14 a. 4 Al(s) + 3 O$_2$(g) → 2 Al$_2$O$_3$(s)
  b. Ca(s) + F$_2$(g) → CaF$_2$(s)
  c. 2 Ba(s) + O$_2$(g) → 2 BaO(s)
  d. 2 Na(s) + S(s) → Na$_2$S(s)
5.15 a. 6 K(s) + N$_2$(g) → 2 K$_3$N(s)
  b. Zn(s) + Br$_2$(l) → ZnBr$_2$(s)
  c. 2 Ni(s) + O$_2$(g) → 2 NiO(s)
  d. 2 Ag(s) + Cl$_2$(g) → 2 AgCl(s)
5.16 a. C$_2$H$_4$(g) + 3 O$_2$(g) → 2 CO$_2$(g) + 2 H$_2$O(l)
  b. 2 C$_2$H$_2$(g) + 5 O$_2$(g) → 4 CO$_2$(g) + 2 H$_2$O(l)
  c. 2 C$_6$H$_6$(l) + 15 O$_2$(g) → 12 CO$_2$(g) + 6 H$_2$O(l)
5.17 a. formula is O$_2$, not O$_3$
  b. can divide coefficients by two
  c. unbalanced
  d. formula of iron(III) oxide is wrong

5.18 a. $Sr(s) + 2 H_2O(l) \rightarrow Sr(OH)_2(s) + H_2(g)$
b. $(NH_4)_2CO_3(s) \rightarrow 2 NH_3(g) + CO_2(g) + H_2O(l)$
c. $2 Al(s) + 3 H_2C_2O_4(aq) \rightarrow Al_2(C_2O_4)_3(s) + 3 H_2(g)$
d. $Co_3O_4(s) + 4 CO(g) \rightarrow 3 Co(s) + 4 CO_2(g)$
5.19 a. 3.78 mol   b. 1.24 mol   c. 1.30×10^{-3} mol
5.20 a. 0.585 mol   b. 0.878 mol   c. 0.787 mol
5.21 3.28, 3.28; 3.18, 1.59
5.22 0.438, 0.438; 38.8, 7.76
5.23 a. 1.60×10^2 g   b. 222 g   c. 121 g
5.24 a. 976 g   b. 6.40×10^2 g   c. 0.0372 kg
5.25 49.2 g, 58.1 g; 58.1 g, 107 g
5.26 467 g, 603 g; 4.30×10^2 g, 764 g
5.27 a. 0.475 mol   b. 2.14 mol   c. 4.45 mol
5.28 a. 1.79 mol   b. 2.69 mol   c. 1.99 mol
5.29 0.105 mol, 0.105 mol; 0.865 mol, 1.73 mol
5.30 1.83 mol, 9.14 mol; 0.102 mol, 0.102 mol
5.31 0.114 g
5.32 26.0 g
5.33 0.6308 g
5.34 69.0 g, 122 g, 109 g; 42.3 g, 31.7 g, 28.6 g
5.35 3.65 g, 3.64 g, 4.10 g; 86.1 g, 75.4 g, 97.0 g
5.36 a. $2 Al(s) + 6 HCl(g) \rightarrow 2 AlCl_3(s) + 3 H_2(g)$ b. 4.94 g
5.37 a. $4 Al(s) + 3 C(s) \rightarrow Al_4C_3(s)$ b. 3.41×10^4 g
5.38 $2 K(s) + I_2(s) \rightarrow 2 KI(s)$; 7.90 g
5.39 $2 Zn(s) + O_2(g) \rightarrow 2 ZnO(s)$; 1.24×10^3 g
5.40 1.64×10^5 g
5.41 1.7×10^{16} g $C_6H_{12}O_6$, 1.8×10^{16} g O_2
5.42 O_2; 1.20 mol
5.43 O_2; 1.65 g
5.44 12 g
5.45 Fe; 38 g
5.46 12.9 g; 86.8%
5.47 61.0 g; 82.0%
5.48 1.40 g
5.49 b   **5.52** d   **5.55** a   **5.58** b
5.50 c   **5.53** d   **5.56** c   **5.59** c
5.51 c   **5.54** e   **5.57** c   **5.60** a

CHAPTER 6

6.12 a. 775 mm Hg   b. 0.937 atm   c. 0.950 atm
6.13 a. 762.5 mm Hg   b. 1.003 atm
6.14 a. 754.7 mm Hg   b. 0.9931 atm
6.15 2.24 L
6.16 2.10×10^3 atm
6.17 1.88 cm^3
6.18 1.13 atm

Appendix 2

6.19 649 cm³
6.20 −123°C
6.21 1230 K
6.22 9.2%
6.23 2.75 atm
6.24 702 mm Hg
6.25 1.22 L
6.26 3.85 cm³
6.27 3.34 atm
6.28 212 K
6.29 a. 1.25 g/L   b. 0.179 g/L   c. 6.52 g/L
6.30 60.3 g/mol
6.31 3.84 g
6.32 a. 0.800 L   b. 5.60 L   c. 0.153 L
6.33 39.4 L
6.34 0.0131 mol
6.35 1.20 atm
6.36 609 K
6.37 0.184 g
6.38 0.827 L
6.39 0.0822 L·atm/(mol·K)
6.40 2.64 kg
6.41 741 mm Hg
6.42 a. 720 mm Hg   b. 8.60×10^{-3} mol
6.43 a. 725 mm Hg   b. 0.0325 g
6.44 a. 730 mm Hg   b. 116 cm³
6.45 a. 3.66 L   b. 0.597 L
6.46 8.97 L
6.47 2.35 L
6.48 0.700 g
6.49 0.533 g
6.50 0.533 L
6.51 d   6.54 c   6.57 d   6.59 d
6.52 c   6.55 d   6.58 b   6.60 b
6.53 b   6.56 c

Chapter 7

7.23 a. $1s^2 2s^2$ b. $1s^2 2s^2 2p^6 3s^2 3p^6 4s^2 3d^1$ c. $1s^2 2s^2 2p^6 3s^2 3p^4$
7.24 a. $1s^2 2s^2 2p^6 3s^2 3p^6 4s^2 3d^8$ b. $1s^2 2s^2 2p^6$ c. $1s^2 2s^2 2p^3$
7.25 a. F   b. Fe   c. Ti
7.26 a. O   b. Al   c. As
7.27 a. Li   b. Al   c. Sc
7.28 13%, 53%, 33%
7.29 a. 36   b. 32   c. 16
7.30 8, 12, 5

7.31

	1s	2s	2p	3s	3p	4s	3d
a. Mg	(↑↓)	(↑↓)	(↑↓)(↑↓)(↑↓)	(↑↓)			
b. Cu	(↑↓)	(↑↓)	(↑↓)(↑↓)(↑↓)	(↑↓)	(↑↓)(↑↓)(↑↓)	(↑)	(↑↓)(↑↓)(↑↓)(↑↓)(↑↓)
c. Cl	(↑↓)	(↑↓)	(↑↓)(↑↓)(↑↓)	(↑↓)	(↑↓)(↑↓)(↑)		

7.32

	4s	3d	4p
a. V	(↑↓)	(↑)(↑)(↑)( )( )	
b. Br	(↑↓)	(↑↓)(↑↓)(↑↓)(↑↓)(↑↓)	(↑↓)(↑↓)(↑)
c. Co	(↑↓)	(↑↓)(↑↓)(↑)(↑)(↑)	

7.33 a. S b. Mg c. K
7.34 a. Ga b. V c. Zn
7.35 a. 1 b. 2 c. 2
7.36 b, c, d
7.37 a. $5s^25p^1$ b. $5s^25p^3$ c. $5s^25p^5$
7.38 a. P b. Bi c. Xe
7.39 a. 4d b. 6s c. 5d
7.40 a. 4f b. 5f c. 5d
7.41 a. $5s^2$ b. $7s^1$ c. $6s^26p^6$
7.42 a. Cs b. At c. As
7.43 a. Ra b. Be c. Be
7.44 a. At b. F c. F
7.45 Group 1
7.46 Si < Al < Ga
7.47 Al > Mg > Na > K
7.48 Se < S < O < F

7.49 d	**7.52** d	**7.55** b	**7.58** c
7.50 d	**7.53** b	**7.56** c	**7.59** b
7.51 c	**7.54** b	**7.57** c	**7.60** c

Chapter 8

8.21 a. $1s^22s^22p^6$ b. $1s^22s^22p^63s^23p^6$ c. $1s^22s^22p^63s^23p^64s^23d^{10}4p^6$
8.22 a. $1s^22s^22p^63s^23p^6$ b. $1s^22s^22p^63s^23p^64s^23d^{10}4p^6$ c. $1s^22s^22p^6$
8.23 S^{2-}, Cl^-, K^+, Ca^{2+}
8.24 a. $3d^4$ b. $3d^3$ c. $3d^7$ d. $3d^6$
8.25 a. $3d^5$ b. $3d^8$ c. $3d^{10}$ d. $3d^{10}$
8.26 a. $4d^2$ b. $4d^4$ d. $4d^6$
8.27 a, c
8.28 a. Ar b. Xe c. Ne d. Ar e. He
8.29 a. ·Si· b. ·S̈· c. :C̈l· d. H·
8.30 a. ·P̈· b. :Är: c. ·Ö· d. :F̈·
8.31 a. 14 b. 20 c. 8 d. 32

8.32 a. H—C(H)(H)—Cl: b. :Cl—S—Cl: c. H—As(H)—H d. :Cl—Si(:Cl:)(:Cl:)—Cl:

Appendix 2

8.33 a. :F—C̈—F: b. :Ï—Ï: c. H—N̈—H   d. :C̈l—N̈—C̈l:
       with Cl above and Cl below C   H   :C̈l: below

(Lewis structures:)
8.33 a. :F̈—C̈—F̈: with :C̈l: on top and :C̈l: on bottom
     b. :Ï—Ï:
     c. H—N̈—H with H below
     d. :C̈l—N̈—C̈l: with :C̈l: below

8.34 a. :N≡N: b. :S̈=C=S̈:   c. :Ö—S̈e=Ö

8.35 a. :C̈l—P̈—C̈l: with :C̈l: below
     b. :C≡O:
     c. :Ö—S̈=Ö with :Ö: below

8.36 a. :F̈—Be—F̈:   b. :F̈—B—F̈: with :F̈: below   c. ·N̈=Ö

8.37 a. 8   b. 20   c. 24   d. 24

8.38 a. (:Ö—H)⁻   b. (:Ö—C̈l—Ö:)⁻   c. (:Ö—C—Ö:)²⁻ with :Ö: double-bonded below

     d. (:Ö—N—Ö:)⁻ with :Ö: double-bonded below

8.39 a. linear   b. linear   c. equilateral triangle
8.40 a. tetrahedral   b. pyramidal   c. bent   d. linear
8.41 a. tetrahedral   b. bent   c. pyramidal   d. tetrahedral
8.42 a. pyramidal   b. tetrahedral   c. bent
8.43 a. nonpolar   b. polar   c. nonpolar   d. polar
8.44 b, c, d
8.45 a, b, c
8.46 a, b, c
8.47 b   8.51 d   8.55 b   8.58 c
8.48 d   8.52 b   8.56 b   8.59 d
8.49 c   8.53 c   8.57 c   8.60 c
8.50 c   8.54 d

Chapter 9

9.16 a. $PbCl_2$ b. $PbSO_4$ c. PbS
9.17 a. Ag_2CO_3 b. As_2S_3 d. $Cu(OH)_2$
9.18 a. $Pb^{2+}(aq) + 2\ Cl^-(aq) \rightarrow PbCl_2(s)$
     b. $Pb^{2+}(aq) + SO_4^{2-}(aq) \rightarrow PbSO_4(s)$
     c. $Pb^{2+}(aq) + S^{2-}(aq) \rightarrow PbS(s)$
9.19 a. $2\ Ag^+(aq) + CO_3^{2-}(aq) \rightarrow Ag_2CO_3(s)$
     b. $2\ As^{3+}(aq) + 3\ S^{2-}(aq) \rightarrow As_2S_3(s)$
     d. $Cu^{2+}(aq) + 2\ OH^-(aq) \rightarrow Cu(OH)_2(s)$
9.20 a. $Ba^{2+}(aq) + SO_4^{2-}(aq) \rightarrow BaSO_4(s)$
     b. $2\ Ag^+(aq) + S^{2-}(aq) \rightarrow Ag_2S(s)$
     c. $Ni^{2+}(aq) + 2\ OH^-(aq) \rightarrow Ni(OH)_2(s)$
     d. $Al^{3+}(aq) + 3\ OH^-(aq) \rightarrow Al(OH)_3(s)$
9.21 b. $Ni^{2+}(aq) + 2\ OH^-(aq) \rightarrow Ni(OH)_2(s)$
     c. $Ni^{2+}(aq) + 2\ OH^-(aq) \rightarrow Ni(OH)_2(s)$

Appendix 2

$$Ba^{2+}(aq) + SO_4^{2-}(aq) \rightarrow BaSO_4(s)$$
    d. $Zn^{2+}(aq) + CO_3^{2-}(aq) \rightarrow ZnCO_3(s)$

9.22 a. $NaNO_3$, Ag_2CrO_4 b. Ag_2CrO_4
    c. $2\,Ag^+(aq) + CrO_4^{2-}(aq) \rightarrow Ag_2CrO_4(s)$

9.23 a. $FePO_4$, $NaNO_3$ b. $Fe^{3+}(aq) + PO_4^{3-}(aq) \rightarrow FePO_4(s)$

9.24 a. $Pb^{2+}(aq) + S^{2-}(aq) \rightarrow PbS(s)$
    b. $Fe^{3+}(aq) + 3\,OH^-(aq) \rightarrow Fe(OH)_3(s)$
    c. $Hg_2^{2+}(aq) + 2\,Cl^-(aq) \rightarrow Hg_2Cl_2(s)$

9.25 a. $2\,Ag^+(aq) + CrO_4^{2-}(aq) \rightarrow Ag_2CrO_4(s)$
    b. $Ca^{2+}(aq) + 2\,F^-(aq) \rightarrow CaF_2(s)$
    c. $Ba^{2+}(aq) + SO_4^{2-}(aq) \rightarrow BaSO_4(s)$

9.26 a. 1.83   b. 0.216   c. 7.34
9.27 a. 3.84   b. 0.0167   c. 0.0525
9.28 a. 1.00 L   b. 6.00 cm^3 c. 5.15 cm^3
9.29 a. 3.22   b. 0.0311 L   c. 3.22×10^{-3}
9.30 a. 353 g   b. 9.62 L
9.31 Dissolve 3.839 g (volumetric flask) to form 500.0 cm^3 solution
9.32 4.2 cm^3
9.33 3.75 cm^3
9.34 a. 1.02×10^3 g   b. 16.2
9.35 60.0 g; 440 g
9.36 38.8
9.37 56 g H_2SO_4, 2 g H_2O
9.38 a. 3.00 g   b. 66.0 g   c. 50.0 g
9.39 a. 0.832   b. 2.42
9.40 6.85 g
9.41 a. 7.24   b. 25.0
9.42 6.86
9.43 $-1.61°C$, $100.45°C$
9.44 $-1.48°C$, $100.41°C$
9.45 a. 3.23 g   b. 5.76 g
9.46 576 g
9.47 a. 4.03   b. $-7.50°C$
9.48 2.00×10^2 g/mol
9.49 79.3 g/mol
9.50 174 g/mol

9.51 d	**9.54** d	**9.57** b	**9.59** a
9.52 a	**9.55** d	**9.58** b	**9.60** c
9.53 a	**9.56** d		

Chapter 10

10.21 a. increase   b. no effect   c. decrease   d. increase   e. decrease
10.22 add CO or NO_2, compress, raise T, use catalyst
10.23 reduce T and P
10.24 a. $\dfrac{[H_2O]^2}{[H_2]^2 \times [O_2]}$ b. $\dfrac{[SO_2] \times [O_2]^{1/2}}{[SO_3]}$ c. $\dfrac{[NH_3]^2}{[N_2] \times [H_2]^3}$

Appendix 2

10.25 a. $\dfrac{[H_2] \times [I_2]}{[HI]^2}$ b. $\dfrac{[HI]^2}{[H_2] \times [I_2]}$ c. $\dfrac{[HI]}{[H_2]^{1/2} \times [I_2]^{1/2}}$

10.26 a. $4\ NH_3(g) + 5\ O_2(g) \rightleftharpoons 4\ NO(g) + 6\ H_2O(g)$
   b. $2\ NH_3(g) + 5/2\ O_2(g) \rightleftharpoons 2\ NO(g) + 3\ H_2O(g)$
   c. $4\ NO(g) + 6\ H_2O(g) \rightleftharpoons 4\ NH_3(g) + 5\ O_2(g)$

10.27 a. $N_2(g) + O_2(g) \rightleftharpoons 2\ NO(g)$
   b. $N_2O_4(g) \rightleftharpoons 2\ NO_2(g)$
   c. $S(g) + 3\ F_2(g) \rightleftharpoons SF_6(g)$

10.28 0.24
10.29 51
10.30 0.13
10.31 0.36
10.32 a. 0.18 M b. 0.82 M c. 0.82 M d. 3.7
10.33 a. 0.0500 M b. 0.050 M c. 0.025 M d. 0.025
10.34 0.082 M
10.35 0.050 M
10.36 a. 1×10^{-2} M b. 1×10^{-20} M
10.37 a. 0.19 M b. 0.028 M
10.38 7.1×10^{-5} M
10.39 a. → b. ← c. ← d. →
10.40 compressed ←; expanded →
10.41 a. → b. → c. →
10.42 a. ← b. ← c. →
10.43 a. no effect b. → c. → d. →
10.44 a. ← b. decrease
10.45 a. → b. increase
10.46 temperature →; pressure, no effect
10.47 low temperature, high pressure

10.48 c	10.52 d	10.55 c	10.58 d
10.49 c	10.53 c	10.56 a	10.59 e
10.50 d	10.54 c	10.57 b	10.60 d
10.51 c			

Chapter 11

11.15 a. 1.0×10^{-11} M b. 5.0×10^{-9} M c. 1.9×10^{-6} M
11.16 a. 1.0×10^{-9} M b. 4.0×10^{-8} M c. 2.9×10^{-5} M
11.17 acidic, acidic, basic; basic, basic, acidic
11.18 a. $[OH^-] = 0.12$ M, $[H^+] = 8.3 \times 10^{-14}$ M
   b. $[H^+] = 0.023$ M, $[OH^-] = 4.3 \times 10^{-13}$ M
11.19 a. 0.50 M b. 2.0×10^{-14} M
11.20 a. 2.05×10^{-2} M b. 4.9×10^{-13} M
11.21 a. 6.00 b. 9.0 c. 3.70 d. 2.27
11.22 a. 13.0 b. -1.00 c. 4.48 d. 8.09
11.23 a. 10^{-6} M b. 1×10^{-12} M c. 5×10^{-5} M d. 2×10^{-7} M
11.24 a. 1×10^{-9} M b. 10 M c. 6×10^{-6} M d. 2×10^{-12} M
11.25 acidic, basic, acidic, acidic; basic, acidic, acidic, basic

Appendix 2   475

11.26 A more acidic, B more basic; 10^3
11.27 $[H^+] = 4.2 \times 10^{-9}$ M; pH = 8.38
11.28 1.30
11.29 a. $HNO_3(aq) \rightarrow H^+(aq) + NO_3^-(aq)$
 b. $HBr(aq) \rightarrow H^+(aq) + Br^-(aq)$
 c. $H_2SO_4(aq) \rightarrow H^+(aq) + HSO_4^-(aq)$
11.30 a. $HF(aq) \rightleftharpoons H^+(aq) + F^-(aq)$
 b. $H_3PO_4(aq) \rightleftharpoons H^+(aq) + H_2PO_4^-(aq)$
 c. $HC_2H_3O_2(aq) \rightleftharpoons H^+(aq) + C_2H_3O_2^-(aq)$
11.31 a. $HCl(aq) \rightarrow H^+(aq) + Cl^-(aq)$
 b. $H_2CO_3(aq) \rightleftharpoons H^+(aq) + HCO_3^-(aq)$
 c. $HNO_2(aq) \rightleftharpoons H^+(aq) + NO_2^-(aq)$
 d. $HClO_4(aq) \rightarrow H^+(aq) + ClO_4^-(aq)$
11.32 8.0×10^{-5}
11.33 4.0×10^{-7}
11.34 4.0×10^{-10}
11.35 a. 1×10^{-4} b. 1×10^{-7}
11.36 0.10 M
11.37 1.3×10^{-3} M
11.38 5.7×10^{-5} M
11.39 a. 2.1×10^{-4} M   b. 3.68   c. 0.21%
11.40 a. $H^+(aq) + OH^-(aq) \rightarrow H_2O$
 b. $H^+(aq) + OH^-(aq) \rightarrow H_2O$
 c. $H^+(aq) + OH^-(aq) \rightarrow H_2O$
11.41 a. $HC_2H_3O_2(aq) + OH^-(aq) \rightarrow C_2H_3O_2^-(aq) + H_2O$
 b. $HNO_2(aq) + OH^-(aq) \rightarrow NO_2^-(aq) + H_2O$
 c. $HF(aq) + OH^-(aq) \rightarrow F^-(aq) + H_2O$
11.42 a. $HC_2H_3O_2(aq) + OH^-(aq) \rightarrow C_2H_3O_2^-(aq) + H_2O$
 b. $H^+(aq) + OH^-(aq) \rightarrow H_2O$
 c. $H^+(aq) + NH_3(aq) \rightarrow NH_4^+(aq)$
11.43 a. $H^+(aq) + NH_3(aq) \rightarrow NH_4^+(aq)$
 b. $H^+(aq) + OH^-(aq) \rightarrow H_2O$
 c. $HCN(aq) + OH^-(aq) \rightarrow CN^-(aq) + H_2O$
11.44 184 cm^3
11.45 0.0857 M
11.46 0.171 M
11.47 208 cm^3
11.48 a. HNO_3 acid, H_2O base   b. H_2O acid, F^- base
11.49 H_2CO_3, HCO_3^- are acids; CO_3^{2-}, HCO_3^- are bases
11.50 a. NH_4^+ b. NO_2^- c. OH^- d. H_3O^+
11.51 a   11.54 c   11.57 a   11.59 b
11.52 b   11.55 c   11.58 e   11.60 b
11.53 b   11.56 c

Chapter 12

12.16 a. 0   b. −1   c. +5   d. +7
12.17 a. 0   b. +3   c. +6   d. +6
12.18 a. −3   b. 0   c. −2   d. +5

Appendix 2

12.19 a. K = +1, O = −2, Mn = +7   b. O = −2, S = +6
c. Na = +1, O = −1   d. C = −4, H = +1

12.20 a. Na = +1, S = −2   b. Al = +3, O = −2   c. P = +5, O = −2
d. H = +1, O = −2, S = +6

12.21 a. H_2 b. HCl, H_2O, etc.   c. H^-

12.22 a. Al oxidized, Pb^{2+} reduced   b. Zn oxidized, NO_3^- reduced
c. Mn^{2+} oxidized, BiO_3^- reduced

12.23 a. Pb^{2+} oxidizing agent, Al reducing agent
b. NO_3^- oxidizing agent, Zn reducing agent
c. BiO_3^- oxidizing agent, Mn^{2+} reducing agent

12.24 a. NO_3^- oxidizing agent, Ag reducing agent
b. $Cr_2O_7^{2-}$ oxidizing agent, I^- reducing agent
c. PbO_2 oxidizing agent, Cl^- reducing agent

12.25 a. MnO_2 oxidizing agent, Cl^- reducing agent
b. Ag^+ oxidizing agent, Cu reducing agent
c. NO_3^- oxidizing agent, Br^- reducing agent

12.26 a. OA   b. OA   c. RA   d. OA

12.27 a. RA   b. OA   c. OA   d. OA

12.28 a. $2\,Al(s) + 3\,Pb^{2+}(aq) \rightarrow 2\,Al^{3+}(aq) + 3\,Pb(s)$
b. $4\,Zn(s) + NO_3^-(aq) + 10\,H^+(aq) \rightarrow 4\,Zn^{2+}(aq) + NH_4^+(aq) + 3\,H_2O$
c. $2\,Mn^{2+}(aq) + 5\,BiO_3^-(aq) + 14\,H^+(aq) \rightarrow 2\,MnO_4^-(aq) + 5\,Bi^{3+}(aq) + 7\,H_2O$

12.29 a. $3\,Ag(s) + NO_3^-(aq) + 4\,H^+(aq) \rightarrow 3\,Ag^+(aq) + NO(g) + 2\,H_2O$
b. $Cr_2O_7^{2-}(aq) + 14\,H^+(aq) + 6\,I^-(aq) \rightarrow 2\,Cr^{3+}(aq) + 7\,H_2O + 3\,I_2(s)$
c. $PbO_2(s) + 4\,H^+(aq) + 2\,Cl^-(aq) \rightarrow Pb^{2+}(aq) + Cl_2(g) + 2\,H_2O$

12.30 a. $MnO_2(s) + 4\,H^+(aq) + 2\,Cl^-(aq) \rightarrow Mn^{2+}(aq) + Cl_2(g) + 2\,H_2O$
b. $2\,Ag^+(aq) + Cu(s) \rightarrow 2\,Ag(s) + Cu^{2+}(aq)$
c. $2\,NO_3^-(aq) + 8\,H^+(aq) + 6\,Br^-(aq) \rightarrow 2\,NO(g) + 4\,H_2O + 3\,Br_2(l)$

12.31 a. $Cl_2(g) + 2\,I^-(aq) \rightarrow 2\,Cl^-(aq) + I_2(s)$
b. $2\,Al(s) + 3\,Ni^{2+}(aq) \rightarrow 2\,Al^{3+}(aq) + 3\,Ni(s)$
c. $Sn^{2+}(aq) + 2\,Fe^{3+}(aq) \rightarrow Sn^{4+}(aq) + 2\,Fe^{2+}(aq)$

12.32 a. $Cu(s) + 2\,NO_3^-(aq) + 4\,H^+(aq) \rightarrow Cu^{2+}(aq) + 2\,NO_2(g) + 2\,H_2O$
b. $2\,ClO_3^-(aq) + SO_2(g) \rightarrow 2\,ClO_2(g) + SO_4^{2-}(aq)$
c. $2\,MnO_4^-(aq) + 6\,H^+(aq) + 5\,H_2S(aq) \rightarrow 2\,Mn^{2+}(aq) + 8\,H_2O + 5\,S(s)$

12.33 a. $BrO_3^-(aq) + 6\,H^+(aq) + 6\,I^-(aq) \rightarrow Br^-(aq) + 3\,H_2O + 3\,I_2(s)$
b. $5\,Fe^{2+}(aq) + MnO_4^-(aq) + 8\,H^+(aq) \rightarrow 5\,Fe^{3+}(aq) + Mn^{2+}(aq) + 4\,H_2O$
c. $2\,Cr_2O_7^{2-}(aq) + 16\,H^+(aq) + C_2H_4(g) \rightarrow 4\,Cr^{3+}(aq) + 10\,H_2O + 2\,CO_2(g)$

12.34 $Sn(s) + 4\,NO_3^-(aq) + 4\,H^+(aq) \rightarrow SnO_2(s) + 4\,NO_2(g) + 2\,H_2O$

12.35 a. $2\,Cl^-(aq) \rightarrow Cl_2(g) + 2\,e^-$
b. $NO_3^-(aq) + 2\,H^+(aq) + e^- \rightarrow NO_2(g) + H_2O$
c. $2\,Cl^-(aq) + 2\,NO_3^-(aq) + 4\,H^+(aq) \rightarrow Cl_2(g) + 2\,NO_2(g) + 2\,H_2O$

12.36 a. OA   b. OA, RA   c. RA   d. OA   e. OA   f. RA

12.37 $Ag(s) < Fe^{2+}(aq) < Sn(s)$; $Zn^{2+}(aq) < Fe^{2+}(aq) < NO_3^-(aq) < Br_2(l)$

12.38 a. 21   b. 11   c. 4   d. 0

12.39 a. 17   b. 2   c. 6   d. 23

12.40 a. no   b. yes   c. yes

12.41 a. yes   b. no   c. no

12.42 Cu^{2+}, Ag^+, Cl_2

Appendix 2   477

12.43 Zn, Co
12.44 a. $2\,Cl^-(aq) + Ni^{2+}(aq) \rightarrow Cl_2(g) + Ni(s)$
       b. $2\,Br^-(aq) + Co^{2+}(aq) \rightarrow Br_2(l) + Co(s)$
       c. $2\,I^-(aq) + Fe^{2+}(aq) \rightarrow I_2(s) + Fe(s)$
12.45 a. $2\,Cl^-(aq) + 2\,H_2O \rightarrow Cl_2(g) + H_2(g) + 2\,OH^-(aq)$
       b. $2\,I^-(aq) + 2\,H_2O \rightarrow I_2(s) + H_2(g) + 2\,OH^-(aq)$
       c. $2\,Br^-(aq) + 2\,H_2O \rightarrow Br_2(l) + H_2(g) + 2\,OH^-(aq)$
12.46 [diagram: electrolytic cell with electrodes A and C, e^- flow, Cl^- and Ni^{2+} ions]

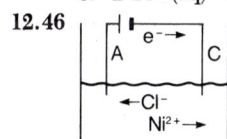

12.47 a. $Zn(s) + 2\,Ag^+(aq) \rightarrow Zn^{2+}(aq) + 2\,Ag(s)$
       b. $Zn(s) \rightarrow Zn^{2+}(aq) + 2\,e^-$
       c. $2\,Ag^+(aq) + 2\,e^- \rightarrow 2\,Ag(s)$
12.48 a. $Cl_2(g) + 2\,I^-(aq) \rightarrow 2\,Cl^-(aq) + I_2(s)$
       b. $2\,I^-(aq) \rightarrow I_2(s) + 2\,e^-$
       c. $Cl_2(g) + 2\,e^-(aq) \rightarrow 2\,Cl^-(aq)$
12.49 [diagram: galvanic cell with Cu electrode A and Ag electrode C, e^- flow, Cu^{2+} and Ag^+ ions]

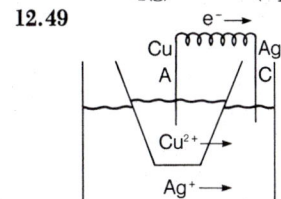

12.50 b   12.53 d   12.56 d   12.59 d
12.51 b   12.54 d   12.57 c   12.60 c
12.52 e   12.55 c   12.58 b

Chapter 13

13.27 a. $2\,K(s) + H_2(g) \rightarrow 2\,KH(s)$
       b. $2\,K(s) + Cl_2(g) \rightarrow 2\,KCl(s)$
       c. $2\,K(s) + S(s) \rightarrow K_2S(s)$
       d. $2\,K(s) + 2\,H_2O \rightarrow 2\,K^+(aq) + 2\,OH^-(aq) + H_2(g)$
13.28 a. $Sr(s) + H_2(g) \rightarrow SrH_2(s)$
       b. $Sr(s) + Br_2(l) \rightarrow SrBr_2(s)$
       c. $3\,Sr(s) + N_2(g) \rightarrow Sr_3N_2(s)$
       d. $Sr(s) + S(s) \rightarrow SrS(s)$
13.29 a. $4\,Li(s) + O_2(g) \rightarrow 2\,Li_2O(s)$
       b. $2\,Na(s) + O_2(g) \rightarrow Na_2O_2(s)$
       c. $Rb(s) + O_2(g) \rightarrow RbO_2(s)$
       d. $2\,Mg(s) + O_2(g) \rightarrow 2\,MgO(s)$
       e. $2\,Ca(s) + O_2(g) \rightarrow 2\,CaO(s)$
13.30 a. Li_3N b. K_2O_2 c. K_2O d. KO_2
13.31 a. BaO b. BaO_2 c. Sr_3N_2 d. CaH_2

Appendix 2

13.32 $Zn(NH_3)_4^{2+}$, $Zn(NH_3)_3Cl^+$, $Zn(NH_3)_2Cl_2$, $Zn(NH_3)Cl_3^-$, $ZnCl_4^{2-}$
13.33 a. $Co(NH_3)_6^{3+}$ b. $Co(NH_3)_3(H_2O)_3^{3+}$ c. $Co(NH_3)_4Cl_2^+$
   d. $Co(NH_3)_2Cl_4^-$
13.34 a. +3   b. +2   c. +2   d. +4   e. +1
13.35 6, 6, 6, 6; 6, 4, 4, 4, 2
13.36 a. $2\ C(s) + O_2(g) \rightarrow 2\ CO(g)$
   b. $Fe_2O_3(s) + 3\ CO(g) \rightarrow 2\ Fe(l) + 3\ CO_2(g)$
13.37 a. $CaCO_3(s) + SiO_2(s) \rightarrow CaSiO_3(l) + CO_2(g)$
   b. $C(s) + O_2(g) \rightarrow CO_2(g)$
13.38 a. $Al_2O_3(s) + 2\ OH^-(aq) + 3\ H_2O \rightarrow 2\ Al(OH)_4^-(aq)$
   b. $2\ Al_2O_3(l) \rightarrow 4\ Al(l) + 3\ O_2(g)$
13.39 a. $SnO_2(s) + 2\ CO(g) \rightarrow Sn(s) + 2\ CO_2(g)$
   b. $ZnO(s) + CO(g) \rightarrow Zn(s) + CO_2(g)$
   c. $Cr_2O_3(s) + 3\ CO(g) \rightarrow 2\ Cr(s) + 3\ CO_2(g)$
   d. $Co_3O_4(s) + 4\ CO(g) \rightarrow 3\ Co(s) + 4\ CO_2(g)$
13.40 a. $2\ H_2O(l) \rightarrow 2\ H_2(g) + O_2(g)$; $Zn(s) + 2\ H^+(aq) \rightarrow Zn^{2+}(aq) + H_2(g)$
   b. $CH_4(g) + H_2O(g) \rightarrow CO(g) + 3\ H_2(g)$
13.41 a. $2\ Na(s) + H_2(g) \rightarrow 2\ NaH(s)$
   b. $H_2(g) + F_2(g) \rightarrow 2\ HF(g)$
   c. $3\ H_2(g) + N_2(g) \rightarrow 2\ NH_3(g)$
   d. $Ba(s) + H_2(g) \rightarrow BaH_2(s)$
13.42 $2\ H_2O(l) \rightarrow 2\ H_2(g) + O_2(g)$; $2\ H_2O_2(aq) \rightarrow 2\ H_2O + O_2(g)$
13.43 a. $4\ Al(s) + 3\ O_2(g) \rightarrow 2\ Al_2O_3(s)$
   b. $S(s) + O_2(g) \rightarrow SO_2(g)$
   c. $N_2(g) + O_2(g) \rightarrow 2\ NO(g)$
13.44 NaH, BaH_2, Al_2O_3 ionic; others molecular
13.45 a. N_2O_5, N_2O_4, N_2O_3, N_2O_2, NO_2, NO, N_2O b. SO_2, SO_3
   c. P_4O_6, P_4O_{10}
13.46 a. CaO b. Li_2O c. Cr_2O_3
13.47 a. $CaO(s) + H_2O(l) \rightarrow Ca(OH)_2(s)$
   b. $Li_2O(s) + H_2O(l) \rightarrow 2\ LiOH(s)$
   c. $Cr_2O_3(s) + 3\ H_2O(l) \rightarrow 2\ Cr(OH)_3(s)$
13.48 a. 4   b. CO_2 c. $CO_2(g) + H_2O \rightarrow H_2CO_3(aq)$
13.49 a. SO_2 b. SO_3 c. P_4O_{10} d. N_2O_5
13.50 $Ca(OH)_2$, $LiOH$, $Cr(OH)_3$ ionic; others molecular
13.51 b   13.54 b   13.57 d   13.59 d
13.52 b   13.55 e   13.58 c   13.60 b
13.53 e   13.56 d

Chapter 14

14.25 a. C_7H_{16} b. C_7H_{14} c. C_7H_{12}
14.26 a. C_6H_{14} b. C_7H_{14} c. C_8H_{14}
14.27 a. alkane   b. alkene   c. alkyne   d. alkyne   e. alkane
14.28 a. alkene   b. alkane   c. alkyne   d. alkyne   e. alkene
14.29 a. C_6H_{14} b. C_3H_6 c. C_4H_6 d. C_4H_{10}

Appendix 2

14.30 CH$_3$—CH$_2$—CH$_2$—CH$_2$—CH$_2$—CH$_3$,   CH$_3$—C(CH$_3$)(H)—CH$_2$—CH$_2$—CH$_3$,

CH$_3$—CH$_2$—C(CH$_3$)(H)—CH$_2$—CH$_3$,   CH$_3$—C(CH$_3$)(H)—C(CH$_3$)(H)—CH$_3$,   CH$_3$—C(CH$_3$)(CH$_3$)—CH$_2$—CH$_3$

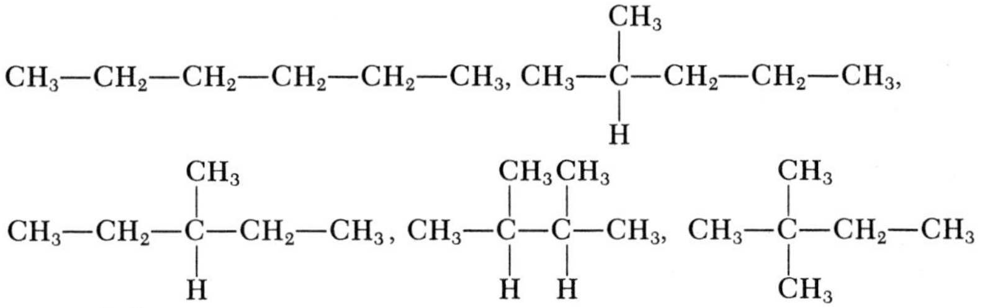

14.31 a, b, and d

14.32 CH$_3$—CH$_2$—CH$_2$—CH$_2$—Cl,   CH$_3$—CH$_2$—C(H)(Cl)—CH$_3$,   CH$_3$—C(CH$_3$)(Cl)—CH$_3$,   CH$_3$—C(CH$_3$)(H)—CH$_2$Cl

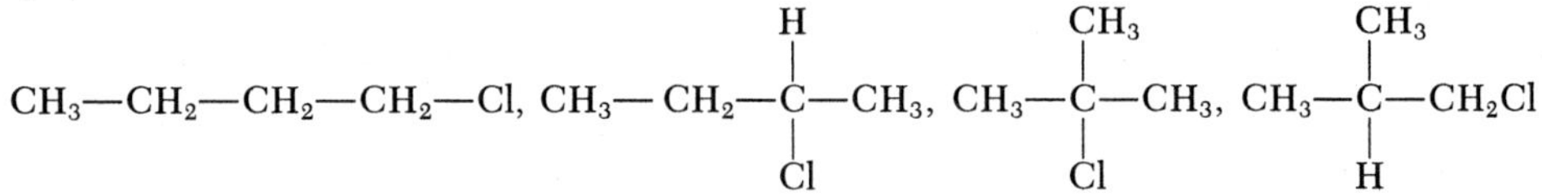

14.33 H—C≡C—CH$_2$—CH$_3$,   CH$_3$—C≡C—CH$_3$

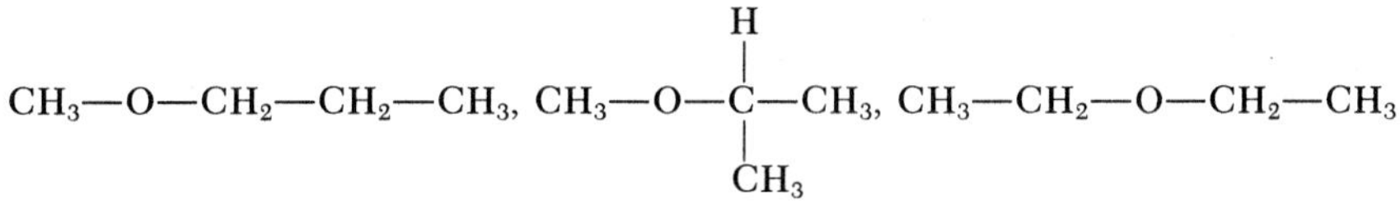

14.34 CH$_3$—CH$_2$—C(Cl)(Cl)—Cl,   CH$_3$—C(Cl)(H)—C(Cl)(H)—Cl,   H—C(H)(Cl)—CH$_2$—C(Cl)(H)—Cl,

CH$_3$—C(Cl)(Cl)—CH$_2$Cl,   H—C(H)(Cl)—C(H)(Cl)—C(H)(Cl)—H

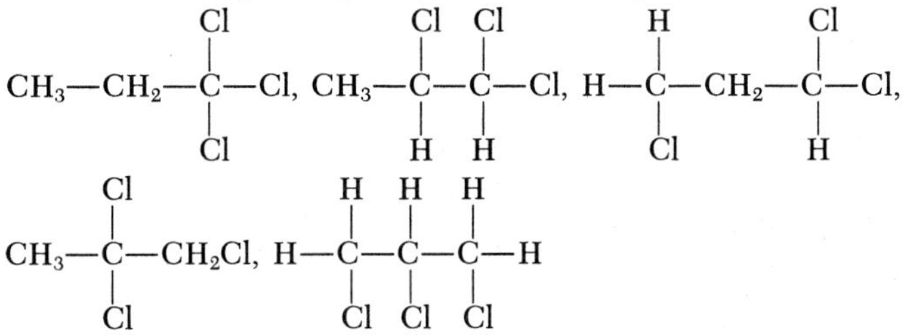

14.35 C$_4$H$_8$O$_2$

14.36 CH$_3$—O—CH$_2$—CH$_2$—CH$_3$,   CH$_3$—O—C(H)(CH$_3$)—CH$_3$,   CH$_3$—CH$_2$—O—CH$_2$—CH$_3$

14.37 (1,2,3-trichlorobenzene), (1,2,4-trichlorobenzene), (1,3,5-trichlorobenzene)

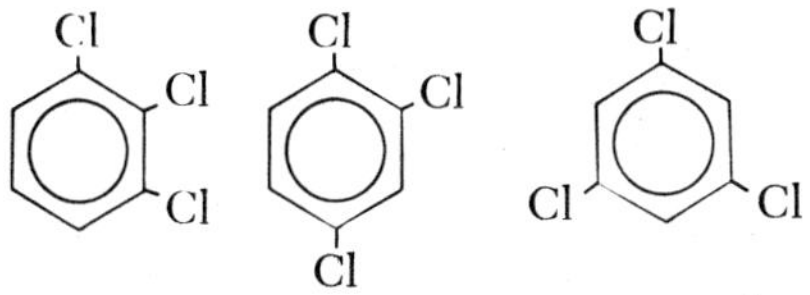

14.38 hexane, 2-methylpentane, 3-methylpentane, 2,3-dimethylbutane, 2,2-dimethylbutane

14.39 2-methylbutane, 2,2-dimethylpropane

14.40 a. CH$_3$—C(H)(CH$_3$)—CH$_2$—CH$_2$—CH$_2$—CH$_3$ b. CH$_3$—C(CH$_3$)(CH$_3$)—CH$_2$—CH$_2$—CH$_3$

c. CH$_3$—C(H)(CH$_3$)—C(H)(CH$_2$CH$_3$)—CH$_2$—CH$_2$—CH$_3$

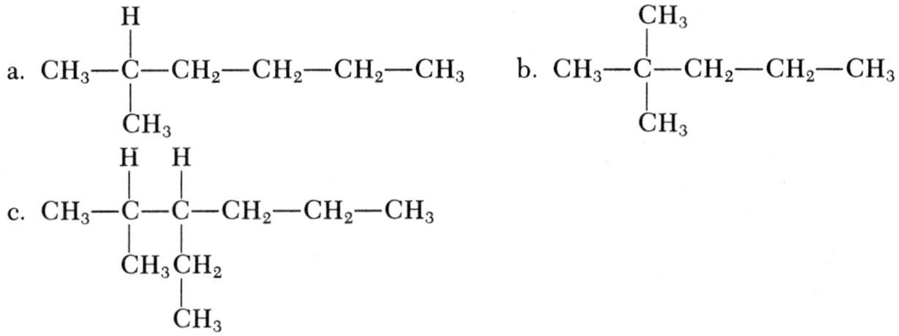

d. CH₃—CH₂—CH₂—CH₂—CH₂—CH₂—CH₂—CH₃

e. CH₃—CH₂—CH₂—CH(CH₂CH₃)—CH₂—CH₂—CH₃ (with CH₂—CH₃ branch on middle C)

14.41 a. 2-chlorobutane   b. 2,3-dichlorobutane   c. 1,1,1-trichlorobutane
d. 1,3-dichloropropane

14.42 H₂C=CH—CN (cis arrangement shown: H,H on one carbon; H,CN on other)

14.43 a. —C(H)(H)—C(H)(CH₃)—C(H)(H)—C(H)(CH₃)—
b. —CF₂—CF₂—CF₂—CF₂—
c. —C(H)(C₆H₅)—C(H)(H)—C(H)(C₆H₅)—C(H)(H)—

14.44 a. CH₂   b. 85.60% C, 14.40% H
14.45 a. aldehyde   b. acid   c. ester   d. ether   e. alcohol
14.46 a. CH₃CH₂CH₂OH   b. CH₃—O—CH₂CH₃   c. CH₃CH₂—C(=O)H
d. CH₃—C(=O)—CH₃   e. CH₃CH₂—C(=O)—OH   f. CH₃—C(=O)—OCH₃

14.47 a. 1   b. 2   c. 1   d. 3   e. 1   f. 2

14.48 a. H—C(=O)—O—CH₂—CH₂—CH₃   b. CH₃—C(=O)—O—CH(H)(CH₃)—CH₃

14.49 H—C(=O)—OCH₃, CH₃—C(=O)—OCH₃, H—C(=O)—OCH₂CH₃, CH₃—C(=O)—CH₂CH₃

14.50 a. formic acid, ethyl alcohol   b. acetic acid, propyl alcohol
c. acetic acid, isopropyl alcohol

14.51 c	14.54 a	14.57 b	14.59 d
14.52 c	14.55 b	14.58 d	14.60 c
14.53 c	14.56 c		

Chapter 15

15.27 a. $^{235}_{92}U \rightarrow {}^{4}_{2}He + {}^{231}_{90}Th$
b. $^{137}_{55}Cs \rightarrow {}^{0}_{-1}e + {}^{137}_{56}Ba$

15.28 a. $^{232}_{90}Th \rightarrow {}^{4}_{2}He + {}^{228}_{88}Ra$
b. $^{90}_{38}Sr \rightarrow {}^{0}_{-1}e + {}^{90}_{39}Y$

15.29 a. $^{222}_{86}Rn$ b. $^{3}_{2}He$

15.30 a. $^{192}_{81}Tl$ b. $^{28}_{14}Si$

15.31 a. $^{1}_{0}n$ b. $^{12}_{6}C$

15.32 a. $^{4}_{2}He$ b. $^{7}_{4}Be$
15.33 a. $^{147}_{57}La$ b. $^{9}_{4}Be$
15.34 eight alphas, six electrons
15.35 seven alphas, four electrons
15.36 a. 0.800 g b. 0.400 g c. 0.100 g
15.37 a. 12 yr b. 36 yr
15.38 a. 1/2 b. 1/256
15.39 a. 106 d b. 212 d
15.40 a. 0.78 b. 0.38
15.41 a. 2400 yr b. 7600 yr
15.42 3600 yr
15.43 0.79
15.44 a. -1.3×10^{3} kJ b. -2.2×10^{5} kJ
15.45 a. -3.8×10^{-5} g b. -9.9×10^{-3} g
15.46 a. -2.2×10^{-5} g b. -2.0×10^{6} kJ
15.47 a. -3.5×10^{-5} g b. -3.2×10^{6} kJ
15.48 -1.78×10^{-10} g
15.49 -3.1×10^{-11} g
15.50 a. -1.3×10^{-2} g b. -1.2×10^{9} kJ
15.51 d   **15.54** a   **15.57** c   **15.59** b
15.52 a   **15.55** d   **15.58** d   **15.60** b
15.53 a   **15.56** b

Index

Page numbers in *italics* indicate a figure; page numbers followed by a t indicate a table.

Absolute atomic mass, 69–71
Absolute temperature, 148
Acetic acid, 397, 403
Acetone, 399
Acetylene, 394
Acid(s), 288–312, 437
   acetic, 403
   and bases, 304–308, *304*
   Bronsted-Lowry model of, 309–311, *310*
   carboxylic, 402–404
   citric, 403
   fatty, 405
   formic, 403
   general models of, 308–312
   Lewis model of, 311–312
   malic, 403
   naming of, 92–94, 93t
   pH value and, 292–296, 293t, *294*
   properties of, 288, 289–292, *291*
   strong, 296–297, *298*
      and strong base reaction, 304–305, *304*
      neutralization, 305, *304*
      and weak base reaction, 306–307
      titration, 307–308, *307*
   weak, 296, 298, *298*
      dissociation constants of, 299–301, 299t
      and strong base reaction, 305–306
Acid anhydride, 372, 437
Acid-base indicator, 307, *307*, 437
Acid dissociation constant. *See* Dissociation constant.
"Acid rain," 374, 437
Acid solution, 292–294, 437
Actinide, 188, 437
Activation energy, 258–260, *259*, 437
Activity series, 328–332, 329t, 437
Actual yield, 124, 437
Addition polymer, 392
Addition, uncertainty in, 25
Age
   of organic material, 425
   of rocks, 424–425
Alabaster, 88t
Alcohol(s), 397–398, *398*, 399t, 400–402, 401t, 437

ethyl, 398, *398*, 400
methyl, 397–398, *398*, 400
Alcoholic beverage, 400–401, 401t
Aldehyde, 398, 399, 399t, 437
Algebraic equations, 460–462
Alkali metal, 185, 186, 188, 348–352
   and alkaline earth metals, 350t
   properties of, 349–351, 349t
Alkaline earth metal, 185, 186, 348–352, 437
   and alkali earth metals, 350t
   properties of, 349–351, 349t
Alkane, 381–389, 437
   formulas for, 381–384
   isomerism in, 384–385, *384*
   naming of, 386–389, 387t
   properties of, 383t
   sources of, 389
Alkene, 390–394, *391*, 437
Alkyl group, 437
Alkyne, 394–395
Allotropy, 363–366, 437
   of carbon, 365–366, *366*
   of nitrogen, 365
   of oxygen, 363
   of phosphorus, 365
   of sulfur, 363
Alpha particle, 413–415, *414*, 437
Alpha radiation, 413–414
Aluminum, extraction of, 361–362, *362*
Ammonia, 216, 369
Analytical balance, 15, *15*, 21
Anhydride, 372, 437
Anion, 66, 88, 333, 437
Anode, 333–340, *334*, *335*, *337*, *338*, *340*, 437
Antilogarithm, on calculator, 453
Aqua fortis, 88t
Aqua regia, 88t
Area, 23
Aristotle, 56–57
Arithmetic operation, on calculator, 448
Aromatic hydrocarbons, 395–397, 437
Arrhenius acid, 308, 437
Arrhenius base, 308, 437
Atmosphere, 141, 143, 437
Atmospheric pressure, 140–141, 143

Atom(s), 56–58, 437
   components of, 61t
   electrons in, 58, *59*, 60
      configuration of, 176–179
   isotopes of, 63
   masses of, 67–71
   mass number of, 62
   moles of, 71–74
   nuclear symbol of, 63
   nucleus of, 60
Atomic mass, 67–71, 437
   absolute, 69–71
   relative, 67–69
Atomic number, 61–62, 168, 437
Atomic spectrum, 171–172, *171*, 437
Atomic theory, 56–58
Avogadro, Amedeo, 149
Avogadro's Law, 149–150, 437
Avogadro's number, 70, 71, 114, 437

Baking soda, 88t
Balance, 14–15
Balanced equation, 109–114, 437
   redox, 324–328
Balmer series, *171*
Barometer, 141, *142*, 437
Base(s), 288–312, 437
   and acid reaction, 304–308, *304*
   general models of, 308–312
      Bronsted-Lowry, 309–311, *310*
      Lewis, 311–312
   pH values and, 292–296, 293t, *294*
   properties of, 288, 289–292, *291*
   strong, 301–303
      and strong acid reaction, 304–305, *304*
      and weak acid reaction, 305–306
   titration in, 307–308, *307*
   weak, 303
      and strong acid reaction, 306–307
Basic anhydride, 372, 438
Basic solution, 438
Battery. *See* Storage battery.
Becquerel, Henri, 413

483

Bent molecule, 214, *217*, 218, *218*, 438
Benzene, 395–397
Beta radiation, 415–416, *414*, 438
Binary molecular compound, 91–92
"Blast" furnace, 359, *359*
Blue vitriol, 88t
Bohr, Niels, 168
Bohr's model, 168–172, 174, 438
   limitations of, 172
Boiling point, 16, 17, 42–44, *43*, 438
   of solutions, 246–249, 247t
Boiling point elevation, 438
Boltzmann, Ludwig, 137
Bombardment reaction, 416–418, *417*
Bond(s), 198–222, 437
   covalent, 198, 203–206, *203*, *204*
   hydrogen, 231–232, *232*
   ionic, 198–202, *200*, 200t
   Lewis structure and, 206–214
   and molecular geometry, 214–218, *215*, *216*, *217*
   polarity and, 218–222, *219*, *220*, *221*
Boyle, Robert, 137
Boyle's Law, 144, *144*, 149, 438
Breeder reactor, 429–430
Bronsted-Lowry acid-base model, 309–311, *310*, 438
Buret, 13
Butane, 384, 383t

Calculator(s), 7, 447–454
   antilogarithms on, 453
   arithmetic operations on, 448
   division on, 449
   exponentials on, 451
   extracting roots on, 450
   logarithms on, 453
   raising power on, 450
   successive multiplication on, 449
Calorie, 45, 438
Carbon, 380–381
   conversion of, to carbon monoxide, 359
   as nonmetal, 365
Carbon-14, 416, 425
Carbonate, 358
Carboxylic acid, 402–404, 399t, 438
Catalyst, reaction rate and, 263–264, *263*, 438
Cathode, 58, *59*, 333–340, *334*, *335*, 337, 338, *340*, 438
Cation, 66, 88, 333, 438
Caustic soda, 88t
Celsius scale, 15–16, *16*, 438
   conversion of, to Fahrenheit, 17
   vs. Kelvin scale, 17–18, *18*
Centi-, 7t, 438
Centigrade scale, 15
Centimeter, 8
Cesium-144, 428
Chameleon mineral, 88t
Charles, Jacques, 139
Charles' Law, 146–147, *147*, 148, 149, 438
Chemical(s)
   bonds, 64. *See also* Bond(s).
   changes, 40, 438

   energy and, 44–48
   common names of, 88t
   equations, 108–113
   equilibrium of, 256, 264–280, *265*, *266*, 267t. *See also* Gaseous system.
   properties, 41, 439
   reactions, 40
   vs. nuclear reaction, 412
Chemist, 2
Chemistry, 2
   definition of, 34
   nuclear, 413–433
   organic, 378–406
Chloride, 358
Citric acid, 403
Coal tar, aromatic hydrocarbons and, 397
Coarse balance, 14–15, *14*, 21
Coarse mixture, 34
Coefficient, 3, 109, 439
"Coke," 359
Colligative property, 247, 439
Collision, elastic, 137, 139, *139*, *140*, 258
Combustion, heat of, 126, 127t
Complex ion, 439
Compound(s), 34, 37–39, 439
   color of, 354, 354t, 355
   Law of Constant Composition for, 38
   mass percent of elements in, 38t
   organic, 380–406
   of oxygen, 397–406
   percent composition and, 94–100
   properties of, 38
Compression, of gases, 136, 137, 277–278
Concentration of solution, 240–246, 439
   mass percent and, 244–246
   molarity and, 240–246, *243*,
Condensation, 40
Condensed structural formula, 83
   of carbon compounds, 383
Constant, 144
Conversion, of exponentials to ordinary numbers, 3–5
Conversion factors, 9–10, 439
   procedure for using, 10
   properties of, 9–10
Coordination number, 356, 439
Corrosive sublimate, 80t
Covalent bond, 198, 203–206, *203*, *204*
Cracking, 391
Cubic centimeter, 439
Curie, Marie and Pierre, 413
Cyclotron, 417, *417*, 439

Dalton, John, 57, 58, 154
Dalton's Law, 154–156, *155*, 155t, 439
Decane, properties of, 383t
Decay of nuclei, 413
   radioactive, rate of, 420–425, 421t, *423*
      age of organic materials and, 425
      age of rocks and, 424–425
      calculation of, 423
Deci-, 7t, 439
Democritus, 56
Density, 23, 439

   measurement of, 18–21, *18*, *21*
   of gases, 136, 137, 138
Deuterium, 439
Diamond, as carbon allotrope, 365, 366, *366*
Diatomic molecule, 219–221, *230*
   bonding of, 214
Diffusion of gases, 136, 137, 138
Dilute solution, 240–246, *243*, 439
Dimensional analysis, 11
Direct proportionality, 456–459
Dissociation constant, 299–301, 299t, 439
Distillation, 370, 439
Division
   on calculator, 449
   of exponential numbers, 6–7
   significant figures in, 24
Double bonds, 205, 439
Double-pan balance, 14, 14t
d Sublevel, 175
Dynamic equilibrium, 266

EE key, 7
EEX key, 7
Eicosane, properties of, 383t
Einstein, Albert, 46
Einstein equation, 46
Elastic collision, 137, 139, *139*, *140*, 258
Electrical cell, 332–340
   electrolytic, 332–336, 439
   voltaic, 332–333, 445
Electrical energy, 45
Electrode, 333, 439
Electrolysis, 333–336, *334*, *335*, 439
   of water, 37
Electrolyte, 229, 439
Electrolytic cell, 332–336, *334*, *335*, 439
   electrolysis in, 333–336, *334*, 335
Electron(s), 58, *59*, 60, 168, 415, 439
   configurations, 176–179
      of ions, 200–202
      of transition metals, 202
   excited state of, 169
   ground state of, 168
   orbitals of, 179, 180–182, *180*, *181*, 181t, *182*
   spin of, 179, 182–183, *182*
   transfer, 199
Electron cloud, 172, *173*, 180, 439
Electronegativity, 189, 191–192, *191*, 439
Electroneutrality, 84
   principle of, 86
Electronic structure, 168–192
   Bohr model in, 168–172, *170*, *171*
   electron configuration in, 176–179, *178*, *179*
   electron spin in, 179, 182–183, *182*
   energy levels in, 173–175
      principal, 173, 174, 175t
      sublevel, 174–175, 175t
   orbitals in, 179, 180–182, *180*, *181*, 181t
      diagrams of, 183–184
   quantum mechanics and, 172–173, *173*
Element(s), 34–37, 37t, 439
   atomic number of, 61
   definition of, 34

Index

electron configuration of, 176–177
mass percents of, 35, 35t, 38t
metals as, 36, 36t
nonmetals as, 36, 36t
Periodic Table of, 36
symbols for, 35, 36t
Empirical formula, 82
Endothermic reaction, 47, 49, 125, *126*, 439
Energy, 34, 44–49
  consumption of, 46–47
  endothermic reaction and, 47, 49
  exothermic reaction and, 47
  heat flow and, 47–49
  in nuclear reactions, 412
  Law of Conservation of, 46
  units of, 45
Enthalpy change, 47, 440
Enzyme, 401
Epsom salts, 88t
Equation(s), 108–113
  algebraic, 460–462
  balanced, 109–114
  redox, 324–328, *326*
  thermochemical, 125–127, *126*
  unbalanced, 109
Equilateral triangle molecule, 215, *215*, *217*, 390, 440
Equilibrium, 440
  chemical, 264–280, *265*, *266*, 267t
  dynamic, 266
Equilibrium constant, 256, 439
  for gaseous systems, 267–271. *See also* Gaseous system.
  evaluation of, 269
  expression of, 268–271
  and extent of reaction, 271–274
Equilibrium system, 265
  changes in, 274–281, 276t, 277
  compression and, 272–278, *277*
  expansion and, 277, *277*
  pressure and, 278
  temperature changes and, 278–280, 281t
Error of measurement, 21–22
Ester, 399t, 404–406, 440
Ethane, 381, *382*, 383
  properties of, 383t
Ethene, 390
Ether, 399t, 440
Ethyl alcohol, 400
Ethylene, 390
Ethylene glycol, 402
Exact number, 26
Excited state, 440
  of electrons, 169
Exothermic reaction, 47, 49, 125, *126*, 199, 440
Expansion
  of equilibrium systems, 277, *277*
  of gases, 136, 137
EXP key, 7
Exponential(s), 3–7, 440
  on calculator, 451
  conversion of, to ordinary numbers, 3–5
    rule for, 4
  division of, 6–7
  multiplication of, 6–7

ordinary numbers converted to, 5
significant figures as, 23

Factor label method, 11
Fahrenheit scale, 16, *16*, 440
  conversion to Celsius, 17
"Family name" of alkanes, 386
Fat, 405, 440
Fatty acids, 405
Fermentation, 400–401
Fission, 412, 425, 429, 440
  characteristics of, 426
Fission reactor, 426–429, *427*
Fool's gold, 88t
Formation, heat of, 126, 127t
Formic acid, 403
Formula(s), 82–87, 440
  of acids, 92–93
  of alkanes, 381–384
  of ionic compounds, 84–87
  molecular, 100
  of monatomic ions, 84–86
  vs. names, 87
  and percent composition, 94–100
  of polyatomic ions, 86–88, 87t
  simplest, 97–100
  types of, 82–83
Formula mass, 72, 440
Four electron pairs, geometry and, 216, *215*, *216*, *217*
Fractional distillation, 370, 440
Freezing point, 16, 17, 40, 440
  lowering of, 247, 440
  vs. melting point, 42
  of solutions, 246–249, 247t
f Sublevel, 175
Functional group, 397
Fusion, 47–48, 48t, 413, 440
  nuclear, 430–431

Gamma radiation, *414*, 416, 440
Gas(es), 136–161
  Boyle's Law of, 144, *144*
  Charles' Law of, 146–147, 148, *147*
  Dalton's Law of, 154–156, *155*, 155t
  Ideal Gas Law, 152–154
  Kinetic molecular theory of, 137–139, *136*, *139*
  Law of Combining Volumes and, 160–161, *160*
  molar volumes of, 149–152
  pressure of, 140, 141, 143
  properties of, 136
  reactions involving, 157–161
    mass to volume conversions in, 157–158
    volume to mass conversions in, 157, 158–159
    volume to volume conversions in, 157, 159–161
  temperature and, 137–138, *136*, 140, 143
  volume of, 139, 140
    pressure and, 143–146, *143*, *144*, 144t

    pressure, temperature, and, 148–149
    temperature and, 146–148, *147*
Gaseous equilibrium system(s)
  changes in, 274–280, *277*, 276t
  reactants and, 275–276, 276t
  temperature and, 278–280, 281t
  volume and, 277–278, *277*
  constants in, 267–271, *266*
  and extent of reaction, 271–274
Gay-Lussac, Joseph, 137
Gay-Lussac Law of Combining Volumes, 160–161, *161*
GLAUBER'S salt, 88t
Glucose, 400
Glycerol, 402
Graduated cylinder, 12, *12*
Gram, 9, 13, 440
  -mole conversions, 73–74
Graph(s), 462–464
Graphite as carbon allotrope, 365
Green vitriol, 88t
Ground state, 168, 440
Groups of Periodic Table, 36, 440
Gypsum, 88t

Haber, Fritz, 369
Haber process, 369, 440
Hahn, Otto, 425
Half-equation, 324, 440
Half-life, 420–425, 421t, *423*, 440
  ages of organic materials and, 425
  ages of rocks and, 424–425
  calculation of, 423
Half-reaction, 319
Halide, 188
Hall, Charles, 361
Halogens, 65, 188, 440
Heat, 44
  of combustion, 126, 127t
  enthalpy change and, 47
  of formation, 126, 127t
  of fusion, 47–48, 48t
  of vaporization, 48, 441
"Heavy" hydrogen, 62, 63
Heptane, properties of, 383t
Heroult, 361
Heterogeneous mixture, 39, 441
Hexane, properties of, 383t
High-carbon steel, 361
Homogeneous mixture, 39
Hydrate
  formulas of, 83
Hydride ion, 350
Hydrocarbon, 441
  aromatic, 395–397
  saturated, 381–389
    formulas for, 381–384, *382*
  unsaturated, 389–395, *391*, *393*
Hydrogen, 367–370
Hydrogenation, 405
Hydrogen atom
  Bohr's theory of, 168–172
Hydrogen bond, 231–232, *232*, 441
Hydronium ion, 289, 441

Index

Hyperbola, 459
Hypo, 88t

Ideal Gas Law, 152–154, 441
Indicator, acid-base, 437, 441
Induced radioactivity, 413, 416–418, *417*
Inorganic compound(s)
  names of, 87–94
  solubility of, 235–240
    net ionic equations and, 235–240
    precipitation reactions and, 235–237, *236*
    rules for, 235t
International Union of Pure and Applied Chemistry, 386
Inverse proportionality, 458–460
Ion, 64, 65–67
Ionic bond, 198–202, *200*, 200t, 441
Ionic compound(s), formulas of, 84–87
Ionization energy, 171, 189–190, *190*, 441
Isomer(s), 441
  of alkanes, 384–385, *384*
Isotope, 63, 412, 441
IUPAC, 386

Joule, 45, 441

K_a. *See* Dissociation constant.
K_c. *See* Equilibrium constant.
K_w, 441
Kekulé, Friedrich, 395
Kekulé structure, 395–396
Kelvin scale, 17–18, 441
  vs. Celsius scale, 17–18, *18*
Ketone, 399t, 441
Kilo-, 7t, 441
Kilocalorie, 45
Kilogram, 9, 14
Kilojoule, 45, 441
Kilometer, 8
Kilopascal, 140, *141*, 441
Kinetic molecular theory, 137–139, 441

Lanthanides, 187, *187*, 441
Law of Combining Volumes, 160, *160*, 161, 441
Law of Conservation of Energy, 46, 441
Law of Conservation of Mass, 40–41, 58, 109, 441
Law of Constant Composition, 38, 58, 442
LeChatelier, Henri, 275
LeChatelier's Principle, 275, 442
Leucippus, 56
Lewis, G. N., 206
Lewis acid-base model, 311–312, 442

Lewis structure, 206–214, 442
  of atoms, 207–208, 208t
  of molecules, 209
    writing of, 210–211
  multiple bonds in, 212–213
  octet rule for, 209–212
    exceptions to, 213–214
Ligand, 355, 442
Light, 45
Light hydrogen, 63
Lime, 88t
Limiting reactant, 121–125, 442
Linear molecule, 214, 442
Liquid
  boiling point of, 43
  density of, 19
Liter, 12, 442
Lobe, orbital, 180
Logarithm, on calculator, 453
Lunar caustic, 88t
Lye, 88t
Lyman series in atomic spectrum, *171*

Macromolecularity, 442
  of red phosphorus allotrope, 365
Macromolecular substances, covalent bonding and, 206
Main-group elements, 36, 186, 442
  in Periodic Table, 186, 188
Malic acid, 403
Mass, 13–15, 442
  of atoms, 67–71
    absolute, 69–71
    relative, 67–69
Mass and energy relations, 431–433, 433t
Mass number, 62, 442
Mass percent, 94, 442
  relations, 115–116
  solutions and, 244–246
Mass to mole conversion, 119–120
Mass to volume conversion, 157–158
Matter, 34–44, 442
  changes in, 39–41
    chemical, 40
    physical, 39–40
  classification of, 35t
  compounds of, 37–39
  elements of, 34–37t, 35t
  Law of Conservation of Mass for, 40–41
  mixtures as, 39
  particle structure of, 56–74
  properties of, 41–44
Maxwell, James, 137
Measurement, 2–26
  conversion factors in, 9–10
  of density, 18–21, *19*, *21*
  error of, 21–22
  exact numbers in, 26
  exponential numbers in, 3–7
    conversion to ordinary numbers in, 3–5
    division of, 6–7
    multiplication of, 6–7

  of mass, 13–15
  metric system in, 7–11
  ordinary number conversion in, 3–5
  of temperature, 15–18
  uncertainty in, 21–22
    in addition, 25
    in division, 23–24
    in multiplication, 23–24
    significant figures in, 22–25, 24t
    rules for, 25
    in subtraction, 25
  of volume, 12–13
Measuring pipet, 13, *12*
Mechanical energy, 44
Mega-, 7t, 442
Meitner, Lise, 425
Melting point, 40, 42, *42*, 442
Mendeleev, Dmitri, 185
Metal(s), 36, 36t, 348–352, 349t, 350t, 442
  as carbonates, 358
  as chlorides, 358
  extraction of, 358–362, *359*, *362*
  iron extraction and, 358–360, *359*
  naming of, 89
  as oxides, 357–358
  reaction of oxygen with, 371–372
  as sulfides, 358
  transition, 352–357
Metallic character, 189, 442
Metalloid, 36, 442
Metallurgy, 357–362, *358*, *359*, *362*, 442
Meter, 7–8, 442
Methane, 216, *216*
  geometry of, 381, 383, *382*
  properties of, 383t
Methyl alcohol, 397–398, *398*, 400
Metric system, 7–11
  prefixes in, 7t
Micro-, 7t, 442
"Mild" steel, 360
Milk of magnesia, 88t
Milli-, 7t, 442
Milligram, 9, 14
Millikan, Robert, 58
Milliliter, 12
Millimeter, 8
  of mercury, 442
Mixture, 34, 39
  melting point of, 42, *42*
Molality, 245–246, 442
Molarity, 7t, 240–244, *243*, 442
Molar mass, 72–73, 442
Molar volume of gases, 149–152
  standard conditions for, 150, 152
Mole, 71–74, 114–115, 442
Mole to gram conversion, 73–74
Mole to mass conversion, 118–119
Mole to mole conversion, 117–118
Molecular compound, naming of, 91
Molecular formula, 64–65, 82–83, 442
  of carbon compounds, 383
  percent composition and, 94
  from simplest formula, 100
Molecular speed, 138
Molecule, 64–65, *65*, 443

Index

bent, 214, 218, *217*, *218*
diatomic, 219–221, *220*
geometric structure of, 214–218, *215*, *216*, *217*
linear, 214, 215
nonpolar, 219, *219*
planar, 390
polarity of, 218–222, *219*, *220*, *221*
polyatomic, 221–222, *221*
pyramidal, 217, *217*
tetrahedral, 216, *216*
triangular, 215–216
Monatomic ion, 84–86
naming of, 89
Monoclinic sulfur, 363
Monomer, 392, 443
Multiplication, on calculator, 449
of exponential numbers, 6–7
of significant numbers, 24
Muriatic acid, 88t

Names, of acids, 92–94, 93t
of alkanes, 386–389, 387t
of binary molecular compounds, 91–92
common, of chemicals, 88t
of inorganic compounds, 87–94
of ionic compounds, 88–91
Nano-, 7t, 443
Nanometer, 8
Natural gas, as alkane source, 389
Natural radioactivity, 413–416
Net ionic equation, 237–240, 443
Neutralization, 443
in acid-base reaction, 305, *304*
Neutral solution, 292, 294, *294*, 304
Neutron, 60, 443
Neutron bombardment, 417
New Systems in Chemical Philosophy, 57
Nitride ion, 350
Nitrogen, as nonmetal, 365
Noble gases, 64, 443
ions and, 201
monatomic ion charge and, 84, *85*
Nonane, properties of, 383t
Nonelectrolyte, 229, 443
Nonmetal(s), 36, 36t, 362–367, *364*, *366*, 443
hydrogen and, 367–370
reactions of, 368–370
oxygen and, 372–374
Nonpolar molecules, 219, *219*, 443
Nonspontaneous reaction, 332, 443
Normal boiling point, 44, 442
Nuclear chemistry, 413–433. *See also* Nuclear reaction and specific topics.
Nuclear reaction
fission as, 425–429
fusion as, 430–431, *430*
mass and energy relation in, 431–433, 433t
vs. ordinary chemical reaction, 412
radioactivity in, 413–420
Nuclear symbol, 63–64, 443
Nucleus, 168, 443

of atoms, 60
identification of, 63
Normal boiling point, 44

Octadecane, properties of, 383t
Octahedron, 355, 443
Octane, properties of, 383t
Octet rule, 209–212, 443
exceptions to, 213–214
Oil of vitriol, 88t
Opposed spin, 183
Orbital(s), 179, 180–182, *180*, *181*, 181t, *182*, 443
diagrams of, 183–184, 183t, 184t, 443
spherical, 180
Ordinary numbers, conversion to exponentials, 3–5
rule for, 4
Ore, 35
extraction of aluminum from, 361–362, *362*
extraction of iron from, 358–360, *359*
Organic compound(s), 378–406, 443
structural formula of, 83
Organic material, dating of, 425
Oxidation, 319, 443
Oxidation number, 320–323, 443
of transition metals, 353, 353t
Oxidation reduction, 318–340, *319*
electrolysis cell and, 332–336, *334*, *335*
equations, 324–328, *326*
oxidation numbers and, 320–323
predicting reactions of, 328–332
voltaic cell and, 332–333, 336–340, *337*, *338*
Oxide ion, 351
Oxide ore, 357–358
Oxydizing agents, 328–332, 329t, 443
Oxyacids, 92–93, 93t, 443
Oxyanion, 86, 93, 93t, 443
Oxygen, 370–374, *371*
and metals, 371–372
as nonmetal, 363, 372–374, 373t
organic components, 397–406
alcohols as, 400–402, 401t
carboxylic acids as, 402–404
esters as, 404–406
functional groups in, 397–400, *398*, 399t
preparation of, 370–371, *371*
Ozone, 363

Parallel spin, 183
Partial pressure, 154–156, 443
Path of reaction, 257–260, *258*, *259*
Pauling, Linus, 191
Pentane, properties of, 383t
Percent, 454, 456
Percent composition, 443
formulas and, 94–100
Percent yield, 124, 443
Period(s), 36, 443

Periodic Table, 36, 61, 443
electron configuration and, 176–177, *178*, 185–188, *187*
periods in, 36
Peroxide ion, 351
Petroleum as alkane source, 389, 390t
pH, 292–296, 293t, *294*
Phosphorescence, 365
Phosphorus as nonmetal, 365
Physical change, 39–40, 443
energy and, 44–48
Physical property, 41–42, 41t, 444
"Pig iron," 360
Pitchblende, 413
Planar molecule, 390
Planck, Max, 169
"Plastic" sulfur, 363, *364*
Polarity, 443
molecular, 218–222, *219*, *220*, *221*
Polonium, 413
Polyatomic ion, 86–88, 443
covalent bonding and, 205
naming of, 89
Polyatomic molecule, 221–222, *221*
Polyethylene, 392, *393*
Polymer, 391, 444
Polypropylene, 392
Positive ion bombardment, 417
Posttransition metal, 86, 444
Precipitate, 235, 444
Precipitation reaction, 235–237, *236*
Prefix
in compound names, 91, 91t
in IUPAC name, 386
Pressure
and boiling point, 43–44
gaseous systems and, 278
and gases, 137
measurement of, 140–144, *142*
volume and, 143–146, *143*, *144*, 144t
and solubility, 233–234, *234*
Priestley, Joseph, 38
Principal energy level, 173–174, 175, 175t, 444
Products, 40, 444
Propane, 381–382, 383, *382*
properties of, 383t
Propene, 391
Proportionality, 456–460
direct, 456–458
inverse, 458–460
Propylene, 391
Proton, 60, 444
in nuclear reactions, 412
p Sublevel, 173
Pure substance, 34
melting point of, 42, *42*
Pyramidal molecule, 217, *217*, 444
Pyrite, 88t

Quantum mechanics, 172–173, *173*, 174, 444
electron configuration and, 176–179

Index

principal energy levels and, 173–174, 175, 175t
sublevels and, 173, 174–176, 175t
Quantum number, 170, 173, 444
Quantum theory, 169, *171*
Quicklime, 88t

Rad, 418, 444
Radiation
  alpha, 413–414
  beta, 415–416, *414*
  gamma, 416, *414*
Radioactivity, 413–420, 444
  decay, 420–425, 421t, 444
    dating organic material and, 425
    dating rocks and, 424–425
    half-life in, 420–425, 421t, *423*
    calculation of, 423
  effects of, 418–420
  induced, 413, 416–418, *417*
  natural, 413–416, *414*
  sources of, 419, 419t
Radium, 413
Rate of reaction, 256, 444
Reactant, 40, 444
  addition of, 275–278, 276t
  gaseous, 261, *261*
  limiting, 121–125
  removal of, 275–278, 276t
Reaction
  bombardment, 416–418, *417*
  chemical vs. nuclear, 412
  extent of, 256
  fission, 412, 425–429
  oxidation-reduction, 318–340
  rate of, 256–264, 444
    activation energy and, 258–260, *259*
    catalysts and, 263–264, *263*
    concentration of reactant and, 260–261, *260, 261*
    path and, 257–260, *258, 259*
    temperature and, 261–262, *262*
  redox, 324–328, *326*
Reactor(s)
  breeder, 429–430
  fission, 426–429, *427*
  fusion, 430–431, *430*
Red phosphorus, 365
Redox, 318–340, 444. See also Oxidation reduction.
Reducing agent, 323, 444
Reduction, 319, 444
Relative atomic mass, 67–69
Rem, 418, 444
Rhombic sulfur, 363
Root, on calculator, 450
"Rounding off" significant figures, 25
Rutherford, Ernest, 60

Sal ammoniac, 88t
"Salt bridge," 338, *338*
Saltpeter, 88t
"Saturated fat," 367
Saturated fatty acid, 405–406
Saturated hydrocarbon, 381–389, 444
  formulas for, 381–384, *382*
Saturated solution, 230–231, *231*, 444
Scientific calculator, 7
Scientific notation, 3
Semiconductor, 36
Significant figure, 22–24, 24t, 444
  in calculated quantities, 24t
  in division, 24
  as exponential number, 23
  in multiplication, 24
  "rounding off" of, 25
Simplest formula, 82, 83, 97–100, 444
  percent composition and, 94
Single bond, 205, 444
Slag, 360, 444
Slaked lime, 88t
Solid
  density of, 20, *21*
  volume of, 20
Solubility, 230–234, *231, 232, 233, 234*, 230t
  pressure and, 233–234, *234*
  rules for, 235, 235t, 444
  temperature and, 232–233, *234*
Solute, 74, 228–230, 444
Solution, 34, 39, 228–249, 444
  boiling point of, 246–249, 247t
  concentration of, 240–246
  freezing point of, 246–249, 247t
  ionic compounds in, 235–240
  solubility and, 230–234, *231, 232, 233, 234*, 230t
  solutes in, 228–230
Solvent, 74, 445
Sørensen, Peder, 293
Specific gravity, 21, 445
"Spectator ion," 236, 445
Spectrum, atomic, 171–172, *171*
Spherical orbitals, 180, *180*
Spin, electron, 179, 182–183, *182*, 445
Spontaneous reaction, 332, 445
s Sublevel, 175
Stainless steel, 361
Standard atmosphere, 141
"Standard conditions," 150
Steel, 360–361
Stoichiometry, 117–121
  mass to mole conversion in, 119–120
  mole to mass conversion in, 118–119
  mole to mole conversion in, 117–118
Storage battery, 333–339, *334, 335*, 340
STP, 150, 445
"Straight-chain" alkanes, 386
"Straight-chain" hydrocarbons, 381–382
Strassman, Fritz, 425
Strong acid, 296, 297, 445
Strong base, 296, 298, *298*, 445
Strontium-90, 428
Structural formula, 83, 445
  of carbon compound, 381–382, *382*
Structural isomers, 385, 383t, 445
Subatomic particle(s), 58–64

Sublevel(s), 173, 174–176, 175t, 445
  principles of, 175
Sublimation, 40
Subtraction, uncertainty in, 25
Suffix
  in compound names, 91, 91t
  in IUPAC name, 386
Sulfides, 358
Sulfur, allotropy and, 363–364, *364*
Superoxide ion, 351
Supersaturated solution, 231, 445
"Surname" of alkanes, 386
Symbol, 445
  of elements, 35, 36t

Temperature, 15–18
  and gases, 137, 138–139
  gaseous systems and, 278–280, 281t
    measurement of, 140
    and volume, 146–148
  reaction rate and, 261–262, *262*
  and solubility, 232–233, *234*
  and time boiling point plot, 43, *43*
Tetrahedral configuration, 216
  of carbon compound, 381
Tetrahedral molecule, 216, *216*, 445
Tetrahedron, 215, 216, 445
Theoretical yield, 445
Theory of relativity, 46
Thermochemical equation, 125–127, *126*, 445
Thermometer, 15, *16*
Thomson, J. J., 58
Three electron pairs, geometry and, 215, *215*
Three-Mile Island, 428–429
Titration, 307–308, *307*, 445
Transfer pipet, *12*, 13
Transition metal(s), 86, 352–357
  color of, 354, 354t, *355*
  complex ions and, 355–57, *356*
  coordination number of, 356
  oxidation numbers of, 353, 353t
  in Periodic Table, 186–187, *187*
  properties of, 352–353, 352t
Triangular molecule, 215, *215*, 216
Triple bonds, 205, 445
Two electron pairs, geometry and, 215, *215*

Unbalanced equation, 109
Uncertainty
  in addition, 25
  in division, 23–24
  of measurement, 21–22
  in multiplication, 23–24
  in subtraction, 25
"Unsaturated" fat molecule, 367
Unsaturated fatty acid, 405–406
Unsaturated hydrocarbon, 389–395, 389t, *391, 393*, 445
  alkenes as, 389–394
  alkynes as, 394–395
Unsaturated solution, 231, 445
Unshared pair, 209, 445

Index **489**

Valence electrons, 207–208, 445
Vaporization, 40, 445
   heat of, 48, 48t
Vapor pressure, 43, 445
Vitriol, 88t
Volatile, 445
Voltaic cell, 332–333, *337, 338,* 445
Volume, 12–13
   and gaseous equilibrium
      decrease in, 277–278
      increase in, 277, *277*
   of gases, 139, 140
   pressure and, 143–146, *143, 144,* 144t
   temperature and, 146–148, *147*
Volume to mass conversion, 157, 158–159
Volume to volume conversion, 157, 159–161
Volumetric flask, 13, *13*
Volumetric glassware, 13

Washing soda, 88t
Water
   density of, 21
   electrolysis of, 37, 37t
Weak acid, 296, 298, *298,* 445
Weak base, 303, 306–307, 445
Weight vs. mass, 13
White phosphorus, 365
Wohler, Friedrich, 380
"Wood alcohol," 400

X-ray, 419